FINITE ELEMENT ANALYSIS OF ANTENNAS AND ARRAYS

FINITE ELEMENT ANALYSIS OF ANTENNAS AND ARRAYS

Jian-Ming Jin
University of Illinois

Douglas J. Riley
Northrop Grumman Corporation

IEEE PRESS

WILEY
A JOHN WILEY & SONS, INC., PUBLICATION

Published by John Wiley & Sons, Inc., Hoboken, New Jersey.
Published simultaneously in Canada.

Library of Congress Cataloging-in-Publication Data:

Jin, Jian-Ming, 1962–
Finite element analysis of antennas and arrays / Jian-Ming Jin, Douglas J. Riley.
p. cm.
Includes bibliographical references and index.
ISBN 978-0-470-40128-6 (cloth)
1. Antenna arrays. 2. Finite element method. I. Riley, Douglas J., 1958– II. Title.
TK7871.6.J56 2008
621.382′4–dc22

2008026770

10 9 8 7 6 5 4 3 2 1

CONTENTS

PREFACE

The idea to write this book came about purely by accident. Neither of the authors even considered this prior to 2005. In the summer of 2005, the authors were invited by Dr. Constantine Balanis to contribute a chapter on "Finite Element Analysis and Modeling of Antennas" for his *Modern Antenna Handbook*. During the preparation of this book chapter, the authors discovered two interesting facts. First, the work by their two research groups combined had touched nearly all aspects of the finite element analysis of antennas, ranging from frequency- and time-domain analysis of narrow- and broadband antennas, to the modeling of antenna feeds and dispersive media, and to the simulation of phased-array antennas and finite arrays. Second, the amount of material collected was by far more than what can be included in a book chapter. The idea of expanding the book chapter into a full book was mentioned, but eventually dropped after submission of the book chapter in the summer of 2006. In May 2007 the authors jointly offered a short course on "Finite Element Analysis of Complex Antennas and Arrays" for the Electromagnetic Code Consortium's Annual Conference. This short course provided an opportunity for the authors to organize the course material into various topics, which resulted in over 200 viewgraphs and formed the framework of this book.

The book is organized into 12 chapters. In Chapter 1 we describe the need for numerical simulation in the design of complex antennas and arrays and compare the finite element method with other numerical techniques from the viewpoint of antenna analysis. Chapter 2 covers the formulations of finite element analysis of antennas in the frequency and time domains. Chapter 3 deals with the problem of mesh truncation in the finite element analysis of antennas. In Chapter 4 we describe a hybrid technique that combines the finite element time-domain method with the finite-difference time-domain method. Chapter 5 deals with the critical problem of modeling antenna feeds for radiation analysis and plane-wave excitation for scattering analysis. In Chapter 6 we consider the finite element modeling of complex structures and circuit components that are widely used in antenna designs. In Chapter 7 we present a variety of antenna simulation examples to demonstrate the capability and versatility of the finite element method. Chapter 8 covers the analysis of axisymmetric antennas using a two-dimensional finite element method. The modeling of infinitely large phased arrays is the topic of Chapter 9, which covers both the frequency- and time-domain analyses. In Chapter 10 we present various domain-decomposition techniques for the analysis of large finite arrays, which is one of the most challenging problems for computational electromagnetics. Chapter 11 deals with another challenging problem, which is modeling of antenna–platform interactions by combining the finite element method with other numerical or asymptotic techniques. Finally, in Chapter 12 we discuss various numerical and practical considerations in the finite element analysis of antennas and arrays.

In the writing of this book, special care was exercised to avoid significant overlap with existing books on the finite element method for electromagnetic analysis, including a book by the first author. The writing assumed that the reader has basic knowledge of antennas, electromagnetics, and the finite element method. For the frequency-domain analysis, the time convention $e^{j\omega t}$ is used and suppressed throughout. References are listed at the end of each chapter. All the numerical examples were obtained using our in-house-developed computer codes and whenever possible they were verified using published data, experimental data, or results computed using commercial software.

The book is the result of a close collaboration between the two authors—both have worked on every chapter. Each author takes responsibility for the accuracy of the material he wrote. The first author (J.M.J.) was mainly responsible for writing Chapters 1 to 3, 5, and 7 to 11. The second author (D.J.R.) was mainly responsible for writing Chapters 4, 6, and 12 and Sections 9.2.3, 10.2.2, and 11.1.2, and he also contributed to the writing of Chapter 7 and Sections 2.3, 5.1.2, and 5.2.3. The authors would appreciate having any errors brought to their attention.

This book can be used as a research reference by graduate students and researchers in the fields of computational electromagnetics and antennas and as a general reference by practicing antenna designers and engineers. It can also be used as a textbook for a graduate course on numerical analysis and modeling of electromagnetic fields. The techniques presented in the book can be extended to deal with problems in other electromagnetics-, RF-, microwave-, and optics-related technical fields.

Acknowledgments

The authors are indebted to many people who have contributed, either directly or indirectly, to the work presented in this book. Foremost, they wish to thank Drs. Zheng Lou and Norma Riley for coauthoring the chapter on "Finite Element Analysis and Modeling of Antennas" for the *Modern Antenna Handbook* edited by Dr. Constantine Balanis. That book chapter laid the foundation for writing this book. They would also like to thank Dr. Balanis for his invitation to contribute to his book.

The first author (J.M.J.) would like to thank all of his former and current graduate students and postdoctoral research fellows for contributing to the work related to the finite element method for antennas and arrays. Dr. Zheng Lou developed the higher-order finite element method for analysis of phased arrays, formulated the time-domain waveguide port boundary condition to model antenna feeds, and proposed the dual-field domain-decomposition algorithm for the time-domain finite element analysis of large antenna and array problems. Mr. Yu-Jia Li developed the dual–primal domain-decomposition method for the large-scale frequency-domain finite element analysis. Dr. Rickard Petersson developed the time-domain finite element method with Floquet absorbing boundary conditions for analysis of periodic structures, including phased-array antennas. Dr. Dan Jiao developed numerical schemes for implementing perfectly matched layers and boundary integral equations in the time-domain finite element method. This work was extended by Dr. Thomas Rylander and applied to antenna analysis. Dr. Jian Liu contributed to the development of the hybrid finite element–boundary integral method for the analysis of antenna and platform interactions. The time-domain counterpart was developed and demonstrated for antenna applications by Dr. Ali Yilmaz. Dr. Andrew Greenwood developed the finite element modeling of axisymmetric antennas using perfectly matched layers and investigated the modeling of antenna–platform interactions by combining the finite element method with an asymptotic technique. The method for axisymmetric antenna modeling was further extended and refined by Drs. Eric Dunn, Jin-Kyu Byun, and Eric Branch. Dr. Matthys Botha developed a symmetric finite element–boundary integral formulation, which was applied to axisymmetric problems by Mr. Rui Wang. Finally, Dr. Davi Correia proposed the concept of second-order perfectly matched layers, which is also covered in this book. Messrs. Jamie Hutchinson, Yu-Jia Li, Rui Wang, Shih-Hao Lee, and Xiao-Lei Li have proofread the manuscript. Some of the work reported in this book was supported by the Air Force Office of Scientific Research under a Multidisciplinary University Research Initiative grant managed by Dr. Arje Nachman. Also gratefully acknowledged is support from a number of industrial organizations, most notably Lockheed Martin and Northrop Grumman corporations, and the visiting professorship from the City University of Hong Kong and the University of Hong Kong, which provided the first author with the time necessary to complete the book.

The second author (D.J.R.) wishes to thank Dr. C. David Turner of Sandia National Laboratories for his initial contributions to the development of subcell algorithms in the FDTD method, as well as his contributions to the development of the interface between unstructured tetrahedral grids and the FDTD algorithm. Provably stable subcell techniques for wires and slots within the time-domain finite element setting were developed further by Dr. Fredrik Edelvik. Drs. Thomas Rylander and Anders Bondeson contributed significantly to the development of the provably stable FETD–FDTD method. Dr. Norma Riley has made important contributions to the design and understanding of ultrawideband phased arrays based on engineered materials, and some of these designs are described in the book. She has also proofread the entire manuscript and made many valuable suggestions. Also gratefully acknowledged is Dr. Riley's contributions to Chapter 12, where she has provided important insight into many of the practical issues associated with the application of advanced numerical methods in computational electromagnetics. The guidance provided by Dr. Robert C. Hansen in wideband phased-array design is also gratefully acknowledged. Finally, long-term support by Sandia National Laboratories and Northrop Grumman Corporation to advance and apply the time-domain finite element method is gratefully acknowledged.

JIAN-MING JIN
DOUGLAS J. RILEY

1 Introduction

In recent years, antenna technologies have received heightened interest because of their importance in wireless communication, remote sensing, space exploration, defense, electronic warfare, and many other electronic systems. Quantitative antenna analysis is critical to the design and optimization of antennas, especially complex antennas that are not easily designed by intuitive approaches. In a typical antenna analysis, the goal is to find the radiated field and input impedance. In the case of multiple antennas, such as antenna arrays, it is also important to quantify the mutual coupling between antennas, which can be characterized by either a mutual impedance matrix or a scattering matrix. The calculation of radiated fields, input impedances, and scattering matrices requires solving Maxwell's equations subject to certain boundary conditions determined by antenna configurations. Unfortunately, Maxwell's equations can be solved analytically only for a very few idealized antenna geometries. For example, when a linear antenna can be approximated as an infinitesimally short current element or a finite wire with a known current distribution, its radiated field can be calculated analytically. When a biconical antenna is assumed to extend to infinity, its radiated field and input impedance can also be obtained analytically. Without an approximation, antennas cannot be analyzed analytically primarily because of their structural configurations. Whereas a variety of approximate analytical techniques have been developed for relatively simple antennas, accurate and complete analysis of complex antennas, especially antenna arrays, can be accomplished only through a numerical method that solves Maxwell's equations numerically with the aid of high-speed computers.

1.1 NUMERICAL SIMULATION OF ANTENNAS

Computational electromagnetics deals with the art and science of solving Maxwell's equations numerically or with the numerical simulation of electromagnetic fields. It has become an indispensable tool for antenna analysis because of the predictive power of Maxwell's equations: If these equations are solved correctly, the solution can predict experimental outcomes and design performances. Because of their high predictive power and capability of dealing with complex structures, numerical simulation tools can support a wide variety of engineering applications, such as designing antennas analytically and predicting the impact of platforms on antenna performance, and

Finite Element Analysis of Antennas and Arrays, By Jian-Ming Jin and Douglas J. Riley

address more complex applications, including calibration of antenna systems, estimating co-site interference of multiple-antenna systems on a platform, and predicting scattering from low-observable antenna installations.

In addition to the capability of analyzing complex antennas, numerical simulation has four more distinctive advantages over traditional antenna design by experiment. The first advantage is low cost. When an antenna can be designed, analyzed, and optimized on a computer, its design cost is reduced significantly compared to that of constructing a prototype physically and measuring it in an anechoic chamber. The second advantage is the short design cycle. It typically takes far less time to simulate an antenna on a computer than to actually build one and measure it in a laboratory. The third advantage is the full exploration of the design space. Because of the low cost and short design cycle, the designer can evaluate a large variety of design parameters systematically to come up with an optimal design through numerical simulation, which is simply impossible with laboratory experiments. The last but not the least advantage of numerical simulation is the enormous amount of physical insight it provides. With a numerical solution to Maxwell's equations, the designer can now use a computer visualization tool to "see" the current flow on an antenna and field distributions around the antenna. Such a capability is extremely useful because it can help to pinpoint the source of design deficiency, such as the source of mutual coupling between antennas and the source of interference for antennas mounted on a platform. All these advantages become much more pronounced when dealing with more complex antennas involving many design parameters. Indeed, in many cases numerical simulation coupled with an appropriate set of validating measurements is the best practical solution to an antenna design problem.

Unfortunately, the great advantages of numerical simulation are also accompanied by a series of challenges. The main challenge is due to improper use of a numerical simulation, such as insufficient discretization and use of a method outside its bounds. Such improper use would yield either a poor or a completely erroneous design while wasting time and resources. Therefore, it is very important to understand the basic principles, solution technologies, and applicability and capabilities of numerical methods behind the numerical simulation tools. Such knowledge can not only reduce the possibility of improper use of a method, but also help in choosing from a suite of tools the technique best suited for a specific problem, thus increasing the designer's productivity.

1.2 FINITE ELEMENT ANALYSIS VERSUS OTHER NUMERICAL METHODS

Among a variety of numerical simulation tools in computational electromagnetics that provide a complete solution to Maxwell's equations, many are based on the method of moments, the finite-difference time-domain method, and the finite element method. Other methods, such as the transmission-line method and the finite integration technique, can be identified as either a variation or an equivalent of one of the first three.

Among the three major numerical techniques, the method of moments [1–6] has the longest history for antenna analysis. The method is based on the formulation of integral equations in terms of Green's functions as the fundamental solution to Maxwell's equations. Early development of the moment method for antenna analysis is natural because certain traditional antennas, such as dipoles and monopoles, can be represented by wires and thus require only a one-dimensional discretization for a numerical solution by the moment method. Furthermore, the Sommerfeld radiation condition, which has to be satisfied by an antenna's radiated fields, is built into the moment-method formulation automatically through the use of an appropriate Green's function; therefore, it requires no special treatment. The moment method is ideally suited for modeling metallic antennas because by using a surface integral equation, the computational domain is confined to the metallic surfaces. It is also highly efficient for antennas consisting of layered substrates, such as microstrip patch antennas, and for antennas comprising bulk homogeneous dielectrics, such as dielectric resonator antennas, because for these cases, the effect of the dielectrics can either be accounted for by a special Green's function or be modeled by equivalent electric and magnetic surface currents. However, the capability of the moment method is challenged when one attempts to model complex antennas designed with complex materials that may be anisotropic and inhomogeneous. Moreover, because of the use of Green's functions, the moment method generates a fully populated matrix whose computation and solution are associated with a high degree of computational complexity. Therefore, the traditional moment method becomes very time consuming and memory intensive for the analysis of large antennas, especially array-type antennas, which are often modeled with millions of unknowns. Fortunately, this challenge has largely been alleviated by the development of a variety of fast solvers, such as the fast multipole method, the adaptive integral method, and other fast fourier transform (FFT)–based methods [7–10]. Despite its drawbacks, the distinctive advantages of the moment method (mainly a surface-only discretization for a three-dimensional problem), coupled with the development of fast solvers, make the method a powerful tool and a preferred choice for the analysis of metallic antennas and antennas mounted on a metallic platform.

The finite-difference time-domain method [11–13], invented in the mid-1960s, solves Maxwell's equations discretized on a rectangular grid directly in the time domain. The method can easily handle material anisotropy and inhomogeneity and has become very powerful and increasingly popular because of its simplicity in formulation, implementation, and grid generation. It is also highly efficient because it does not involve any matrix solutions, and through the Fourier transform it yields a broadband solution with one time-domain calculation. As a method that solves partial differential equations directly, the finite-difference time-domain method requires a grid discretization of a three-dimensional volume to compute the fields in the volume. Since the solution region extends to infinity in an antenna radiation problem, the volume must be truncated and treated specially so that the truncated volume still mimics the original open environment. This was a major limiting factor affecting the accuracy and use of the method for many years; however, this difficulty has been removed successfully with the development of perfectly matched layers for grid truncation [14]. The remaining major challenge for the finite-difference time-domain method is the accurate modeling of complex geometrical structures, especially very fine structures whose sizes are on the order of a few hundredths or even thousandths of a wavelength, by using a rectangular grid. Although the geometrical modeling accuracy can be improved by the use of conformal grids or subgriding techniques, the resulting numerical schemes become either more complicated or much less efficient because in such a case one has to reduce the time-step size to maintain the stability of the numerical solution. Nevertheless, since most antenna geometries can be modeled accurately with a sufficiently fine grid, the finite-difference time-domain method will remain a powerful and popular choice for the modeling of antennas with complex structures and embedded in complex materials.

Compared to the method of moments and the finite-difference time-domain method, the finite element method [15–22] is not as mature and popular for antenna analysis because its formulation is more complicated than that of the finite-difference time-domain method and its use requires sophisticated volumetric mesh generation. However, the finite element method has an unmatched capability for modeling both complex structures and materials. By using unstructured meshes with curvilinear triangular and tetrahedral elements, the method can accurately model curved surfaces, fine structures, and artificial engineered materials. Since the finite element method in the time domain can be formulated to be unconditionally stable, the time-step size does not have to be reduced even for problems containing very small finite elements. This unconditional stability is critical to the analysis of complicated antenna applications such as the one illustrated in Figure 1.1. Although the finite element method requires solving a large matrix equation, the associated matrix is very sparse and

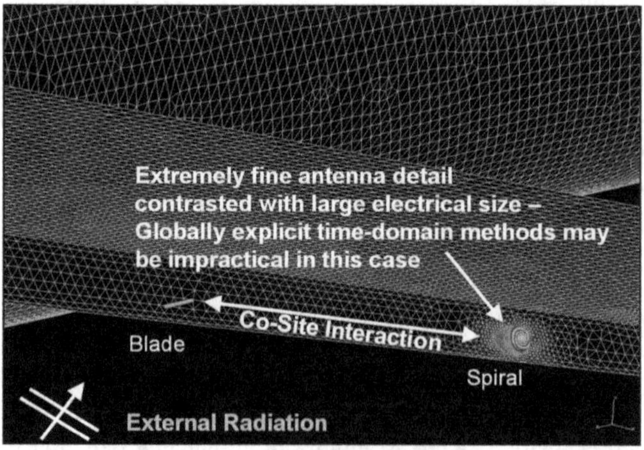

Figure 1.1 Example showing very small finite elements to model fine structures on a large object. Such a problem is challenging for explicit, conditionally stable time-domain methods and can be better handled by either implicit, unconditionally stable time- or frequency-domain techniques. (*See insert for color representation of figure.*)

often symmetrical, and its solution can be obtained efficiently by using advanced sparse solvers. Furthermore, the finite element method is well suited for parallel computation through the use of a variety of domain-decomposition algorithms. Like the finite-difference time-domain method, the finite element method solves partial differential equations directly without using Green's functions. As such, it requires the discretization of the three-dimensional space that surrounds the antenna to be analyzed and the truncation of this open space to make the solution domain finite. Proper treatment of the mesh truncation has been one of the major research subjects for the finite element analysis of antenna problems, and a variety of highly effective techniques have now been developed. The remaining major obstacle that has made the finite element method a less popular choice is the necessity for complicated mesh generation. However, this situation is changing quickly because of tremendous ongoing activities in the development of highly robust mesh generators.

From the discussions above, it can be seen clearly that the three methods have unique strengths and shortcomings. No single method is superior to the other two for every application. The moment method models free space accurately and requires only a surface discretization; thus, it is an attractive choice for modeling large metallic surfaces and homogeneous objects. The finite-difference time-domain method does not require a solution of a matrix equation and thus is highly efficient. Its implementation of perfectly matched layers for grid truncation has been well developed and is highly robust. For the finite element method to be competitive with these two methods, it must absorb their strengths into its formulation to compensate for its deficiencies. For example, the finite element method can be combined with the moment method such that the exterior open space and the antenna platform can be modeled accurately using the moment method, and the finite element method can then focus on the modeling of complex antenna structures. The finite element method can also be combined with the finite-difference time-domain method, with the latter being used to model the surrounding free space and any other homogeneous regions to fully exploit its high efficiency and its robust implementation of perfectly matched layers. These ideas lead to the development of various hybrid techniques, which are much more powerful than their individual components. These hybrid techniques should not simply "bundle" different methods together; rather, they should be formulated based on well-established electromagnetic and mathematical principles, they should be error controllable, and it should be possible to improve their accuracy in a systematic manner. Two such hybrid techniques are covered in this book; one combines the finite element method and the moment method, and the other combines the finite element method and the finite-difference time-domain method.

1.3 FREQUENCY- VERSUS TIME-DOMAIN SIMULATIONS

Since Maxwell's equations can be cast in both the time and frequency domains, a numerical solution to an electromagnetic problem can be sought in either the time or the frequency domain. In principle, it is sufficient to seek a solution in only one domain because the solution in the other domain can always be obtained using the Fourier transform. However, since the solution processes in the two domains are different, the two solutions possess different strengths. For example, when a frequency-domain numerical method is employed to solve Maxwell's equations, we have to solve a system of linear equations (matrix equation) for each frequency. However, for a general electromagnetic problem, the system matrix is independent of the excitation. Once this matrix is inverted or factorized, it becomes trivial to find a solution for a new excitation. This feature makes the frequency-domain method ideally suited for scattering analysis, where one is often interested in scattering due to plane waves from many incident directions, and perhaps less attractive for antenna analysis, where the number of different excitations is usually small and the response over many frequencies is typically required. On the other hand, when a time-domain numerical method is adopted to solve Maxwell's equations, we have to seek a solution by time marching for each excitation. Once the solution in the time domain is obtained, we can find the solution over a wide band of frequency using the Fourier transform. However, the entire solution process must be repeated for a new excitation. Therefore, the time-domain method is ideally suited for antenna analysis, where one is often interested in a solution over a broad frequency band for one or a few excitations, and becomes less efficient for scattering analysis because it requires many solutions to many excitations. Because of its importance to antenna analysis, we have devoted much effort in this book to time-domain techniques.

Although the discussions and the conclusion above are true in a general sense, we have to consider many factors when we evaluate and choose a specific solution method. For example, when a frequency-domain method is equipped with a fast solver and a robust frequency interpolation algorithm, it can become as efficient as, or even more efficient than, a time-domain technique even for the broadband analysis of antennas. Besides the use of a fast solver, a frequency-domain method has three additional unique capabilities. The first is the ability to use different mesh densities at different frequencies. This allows the use of a much coarser mesh at a lower frequency, which can speed up the simulation greatly. In contrast, the mesh density in a time-domain solution has to be determined based on the highest frequency of interest. The second capability is the ease of performing parallel computations for a broadband simulation. One need only assign different processors to carry out computations at different frequencies. This embarrassingly simple parallelization requires no interprocessor communications and hence is highly efficient. The third capability, which is perhaps also the most important, is that in the frequency domain a large discretized electromagnetic problem can be represented by a reduced-order model that contains only a few degrees of freedom. For example, the property of an antenna, which is originally characterized by a matrix having an order of a few thousands, can be represented accurately by a much smaller matrix having an order of a few tens. This feature allows the development of special techniques that can handle very large antenna arrays which originally have to be modeled with millions or even billions of degrees of freedom. The development of such a technique in the time domain is, however, not as straightforward. Therefore, because of these enhancements, the frequency-domain methods will remain important simulation tools for antenna analysis.

The truly unique strength of time-domain methods is their capability to model nonlinear components, devices, and media in an antenna system, similar to the nonlinear circuits treated in Ref. 23. This capability will become more important in the future with the development of advanced antenna systems that integrate active devices such as sources directly into antenna radiating elements. Simulation of such antenna systems in the frequency domain by means of harmonic balancing is cumbersome and very time consuming. Although the time-domain methods discussed in the book can be employed to model nonlinear antenna problems, this topic is discussed only briefly, in the context of nonlinear lumped-circuit components.

1.4 BRIEF REVIEW OF PAST WORK

Since Silvester [24] introduced the finite element method into the field of microwave engineering and electromagnetics in 1969, a tremendous amount of research has been carried out to develop the method for the analysis of electrostatic, magnetostatic, and electrodynamic problems. Most early applications dealt with problems within a bounded region, such as waveguide problems. In 1974, Mei developed a technique that combined the finite element method with a wavefunction expansion to deal with open-region electromagnetic problems such as antenna and scattering analysis [25]. In 1982, Marin developed an alternative method to deal with open-region scattering problems, which combined the finite element method with a boundary integral equation [26]. This work can be considered an extension of early formulations [27,28] for static fields. These developments enabled the application of the finite element method to open-region electromagnetic problems.

An important breakthrough in the finite element analysis of vector electromagnetic field problems occurred in the 1980s with the development of edge-based vector elements [29–31]. These new elements accurately model the nature of the electric and magnetic fields and eliminate many of the challenges associated with traditional node-based scalar elements that were used in the early finite element formulations. Since the development of vector elements, the finite element method has become a very powerful numerical technique for the analysis of three-dimensional electromagnetic fields. Although much research has been carried out and published on the finite element method for electromagnetic analysis, most of it focused on bounded field and open-region scattering problems. The subject of the finite element analysis of antennas has not received as much attention as it deserves. In the following text we review briefly the development and application of the finite element method for antenna analysis.

Application of the finite element method for the analysis and design of various antennas dates back to the 1970s, when Mei developed the first accurate approach that enabled the finite element method to deal with unbounded open-region problems [25]. The method was applied to axisymmetric antennas. For many years, the finite element method was limited to simplified two-dimensional and axisymmetric models of antennas [32–35] because of the difficulty of using the node-based elements to model vector electromagnetic fields, with the exception that Ref. 34 contained an example of calculating the field radiated by an electric current element in free space using edge-based elements. The first full-wave three-dimensional finite element analysis of realistic antennas appeared in the early 1990s [36], where the finite element method was coupled with a boundary integral equation to simulate cavity-backed microstrip antennas on a ground plane. A simple probe feed model was developed to excite antennas, and both radiation patterns and input impedances were calculated and compared with measured data. Thereafter, a variety of finite element–based numerical techniques have been developed for the analysis and simulation of various antennas and antenna arrays.

Most notably, the finite element–based numerical techniques have been developed to analyze infinitely periodic array antennas [37–40] and finite array antennas [41–48]. For the analysis of infinite array antennas, boundary integral equations were developed based on the Floquet theorem to accurately model the radiation condition, and periodic boundary conditions were formulated to confine the analysis to a single unit cell. For the analysis of large finite array antennas, novel domain-decomposition schemes were proposed that exploit the geometric repetition in the array configuration to make the analysis possible. The finite element method has also been used for the analysis of complex horn antennas [49–54] and dielectric lens antennas and radomes [55–58]. For these analyses, the axisymmetric feature of the antenna geometry, except for the excitation, can be utilized to reduce the computational domain from a three-dimensional volume to a two-dimensional slice by expanding the fields in terms of Fourier modes. The finite element method has been found to be ideally suited for modeling conformal antennas, such as cavity-backed aperture, slot, and patch antennas [59–66], because the finite element analysis can be confined to the cavity region, which contains complex antenna geometries, leaving the aperture field to be handled by a boundary integral equation. The excellent material modeling capability of the finite element method enabled the analysis of antennas residing on complex materials, such as those designed with ferrite and chiral substrates [67–69]. Combined with the moment method and a high-frequency asymptotic technique, the finite element method has been employed to analyze antennas mounted on a finite platform [70–77]. In this type of analysis, the finite element method is used to model antennas, and the effect of platforms is modeled either by the moment method based on a surface integral equation or by a high-frequency asymptotic technique such as physical optics, the geometrical theory of diffraction, the uniform theory of diffraction, and the shooting- and bouncing-ray method.

Most of the analyses discussed above were carried out in the frequency domain. To perform a frequency sweep analysis, a model-order reduction technique has been proposed [78,79], which was based on the asymptotic waveform evaluation technique originally developed for circuit analysis. Recently, the finite element method has been developed for antenna analysis directly in the time domain [80–93]. As discussed earlier, such a time-domain analysis is highly efficient for the characterization of broadband responses and is capable of modeling nonlinear materials and devices. In these works, the computational domain was truncated by perfectly matched layers implemented either directly in the finite element method [82,85] or in combination with the finite-difference time-domain method [81,83,84]. Accurate feed models have been developed to provide an excitation to antennas and to extract input impedances or S-parameters [85]. Novel domain-decomposition schemes have been developed for the analysis of large antennas and finite arrays [86–88]. A highly effective approach based on field transformation has been proposed for the analysis of infinitely periodic antenna arrays using time-domain finite element formulations [89–91]. Preliminary studies have also been conducted on incorporating a distributed feed network into the finite element modeling of antenna arrays [92] and on the simulation of antennas installed on a platform by combining the time-domain finite element method with a fast solution of a time-domain surface integral equation for induced currents on the platform [93].

1.5 OVERVIEW OF THE BOOK

The objective of this book is to present the basic formulations and discuss all the technical aspects in the finite element analysis of complex antennas and arrays. The remaining 11 chapters of the book are organized as follows.

In Chapter 2 we describe the formulations of the finite element analysis of antennas in the frequency and time domains. In this description, emphasis is placed on the basic principle of the finite element method instead of its numerical implementation. The modeling of complex anisotropic, dispersive, and lossy materials in the time-domain finite element analysis is discussed in detail. Techniques for solving finite element equations and the use of higher-order curvilinear finite elements are also addressed briefly.

Chapter 3 deals with the fundamental challenge in the partial differential equation–based numerical analysis of open-region electromagnetic radiation and scattering problems, which is the truncation of the infinite solution space into a finite-sized computational domain. The truncation techniques covered include first- and second-order absorbing boundary conditions, various perfectly matched layers, and free-space and half-space boundary integral equations. Their formulation and implementation in the frequency and time domains are discussed in detail.

In Chapter 4 we describe a stable formulation that combines the finite element time-domain method with the highly efficient finite-difference time-domain method. An immediate benefit of this combination is to use the well-established finite-difference time-domain implementation of perfectly matched layers for the truncation of computational domains. Certain equivalence between the finite-difference time-domain and finite element time-domain methods, which provides a theoretical foundation for the stable interface formulation, is illustrated and an accurate near-to-far-field transformation is described.

Chapter 5 deals with another critical challenge specific to the finite element analysis of antennas: the modeling of antenna feeds for radiation analysis and plane-wave excitation for scattering analysis. Described are simplified feed models such as current probes and voltage gaps, as well as a more accurate modeling of waveguide feeds using a waveguide port boundary condition. For scattering analysis, we describe the total- and scattered-field formulations and a novel total- and scattered-field decomposition approach. Far-field computation and near-field visualization are also addressed briefly.

In Chapter 6 we consider the finite element modeling of complex structures and circuit components, such as thin-material layers and sheets, thin wires and slots, lumped-circuit components, and feeding networks. Such modeling is important for practical applications since these types of structures and circuit components are used widely in antenna designs. The chapter includes many practical examples and, in particular, an example of predicting electromagnetic coupling into an electronic subsystem.

Chapter 7 covers a variety of antenna simulation examples, including two scattering examples, to demonstrate the capability and versatility of the finite element method based on Chapters 2 through 6. The narrowband examples include monopole and microstrip patch antennas, and the broadband examples include the horn, spiral, sinuous, Vivaldi, and Vlasov antennas. Whenever possible, the simulation results are validated or verified with published data, experimental data, or results computed using commercial software.

In Chapter 8 we describe the finite element analysis of axisymmetric antennas in conjunction with absorbing boundary conditions, perfectly matched layers, and boundary integral equations. The analysis exploits the rotational symmetry of the problem by expanding fields and excitations in terms of Fourier modes, which reduces the original three-dimensional problem to a two-dimensional problem where the simulation can be carried out more efficiently using a two-dimensional finite element method.

The modeling of infinitely large phased arrays is the topic of Chapter 9, which covers both the frequency- and time-domain analyses. The implementation of periodic boundary conditions, the formulation of mesh truncation techniques specific to this class of problems, and the modeling of general complex materials are discussed in detail. The use of an infinite phased-array solution to approximate a corresponding finite array is also addressed as a fast practical solution to a very complicated problem.

In Chapter 10 we treat one of the more challenging applications in the numerical simulation of antennas and perhaps in the entire field of computational electromagnetics: analysis of large finite arrays. Two major numerical techniques are presented to deal with this problem. One is based on the finite element tearing and interconnecting algorithm in the frequency domain, and the other is based on domain decomposition strategies in the time domain. Both approaches effectively exploit the geometrical repetitions of a finite array to make the problem tractable.

Chapter 11 deals with another highly challenging problem for antenna analysis: modeling of antenna–platform interactions. Two approaches are described. One approach is based on an accurate simultaneously coupled analysis, which simulates the entire problem numerically in one stage. The other approach is first to apply the finite element method to the antenna and its nearby structure and compute the near field, then calculate the far-field radiation according to Huygens' principle using either a numerical method based on surface integral equations or a high-frequency asymptotic technique based on ray tracing.

Finally, in Chapter 12 we discuss various numerical and practical considerations in the application of finite element analysis to antennas and arrays, such as selection of the most suitable analysis tool and solver for a specific problem, finite element discretization and corresponding numerical convergence, fast frequency sweep based on sampled frequency solutions, the application of domain decomposition and parallel computing, and the verification and validation of numerical predictions.

REFERENCES

1. R. F. Harrington, *Field Computation by Moment Methods*. New York: Macmillan, 1968; reprinted by IEEE Press, 1993.
2. R. Mittra, Ed., *Computer Techniques for Electromagnetics*. Elmsford, NY: Permagon, 1973.
3. R. C. Hansen, Ed., *Moment Methods in Antennas and Scattering*. Norwood, MA: Artech House, 1990.
4. J. J. H. Wang, *Generalized Moment Methods in Electromagnetics*. New York: Wiley, 1991.
5. E. K. Miller, L. Medgyesi-Mitschang, and E. H. Newman, Eds., *Computational Electromagnetics: Frequency-Domain Method of Moments*. New York: IEEE Press, 1992.
6. A. F. Peterson, S. L. Ray, and R. Mittra, *Computational Methods for Electromagnetics*. New York: IEEE Press, 1998.
7. N. N. Bojarski, "*k*-Space formulation of the electromagnetic scattering problem," Tech. Rep. AFAL-TR-71-75, Air Force Avionics Laboratory, Mar. 1971.
8. R. Coifman, V. Rokhlin, and S. Wandzura, "The fast multipole method for the wave equation: a pedestrian prescription," *IEEE Antennas Propagat. Mag.*, vol. 35, pp. 7–12, June 1993.
9. E. Bleszynski, M. Bleszynski, and T. Jaroszewicz, "AIM: adaptive integral method for solving large-scale electromagnetic scattering and radiation problems," *Radio Sci.*, vol. 31, pp. 1225–1251, Sept.–Oct. 1996.
10. W. C. Chew, J. M. Jin, E. Michielssen, and J. M. Song, Eds., *Fast and Efficient Algorithms in Computational Electromagnetics*. Norwood, MA: Artech House, 2001.
11. K. S. Yee, "Numerical solution of initial boundary value problems involving Maxwell's equations in isotropic media," *IEEE Trans. Antennas Propagat.*, vol. 14, pp. 302–307, May 1966.
12. K. S. Kunz and R. J. Luebbers, *The Finite Difference Time Domain Method for Electromagnetics*. Boca Raton, FL: CRC Press, 1994.
13. A. Taflove and S. C. Hagness, *Computational Electrodynamics: The Finite Difference Time Domain Method*, 3rd ed. Norwood, MA: Artech House, 2005.
14. J.-P. Berenger, "A perfectly matched layer for the absorption of electromagnetic waves," *J. Comput. Phys.*, vol. 114, no. 2, pp. 185–200, 1994.
15. P. P. Silvester and R. L. Ferrari, *Finite Elements for Electrical Engineers*, 3rd ed. Cambridge, UK: Cambridge University Press, 1996.
16. J.-M. Jin, *The Finite Element Method in Electromagnetics*, 2nd ed. Hoboken, NJ: Wiley, 2002.

17. P. P. Silvester and G. Pelosi, Eds., *Finite Elements for Wave Electromagnetics*. New York: IEEE Press, 1994.
18. T. Itoh, G. Pelosi, and P. P. Silvester, Eds., *Finite Element Software for Microwave Engineering*. New York: Wiley, 1996.
19. J. L. Volakiş, A. Chatterjee, and L. C. Kempel, *Finite Element Method for Electromagnetics: Antennas, Microwave Circuits and Scattering Applications*. New York: IEEE Press, 1998.
20. M. Salazar-Palma, T. K. Sarkar, L. E. Garcia-Castillo, T. Roy, and A. R. Djordjevic, *Iterative and Self-Adaptive Finite Elements in Electromagnetic Modeling*. Norwood, MA: Artech House, 1998.
21. A. Bossavit, *Computational Electromagnetism: Variational Formulations, Complementarity, Edge Elements*. San Diego, CA: Academic Press, 1998.
22. Y. Zhu and A. C. Cangellaris, *Multigrid Finite Element Methods for Electromagnetic Field Modeling*. New York: IEEE Press, 2006.
23. S.-H. Chang, R. Coccioli, Y. Qian, and T. Itoh, "Global finite-element time-domain analysis of active nonlinear microwave circuits," *IEEE Trans. Microwave Theory Tech.*, vol. 47, no. 12, pp. 2410–2416, Dec. 1999.
24. P. P. Silvester, "Finite element solution of homogeneous waveguide problems," *Alta Freq.*, vol. 38, pp. 313–317, May 1969.
25. K. K. Mei, "Unimoment method of solving antenna and scattering problems," *IEEE Trans. Antennas Propagat.*, vol. 22, pp. 760–766, Nov. 1974.
26. S. P. Marin, "Computing scattering amplitudes for arbitrary cylinders under incident plane waves," *IEEE Trans. Antennas Propagat.*, vol. 30, pp. 1045–1049, Nov. 1982.
27. P. P. Silvester and M. S. Hsieh, "Finite-element solution of 2-dimensional exterior field problems," *IEE Proc. H*, vol. 118, pp. 1743–1747, Dec. 1971.
28. B. H. McDonald and A. Wexler, "Finite-element solution of unbounded field problems," *IEEE Trans. Microwave Theory Tech.*, vol. 20, pp. 841–847, Dec. 1972.
29. J. C. Nedelec, "Mixed finite elements in R^3," *Numer. Math.*, vol. 35, pp. 315–341, 1980.
30. A. Bossavit and J. C. Verite, "A mixed FEM–BIEM method to solve 3-D eddy current problems," *IEEE Trans. Magn.*, vol. 18, pp. 431–435, Mar. 1982.
31. M. L. Barton and Z. J. Cendes, "New vector finite elements for three-dimensional magnetic field computation," *J. Appl. Phys.*, vol. 61, pp. 3919–3921, Apr. 1987.
32. T. Orikasa, S. Washisu, T. Honma, and I. Fukai, "Finite element method for unbounded field problems and application to two-dimensional taper," *Int. J. Numer. Methods Eng.*, vol. 19, pp. 157–168, 1983.
33. J. D'Angelo, M. J. Povinelli, and M. A. Palmo, "Hybrid finite element/boundary element analysis of a strip line notch array," *IEEE AP-S Int. Symp. Dig.*, vol. 3, pp. 1126–1129, 1988.
34. J. D'Angelo and I. D. Mayergoyz, "Finite element methods for the solution of RF radiation and scattering problems," *Electromagnetics*, vol. 10, pp. 177–199, 1990.
35. H. Ali and G. Costache, "Finite-element time-domain analysis of axisymmetrical radiators," *IEEE Trans. Antennas Propagat.*, vol. 42, no. 2, pp. 272–275, Feb. 1994.
36. J. M. Jin and J. L. Volakis, "A hybrid finite element method for scattering and radiation by microstrip patch antennas and arrays residing in a cavity," *IEEE Trans. Antennas Propagat.*, vol. 39, pp. 1598–1604, Nov. 1991.

37. J. M. Jin and J. L. Volakis, "Scattering and radiation analysis of three-dimensional cavity arrays via a hybrid finite element method," *IEEE Trans. Antennas Propagat.*, vol. 41, pp. 1580–1586, Nov. 1993.

38. D. T. McGrath and V. P. Pyati, "Phased array antenna analysis with the hybrid finite element method," *IEEE Trans. Antennas Propagat.*, vol. 42, pp. 1625–1630, Dec. 1994.

39. E. W. Lucas and T. P. Fontana, "A 3-D hybrid finite element/boundary element method for the unified radiation and scattering analysis of general infinite periodic arrays," *IEEE Trans. Antennas Propagat.*, vol. 43, no. 2, pp. 145–153, Feb. 1995.

40. Z. Lou and J. M. Jin, "Finite element analysis of phased array antennas," *Microwave Opt. Tech. Lett.*, vol. 40, no. 6, pp. 490–496, Mar. 2004.

41. R. Kindt, K. Sertel, E. Topsakal, and J. L. Volakis, "Array decomposition method for the accurate analysis of finite arrays," *IEEE Trans. Antennas Propagat.*, vol. 51, pp. 1364–1372, June 2003.

42. M. N. Vouvakis, S.-C. Lee, K. Zhao, and J.-F. Lee, "A symmetric FEM-IE formulation with a single-level IE-QR algorithm for solving electromagnetic radiation and scattering problems," *IEEE Trans. Antennas Propagat.*, vol. 52, no. 11, pp. 3060–3070, Nov. 2004.

43. S.-C. Lee, M. N. Vouvakis, and J.-F. Lee, "A non-overlapping domain decomposition method with non-matching grids for modeling large finite antenna arrays," *J. Comput. Phys.*, vol. 203, pp. 1–21, Feb. 2005.

44. J. Rubio, M. A. Gonzalez, and J. Zapata, "Generalized-scattering-matrix analysis of a class of finite arrays of coupled antennas by using 3-D FEM and spherical mode expansion," *IEEE Trans. Antennas Propagat.*, vol. 53, no. 3, pp. 1133–1144, Mar. 2005.

45. Y. J. Li and J. M. Jin, "A vector dual–primal finite element tearing and interconnecting method for solving 3-D large-scale electromagnetic problems," *IEEE Trans. Antennas Propagat.*, vol. 54, no. 10, pp. 3000–3009, Oct. 2006.

46. K. Zhao, V. Rawat, S.-C. Lee, and J.-F. Lee, "A domain decomposition method with nonconformal meshes for finite periodic and semi-periodic structures," *IEEE Trans. Antennas Propagat.*, vol. 55, no. 9, pp. 2559–2570, Sept. 2007.

47. Y. J. Li and J. M. Jin, "A new dual–primal domain decomposition approach for finite element simulation of 3D large-scale electromagnetic problems," *IEEE Trans. Antennas Propagat.*, vol. 55, no. 10, pp. 2803–2810, Oct. 2007.

48. C. Smith, M. Little, B. Porter, and M. N. Vouvakis, "Analysis of co-planar phased array coupling using finite element domain decomposition," *IEEE AP-S Int. Symp. Dig.*, pp. 3524–3527, 2007.

49. G. C. Chinn, L. W. Epp, and D. J. Hoppe, "A hybrid finite-element method for axisymmetric waveguide-fed horns," *IEEE Trans. Antennas Propagat.*, vol. 44, no. 3, pp. 280–285, Mar. 1996.

50. C. Zuffada, T. Cwik, and V. Jamnejad, "Modeling radiation with an efficient hybrid finite-element integral-equation waveguide mode-matching technique," *IEEE Trans. Antennas Propagat.*, vol. 45, no. 1, pp. 34–39, Jan. 1997.

51. D. T. McGrath and C. E. Baum, "Scanning and impedance properties of TEM horn arrays for transient radiation," *IEEE Trans. Antennas Propagat.*, vol. 47, no. 3, pp. 469–473, Mar. 1999.

52. A. D. Greenwood and J. M. Jin, "Finite element analysis of complex axisymmetric radiating structures," *IEEE Trans. Antennas Propagat.*, vol. 47, no. 8, pp. 1260–1266, Aug. 1999.

53. J. M. Gil, J. Monge, J. Rubio, and J. Zapata, "A CAD-oriented method to analyze and design radiating structures based on bodies of revolution by using finite elements and generalized scattering matrix," *IEEE Trans. Antennas Propagat.*, vol. 54, no. 3, pp. 899–907, Mar. 2006.

54. G. G. Gentili, P. Bolli, R. Nesti, G. Pelosi, and L. Toso, "High-order FEM mode matching analysis of circular horns with rotationally symmetric dielectrics," *IEEE Trans. Antennas Propagat.*, vol. 55, no. 10, pp. 2915–2918, Oct. 2007.

55. A. D. Greenwood and J. M. Jin, "A field picture of wave propagation in inhomogeneous dielectric lenses," *IEEE Antennas Propagat. Mag.*, vol. 41, no. 5, pp. 9–18, Oct. 1999.

56. C. S. Liang, D. A. Streater, J. M. Jin, E. Dunn, and T. Rozendal, "A quantitative study of Luneberg lens reflectors," *IEEE Antennas Propagat. Mag.*, vol. 47, no. 2, pp. 30–42, Apr. 2005.

57. R. K. Gordon and R. Mittra, "Finite element analysis of axisymmetric radomes," *IEEE Trans. Antennas Propagat.*, vol. 41, no. 7, pp. 975–981, July 1993.

58. E. Dunn, J. K. Byun, E. Branch, and J. M. Jin, "Numerical simulation of BOR scattering and radiation using a higher-order FEM," *IEEE Trans. Antennas Propagat.*, vol. 54, no. 3, pp. 945–952, Mar. 2006.

59. L. C. Kempel, J. L. Volakis, and R. J. Sliva, "Radiation by cavity-backed antennas on a circular cylinder," *IEE Proc. Microwaves Antennas Propagat.*, vol. 142, no. 3, pp. 233–239, June 1995.

60. G. E. Antilla and N. G. Alexopoulos, "Radiation and scattering from complex 3D curvilinear geometries using a hybrid finite element–integral equation method," *IEEE AP-S Int. Symp. Dig.*, pp. 1758–1761, 1992.

61. M. A. Gonzalez de Aza, J. A. Encinar, J. Zapata, and M. Lambea, "Full-wave analysis of cavity-backed and probe-fed microstrip patch arrays by a hybrid mode-matching generalized scattering matrix and finite-element method," *IEEE Trans. Antennas Propagat.*, vol. 46, no. 2, pp. 234–242, Feb. 1998.

62. T. Ozdemir, J. L. Volakis, and M. W. Nurnberger, "Analysis of thin multioctave cavity-backed slot spiral antennas," *IEE Proc. Microwaves Antennas Propagat.*, vol. 146, pp. 447–454, Dec. 1999.

63. C. A. Macon, L. C. Kempel, and S. W. Schneider, "Radiation and scattering by complex conformal antennas on a circular cylinder," *Adv. Comput. Math.*, vol. 16, pp. 191–209, 2002.

64. M. N. Vouvakis, C. A. Balanis, C. Birtcher, and A. C. Polycarpou, "Multilayer effects on cavity-backed slot antennas," *IEEE Trans. Antennas Propagat.*, vol. 52, no. 3, pp. 880–887, Mar. 2004.

65. C. A. Macon, L. C. Kempel, S. W. Schneider, and K. D. Trott, "Modeling conformal antennas on metallic prolate spheroid surfaces using a hybrid finite element method," *IEEE Trans. Antennas Propagat.*, vol. 52, no. 3, pp. 750–758, Mar. 2004.

66. K. Mao, J. K. Byun, and J. M. Jin, "Enhancing the modeling capability of the FE-BI method for simulation of cavity-backed antennas and arrays," *Electromagnetics*, vol. 26, no. 7, pp. 503–515, Oct. 2006.

67. A. C. Polycarpou and C. A. Balanis, "Finite-element investigation of scan performance characteristics of probe-fed phased arrays on magnetized ferrite substrates," *IEEE AP-S Int. Symp. Dig.*, vol. 1, pp. 666–669, July 1999.

68. M. N. Vouvakis, C. A. Balanis, C. R. Birtcher, and A. C. Polycarpou, "Ferrite-loaded cavity-backed antennas including nonuniform and nonlinear magnetization effects," *IEEE Trans. Antennas Propagat.*, vol. 51, no. 5, pp. 1000–1010, May 2003.

69. F. Bilotti, A. Toscano, and L. Vegni, "FEM–BEM formulation for the analysis of cavity-backed patch antennas on chiral substrates," *IEEE Trans. Antennas Propagat.*, vol. 51, no. 2, pp. 306–311, Feb. 2003.

70. T. Özdemir, M. W. Nurnberger, J. L. Volakis, R. Kipp, and J. Berrie, "A hybridization of finite-element and high-frequency methods for pattern prediction for antennas on aircraft structures," *IEEE Antennas Propagat. Mag.*, vol. 38, pp. 28–37, June 1996.

71. C. J. Reddy, M. D. Deshpande, C. R. Cockrell, and F. B. Beck, "Radiation characteristics of cavity-backed aperture antennas in finite ground plane using the hybrid FEM/MoM technique and geometrical theory of diffraction," *IEEE Trans. Antennas Propagat.*, vol. 44, no. 10, pp. 1327–1333, Oct. 1996.

72. A. D. Greenwood, S. S. Ni, J. M. Jin, and S. W. Lee, "Hybrid FEM/SBR method to compute the radiation pattern from a microstrip patch antenna in a complex geometry," *Microwave Opt. Tech. Lett.*, vol. 13, no. 2, pp. 84–87, Oct. 1996.

73. J. M. Jin, J. A. Berrie, R. Kipp, and S. W. Lee, "Calculation of radiation patterns of microstrip antennas on cylindrical bodies of arbitrary cross section," *IEEE Trans. Antennas Propagat.*, vol. 45, no. 1, pp. 126–132, Jan. 1997.

74. M. Alaydrus, V. Hansen, and T. F. Eibert, "Hybrid^2: Combining the three-dimensional hybrid finite element–boundary integral technique for planar multilayered media with the uniform geometrical theory of diffraction," *IEEE Trans. Antennas Propagat.*, vol. 50, no. 1, pp. 67–74, Jan. 2002.

75. J. Liu and J. M. Jin, "Analysis of conformal antennas on a complex platform," *Microwave Opt. Tech. Lett.*, vol. 36, no. 2, pp. 139–142, Jan. 2003.

76. R. Fernandez-Recio, L. E. Garcia-Castillo, I. Gomez-Revuelto, and M. Salazar-Palma, "Fully coupled multi-hybrid FEM-PO/PTD-UTD method for the analysis of radiation problems," *IEEE Trans. Magn.*, vol. 43, pp. 1341–1344, Apr. 2007.

77. A. Barka and P. Caudrillier, "Domain decomposition method based on generalized scattering matrix for installed performance of antennas on aircraft," *IEEE Trans. Antennas Propagat.*, vol. 55, no. 6, pp. 1833–1842, June 2007.

78. D. Jiao and J. M. Jin, "Fast frequency-sweep analysis of cavity-backed microstrip patch antennas," *Microwave Opt. Tech. Lett.*, vol. 22, no. 6, pp. 389–393, Sept. 1999.

79. D. Jiao and J. M. Jin, "Fast frequency-sweep analysis of microstrip antennas on a dispersive substrate," *Electron. Lett.*, vol. 35, no. 14, pp. 1122–1123, July 1999.

80. C. Geuzaine, B. Meys, V. Beauvois, and W. Legros, "An FETD approach for the modeling of antennas," *IEEE Trans. Magn.*, vol. 36, no. 4, pp. 892–896, July 2000.

81. N. Montgomery, R. Hutchins, and D. J. Riley, "Thin-wire hybrid FETD/FDTD broadband antenna predictions," *USNC/URSI Natl. Radio Sci. Mtg. Dig.*, p. 194, July 2001.

82. D. Jiao and J. M. Jin, "Time-domain finite element simulation of cavity-backed microstrip patch antennas," *Microwave Opt. Tech. Lett.*, vol. 32, no. 4, pp. 251–254, Feb. 2002.

83. F. Edelvik, G. Ledfelt, P. Lotstedt, and D. J. Riley, "An unconditionally stable subcell model for arbitrarily oriented thin wires in the FETD method," *IEEE Trans. Antennas Propagat.*, vol. 51, no. 8, pp. 1797–1805, Aug. 2003.

84. T. Rylander, F. Edelvik, A. Bondeson, and D. Riley, "Advances in hybrid FDTD-FE techniques," in *Computational Electrodynamics: The Finite-Difference Time-Domain Method*,

3rd ed., A. Taflove and S. C. Hagness, Eds. Norwood, MA: Artech House, pp. 907–953, 2005.

85. Z. Lou and J. M. Jin, "Modeling and simulation of broadband antennas using the time-domain finite element method," *IEEE Trans. Antennas Propagat.*, vol. 53, no. 12, pp. 4099–4110, Dec. 2005.

86. Z. Lou and J. M. Jin, "A novel dual-field time-domain finite-element domain-decomposition method for computational electromagnetics," *IEEE Trans. Antennas Propagat.*, vol. 54, no. 6, pp. 1850–1862, June 2006.

87. Z. Lou and J. M. Jin, "A new explicit time-domain finite-element method based on element-level decomposition," *IEEE Trans. Antennas Propagat.*, vol. 54, no. 10, pp. 2990–2999, Oct. 2006.

88. Z. Lou and J. M. Jin, "A dual-field domain-decomposition method for the time-domain finite-element analysis of large finite arrays," *J. Comput. Phys.*, vol. 222, no. 1, pp. 408–427, Mar. 2007.

89. L. E. R. Petersson and J. M. Jin, "A three-dimensional time-domain finite element formulation for periodic structures," *IEEE Trans. Antennas Propagat.*, vol. 54, no. 1, pp. 12–19, Jan. 2006.

90. L. E. R. Petersson and J. M. Jin, "Analysis of periodic structures via a time-domain finite element formulation with a Floquet ABC," *IEEE Trans. Antennas Propagat.*, vol. 54, no. 3, pp. 933–944, Mar. 2006.

91. D. J. Riley and J. M. Jin, "Time-domain finite-element analysis of electrically and magnetically dispersive periodic structures," *USNC/URSI Natl. Radio Sci. Mtg. Dig.*, p. 212.4, San Diego, CA, July 2008.

92. R. Wang, H. Wu, A. C. Cangellaris, and J.-M. Jin, "Time-domain finite-element modeling of antenna arrays with distributed feed network," *IEEE AP-S Int. Symp. Dig.*, p. 132.1, San Diego, CA, July 2008.

93. A. E. Yilmaz, Z. Lou, E. Michielssen, and J. M. Jin, "A single-boundary, implicit, and FFT-accelerated time-domain finite element–boundary integral solver," *IEEE Trans. Antennas Propagat.*, vol. 55, no. 5, pp. 1382–1397, May 2007.

2 Finite Element Formulation

The finite element method is a numerical procedure used to obtain approximate solutions to boundary-value problems of mathematical physics with the aid of an electronic computer. The method was proposed by Courant in 1943 to solve variational problems in potential theory [1]. Thereafter, the method has been developed and applied extensively to problems of structural analysis and increasingly to problems in other fields. Today, the finite element method is recognized as a general preeminent method applicable to a wide variety of engineering and mathematical problems, including those in antenna and microwave engineering.

In this chapter we describe the general formulation of the finite element method for solving the vector wave equations arising from Maxwell's equations for electric and magnetic fields. Our presentation assumes that the reader is familiar with the basic concepts in the finite element method, such as subdivision of a computational domain into finite elements, formulation of basis or interpolation functions, variational formulation or the weak-form representation of partial differential equations, and finite element matrix assembly. The reader is referred to a textbook on the finite element method for electromagnetic analysis, such as the one by Jin [2], to acquire the necessary background material. In this chapter the emphasis is on the general principle of the finite element method and the modeling of general materials in the time-domain finite element method since such modeling has not been well developed elsewhere.

2.1 FINITE ELEMENT FORMULATION IN THE FREQUENCY DOMAIN

We first formulate finite element analysis in the frequency domain. Consider a generic antenna, sketched in Figure 2.1. The antenna is excited by a current source with electric current density denoted by $\mathbf{J}_{\text{imp}}$. This current radiates an electromagnetic field that is modified by the structure of the antenna. The antenna may contain or may be embedded in anisotropic materials characterized by the permittivity and permeability tensors, denoted as $\overleftrightarrow{\varepsilon}$ and $\overleftrightarrow{\mu}$, respectively. The main objective of an antenna analysis is to predict the performance characteristics of the antenna, which

Finite Element Analysis of Antennas and Arrays, By Jian-Ming Jin and Douglas J. Riley

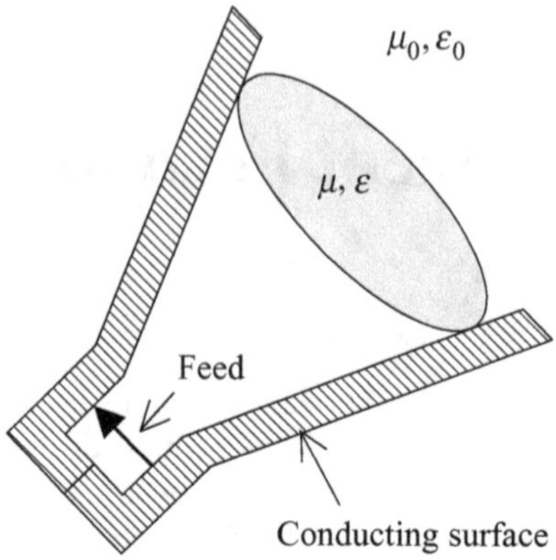

Figure 2.1 Original boundary-value problem: a generic antenna consisting of a conducting surface, a dielectric, and a feed.

include the input impedance and radiation patterns. To this end, we have to solve Maxwell's equations:

$$\nabla \times \mathbf{E} = -j\omega\overleftrightarrow{\mu} \cdot \mathbf{H} - \mathbf{M}_{\text{imp}} \tag{2.1}$$

$$\nabla \times \mathbf{H} = j\omega\overleftrightarrow{\varepsilon} \cdot \mathbf{E} + \mathbf{J}_{\text{imp}} \tag{2.2}$$

$$\nabla \cdot (\overleftrightarrow{\varepsilon} \cdot \mathbf{E}) = -\frac{1}{j\omega}\nabla \cdot \mathbf{J}_{\text{imp}} \tag{2.3}$$

$$\nabla \cdot (\overleftrightarrow{\mu} \cdot \mathbf{H}) = -\frac{1}{j\omega}\nabla \cdot \mathbf{M}_{\text{imp}} \tag{2.4}$$

together with the boundary condition

$$\hat{n} \times \mathbf{E} = 0 \qquad \mathbf{r} \in S_{\text{PEC}} \tag{2.5}$$

where S_{PEC} denotes the perfect electrically conducting (PEC) surface of the antenna. In (2.1), $\mathbf{M}_{\text{imp}}$ denotes the magnetic current density of an impressed magnetic current. Although a magnetic current does not exist in reality, it is a useful quantity to model certain antenna feeds, such as a magnetic frill generator. In addition to (2.5), the electric and magnetic fields have to satisfy the Sommerfeld radiation condition at infinity:

$$\lim_{r\to\infty} r\left[\nabla \times \begin{pmatrix}\mathbf{E}\\ \mathbf{H}\end{pmatrix} + jk_0\hat{r} \times \begin{pmatrix}\mathbf{E}\\ \mathbf{H}\end{pmatrix}\right] = 0 \tag{2.6}$$

where k_0 is the free-space wavenumber.

The electromagnetic problem defined by (2.1)–(2.6) can be solved analytically for only a very few cases where S_{PEC} has a very simple shape. For most practical

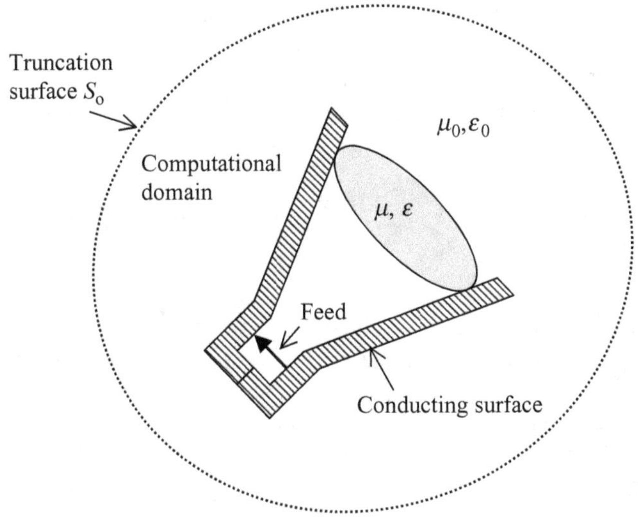

Figure 2.2 Approximate boundary-value problem: computational domain truncated by an artificial surface for numerical finite element analysis.

problems, one has to resort to a numerical method such as the finite element method for an approximate solution. Since the antenna radiates an electromagnetic field to infinity, the problem has an unbounded solution space. To use the finite element method, this unbounded space must be truncated into a finite space. This can be accomplished by introducing a fictitious surface to enclose the antenna, which is denoted as S_o in Figure 2.2. To define the electromagnetic problem bounded by S_o uniquely, one has to specify a boundary condition on S_o. This boundary condition should make S_o as transparent as possible to the radiated field. The ideal boundary condition is to make S_o completely transparent, such that the radiated field can pass through it without any distortion or reflection. This is, unfortunately, not possible in practice. Therefore, instead, we employ an approximate boundary condition,

$$\hat{n} \times \nabla \times \begin{pmatrix} \mathbf{E} \\ \mathbf{H} \end{pmatrix} + jk_0 \hat{n} \times \hat{n} \times \begin{pmatrix} \mathbf{E} \\ \mathbf{H} \end{pmatrix} \approx 0 \qquad \mathbf{r} \in S_o \tag{2.7}$$

which is similar to the Sommerfeld radiation condition in (2.6), where $\hat{n}$ denotes the unit vector normal to S_o and pointing toward the exterior space. Here we assume that S_o resides in air. The approximate radiation condition (2.7) is also called the *first-order absorbing boundary condition*. For this absorbing boundary condition to be reasonably accurate, S_o must be placed some distance (practically on the order of a half-wavelength, but theoretically, the larger the better) away from the antenna. More accurate treatment of the truncation surface S_o is discussed in Chapters 3 and 4.

The electromagnetic problem defined by (2.1)–(2.5) and (2.7) can be solved in terms of either the electric field **E** or the magnetic field **H**. Here we describe the solution procedure for the electric field; the magnetic field can be solved in a similar manner. By eliminating **H** in (2.1)–(2.4), we can derive the vector wave equation for **E** as

$$\nabla \times \left(\overleftrightarrow{\mu}_r^{-1} \cdot \nabla \times \mathbf{E}\right) - k_0^2 \overleftrightarrow{\varepsilon}_r \cdot \mathbf{E} = -jk_0 Z_0 \mathbf{J}_{\text{imp}} - \nabla \times \left(\overleftrightarrow{\mu}_r^{-1} \cdot \mathbf{M}_{\text{imp}}\right) \qquad \mathbf{r} \in V \tag{2.8}$$

where $\overleftrightarrow{\mu}_r = \overleftrightarrow{\mu}/\mu_0$ and $\overleftrightarrow{\varepsilon}_r = \overleftrightarrow{\varepsilon}/\varepsilon_0$ are the relative permeability and permittivity tensors, respectively; $k_0 = \omega\sqrt{\mu_0\varepsilon_0}$ and $Z_0 = \sqrt{\mu_0/\varepsilon_0}$ are the free-space wavenumber and intrinsic impedance, respectively; and V denotes the volume enclosed by S_o.

To solve the approximate boundary-value problem defined in Figure 2.2 with (2.7) and (2.8), we can multiply (2.8) by an appropriate testing function **T** and integrate over V to find

$$\begin{aligned}\iiint_V \mathbf{T} \cdot \left[\nabla \times \left(\overleftrightarrow{\mu}_r^{-1} \cdot \nabla \times \mathbf{E}\right) - k_0^2 \overleftrightarrow{\varepsilon}_r \cdot \mathbf{E}\right] dV \\ = -\iiint_V \mathbf{T} \cdot \left[jk_0 Z_0 \mathbf{J}_{\text{imp}} + \nabla \times \left(\overleftrightarrow{\mu}_r^{-1} \cdot \mathbf{M}_{\text{imp}}\right)\right] dV.\end{aligned} \tag{2.9}$$

By invoking the vector identity,

$$\mathbf{T} \cdot \left[\nabla \times \left(\overleftrightarrow{\mu}_r^{-1} \cdot \nabla \times \mathbf{E}\right)\right] = (\nabla \times \mathbf{T}) \cdot \overleftrightarrow{\mu}_r^{-1} \cdot (\nabla \times \mathbf{E}) - \nabla \cdot \left[\mathbf{T} \times \left(\overleftrightarrow{\mu}_r^{-1} \cdot \nabla \times \mathbf{E}\right)\right] \tag{2.10}$$

and Gauss's theorem,

$$\iiint_V \nabla \cdot \left[\mathbf{T} \times \left(\overleftrightarrow{\mu}_r^{-1} \cdot \nabla \times \mathbf{E}\right)\right] dV = \oiint_S \hat{n} \cdot \left[\mathbf{T} \times \left(\overleftrightarrow{\mu}_r^{-1} \cdot \nabla \times \mathbf{E}\right)\right] dS \tag{2.11}$$

we obtain

$$\begin{aligned}\iiint_V \left[(\nabla \times \mathbf{T}) \cdot \overleftrightarrow{\mu}_r^{-1} \cdot (\nabla \times \mathbf{E}) - k_0^2 \mathbf{T} \cdot \overleftrightarrow{\varepsilon}_r \cdot \mathbf{E}\right] dV \\ = \oiint_{S_o \cup S_{\text{PEC}}} \hat{n} \cdot \left[\mathbf{T} \times \left(\overleftrightarrow{\mu}_r^{-1} \cdot \nabla \times \mathbf{E}\right)\right] dS \\ - \iiint_V \mathbf{T} \cdot \left[jk_0 Z_0 \mathbf{J}_{\text{imp}} + \nabla \times \left(\overleftrightarrow{\mu}_r^{-1} \cdot \mathbf{M}_{\text{imp}}\right)\right] dV.\end{aligned} \tag{2.12}$$

In the formulas above, S denotes the surface that encloses the volume V. Since there exists no electric field in a perfect conductor, the volume V is bounded by the outer surface S_o and the surface of any conductor inside S_o. Application of the first-order absorbing boundary condition in (2.7) to (2.12) yields

$$\iiint_V \left[(\nabla \times \mathbf{T}) \cdot \overleftrightarrow{\mu}_r^{-1} \cdot (\nabla \times \mathbf{E}) - k_0^2 \mathbf{T} \cdot \overleftrightarrow{\varepsilon}_r \cdot \mathbf{E}\right] dV$$
$$= \oiint_{S_{\mathrm{PEC}}} (\hat{n} \times \mathbf{T}) \cdot \overleftrightarrow{\mu}_r^{-1} \cdot (\nabla \times \mathbf{E})\, dS - jk_0 \oiint_{S_o} (\hat{n} \times \mathbf{T}) \cdot (\hat{n} \times \mathbf{E})\, dS$$
$$- \iiint_V \mathbf{T} \cdot \left[jk_0 Z_0 \mathbf{J}_{\mathrm{imp}} + \nabla \times \left(\overleftrightarrow{\mu}_r^{-1} \cdot \mathbf{M}_{\mathrm{imp}}\right)\right] dV. \quad (2.13)$$

To find a numerical solution of (2.13) using the finite element method, the entire volume V is first divided into small finite elements, such as hexahedral, tetrahedral, prism, and/or pyramidal cells (Figure 2.3). The size of finite elements is determined primarily by the wavelength and the geometrical structure of the antenna. Typically, the elements should be smaller than one-tenth to one-twentieth of a wavelength when one uses the first-order basis functions discussed below. Local to the fine structure of

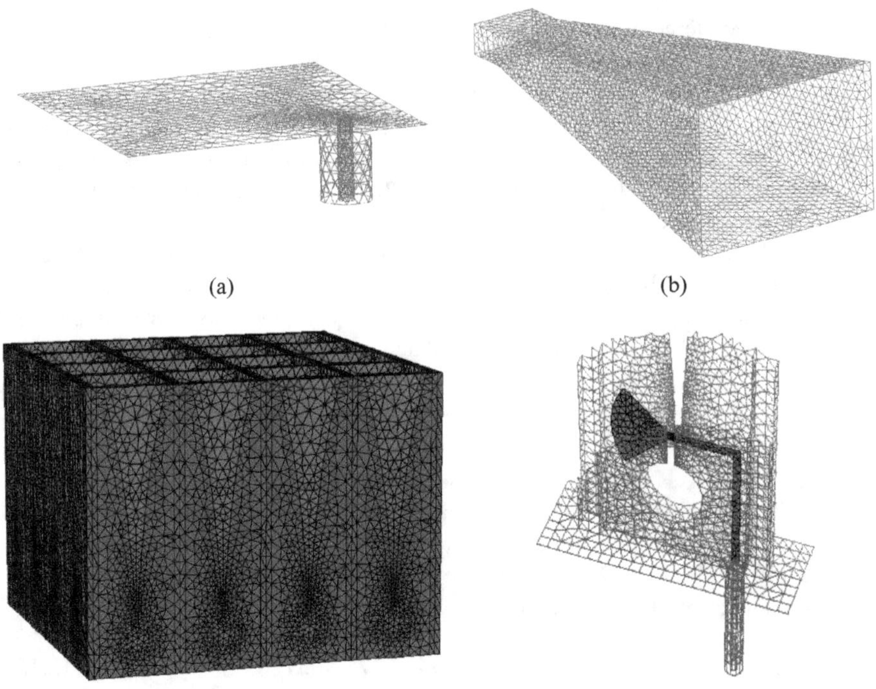

Figure 2.3 Examples of finite element meshes (Only surface meshes are shown for clarity). (a) A microstrip patch antenna fed by a coaxial line with the substrate and ground plane removed. (b) A horn antenna. (c) A 4 × 4 dual-polarized Vivaldi antenna array. (d) The feed structure of a Vivaldi antenna. (*See insert for color representation of figure.*)

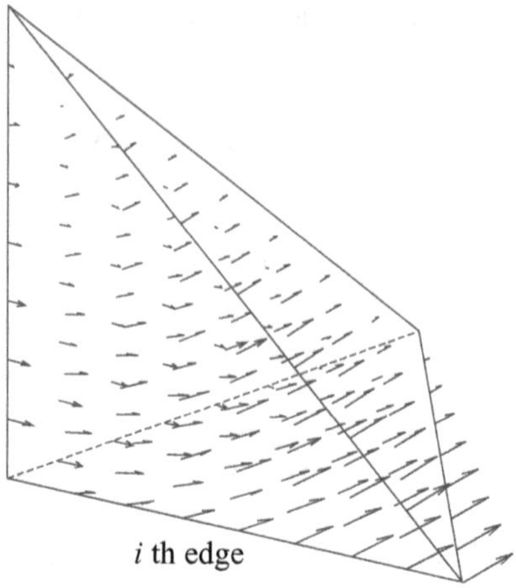

Figure 2.4 Vector plot of the first-order vector basis function associated with the ith edge. There are six such first-order vector basis functions associated with the six edges of a tetrahedral element.

the antenna, the elements have to be much smaller in order to resolve the geometry as well as the fast variation of the fields.

Once the finite element mesh is generated, within each small finite element volume, **E** can be interpolated using a set of discrete values. One approach is to assign **E** at a few points on the element and then interpolate **E** elsewhere using a set of scalar interpolation functions. This approach turns out to be very problematic because of the difficulty in applying correct boundary conditions (boundary conditions at conducting surfaces and interfaces between different media) to the interpolated **E**-field. A better approach is to assign the tangential component of **E** at each edge of the element and then interpolate **E** elsewhere using a set of vector basis functions. For example, the field in a tetrahedral element can be interpolated as

$$\mathbf{E}^e(x, y, z) = \sum_{i=1}^{6} \mathbf{N}_i^e(x, y, z) E_i^e \tag{2.14}$$

where E_i^e denotes the tangential component of **E** at edge i of element e, and $\mathbf{N}_i^e$ is the corresponding interpolation or basis function. Denoting the simplex coordinates of a tetrahedron as $\lambda_l (l = 1, 2, 3, 4)$ [2], the vector basis function associated with the edge that connects nodes l and k can be written as

$$\mathbf{N}_{lk}^e(\mathbf{r}) = \ell_{lk}(\lambda_l \nabla\lambda_k - \lambda_k \nabla\lambda_l) \tag{2.15}$$

where ℓ_{lk} denotes the length of the edge and $l, k = 1, 2, 3, 4$ with $l < k$. Figure 2.4 shows one of the six first-order vector basis functions for a tetrahedral element.[†] To

[†]Note that the order of vector basis functions here is defined differently from that in Ref. 2. The basis functions defined in (2.15) are referred to here as the first order instead of the zeroth order as in Ref. 2.

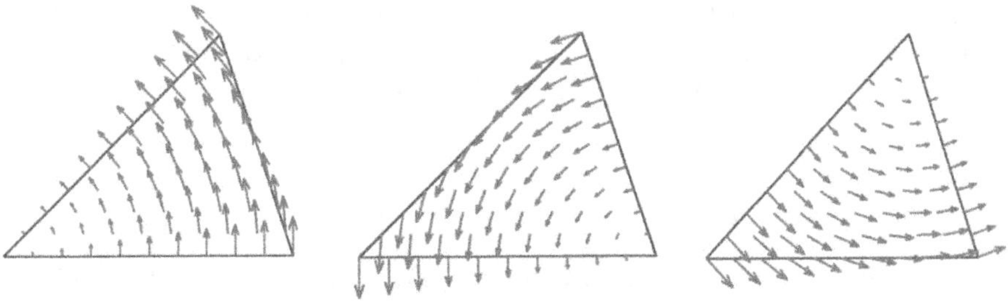

Figure 2.5 Vector basis functions for a linear triangular element, which can be considered as one of the four facets of a tetrahedral element.

visualize the vector basis functions more clearly, Figure 2.5 plots the three vector basis functions on one of the four facets of a tetrahedral element. Clearly, such basis functions have a tangential component only along the associated edge, and as such they ensure the tangential continuity of the interpolated field while allowing the normal component to be discontinuous at a material discontinuity. Hence, they accurately model the nature of the vector field **E**. Higher-order vector basis functions can also be constructed to achieve better interpolation accuracy [2].

When the **E**-field is interpolated in each element using its tangential values at the edges of the element, the **E**-field in the entire volume V can be expressed as

$$\mathbf{E} = \sum_{i=1}^{N_{\text{edge}}} \mathbf{N}_i E_i \tag{2.16}$$

where N_{edge} denotes the total number of edges excluding those on S_{PEC}, E_i denotes the tangential component of **E** at the ith edge, and $\mathbf{N}_i$ is the vector basis function corresponding to the ith edge. Obviously, for an edge inside V, $\mathbf{N}_i$ spans several neighboring elements that share the common edge. Also note that by excluding the edges on S_{PEC} in (2.16), the interpolated field is guaranteed to satisfy the required boundary condition in (2.5).

By substituting (2.16) into (2.13) and using the same $\mathbf{N}_i$ as the weighting function **T**, we obtain

$$\sum_{j=1}^{N_{\text{edge}}} K_{ij} E_j = b_i \qquad i = 1, 2, \ldots, N_{\text{edge}} \tag{2.17}$$

where

$$K_{ij} = \iiint_V \left[(\nabla \times \mathbf{N}_i) \cdot \overleftrightarrow{\mu}_r^{-1} \cdot (\nabla \times \mathbf{N}_j) - k_0^2 \mathbf{N}_i \cdot \overleftrightarrow{\varepsilon}_r \cdot \mathbf{N}_j \right] dV$$

$$+ jk_0 \oiint_{S_o} (\hat{n} \times \mathbf{N}_i) \cdot (\hat{n} \times \mathbf{N}_j)\, dS \tag{2.18}$$

$$b_i = -\iiint_V \mathbf{N}_i \cdot \left[jk_0 Z_0 \mathbf{J}_{\text{imp}} + \nabla \times \left(\overleftrightarrow{\mu}_r^{-1} \cdot \mathbf{M}_{\text{imp}} \right) \right] dV. \tag{2.19}$$

This formulation is called *Galerkin's formulation* since the same functions are used for both expansion and testing. Note that the integral over S_{PEC} in (2.13) vanishes here since $\hat{n} \times \mathbf{N}_i = 0$ on S_{PEC}. Equation (2.17) can be written compactly as

$$[K]\{E\} = \{b\} \tag{2.20}$$

which can be solved for $\{E\}$. Because the elemental interactions in (2.18) are local in nature, $[K]$ is a sparse and symmetric (if $\overleftrightarrow{\varepsilon}_r$ and $\overleftrightarrow{\mu}_r$ are symmetric) matrix that can be solved efficiently using a sparse matrix solver. Once $\{E\}$ is obtained, the field everywhere in V can be calculated using (2.16), from which other parameters, such as the input impedance and radiation patterns, can be computed.

The formulation described above can easily be extended to the case where materials are lossy and/or dispersive because $\overleftrightarrow{\varepsilon}_r$ and $\overleftrightarrow{\mu}_r$ can be made complex (to include the electric and magnetic losses) and can have different values for different frequencies since the analysis is performed at a single frequency each time.

2.2 FINITE ELEMENT FORMULATION IN THE TIME DOMAIN

The finite element formulation described in Section 2.1 operates in the frequency domain. It takes the frequency as an input parameter and solves for the electric field at that specific frequency. However, for most antenna analyses, one is interested in the characteristics of an antenna over a frequency band. In such a case, one often has to repeat a frequency-dependent analysis at many frequencies, which can be quite time consuming for both narrow- and broadband antennas. For narrowband antennas, the antenna characteristics change rapidly with frequency due to resonance. A very small frequency step has to be used to capture impedance variations accurately near resonance. For broadband antennas, the frequency step can be made larger; however, the entire frequency band may be very wide, thus requiring many frequency samples as well. Furthermore, it is found to be difficult to model nonlinear devices/media accurately using a frequency-domain-based method. These two challenges can be alleviated by using the finite element method formulated in the time domain.

In the time domain, the first two of Maxwell's equations (2.1)–(2.4) become

$$\nabla \times \mathbf{E}(t) = -\overleftrightarrow{\mu} \cdot \frac{\partial \mathbf{H}(t)}{\partial t} - \mathbf{M}_{\text{imp}}(t) \tag{2.21}$$

$$\nabla \times \mathbf{H}(t) = \overleftrightarrow{\varepsilon} \cdot \frac{\partial \mathbf{E}(t)}{\partial t} + \overleftrightarrow{\sigma}_e \cdot \mathbf{E}(t) + \mathbf{J}_{\text{imp}}(t) \tag{2.22}$$

where $\overleftrightarrow{\sigma}_e$ denotes the electrical conductivity tensor. The boundary condition on the perfectly conducting surface remains the same as in (2.5), and the boundary condition corresponding to (2.7) becomes

$$\hat{n} \times \left(\frac{1}{\mu_0} \nabla \times \mathbf{E} \right) + Y_0 \hat{n} \times \left(\hat{n} \times \frac{\partial}{\partial t} \mathbf{E} \right) \approx 0 \qquad \mathbf{r} \in S_o \tag{2.23}$$

where $Y_0 = 1/Z_0$ and S_o is assumed to reside in air. By eliminating the magnetic field from (2.21) and (2.22), we obtain the vector wave equation for the electric field as

$$\nabla \times [\overleftrightarrow{\mu}^{-1} \cdot \nabla \times \mathbf{E}(t)] + \overleftrightarrow{\varepsilon} \cdot \frac{\partial^2 \mathbf{E}(t)}{\partial t^2} + \overleftrightarrow{\sigma}_e \cdot \frac{\partial \mathbf{E}(t)}{\partial t} = -\frac{\partial \mathbf{J}_{\text{imp}}}{\partial t} - \nabla \times (\overleftrightarrow{\mu}^{-1} \cdot \mathbf{M}_{\text{imp}}) \tag{2.24}$$

and its weak-form representation as

$$\begin{aligned} &\iiint_V \left[(\nabla \times \mathbf{T}) \cdot \overleftrightarrow{\mu}^{-1} \cdot (\nabla \times \mathbf{E}) + \mathbf{T} \cdot \overleftrightarrow{\varepsilon} \cdot \frac{\partial^2 \mathbf{E}}{\partial t^2} + \mathbf{T} \cdot \overleftrightarrow{\sigma}_e \cdot \frac{\partial \mathbf{E}}{\partial t} \right] dV \\ &\quad + Y_0 \oiint_{S_o} (\hat{n} \times \mathbf{T}) \cdot \left(\hat{n} \times \frac{\partial \mathbf{E}}{\partial t} \right) dS \\ &= -\iiint_V \mathbf{T} \cdot \left[\frac{\partial \mathbf{J}_{\text{imp}}}{\partial t} + \nabla \times (\overleftrightarrow{\mu}^{-1} \cdot \mathbf{M}_{\text{imp}}) \right] dV. \end{aligned} \tag{2.25}$$

The derivation follows that in the frequency domain. Note that the vector testing function is assumed to satisfy the boundary condition $\hat{n} \times \mathbf{T} = 0$ on S_{PEC} since (2.5) is enforced in the solution of (2.25).

To seek the finite element solution of (2.25), we first perform the spatial discretization in exactly the same manner as in the frequency domain. To be more specific, by subdividing the solution volume into small finite elements and expanding the electric field within each element using the vector basis functions, we can express the electric field as

$$\mathbf{E}(\mathbf{r}, t) = \sum_{i=1}^{N_{\text{edge}}} \mathbf{N}_i(\mathbf{r}) E_i(t). \tag{2.26}$$

Substituting (2.26) into (2.25) yields the second-order ordinary differential equation

$$[T] \frac{d^2\{E\}}{dt^2} + [R] \frac{d\{E\}}{dt} + [S]\{E\} = \{f\} \tag{2.27}$$

where $[T]$, $[R]$, and $[S]$ represent sparse, symmetric (if $\overleftrightarrow{\varepsilon}$, $\overleftrightarrow{\mu}$, and $\overleftrightarrow{\sigma}_e$ are all symmetric) matrices whose elements are given by

$$T_{ij} = \iiint_V \mathbf{N}_i \cdot \overleftrightarrow{\varepsilon} \cdot \mathbf{N}_j \, dV \tag{2.28}$$

$$R_{ij} = \iiint_V \mathbf{N}_i \cdot \overleftrightarrow{\sigma}_e \cdot \mathbf{N}_j \, dV + Y_0 \oiint_{S_o} (\hat{n} \times \mathbf{N}_i) \cdot (\hat{n} \times \mathbf{N}_j) \, dS \tag{2.29}$$

$$S_{ij} = \iiint_V (\nabla \times \mathbf{N}_i) \cdot \overleftrightarrow{\mu}^{-1} \cdot (\nabla \times \mathbf{N}_j) \, dV. \tag{2.30}$$

Furthermore, $\{E\} = [E_1, E_2, \ldots, E_{N_{\text{edge}}}]^{\text{T}}$ (where the superscript T denotes the transpose) and the elements of the excitation vector $\{f\}$ are given by

$$f_i(t) = -\iiint_V \mathbf{N}_i \cdot \left[\frac{\partial \mathbf{J}_{\text{imp}}}{\partial t} + \nabla \times (\overleftrightarrow{\mu}^{-1} \cdot \mathbf{M}_{\text{imp}}) \right] dV. \tag{2.31}$$

Having resolved the spatial variation by the finite element procedure, the temporal variation of (2.27) can be solved by using direct time integration or by the finite-difference method [2]. In the finite-difference method, the time variable t is uniformly discretized such that it is represented by $t = n\,\Delta t$ $(n = 0, 1, \ldots)$, where Δt is denoted as the time step. The continuous time derivatives are then approximated by finite differences, which yields an equation that allows for calculation of the unknown vector $\{E\}$ based on its previous values in time. This process is called *time marching*. As shown in Ref. 2, the use of forward differencing will result in an *unstable* time-marching equation (the solution grows exponentially and becomes completely erroneous). The use of backward differencing will result in an *unconditionally stable* (the time-step size Δt is not constrained by the spatial discretization) time-marching equation, which unfortunately is only first-order accurate—the accuracy of the solution is proportional to Δt. The use of central differencing will yield a second-order-accurate time-marching equation, which is *conditionally stable*—the time marching is stable only when Δt is smaller than a certain value dictated by the spatial discretization. For (2.27), a preferred choice is to use a differencing formula derived from the Newmark-beta time integration method [3–5], which is equivalent to using central differencing for the first- and second-order time derivatives

$$\frac{d\{E\}}{dt} \approx \frac{\{E\}^{n+1} - \{E\}^{n-1}}{2\Delta t} \tag{2.32}$$

$$\frac{d^2\{E\}}{dt^2} \approx \frac{\{E\}^{n+1} - 2\{E\}^n + \{E\}^{n-1}}{(\Delta t)^2} \tag{2.33}$$

and using a weighted average for the undifferentiated quantities

$$\{E\} \approx \beta\{E\}^{n+1} + (1-2\beta)\{E\}^{n} + \beta\{E\}^{n-1} \tag{2.34}$$

$$\{f\} \approx \beta\{f\}^{n+1} + (1-2\beta)\{f\}^{n} + \beta\{f\}^{n-1} \tag{2.35}$$

where β is a parameter that takes a value between 0 and 1. In (2.32)–(2.35), the superscript denotes the time at which the associated quantity is evaluated; for example, $\{E\}^n$ denotes the vector $\{E\}$ evaluated at $t = n\,\Delta t$. Substituting these into (2.27), we obtain the time-marching or time-stepping equation

$$\begin{aligned}\left\{\frac{1}{(\Delta t)^2}[T] + \frac{1}{2\Delta t}[R] + \beta[S]\right\}\{E\}^{n+1} &= \left\{\frac{2}{(\Delta t)^2}[T] - (1-2\beta)[S]\right\}\{E\}^{n} \\ &- \left\{\frac{1}{(\Delta t)^2}[T] - \frac{1}{2\Delta t}[R] + \beta[S]\right\}\{E\}^{n-1} + \beta\{f\}^{n+1} + (1-2\beta)\{f\}^{n} + \beta\{f\}^{n-1}.\end{aligned} \tag{2.36}$$

When $\beta = 0$, this equation reduces to the formula obtained using central differencing. However, it can be shown that when $\beta \geq 1/4$, this equation is unconditionally stable [4] while maintaining second-order accuracy. Given the initial values of $\{E\}$, that is, $\{E\}^0$ and $\{E\}^1$, and the values of the excitation vector $\{f\}$, (2.36) can be employed to compute all subsequent values for $\{E\}$. To compute each new $\{E\}$, we have to solve a matrix equation at each time step. The unconditional stability of (2.36) is very important for problems involving very small finite elements in order to model fine structures such as the example shown in Figure 1.1. With this unconditional stability, Δt can be chosen based entirely on the temporal variation of the field instead of being constrained by the smallest finite elements in the solution domain.

2.3 MODELING OF COMPLEX MATERIALS

The formulation described in Section 2.2 assumes that the materials involved are dispersion free. In other words, both the permeability $\overleftrightarrow{\mu}$ and the permittivity $\overleftrightarrow{\varepsilon}$ are invariant with respect to frequency. Furthermore, the magnetic loss is ignored. Because advanced antenna designs often take advantage of engineered materials to achieve desired antenna characteristics, it is critical that any analysis tool be able to model complex materials accurately. In this section we describe the modeling of materials in the time-domain finite element method that can be simultaneously electrically and magnetically lossy and dispersive. (As mentioned earlier, modeling these types of materials in the frequency domain is rather straightforward.) To improve understanding, we first treat nondispersive lossy materials, which is followed by the treatment of electrically and magnetically dispersive materials, and finally, we combine these results to treat the case of general materials [6–11].

2.3.1 Modeling of Electrically and Magnetically Lossy Materials

Consider a nondispersive, anisotropic, lossy volume, which is characterized by permeability $\overset{\leftrightarrow}{\mu}$, permittivity $\overset{\leftrightarrow}{\varepsilon}$, electrical conductivity $\overset{\leftrightarrow}{\sigma}_e$, and magnetic conductivity $\overset{\leftrightarrow}{\sigma}_m$. Maxwell's equations in such a domain can be written as

$$\nabla \times \mathbf{E} = -\overset{\leftrightarrow}{\mu} \cdot \frac{\partial \mathbf{H}}{\partial t} - \overset{\leftrightarrow}{\sigma}_m \cdot \mathbf{H} - \mathbf{M}_{\text{imp}} \tag{2.37}$$

$$\nabla \times \mathbf{H} = \overset{\leftrightarrow}{\varepsilon} \cdot \frac{\partial \mathbf{E}}{\partial t} + \overset{\leftrightarrow}{\sigma}_e \cdot \mathbf{E} + \mathbf{J}_{\text{imp}}. \tag{2.38}$$

Taking the curl of (2.37), we obtain

$$\nabla \times (\overset{\leftrightarrow}{\mu}^{-1} \cdot \nabla \times \mathbf{E}) = -\frac{\partial}{\partial t}(\nabla \times \mathbf{H}) - \nabla \times (\overset{\leftrightarrow}{\mu}^{-1} \cdot \overset{\leftrightarrow}{\sigma}_m \cdot \mathbf{H}) - \nabla \times (\overset{\leftrightarrow}{\mu}^{-1} \cdot \mathbf{M}_{\text{imp}}). \tag{2.39}$$

Substituting (2.38) into (2.39) yields

$$\begin{aligned}\nabla \times (\overset{\leftrightarrow}{\mu}^{-1} \cdot \nabla \times \mathbf{E}) &= -\overset{\leftrightarrow}{\varepsilon} \cdot \frac{\partial^2 \mathbf{E}}{\partial t^2} - \overset{\leftrightarrow}{\sigma}_e \cdot \frac{\partial \mathbf{E}}{\partial t} - \frac{\partial \mathbf{J}_{\text{imp}}}{\partial t} \\ &\quad - \nabla \times (\overset{\leftrightarrow}{\mu}^{-1} \cdot \overset{\leftrightarrow}{\sigma}_m \cdot \mathbf{H}) - \nabla \times (\overset{\leftrightarrow}{\mu}^{-1} \cdot \mathbf{M}_{\text{imp}})\end{aligned} \tag{2.40}$$

which can also be written as

$$\begin{aligned}&\nabla \times (\overset{\leftrightarrow}{\mu}^{-1} \cdot \nabla \times \mathbf{E}) + \overset{\leftrightarrow}{\varepsilon} \cdot \frac{\partial^2 \mathbf{E}}{\partial t^2} + \overset{\leftrightarrow}{\sigma}_e \cdot \frac{\partial \mathbf{E}}{\partial t} + \nabla \times (\overset{\leftrightarrow}{\mu}^{-1} \cdot \overset{\leftrightarrow}{\sigma}_m \cdot \mathbf{H}) \\ &= -\frac{\partial \mathbf{J}_{\text{imp}}}{\partial t} - \nabla \times (\overset{\leftrightarrow}{\mu}^{-1} \cdot \mathbf{M}_{\text{imp}}).\end{aligned} \tag{2.41}$$

Testing (2.41) with $\mathbf{T}$ and integrating over a volume V yields

$$\begin{aligned}&\iiint_V \mathbf{T} \cdot \left[\nabla \times (\overset{\leftrightarrow}{\mu}^{-1} \cdot \nabla \times \mathbf{E})\right] dV + \iiint_V \mathbf{T} \cdot \left(\overset{\leftrightarrow}{\varepsilon} \cdot \frac{\partial^2 \mathbf{E}}{\partial t^2} + \overset{\leftrightarrow}{\sigma}_e \cdot \frac{\partial \mathbf{E}}{\partial t}\right) dV \\ &+ \iiint_V \mathbf{T} \cdot \nabla \times (\overset{\leftrightarrow}{\mu}^{-1} \cdot \overset{\leftrightarrow}{\sigma}_m \cdot \mathbf{H})\, dV = \iiint_V \mathbf{T} \cdot \mathbf{g}_{\text{imp}}\, dV\end{aligned} \tag{2.42}$$

where

$$\mathbf{g}_{\text{imp}} = -\frac{\partial \mathbf{J}_{\text{imp}}}{\partial t} - \nabla \times (\overset{\leftrightarrow}{\mu}^{-1} \cdot \mathbf{M}_{\text{imp}}). \tag{2.43}$$

The finite element discretization of the second and last integrals in (2.42) is straightforward. Let us focus on the first integral, which can be written as

$$
\begin{aligned}
\iiint_V \mathbf{T}\cdot\left[\nabla\times(\overleftrightarrow{\mu}^{-1}\cdot\nabla\times\mathbf{E})\right]dV &= -\iiint_V \nabla\cdot\left[\mathbf{T}\times(\overleftrightarrow{\mu}^{-1}\cdot\nabla\times\mathbf{E})\right]dV \\
&\quad + \iiint_V (\nabla\times\mathbf{T})\cdot\overleftrightarrow{\mu}^{-1}\cdot(\nabla\times\mathbf{E})\,dV \\
&= -\oiint_S \left[\mathbf{T}\times(\overleftrightarrow{\mu}^{-1}\cdot\nabla\times\mathbf{E})\right]\cdot\hat{n}\,dS \\
&\quad + \iiint_V (\nabla\times\mathbf{T})\cdot\overleftrightarrow{\mu}^{-1}\cdot(\nabla\times\mathbf{E})\,dV
\end{aligned}
\tag{2.44}
$$

where we applied a common vector identity and Gauss's divergence theorem. Substituting (2.37) into (2.44), we obtain

$$
\begin{aligned}
\iiint_V \mathbf{T}\cdot\left[\nabla\times(\overleftrightarrow{\mu}^{-1}\cdot\nabla\times\mathbf{E})\right]dV &= \oiint_S \left[\mathbf{T}\times\frac{\partial\mathbf{H}}{\partial t}+\mathbf{T}\times(\overleftrightarrow{\mu}^{-1}\cdot\overleftrightarrow{\sigma}_m\cdot\mathbf{H})\right]\cdot\hat{n}\,dS \\
&\quad + \iiint_V (\nabla\times\mathbf{T})\cdot\overleftrightarrow{\mu}^{-1}\cdot(\nabla\times\mathbf{E})\,dV
\end{aligned}
\tag{2.45}
$$

where it is assumed that no surface magnetic current exists on S.

Let us now consider the third term in (2.42), which can be written as

$$
\begin{aligned}
\iiint_V \mathbf{T}\cdot\nabla\times(\overleftrightarrow{\mu}^{-1}\cdot\overleftrightarrow{\sigma}_m\cdot\mathbf{H})\,dV &= \iiint_V \Big\{(\nabla\times\mathbf{T})\cdot(\overleftrightarrow{\mu}^{-1}\cdot\overleftrightarrow{\sigma}_m\cdot\mathbf{H}) \\
&\quad -\nabla\cdot\left[\mathbf{T}\times(\overleftrightarrow{\mu}^{-1}\cdot\overleftrightarrow{\sigma}_m\cdot\mathbf{H})\right]\Big\}\,dV \\
&= \iiint_V (\nabla\times\mathbf{T})\cdot(\overleftrightarrow{\mu}^{-1}\cdot\overleftrightarrow{\sigma}_m\cdot\mathbf{H})\,dV \\
&\quad - \oiint_S \left[\mathbf{T}\times(\overleftrightarrow{\mu}^{-1}\cdot\overleftrightarrow{\sigma}_m\cdot\mathbf{H})\right]\cdot\hat{n}\,dS.
\end{aligned}
\tag{2.46}
$$

Surprisingly, the second term on the right-hand side of (2.46) cancels the second term in the surface integral of (2.45). Hence, (2.42) becomes

$$
\begin{aligned}
&\iiint_V (\nabla\times\mathbf{T})\cdot\overleftrightarrow{\mu}^{-1}\cdot(\nabla\times\mathbf{E})\,dV + \iiint_V \mathbf{T}\cdot\left(\overleftrightarrow{\varepsilon}\cdot\frac{\partial^2\mathbf{E}}{\partial t^2}+\overleftrightarrow{\sigma}_e\cdot\frac{\partial\mathbf{E}}{\partial t}\right)dV \\
&+\iiint_V (\nabla\times\mathbf{T})\cdot(\overleftrightarrow{\mu}^{-1}\cdot\overleftrightarrow{\sigma}_m\cdot\mathbf{H})\,dV + \oiint_S \left(\mathbf{T}\times\frac{\partial\mathbf{H}}{\partial t}\right)\cdot\hat{n}\,dS = \iiint_V \mathbf{T}\cdot\mathbf{g}_{\text{imp}}\,dV.
\end{aligned}
\tag{2.47}
$$

The surface integral in this equation facilitates the application of the boundary condition over S.

We note that in the finite element implementation of (2.47), the magnetic field $\mathbf{H}$ should not be treated as an independent unknown function. Rather, it can be moved to the right-hand side and treated as the source for the electric field. In particular, with the electric and magnetic fields $\mathbf{E}^n$ and $\mathbf{H}^{n-1/2}$, where n denotes the time step, we can calculate the magnetic field at the next step using (2.37) via central differencing,

$$\begin{aligned}\mathbf{H}^{n+1/2} &= \left[2\overleftrightarrow{\mu} + \overleftrightarrow{\sigma}_m \, \Delta t\right]^{-1} \cdot \left[2\overleftrightarrow{\mu} - \overleftrightarrow{\sigma}_m \, \Delta t\right] \cdot \mathbf{H}^{n-1/2} \\ &\quad - 2\Delta t \left[2\overleftrightarrow{\mu} + \overleftrightarrow{\sigma}_m \, \Delta t\right]^{-1} \cdot \left[(\nabla \times \mathbf{E})^n + \mathbf{M}^n_{\text{imp}}\right] \end{aligned} \tag{2.48}$$

which can be used to update the right-hand side of the time-marching equation. When the first-order edge elements are used to represent the electric field, $\nabla \times \mathbf{E}$—and, hence, the discretized magnetic field—is constant within each element.

To verify the formulation presented above, we consider the problem of scattering from a rectangular slab for the case of plane-wave incidence along the z-axis. The rectangular slab has dimensions of 1 m $\times$ 1 m $\times$ 0.1 m in the x-, y-, and z-directions, respectively. We analyze the problem for both a magnetically and an electrically lossy slab [9]. For the magnetically lossy case, $\sigma_m = 1.047 \times 10^4$ Ω/m, and for the electrically lossy case, $\sigma_e = 0.0738$ S/m. With these parameters, both slabs would produce the same normalized backscattered field. For the finite element time-domain (FETD) solution, the geometry was discretized with first-order tetrahedral edge elements with an average edge length of 1.25 cm external to the slab and 6.25 mm within the slab. For the reference finite-difference time-domain (FDTD) solution, 6.25-mm cubes were used throughout the computational domain. The transient result for the case of the magnetically lossy slab for a unit-amplitude Gaussian pulse excitation is shown in Figure 2.6(a), which shows excellent agreement between the finite element and FDTD solutions. The corresponding backscattered electric field as a function of frequency is shown in Figure 2.6(b), where the finite element and FDTD curves for the magnetically lossy case coincide with each other and a small deviation is observed for the finite element solution of the electrically lossy slab.

2.3.2 Modeling of Electrically Dispersive Materials

In an anisotropic, electrically dispersive material, Maxwell's equations (neglecting the source terms for simplicity) become

$$\nabla \times \mathbf{E}(t) = -\overleftrightarrow{\mu} \cdot \frac{\partial \mathbf{H}(t)}{\partial t} \tag{2.49}$$

$$\nabla \times \mathbf{H}(t) = \frac{\partial \mathbf{D}(t)}{\partial t} \tag{2.50}$$

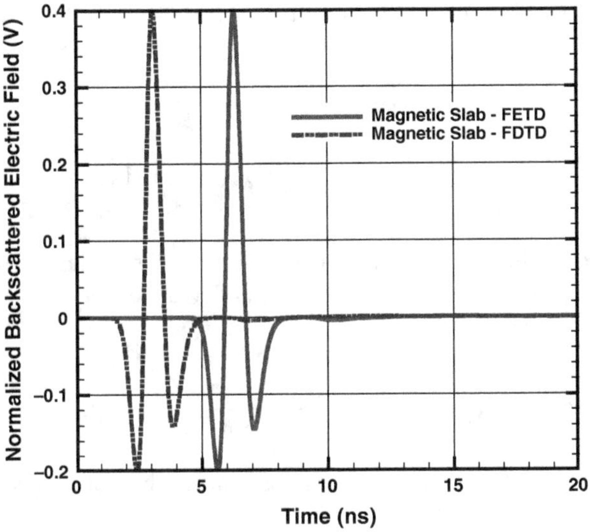

(a)

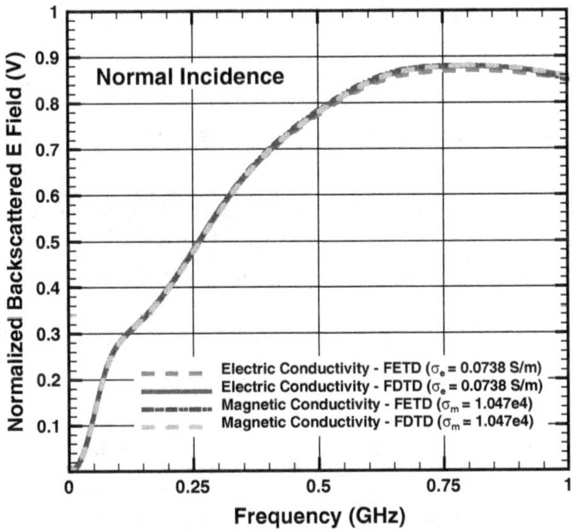

(b)

Figure 2.6 Backscattered electric field from a rectangular slab of dimensions 1 m × 1 m × 0.1 m. Comparison between the FETD magnetic loss formulation based on (2.47) and a traditional FDTD formulation. (a) Field versus time (The FETD curve is shifted for clarity). (b) Field versus frequency (Similar results for an electrically conductive slab are provided for reference). (After Riley and Jin [9], Copyright © Wiley 2005.)

where $\overleftrightarrow{\mu}$ is a time-invariant permeability tensor and $\mathbf{D}(t)$ is related to $\mathbf{E}(t)$ by the constitutive relation

$$\begin{aligned}\mathbf{D}(t) &= \varepsilon_0 \overleftrightarrow{\varepsilon}_\infty \cdot \mathbf{E}(t) + \varepsilon_0 \overleftrightarrow{\chi}_e(t) * \mathbf{E}(t) \\ &= \varepsilon_0 \overleftrightarrow{\varepsilon}_\infty \cdot \mathbf{E}(t) + \varepsilon_0 \int_0^t \overleftrightarrow{\chi}_e(t-\tau) \cdot \mathbf{E}(\tau)\, d\tau \end{aligned} \tag{2.51}$$

where $\overleftrightarrow{\varepsilon}_\infty$ denotes the relative permittivity tensor at the optical frequency, $\overleftrightarrow{\chi}_e(t)$ represents the electrical susceptibility tensor, and $*$ denotes the time convolution. From (2.49) and (2.50), we obtain

$$\nabla \times \left[\overleftrightarrow{\mu}^{-1} \cdot \nabla \times \mathbf{E}(t)\right] + \varepsilon_0 \overleftrightarrow{\varepsilon}_\infty \cdot \frac{\partial^2 \mathbf{E}(t)}{\partial t^2} + \varepsilon_0 \frac{\partial^2}{\partial t^2}\left[\overleftrightarrow{\chi}_e(t) * \mathbf{E}(t)\right] = 0. \tag{2.52}$$

The last term on the left-hand side in (2.52) has two equivalent expressions,

$$\frac{\partial^2}{\partial t^2}\left[\overleftrightarrow{\chi}_e(t) * \mathbf{E}(t)\right] = \frac{\partial^2 \overleftrightarrow{\chi}_e(t)}{\partial t^2} * \mathbf{E}(t) \equiv \ddot{\overleftrightarrow{\chi}}_e(t) * \mathbf{E}(t) \tag{2.53}$$

and

$$\frac{\partial^2}{\partial t^2}\left[\overleftrightarrow{\chi}_e(t) * \mathbf{E}(t)\right] = \overleftrightarrow{\chi}_e(t) * \frac{\partial^2 \mathbf{E}(t)}{\partial t^2} \equiv \overleftrightarrow{\chi}_e(t) * \ddot{\mathbf{E}}(t). \tag{2.54}$$

Therefore, there are two approaches to dealing with the convolution term in (2.52). One is based on (2.53), which requires the second derivative of the electrical susceptibility; the other employs (2.54), which involves the second derivative of the electric field.

Let us first consider the approach based on the use of (2.53), for which (2.52) can be written as

$$\nabla \times \left[\overleftrightarrow{\mu}^{-1} \cdot \nabla \times \mathbf{E}(t)\right] + \varepsilon_0 \overleftrightarrow{\varepsilon}_\infty \cdot \frac{\partial^2 \mathbf{E}(t)}{\partial t^2} + \varepsilon_0 \ddot{\overleftrightarrow{\chi}}_e(t) * \mathbf{E}(t) = 0. \tag{2.55}$$

The weak form of (2.55) can be obtained by taking the dot product with a testing function $\mathbf{T}$ and then invoking Gauss's divergence theorem, yielding

$$\begin{aligned}&\iiint_V \left[(\nabla \times \mathbf{T}) \cdot \overleftrightarrow{\mu}^{-1} \cdot \nabla \times \mathbf{E}(t) + \varepsilon_0 \mathbf{T} \cdot \overleftrightarrow{\varepsilon}_\infty \cdot \frac{\partial^2 \mathbf{E}(t)}{\partial t^2} + \varepsilon_0 \mathbf{T} \cdot \ddot{\overleftrightarrow{\chi}}_e(t) * \mathbf{E}(t)\right] dV \\ &+ \oiint_S \mathbf{T} \cdot \hat{n} \times \left[\overleftrightarrow{\mu}^{-1} \cdot \nabla \times \mathbf{E}(t)\right] dS = 0. \end{aligned} \tag{2.56}$$

The last term (surface integral) can be used to incorporate the boundary condition on the surface that encloses the volume of interest. Before we consider the finite element discretization of (2.56), let us assume that the electrical susceptibility can be represented by

$$\overleftrightarrow{\chi}_e(\omega) = \frac{\overleftrightarrow{a}_e}{j\omega + b_e} \tag{2.57}$$

with the corresponding time-domain expression given by

$$\overleftrightarrow{\chi}_e(t) = \overleftrightarrow{a}_e e^{-b_e t} u(t) \tag{2.58}$$

where $u(t)$ represents the unit step function and $\overleftrightarrow{a}_e$ and b_e are parameters related to specific materials, which for the time being are taken to be real (such as the case of Debye material). Using (2.58), we find that

$$\ddot{\overleftrightarrow{\chi}}_e(t) * \mathbf{E}(t) = \overleftrightarrow{a}_e \cdot \frac{\partial \mathbf{E}(t)}{\partial t} - b_e \overleftrightarrow{a}_e \cdot \mathbf{E}(t) + \overleftrightarrow{a}_e \varphi_e(t) * \mathbf{E}(t) \tag{2.59}$$

where

$$\varphi_e(t) = b_e^2 e^{-b_e t} u(t). \tag{2.60}$$

The finite element discretization of (2.56) together with (2.59) is straightforward, and the resulting second-order ordinary differential equation contains a convolution term, which can be denoted by $[Z]\{\psi\}$. By using Galerkin's approach, the matrix $[Z]$ is given by

$$Z_{ij} = \varepsilon_0 \iiint_V \mathbf{N}_i \cdot \overleftrightarrow{a}_e \cdot \mathbf{N}_j \, dV \tag{2.61}$$

and the vector $\{\psi\}$ is given by

$$\psi_i(t) = \varphi_e(t) * E_i(t) \tag{2.62}$$

where $E_i(t)$ is the finite element expansion coefficient for $\mathbf{E}(t)$. Since $\varphi_e(t)$ has the assumed form of an exponential function, the convolution in (2.62) can be evaluated recursively as

$$\psi_i^n = e^{-b_e \Delta t} \psi_i^{n-1} + b_e^2 e^{-b_e n \Delta t} \int_{(n-1)\Delta t}^{n\Delta t} e^{b_e \tau} E_i(\tau) \, d\tau \tag{2.63}$$

where n denotes the time-step index and Δt denotes the length of each time step. To evaluate (2.63) accurately, we can employ a linear interpolation for $E_i(t)$ in the time interval $[(n-1)\,\Delta t, n\,\Delta t]$ to obtain the second-order accurate recursive formula [6]

$$\psi_i^n = e^{-b_e \Delta t}\psi_i^{n-1} + b_e^2 \frac{\Delta t}{2}\left(E_i^n + e^{-b_e \Delta t} E_i^{n-1}\right). \tag{2.64}$$

The use of this recursive convolution saves significant computation time and memory [6,7]. Obviously, the value of $\{\psi\}^n$ depends only on $\{E\}^n$ and its previous values; hence, $\{\psi\}^n$ can be calculated once $\{E\}^n$ is computed. However, to obtain an unconditionally stable solution using the Newmark-beta method, we have to use a weighted average for $\{\psi\}$, which would require the value of $\{\psi\}^{n+1}$ to calculate the right-hand side of the time-marching equation for computing $\{E\}^{n+1}$. Since $\{\psi\}^{n+1}$ contains $\{E\}^{n+1}$, the unknown term must first be moved to the left-hand side of the time-marching equation. Otherwise, implementation of the Newmark-beta method is incomplete and the time-marching solution is only conditionally stable.

Now, we consider the second approach based on the use of (2.54). In this case, (2.52) can be written as

$$\nabla \times \left[\overleftrightarrow{\mu}^{-1} \cdot \nabla \times \mathbf{E}(t)\right] + \varepsilon_0 \overleftrightarrow{\varepsilon}_\infty \cdot \frac{\partial^2 \mathbf{E}(t)}{\partial t^2} + \varepsilon_0 \overleftrightarrow{\chi}_e(t) * \frac{\partial^2 \mathbf{E}(t)}{\partial t^2} = 0. \tag{2.65}$$

The convolution can be written in semidiscrete form as [8]

$$\begin{aligned}\overleftrightarrow{\chi}_e(t) * \left.\frac{\partial^2 \mathbf{E}(t)}{\partial t^2}\right|_{t=n\Delta t} &\cong \int_0^{\Delta t/2} \overleftrightarrow{\chi}_e(\tau) \cdot \ddot{\mathbf{E}}(n\Delta t - \tau)\, d\tau \\ &\quad + \sum_{k=0}^{n-1} \int_{(k+1/2)\Delta t}^{(k+3/2)\Delta t} \overleftrightarrow{\chi}_e(\tau) \cdot \ddot{\mathbf{E}}(n\Delta t - t)\, d\tau \end{aligned} \tag{2.66}$$

and it is understood that $\mathbf{E}(t) = 0$ for $t \le 0$. Assuming that $\ddot{\mathbf{E}}$ is constant over the time integration intervals and the electric susceptibility can be written as (2.58), (2.66) becomes

$$\begin{aligned}\overleftrightarrow{\chi}_e * \left.\frac{\partial^2 \mathbf{E}}{\partial t^2}\right|_{t=n\Delta t} &\cong \left[\int_0^{\Delta t/2} \overleftrightarrow{\chi}_e(\tau)\, d\tau\right] \cdot \ddot{\mathbf{E}}^n + \sum_{k=0}^{n-1} \left[\int_{(k+1/2)\Delta t}^{(k+3/2)\Delta t} \overleftrightarrow{\chi}_e(\tau)\, d\tau\right] \cdot \ddot{\mathbf{E}}^{n-k-1} \\ &= \left[\frac{1}{b_e}(1 - e^{-b_e\,\Delta t/2})\overleftrightarrow{a}_e\right] \cdot \ddot{\mathbf{E}}^n \\ &\quad + \sum_{k=0}^{n-1} \left[\frac{1}{b_e}(1 - e^{-b_e\,\Delta t}) e^{-b_e\,(k+1/2)\Delta t}\, \overleftrightarrow{a}_e\right] \cdot \ddot{\mathbf{E}}^{n-k-1}. \end{aligned} \tag{2.67}$$

We can now construct the finite element discretization by first noting that the weak-form representation of (2.65) is similar to (2.56), with the exception of the placement of the time derivatives in the convolution. By using Galerkin's approach and applying the Newmark-beta method with $\beta = 1/4$, we obtain the following equation for the expansion coefficients of the electric field:

$$\begin{aligned}
&\left\{\frac{1}{(\Delta t)^2}[T] + \frac{1}{2\,\Delta t}[R] + \frac{1}{4}[S]\right\}\{E\}^{n+1} \\
&\quad = 2\left\{\frac{1}{(\Delta t)^2}[T] - \frac{1}{4}[S]\right\}\{E\}^n - \left\{\frac{1}{(\Delta t)^2}[T] - \frac{1}{2\,\Delta t}[R] + \frac{1}{4}[S]\right\}\{E\}^{n-1} \\
&\qquad - \frac{1}{(\Delta t)^2}\{\Psi_e\}^n
\end{aligned} \tag{2.68}$$

where the differences from the nondispersive case are given by

$$T_{ij} = \varepsilon_0 \iiint_V \mathbf{N}_i \cdot \overleftrightarrow{\varepsilon}_\infty \cdot \mathbf{N}_j \, dV + \tilde{\Phi}^0_{ij} \tag{2.69}$$

$$\{\Psi_e\}^n = \sum_{k=0}^{n-1} [\Phi]^k (\{E\}^{n-k} - 2\{E\}^{n-k-1} + \{E\}^{n-k-2}) \tag{2.70}$$

with

$$\begin{aligned}
\tilde{\Phi}^0_{ij} &= \varepsilon_0 \left[\frac{1}{b_e}(1 - e^{-b_e\,\Delta t/2})\right] \iiint_V \mathbf{N}_i \cdot \overleftrightarrow{a}_e \cdot \mathbf{N}_j \, dV \\
\Phi^k_{ij} &= \varepsilon_0 \left[\frac{1}{b_e}(1 - e^{-b_e\,\Delta t}) e^{-b_e\,(k+1/2)\Delta t}\right] \iiint_V \mathbf{N}_i \cdot \overleftrightarrow{a}_e \cdot \mathbf{N}_j \, dV \qquad k \geq 0.
\end{aligned} \tag{2.71}$$

The recursive relationship $[\Phi]^k = e^{-b_e\,\Delta t}\,[\Phi]^{k-1}$ enables (2.70) to be written recursively as

$$\{\Psi_e\}^n = [\Phi]^0(\{E\}^n - 2\{E\}^{n-1} + \{E\}^{n-2}) + e^{-b_e \Delta t}\{\Psi_e\}^{n-1}. \tag{2.72}$$

Comparing the two approaches presented above, we find that the first approach requires an analytical expression for $\overleftrightarrow{\chi}_e(t)$ to find its second time derivative, whereas the second approach only requires the value of $\overleftrightarrow{\chi}_e(t)$. The numerical efficiency of the two approaches is expected to be very similar, and when implemented carefully, both approaches are unconditionally stable.

The formulation described in this section can easily be extended to model materials, such as Lorentz materials, with an electrical susceptibility that is characterized by a simultaneous decay and oscillation according to $\overleftrightarrow{\chi}_e(t) = \overleftrightarrow{a}_e\, e^{-\delta_e t}\cos(\alpha_e t)\, u(t)$. All that is required is to rewrite this as $\overleftrightarrow{\chi}_e(t) = \overleftrightarrow{a}_e\, \mathrm{Re}[e^{-b_e t}]\, u(t)$, where $b_e = \delta_e - j\alpha_e$, and then modify the equations that contain $e^{-b_e t}$ by taking their real part [6]. The formulation can further be extended to model materials that have an electrical susceptibility represented by a multipole expansion

$$\overleftrightarrow{\chi}_e(\omega) = \sum_{p=1}^{N_e} \frac{\overleftrightarrow{a}_{e,p}}{j\omega + b_{e,p}} \leftrightarrow \overleftrightarrow{\chi}_e(t) = \sum_{p=1}^{N_e} \text{Re}\left\{\overleftrightarrow{a}_{e,p}\, e^{-b_{e,p}t}\right\} u(t) \tag{2.73}$$

where N_e denotes the number of poles in the expansion. Such an expansion is useful for simulation over a broad frequency range because it provides increased generality with regard to characterizing specific material behavior as a function of frequency.

2.3.3 Modeling of Magnetically Dispersive Materials

In an anisotropic, magnetically dispersive material, Maxwell's equations become

$$\nabla \times \mathbf{E}(t) = -\frac{\partial \mathbf{B}(t)}{\partial t} \tag{2.74}$$

$$\nabla \times \mathbf{H}(t) = \overleftrightarrow{\varepsilon} \cdot \frac{\partial \mathbf{E}(t)}{\partial t} \tag{2.75}$$

where $\overleftrightarrow{\varepsilon}$ is a time-invariant permittivity tensor and $\mathbf{B}(t)$ is related to $\mathbf{H}(t)$ by the constitutive relation

$$\begin{aligned} \mathbf{B}(t) &= \mu_0 \overleftrightarrow{\mu}_\infty \cdot \mathbf{H}(t) + \mu_0 \overleftrightarrow{\chi}_m(t) * \mathbf{H}(t) \\ &= \mu_0 \overleftrightarrow{\mu}_\infty \cdot \mathbf{H}(t) + \mu_0 \int_0^t \overleftrightarrow{\chi}_m(t-\tau) \cdot \mathbf{H}(\tau)\, d\tau \end{aligned} \tag{2.76}$$

where $\overleftrightarrow{\mu}_\infty$ denotes the relative permeability tensor at the optical frequency and $\overleftrightarrow{\chi}_m(t)$ represents the magnetic susceptibility tensor. This magnetically dispersive material case can be considered analogous to the case treated in the preceding section; hence, we can easily develop a dual solution, which solves for the magnetic field. However, such a solution cannot be combined with that in the preceding section to deal with both electrically and magnetically dispersive materials. Consequently, in this subsection, we describe a different method that still solves for the electric field.

First, we note that the time derivative of (2.76) can be expressed as either

$$\frac{\partial \mathbf{B}(t)}{\partial t} = \mu_0 \overleftrightarrow{\mu}_\infty \cdot \frac{\partial \mathbf{H}(t)}{\partial t} + \mu_0 \frac{\partial \overleftrightarrow{\chi}_m(t)}{\partial t} * \mathbf{H}(t) \tag{2.77}$$

or

$$\frac{\partial \mathbf{B}(t)}{\partial t} = \mu_0 \overleftrightarrow{\mu}_\infty \cdot \frac{\partial \mathbf{H}(t)}{\partial t} + \mu_0 \overleftrightarrow{\chi}_m(t) * \frac{\partial \mathbf{H}(t)}{\partial t}. \tag{2.78}$$

To avoid the use of a time-consuming convolution while still employing the second-order wave equation, we take the time derivative of (2.75) to obtain

$$\nabla \times \frac{\partial \mathbf{H}(t)}{\partial t} = \overleftrightarrow{\varepsilon} \cdot \frac{\partial^2 \mathbf{E}(t)}{\partial t^2} \tag{2.79}$$

whose weak-form representation can be written as

$$\iiint_V \mathbf{T} \cdot \left[-\nabla \times \frac{\partial \mathbf{H}(t)}{\partial t} + \overleftrightarrow{\varepsilon} \cdot \frac{\partial^2 \mathbf{E}(t)}{\partial t^2} \right] dV = 0 \tag{2.80}$$

or

$$\iiint_V \left[-(\nabla \times \mathbf{T}) \cdot \frac{\partial \mathbf{H}(t)}{\partial t} + \mathbf{T} \cdot \overleftrightarrow{\varepsilon} \cdot \frac{\partial^2 \mathbf{E}(t)}{\partial t^2} \right] dV - \oiint_S \mathbf{T} \cdot \left[\hat{n} \times \frac{\partial \mathbf{H}(t)}{\partial t} \right] dS = 0 \tag{2.81}$$

where we applied a simple vector identity and Gauss's divergence theorem. The surface integral term can be used to incorporate an appropriate boundary condition on S. The finite element discretization of (2.81) yields

$$[T]\frac{d^2\{E\}}{dt^2} + [W]\{\dot{\mathbf{H}}\} + \text{other terms} = 0 \tag{2.82}$$

where $\{E\}$ is a vector whose entries are the coefficients of the finite element expansion of $\mathbf{E}(t)$ and $\{\dot{\mathbf{H}}\}$ is a vector whose entries are the coefficients of the compatible finite element expansion of $\partial \mathbf{H}(t)/\partial t$. If we use the first-order vector basis functions to expand $\mathbf{E}(t)$, then $\partial \mathbf{H}(t)/\partial t$ is constant within each element and the entries of $\{\dot{\mathbf{H}}\}$ represent the values of the x-, y-, and z-components of $\partial \mathbf{H}(t)/\partial t$ within each element. In this case, the entries for the matrices $[T]$ and $[W]$ are given by

$$T_{ij} = \iiint_V \mathbf{N}_i \cdot \overleftrightarrow{\varepsilon} \cdot \mathbf{N}_j \, dV \tag{2.83}$$

$$W_{ik} = -\iiint_V (\nabla \times \mathbf{N}_i) \cdot \mathbf{P}_k \, dV \tag{2.84}$$

where $\mathbf{P}_k = (\hat{x}, \hat{y}, \hat{z})$ within element k and is zero elsewhere. Using central differencing for the second-order derivative in (2.82), we obtain a time-marching equation to solve for $\{E\}$. To calculate $\{E\}^{n+1}$, we need the value of $\{\dot{\mathbf{H}}\}^n$, which can be calculated using either (2.77) or (2.78) as described below.

Let us consider first the evaluation of $\{\dot{\mathbf{H}}\}^n$ based on (2.77). By substituting (2.74) into (2.77), we obtain

$$\overleftrightarrow{\mu}_\infty \cdot \frac{\partial \mathbf{H}(t)}{\partial t} = -\frac{1}{\mu_0}\nabla \times \mathbf{E}(t) - \overleftrightarrow{\dot{\chi}}_m(t) * \mathbf{H}(t) \tag{2.85}$$

which can be used to compute $\{\mathbf{H}\}^{n+1/2}$ by using a central difference for $\partial \mathbf{H}(t)/\partial t$ and an average for $\mathbf{H}(t)$. If the time dependence of $\overleftrightarrow{\chi}_m(t)$ can be expressed in terms of the exponential function

$$\overleftrightarrow{\chi}_m(t) = \overleftrightarrow{a}_m e^{-b_m t} u(t) \tag{2.86}$$

the convolution in (2.85) can be evaluated recursively as described earlier. To be more specific, if we denote

$$\left[\dot{\overleftrightarrow{\chi}}_m(t) * \mathbf{H}(t)\right]^n = \overleftrightarrow{a}_m \cdot \mathbf{H}^n - \mathbf{\Psi}^n \tag{2.87}$$

$$\mathbf{\Psi}^n = e^{-b_m \Delta t} \mathbf{\Psi}^{n-1} + \frac{b_m \Delta t}{2} \overleftrightarrow{a}_m \cdot \left[\mathbf{H}^n + e^{-b_m \Delta t} \mathbf{H}^{n-1}\right] \tag{2.88}$$

(2.85) can be written as

$$\begin{aligned}\frac{1}{\Delta t} \overleftrightarrow{\mu}_\infty \cdot \left[\mathbf{H}^{n+1/2} - \mathbf{H}^{n-1/2}\right] = &-\frac{1}{\mu_0} (\nabla \times \mathbf{E})^n - \frac{1}{2} \overleftrightarrow{a}_m \cdot \left[\mathbf{H}^{n+1/2} + \mathbf{H}^{n-1/2}\right] \\ &+ e^{-b_m \Delta t} \mathbf{\Psi}^{n-1} + \frac{\Delta t}{4} b_m \overleftrightarrow{a}_m \cdot \left[\mathbf{H}^{n+1/2} + \mathbf{H}^{n-1/2}\right] \\ &+ \frac{\Delta t}{4} b_m e^{-b_m \Delta t} \overleftrightarrow{a}_m \cdot \left[\mathbf{H}^{n-1/2} + \mathbf{H}^{n-3/2}\right]\end{aligned} \tag{2.89}$$

from which $\mathbf{H}^{n+1/2}$ can be computed. Once $\mathbf{H}^{n+1/2}$ is computed, $\dot{\mathbf{H}}^n$ can be calculated as

$$\dot{\mathbf{H}}^n = \frac{\mathbf{H}^{n+1/2} - \mathbf{H}^{n-1/2}}{\Delta t}. \tag{2.90}$$

The evaluation above requires an analytical expression for $\overleftrightarrow{\chi}_m(t)$ to calculate its time derivative. This requirement can be avoided by using (2.78), where the convolution term can be written as

$$\begin{aligned}\overleftrightarrow{\chi}_m(t) * \left.\frac{\partial \mathbf{H}(t)}{\partial t}\right|_{t=n\Delta t} \cong &\int_0^{\Delta t/2} \overleftrightarrow{\chi}_m(\tau) \cdot \dot{\mathbf{H}}(n\Delta t - \tau)\, d\tau \\ &+ \sum_{k=1}^{n} \int_{(k-1/2)\Delta t}^{(k+1/2)\Delta t} \overleftrightarrow{\chi}_m(\tau) \cdot \dot{\mathbf{H}}(n\Delta t - \tau)\, d\tau\end{aligned} \tag{2.91}$$

and it is understood that $\mathbf{H}(t) = 0$ for $t \leq 0$. Using central differencing to approximate the time derivative in (2.91) and assuming that the magnetic field is constant within the time-integration intervals yield

$$\mu_0 \overleftrightarrow{\chi}_m(t) * \left.\frac{\partial \mathbf{H}(t)}{\partial t}\right|_{t=n\Delta t} \cong \frac{\mu_0}{\Delta t} \overleftrightarrow{\chi}_m^0 \cdot (\mathbf{H}^{n+1/2} - \mathbf{H}^{n-1/2}) + \frac{\mu_0}{\Delta t} \mathbf{\Phi}^n \tag{2.92}$$

where

$$\mathbf{\Phi}^n = \sum_{k=1}^{n} \overleftrightarrow{\chi}_m^k \cdot (\mathbf{H}^{n-k+1/2} - \mathbf{H}^{n-k-1/2}) \tag{2.93}$$

and

$$\overleftrightarrow{\chi}_m^k = \int_{(k-1/2)\Delta t}^{(k+1/2)\Delta t} \overleftrightarrow{\chi}_m(\tau)\, d\tau. \tag{2.94}$$

In (2.94), causality of the susceptibility tensor is assumed so that the lower limit of the integral is zero for the case $k = 0$.

Substituting (2.92) and (2.78) into (2.74), we obtain

$$\mu_0\big(\overleftrightarrow{\mu}_\infty + \overleftrightarrow{\chi}_m^0\big) \cdot \mathbf{H}^{n+1/2} = \mu_0\big(\overleftrightarrow{\mu}_\infty + \overleftrightarrow{\chi}_m^0\big) \cdot \mathbf{H}^{n-1/2} - \Delta t\, (\nabla \times \mathbf{E})^n - \mu_0\, \mathbf{\Phi}^n \tag{2.95}$$

from which $\mathbf{H}^{n+1/2}$ can be computed, and subsequently $\dot{\mathbf{H}}^n$ using (2.90). If $\overleftrightarrow{\chi}_m(t)$ can be expressed as (2.86), then (2.93) can be evaluated efficiently using the recursive relation

$$\mathbf{\Phi}^n = \overleftrightarrow{\chi}_m^1 \cdot (\mathbf{H}^{n-1/2} - \mathbf{H}^{n-3/2}) + e^{-b_m \Delta t}\, \mathbf{\Phi}^{n-1}. \tag{2.96}$$

It is evident that both approaches to calculating $\dot{\mathbf{H}}^n$ can be extended to model more general materials with magnetic susceptibility represented by the pole expansion

$$\overleftrightarrow{\chi}_m(\omega) = \sum_{p=1}^{N_m} \frac{\overleftrightarrow{a}_{m,p}}{j\omega + b_{m,p}} \leftrightarrow \overleftrightarrow{\chi}_m(t) = \sum_{p=1}^{N_m} \mathrm{Re}\big\{\overleftrightarrow{a}_{m,p}\, e^{-b_{m,p} t}\big\} u(t) \tag{2.97}$$

where N_m denotes the number of poles in the expansion.

The time-marching procedure derived from (2.82) using central differencing is only conditionally stable. Application of the Newmark-beta method to (2.82) is complicated since it requires a weighted average for the second term in the form of (2.35). To facilitate the application of the Newmark-beta method and obtain an unconditionally stable solution, we can rewrite (2.81) as

$$\iiint_V \left\{ \frac{1}{\mu_0} (\nabla \times \mathbf{T}) \cdot \overleftrightarrow{\mu}_\infty^{-1} \cdot [\nabla \times \mathbf{E}(t)] - (\nabla \times \mathbf{T}) \cdot \mathbf{Q}(t) + \mathbf{T} \cdot \overleftrightarrow{\varepsilon} \cdot \frac{\partial^2 \mathbf{E}(t)}{\partial t^2} \right\} dV$$
$$- \oiint_S \mathbf{T} \cdot \left[\hat{n} \times \frac{\partial \mathbf{H}(t)}{\partial t} \right] dS = 0 \tag{2.98}$$

where

$$\mathbf{Q}(t) = \frac{\partial \mathbf{H}(t)}{\partial t} + \frac{1}{\mu_0} \overleftrightarrow{\mu}_\infty^{-1} \cdot [\nabla \times \mathbf{E}(t)]. \tag{2.99}$$

The finite element discretization of (2.98) yields

$$[T]\frac{d^2\{E\}}{dt^2} + [S]\{E\} + [W]\{\mathbf{Q}\} + \text{other terms} = 0 \tag{2.100}$$

where $[T]$ and $[W]$ are given in (2.83) and (2.84) and the entries of $[S]$ are given by

$$S_{ij} = \frac{1}{\mu_0}\iiint_V (\nabla \times \mathbf{N}_i) \cdot \overleftrightarrow{\mu}_{\infty}^{-1} \cdot (\nabla \times \mathbf{N}_j)\, dV. \tag{2.101}$$

The vector $\{\mathbf{Q}\}$ stores the value of $\mathbf{Q}(t)$ in each element (assuming the use of the first-order edge elements), which can be calculated using (2.99) once $\partial\mathbf{H}(t)/\partial t$ is evaluated using one of the two approaches described earlier. Applying the Newmark-beta method with $\beta = 1/4$ to (2.98) yields an unconditionally stable time-marching equation

$$\begin{aligned}\left\{\frac{1}{(\Delta t)^2}[T] + \frac{1}{4}[S]\right\}\{E\}^{n+1} &= 2\left\{\frac{1}{(\Delta t)^2}[T] - \frac{1}{4}[S]\right\}\{E\}^{n} \\ &\quad - \left\{\frac{1}{(\Delta t)^2}[T] + \frac{1}{4}[S]\right\}\{E\}^{n-1} - [W]\{\mathbf{Q}\}^n - \text{other terms}.\end{aligned} \tag{2.102}$$

On the surface, (2.102) appears to be an incomplete Newmark-beta implementation since $\{\mathbf{Q}\}^n$ has been used instead of its weighted average. However, because of the extraction of the term $[S]\{E\}$, it is actually equivalent to a complete Newmark-beta implementation with an effective β greater than 1/4. Because an increased value in β does not compromise the stability of a Newmark-beta formulation, (2.102) remains unconditionally stable, although its accuracy is slightly compromised due to the use of a larger value of the effective β. The practical consequence is an increase in the discretization error associated with the time derivatives, which can be offset by defining a reduced time step.

2.3.4 Modeling of Doubly Dispersive Lossy Materials

With the modeling techniques described so far, we are now ready to consider the modeling of a general anisotropic, dispersive, lossy, electric and magnetic material. For this case, Maxwell's equations can be written as

$$\nabla \times \mathbf{E}(t) = -\frac{\partial \mathbf{B}(t)}{\partial t} - \overleftrightarrow{\sigma}_m \cdot \mathbf{H}(t) - \mathbf{M}_{\text{imp}}(t) \tag{2.103}$$

$$\nabla \times \mathbf{H}(t) = \frac{\partial \mathbf{D}(t)}{\partial t} + \overleftrightarrow{\sigma}_e \cdot \mathbf{E}(t) + \mathbf{J}_{\text{imp}}(t) \tag{2.104}$$

where $\mathbf{D}(t)$ is related to $\mathbf{E}(t)$ by the constitutive relation

$$\mathbf{D}(t) = \varepsilon_0 \overleftrightarrow{\varepsilon}_\infty \cdot \mathbf{E}(t) + \varepsilon_0 \overleftrightarrow{\chi}_e(t) * \mathbf{E}(t) \tag{2.105}$$

and $\mathbf{B}(t)$ is related to $\mathbf{H}(t)$ by the constitutive relation

$$\mathbf{B}(t) = \mu_0 \overleftrightarrow{\mu}_\infty \cdot \mathbf{H}(t) + \mu_0 \overleftrightarrow{\chi}_m(t) * \mathbf{H}(t). \tag{2.106}$$

Note that $\overleftrightarrow{\sigma}_e$ and $\overleftrightarrow{\sigma}_m$ represent the electrical and magnetic conductivities as the frequency approaches zero. The dispersion of the material is also accompanied by a loss mechanism where the equivalent conductivities are given by $-\omega\varepsilon_0 \operatorname{Im}[\overleftrightarrow{\chi}_e(\omega)]$ and $-\omega\mu_0 \operatorname{Im}[\overleftrightarrow{\chi}_m(\omega)]$. The effect of this loss is included in the modeling of the material dispersion.

To formulate a weak-form solution of (2.103) and (2.104), we first take the time derivative of (2.104) to obtain

$$\nabla \times \frac{\partial \mathbf{H}(t)}{\partial t} = \frac{\partial^2 \mathbf{D}(t)}{\partial t^2} + \overleftrightarrow{\sigma}_e \cdot \frac{\partial \mathbf{E}(t)}{\partial t} + \frac{\partial \mathbf{J}_{\text{imp}}(t)}{\partial t}. \tag{2.107}$$

Substituting the equation

$$\frac{\partial^2 \mathbf{D}(t)}{\partial t^2} = \varepsilon_0 \overleftrightarrow{\varepsilon}_\infty \cdot \frac{\partial^2 \mathbf{E}(t)}{\partial t^2} + \varepsilon_0 \overleftrightarrow{\chi}_e(t) * \frac{\partial^2 \mathbf{E}(t)}{\partial t^2} \tag{2.108}$$

into (2.107), we obtain

$$-\nabla \times \frac{\partial \mathbf{H}(t)}{\partial t} + \varepsilon_0 \overleftrightarrow{\varepsilon}_\infty \cdot \frac{\partial^2 \mathbf{E}(t)}{\partial t^2} + \overleftrightarrow{\sigma}_e \cdot \frac{\partial \mathbf{E}(t)}{\partial t} + \varepsilon_0 \overleftrightarrow{\chi}_e(t) * \frac{\partial^2 \mathbf{E}(t)}{\partial t^2} = -\frac{\partial \mathbf{J}_{\text{imp}}(t)}{\partial t}. \tag{2.109}$$

To find the weak-form solution of (2.109), we take the dot product with a testing function $\mathbf{T}$ and integrate over a volume V to obtain

$$\begin{aligned} &\iiint_V \left\{ -(\nabla \times \mathbf{T}) \cdot \frac{\partial \mathbf{H}(t)}{\partial t} + \varepsilon_0 \mathbf{T} \cdot \overleftrightarrow{\varepsilon}_\infty \cdot \frac{\partial^2 \mathbf{E}(t)}{\partial t^2} + \mathbf{T} \cdot \overleftrightarrow{\sigma}_e \cdot \frac{\partial \mathbf{E}(t)}{\partial t} \right. \\ &\qquad \left. + \varepsilon_0 \mathbf{T} \cdot \overleftrightarrow{\chi}_e(t) * \frac{\partial^2 \mathbf{E}(t)}{\partial t^2} \right\} dV - \oiint_S \mathbf{T} \cdot \left[\hat{n} \times \frac{\partial \mathbf{H}(t)}{\partial t} \right] dS \\ &= -\iiint_V \mathbf{T} \cdot \frac{\partial \mathbf{J}_{\text{imp}}(t)}{\partial t}\, dV. \end{aligned} \tag{2.110}$$

To cast the final discrete finite element system in a form similar to the standard one that contains the $[S]$ matrix with the intention of obtaining an unconditionally stable solution, as was done in Section 2.3.3, we rewrite (2.110) as

$$\iiint_V \left\{ \frac{1}{\mu_0} (\nabla \times \mathbf{T}) \cdot \overleftrightarrow{\mu}_\infty^{-1} \cdot [\nabla \times \mathbf{E}(t)] - (\nabla \times \mathbf{T}) \cdot \mathbf{Q}(t) + \varepsilon_0 \mathbf{T} \cdot \overleftrightarrow{\varepsilon}_\infty \cdot \frac{\partial^2 \mathbf{E}(t)}{\partial t^2} \right.$$
$$\left. + \mathbf{T} \cdot \overleftrightarrow{\sigma}_e \cdot \frac{\partial \mathbf{E}(t)}{\partial t} + \varepsilon_0 \mathbf{T} \cdot \overleftrightarrow{\chi}_e(t) * \frac{\partial^2 \mathbf{E}(t)}{\partial t^2} \right\} dV - \oiint_S \mathbf{T} \cdot \left[\hat{n} \times \frac{\partial \mathbf{H}(t)}{\partial t} \right] dS$$
$$= - \iiint_V \mathbf{T} \cdot \frac{\partial \mathbf{J}_{\text{imp}}(t)}{\partial t} dV \tag{2.111}$$

where $\mathbf{Q}(t)$ is given by (2.99). The convolution term and $\mathbf{Q}(t)$ can be evaluated by using various approaches discussed in the preceding sections. Here, we summarize their evaluations by using the approach that renders the time-marching solution unconditionally stable. The convolution in the fifth term of (2.111) can be discretized as

$$\left[\overleftrightarrow{\chi}_e(t) * \frac{\partial^2 \mathbf{E}(t)}{\partial t^2} \right]^n = \overleftrightarrow{\chi}_e^0 \cdot \left[\frac{\partial^2 \mathbf{E}(t)}{\partial t^2} \right]^n + \frac{1}{(\Delta t)^2} \sum_{k=1}^{n} \overleftrightarrow{\chi}_e^k \cdot (\mathbf{E}^{n-k+1} - 2\mathbf{E}^{n-k} + \mathbf{E}^{n-k-1}) \tag{2.112}$$

where

$$\overleftrightarrow{\chi}_e^0 = \int_0^{\Delta t/2} \overleftrightarrow{\chi}_e(\tau)\, d\tau \tag{2.113}$$

$$\overleftrightarrow{\chi}_e^k = \int_{(k-1/2)\Delta t}^{(k+1/2)\Delta t} \overleftrightarrow{\chi}_e(\tau)\, d\tau \qquad k > 0. \tag{2.114}$$

As shown earlier, if $\overleftrightarrow{\chi}_e$ can be represented by a pole expansion, this convolution can be evaluated efficiently by using a recursive relation.

To evaluate $\mathbf{Q}(t)$ at the time instant $t = n\,\Delta t$, which requires the value of $\partial \mathbf{H}(t)/\partial t$, we first take the time derivative of (2.106) to find

$$\frac{\partial \mathbf{B}(t)}{\partial t} = \mu_0 \overleftrightarrow{\mu}_\infty \cdot \frac{\partial \mathbf{H}(t)}{\partial t} + \mu_0 \overleftrightarrow{\chi}_m(t) * \frac{\partial \mathbf{H}(t)}{\partial t} \tag{2.115}$$

and then substitute it into (2.103) to obtain

$$\mu_0 \overleftrightarrow{\mu}_\infty \cdot \frac{\partial \mathbf{H}(t)}{\partial t} = -\nabla \times \mathbf{E}(t) - \overleftrightarrow{\sigma}_m \cdot \mathbf{H}(t) - \mu_0 \overleftrightarrow{\chi}_m(t) * \frac{\partial \mathbf{H}(t)}{\partial t} - \mathbf{M}_{\text{imp}}(t). \tag{2.116}$$

Using the central-difference approximation for $\partial \mathbf{H}(t)/\partial t$ and the time average for $\mathbf{H}(t)$, we obtain

$$
\begin{aligned}
\frac{\mu_0}{\Delta t}\overleftrightarrow{\mu}_\infty \cdot (\mathbf{H}^{n+1/2} - \mathbf{H}^{n-1/2}) = &-(\nabla \times \mathbf{E})^n - \tfrac{1}{2}\overleftrightarrow{\sigma}_m \cdot (\mathbf{H}^{n+1/2} + \mathbf{H}^{n-1/2}) \\
&- \frac{\mu_0}{\Delta t}\overleftrightarrow{\chi}_m^0 \cdot (\mathbf{H}^{n+1/2} - \mathbf{H}^{n-1/2}) \\
&- \frac{\mu_0}{\Delta t}\sum_{k=1}^{n}\overleftrightarrow{\chi}_m^k \cdot (\mathbf{H}^{n-k+1/2} - \mathbf{H}^{n-k-1/2}) - \mathbf{M}_{\mathrm{imp}}^n
\end{aligned}
\tag{2.117}
$$

from which we can solve for $\mathbf{H}^{n+1/2}$, which can then be used to calculate $\mathbf{Q}^n$. Note that if $\overleftrightarrow{\chi}_m$ can be represented by a pole expansion, the summation in (2.117) can be evaluated efficiently by recursion, as discussed earlier.

2.3.5 Validation Examples

To demonstrate the validity of the formulations presented in Sections 2.3.2 through 2.3.4, we consider the problem of plane-wave incidence on an infinitely large, uniform dielectric slab of thickness 10 cm, illustrated in Figure 2.7(a). The plane wave is normally incident on the slab and the slab is both electrically and magnetically dispersive. This problem can be solved analytically in the frequency domain and the results can be used as reference. It can also be solved using a one-dimensional finite element formulation. However, to test our formulations for the three-dimensional case, we employ a rectangular box to define a three-dimensional computational domain, as illustrated in Figure 2.7(b). On the two side surfaces perpendicular to the electric field, the electric field satisfies the boundary condition $\hat{n} \times \mathbf{E} = 0$; hence, these two side surfaces can be modeled as perfect electrically conducting surfaces. On the other two side surfaces parallel to the electric field, the field satisfies the boundary condition $\hat{n} \times \mathbf{H} = 0$ or $\hat{n} \times \nabla \times \mathbf{E} = 0$; hence, these two side surfaces can be modeled as perfect magnetically conducting surfaces. It remains to specify a boundary condition for the top and bottom surfaces. For the top surface, the total field is the superposition of the incident and reflected plane waves,

$$
\mathbf{E} = \mathbf{E}^{\mathrm{inc}} + \mathbf{E}^{\mathrm{ref}} = \hat{x}E_0 e^{jk_0 z} + \hat{x}E_0 \Gamma e^{-jk_0 z} \tag{2.118}
$$

where E_0 denotes the magnitude of the incident field and Γ denotes the reflection coefficient. From (2.118), we can easily find

$$
\hat{n} \times (\nabla \times \mathbf{E}) + jk_0 \hat{n} \times (\hat{n} \times \mathbf{E}) = -2jk_0 \mathbf{E}^{\mathrm{inc}} \tag{2.119}
$$

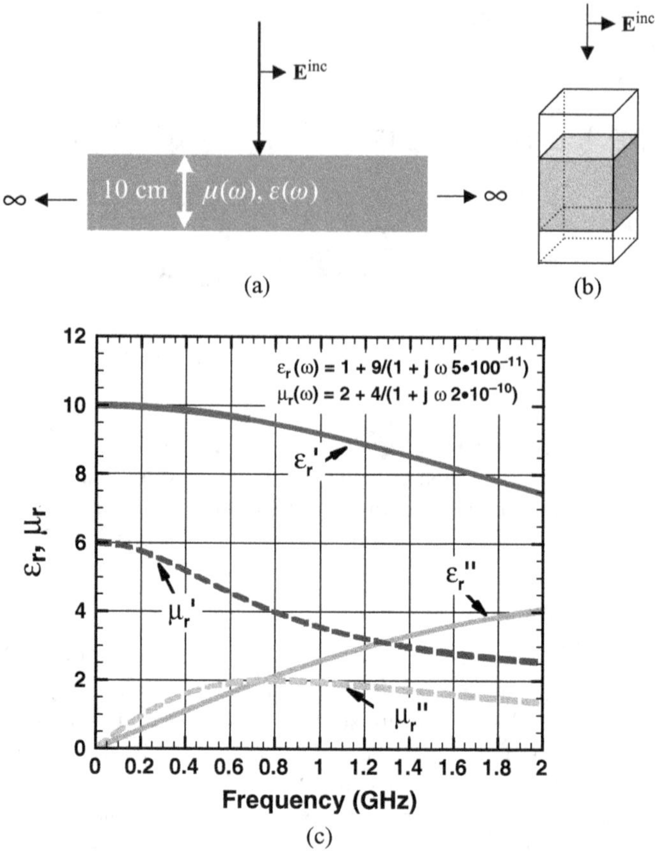

Figure 2.7 Doubly dispersive slab with a normally incident plane wave. (a) Geometry. (b) Computational domain. (c) Dispersive permittivity and permeability frequency profiles of the slab.

which can be transformed into the time domain as

$$\hat{n} \times \left(\frac{1}{\mu_0} \nabla \times \mathbf{E} \right) + Y_0 \, \hat{n} \times \left(\hat{n} \times \frac{\partial}{\partial t} \mathbf{E} \right) = -2Y_0 \frac{\partial}{\partial t} \mathbf{E}^{\text{inc}}. \tag{2.120}$$

This equation can be used as the boundary condition at the top surface. Similarly, for the bottom surface, we can find that the field satisfies the boundary condition

$$\hat{n} \times \left(\frac{1}{\mu_0} \nabla \times \mathbf{E} \right) + Y_0 \, \hat{n} \times \left(\hat{n} \times \frac{\partial}{\partial t} \mathbf{E} \right) = 0. \tag{2.121}$$

With the specification of the boundary conditions on the entire surface of the rectangular box, the field inside and on the surface of this box can be solved for by using

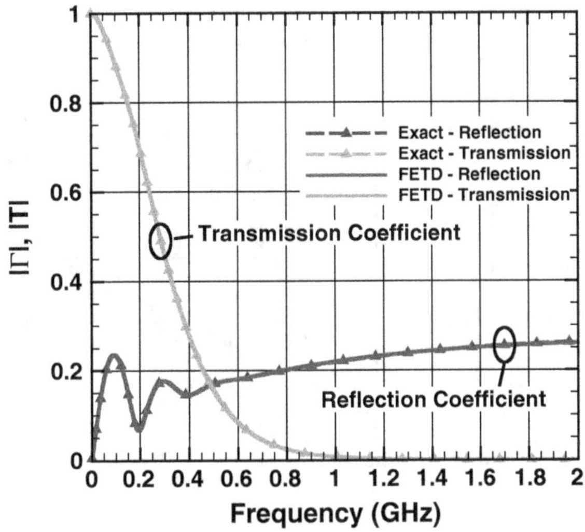

Figure 2.8 Reflection and transmission coefficients of a doubly dispersive slab at normal incidence.

the time-domain finite element method, from which the reflection and transmission coefficients can be calculated. For the relative permittivity and permeability profiles plotted in Figure 2.7(c), the reflection and transmission coefficients of the doubly dispersive slab are given in Figure 2.8 [11]. The results computed are compared with the exact solution, showing excellent agreement and thus verifying the formulations in Sections 2.3.2 through 2.3.4. Figure 2.9 provides another example, where the problem setup is identical to the preceding one, except that the slab is only magnetically dispersive. The Lorentz permeability profile of the slab is given in Figure 2.9(a), which exhibits a sharp resonance at 1 GHz. The reflection and transmission coefficients computed are shown in Figure 2.9(b). The remarkable agreement at the −120-dB level clearly demonstrates the high accuracy of the numerical scheme described in this section.

The formulation presented in this section employs a recursive algorithm for the efficient evaluation of the discrete convolutions introduced by the electric and magnetic dispersion of the medium. To use this algorithm, the electric and magnetic susceptibility functions must first be expressed as a pole expansion, which imposes a restriction on the type of medium that can be modeled. This restriction on the mathematical form of the susceptibility functions can be removed by direct evaluation of the discrete convolutions in (2.66) for the electrically dispersive case, (2.91) for the magnetically dispersive case, or (2.112) and (2.117) for the doubly dispersive case. These discrete convolutions can be written in a general format as

$$\psi^n = \sum_{k=1}^{n} \chi^k u^{n-k} = \sum_{k=0}^{n-1} \chi^{n-k} u^k \tag{2.122}$$

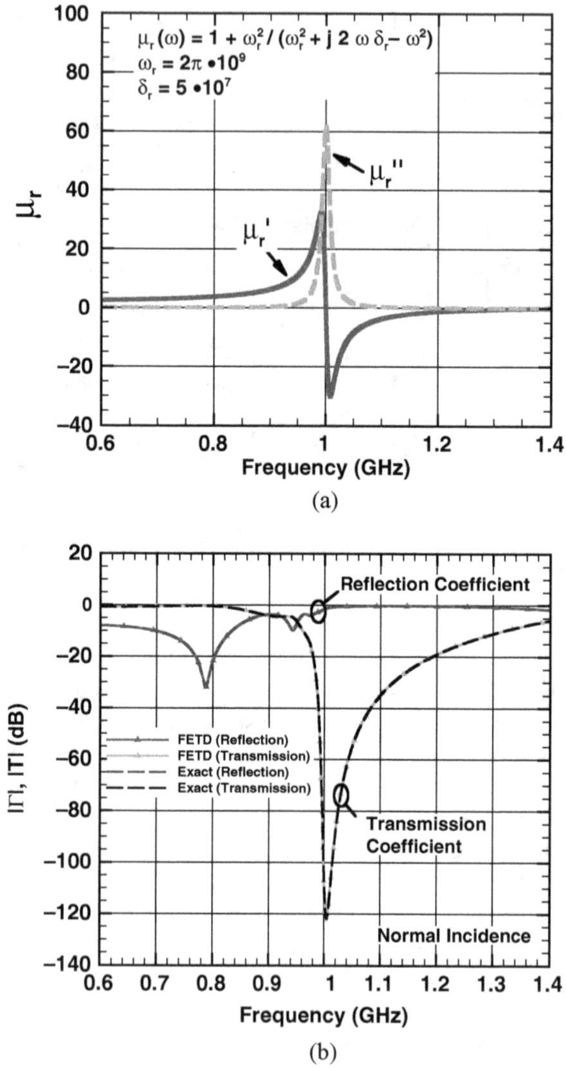

Figure 2.9 Plane-wave reflection by a sharply resonant Lorentz magnetic slab at normal incidence. (a) Permeability profile. (b) Reflection and transmission coefficients.

where χ^k is related to the susceptibility, which is known for all the time steps, and u^k is either the electric or the magnetic field, which is known only for k up to $n-1$. Direct evaluation of (2.122) is very time consuming because it requires $O(n)$ operations in every time step. A better approach is to evaluate (2.122) using a recursive fast Fourier transform (FFT) algorithm [12,13].

The basic idea of the recursive FFT algorithm is first to select an integer M, which is a power of 2, and then evaluate (2.122) using the FFT when n reaches $n = M - 1$.

Specifically, the FFT is used to calulate

$$\{m^k\}_{k=0}^{2M-1} = \{\chi^k\}_{k=0}^{2M-1} * \{\tilde{u}^k\}_{k=0}^{2M-1} \tag{2.123}$$

where $\tilde{u}^k = u^k$ for $k \leq M - 1$ and $\tilde{u}^k = 0$ for $k \geq M$. Subsequent convolutions are then evaluated as

$$\psi^n = m^n + \sum_{k=M}^{n-1} \chi^{n-k} u^k \qquad \text{for } M \leq n < 2M \tag{2.124}$$

where the second term is evaluated directly. Once n reaches $n = 2M - 1$, (2.122) is evaluated again using the FFT to yield

$$\{m^k\}_{k=0}^{4M-1} = \{\chi^k\}_{k=0}^{4M-1} * \{\tilde{u}^k\}_{k=0}^{4M-1} \tag{2.125}$$

where $\tilde{u}^k = u^k$ for $k \leq 2M - 1$ and $\tilde{u}^k = 0$ for $k \geq 2M$. Subsequent convolutions are then evaluated as

$$\psi^n = m^n + \sum_{k=2M}^{n-1} \chi^{n-k} u^k \qquad \text{for } 2M \leq n < 3M \tag{2.126}$$

where the second term is evaluated directly. Once n reaches $n = 3M - 1$, the second term can be evaluated using the FFT as

$$\{l^k\}_{k=0}^{2M-1} = \{\chi^k\}_{k=0}^{2M-1} * \{\tilde{v}^k\}_{k=0}^{2M-1} \tag{2.127}$$

where $\tilde{v}^k = u^{k+2M}$ for $k \leq M - 1$ and $\tilde{v}^k = 0$ for $k \geq M$. Subsequent convolutions are then calculated as

$$\psi^n = m^n + l^{n-2M} + \sum_{k=3M}^{n-1} \chi^{n-k} u^k \qquad \text{for } 3M \leq n < 4M \tag{2.128}$$

where the third term is evaluated directly. This process can continue until n reaches the maximum number of time steps. By doing this, the total cost of evaluating (2.122) for n up to $n = N_t$ is reduced from $O(N_t^2)$ to $O(N_t \log^2 N_t)$. In theory, we can implement the recursive FFT algorithm using $M = 2$. However, because the FFT calculation requires zero padding and operates on complex numbers, the optimal value for M has been found to fall within the interval defined by $M = 64$ to 128 [12].

Since there are no approximations in the use of the recursive FFT algorithm, the accuracy of the final solution is the same as that based on the recursive convolution algorithm. To illustrate this, we again consider the example of plane-wave incidence on a 10-cm-thick dielectric slab; however, we now solve for the reflection and transmission coefficients by using both the recursive FFT algorithm and the recursive convolution approach [13]. For this example, the dielectric slab is made of a fourth-order Lorentz medium, which has two resonances. The relative permittivity is shown in Figure 2.10(a), and the power reflection and transmission coefficients are plotted in Figure 2.10(b) together with the exact solution. As can be seen, the solution obtained using the recursive FFT overlays the solution obtained using the recursive convolution based on the pole expansion of the electrical susceptibility, although use of the recursive FFT makes no specific assumptions with regard to the mathematical form of the electrical susceptibility. However, implementation of the recursive FFT algorithm

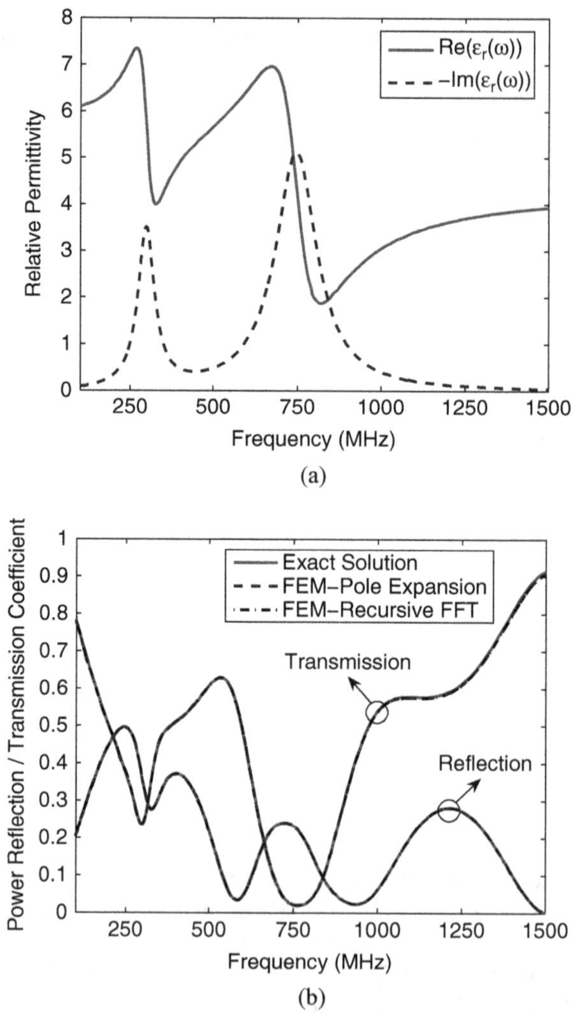

Figure 2.10 Plane-wave reflection by a fourth-order Lorentz medium slab at normal incidence. (a) Permittivity profile. (b) Power reflection and transmission coefficients. (After Li and Jin [13], Copyright © Wiley 2008.)

requires additional memory because all the past field values in the dispersive medium must be stored. In addition, it requires slightly more computational time because of the direct evaluation of a summation in each time step and the overhead associated with the FFT calculations.

2.4 SOLUTION OF THE FINITE ELEMENT EQUATIONS

As described above, when we employ frequency-domain finite element analysis, we have to solve the matrix equation (2.20). Alternatively, when we use the time-domain finite element method, we have to solve a matrix equation similar to (2.36) at each time step. For either formulation, the matrices are usually large, although they are sparse and symmetric. Therefore, an efficient solution of the finite element matrix equation is very important, because this aspect typically dominates the overall computer resource requirements. The important issues are matrix storage schemes, matrix solvers (direct or iterative), and matrix preconditioners (in the iterative case).

The matrices produced by the finite element method are sparse, with only a very small percentage of nonzero elements. By storing only the nonzero entries, the matrix storage requirement is reduced from $O(N^2)$ to $O(N)$. Popular approaches to sparse matrix storage are those based on either a compressed row or a compressed column storage format. In these approaches, the nonzero entries of a sparse matrix are stored in a floating-point vector. In addition, an integer vector is employed to store the row or column indexes of the nonzero entries, and another integer vector is introduced to store the location of the first nonzero entry of each row in the compressed vector. For a symmetric matrix, only the nonzero entries in the upper or lower triangle (including those on the diagonal) need to be stored.

The choice of a matrix solver can have a significant impact on the computational efficiency, and it is therefore important to choose a solver that can best exploit the properties of the finite element matrix. There are two types of matrix solvers. The first, known as a *direct solver*, is based on Gaussian elimination or LU decomposition. These solvers are commonly used for full matrices, although they are also applicable to sparse matrices stored in a banded format, or even a fully sparse format in the case of the frontal and multifrontal methods [14,15]. The alternative to direct solvers is *iterative solvers*, which require significantly less memory than do direct solvers because they are based on calculating successive matrix–vector products according to an iterative algorithm that is designed to converge to the solution [16]. The main drawback of iterative techniques is that they may require a large number of iterations to converge, due primarily to the locations of the eigenvalues of the matrix in the complex plane; however, if the eigenvalues are all clustered around $(1, 0)$, convergence is usually rapid. To improve the convergence of an iterative solver, a preconditioner is typically adopted to move the eigenvalues closer to $(1, 0)$, thereby reducing the iteration count. A preconditioner can be constructed based on physical insight into the problem or on the structure of the original matrix.

During the past two decades, remarkable progress has been made on techniques for solving sparse matrix equations. Today, there are many highly robust and efficient direct and iterative solvers available that deal with sparse matrices. For example, Intel's Math Kernel Library (MKL) [17], SGI's Scientific Computing Software Library (SCSL) [18], and IBM's Watson Sparse Matrix Package (WSMP) [19,20] contain sparse direct solvers for both symmetric and unsymmetric matrices. For unsymmetric sparse matrices, UMFPACK [21], MUMPS [22], and SuperLU [23] provide a scalable parallel solution on distributed memory computing systems. For iterative solvers, PETSc [24] and SPARSKIT [25] provide a variety of Krylov subspace algorithms, such as those based on the stabilized biconjugate gradient squared (BiCGStab) and generalized minimal residual (GMRES) methods, and a variety of preconditioners, such as the incomplete *LU* and successive over relaxation (SSOR) preconditioners, to speed up the iterative convergence. Note that both MKL and SCSL also contain sparse iterative solvers, and that PETSc can also interface with certain external sparse direct solvers noted above.

2.5 HIGHER-ORDER AND CURVILINEAR FINITE ELEMENTS

Most finite element methods use spatial basis functions which vary linearly within every element. However, it is also possible, and often very advantageous, to consider basis functions with higher polynomial orders. Such basis functions are described for vector elements by Graglia et al. [26] and Webb [27].

Higher-order basis functions can be categorized into two classes: interpolatory and hierarchical. *Interpolatory basis functions* relate to a set of elemental points, such that every basis function is of the same order and is equal to unity at one point and zero at all others. On the other hand, *hierarchical basis functions* are formed by adding new higher-order basis functions to the lower-order ones; thus, the elemental solution is expanded in terms of basis functions of differing polynomial orders. The two approaches have the same accuracy, although one may be favored over the other, depending on the application. Interpolatory basis functions generally lead to better conditioned matrices, while hierarchical basis functions permit the use of different elemental orders in a single finite element solution.

Because higher-order basis functions interpolate the solution field much more accurately than do first-order basis functions, the finite element method may be expected to yield much more accurate results with increasing elemental order. Specifically, for smooth functions it can be shown that if p is the order of the basis functions, h denotes the elemental size divided by the order of the element, and λ denotes the excitation wavelength, the interpolation error is of the order $O[(h/\lambda)^{p+1}]$. Therefore, in the case of smooth solutions, the finite element solution error may be reduced accordingly by using higher-order basis functions. If the true solution to the finite element analysis contains a singularity, this interpolation error estimate does not hold and it becomes more advantageous to use smaller elements of low polynomial order local to the singularity.

When the finite element method is applied to the Helmholtz equations (scalar as well as vector cases), the simulated wave propagates at a speed that is slightly different from the exact value. Consequently, there is a numerical error in the phase of the solution called *dispersion error*. Analysis based on the finite element discretization of plane-wave propagation in a uniform medium shows that the dispersion error per wavelength is proportional to $O[(h/\lambda)^{2p}]$.† From this result follows the very important conclusion that phase errors may be decreased exponentially by increasing the order of the basis functions. Therefore, higher-order elements are especially well suited for simulating large-scale wave propagation problems. Note that in the case of time-domain formulations, we have both a finite element discretization in space and, typically, a finite-difference type of discretization in time. Consequently, we have both spatial dispersion errors, which can be reduced by using the same higher-order basis functions as described above, and temporal dispersion errors. Although a higher-order basis function representation can be adopted for the time dependency, a reduced time step is typically used if the temporal discretization errors become unacceptably large.

In the same way that elemental basis functions can be defined to an arbitrary polynomial order, so can the geometric representation of the elements. The elements shown in Figures 2.4 and 2.5 are of linear geometric order because all edges and faces are straight and flat and can thus be described by linear functions of position. Such elements are termed *rectilinear*. Elements of higher geometrical polynomial order may be defined by modeling elemental edges and faces with higher-order polynomial functions of position. For example, one may require the edge of a triangular element to pass through two of its vertex nodes together with an additional node which may not lie on the straight line connecting the vertices. In this case, a second-order geometrical representation would suffice. Such elemental representations are very useful when modeling curved boundaries and are termed *curvilinear*. When the order of geometrical representation is the same as the order of basis functions, the element is called *isoparametric*. Otherwise, it is called either *subparametric* (when the order of geometrical modeling is lower) or *superparametric* (when the order of geometrical modeling is higher).

By using higher-order curvilinear elements, we can obtain a much more accurate solution than is possible with lower-order rectilinear elements. More important, with higher-order curvilinear elements this improvement in accuracy can often be obtained by using spatially larger finite elements (when the problem geometry permits) such that the overall number of unknowns is reduced compared to a lower-order formulation based on rectilinear elements. This can increase the efficiency of the finite element analysis significantly. Our experience indicates that for most applications, we can use four to eight second-order elements per wavelength, or two to four third-order elements per wavelength, or one or two fourth-order elements per wavelength to achieve accuracy similar to that obtained with 10 to 20 first-order elements per wavelength. This results in a reduction in the number of unknowns by a factor of 2, 5, and 15, respectively. Of course, this is simply a guideline, and it ignores the factor of geometrical modeling. The use of higher-order curvilinear elements is particularly beneficial when the computational domain is very large in terms of wavelength and/or when very high accuracy is required, which is a direct result of the exponential convergence discussed earlier.

†Because the order of vector basis fuctions in this book is defined differently from that in Ref. 2, the expression for the grid dispersion error is different.

2.6 SUMMARY

In this chapter we described the finite element formulation used to solve for the electric field surrounding a generic antenna. Starting from the original boundary-value problem for antenna radiation, we first established an approximate representation that truncates the computational domain using a simple, first-order absorbing boundary condition. We then formulated the full three-dimensional finite element solution to solve for the electric field in the frequency domain. This was followed by the finite element formulation in the time domain, where the solution was computed by time marching. Next, we presented various numerical schemes to model electrically and magnetically anisotropic, dispersive, and lossy materials. Finally, we discussed briefly techniques for solving the finite element matrix equations and the benefit of using higher-order curvilinear finite elements.

For the finite element method described in this chapter to be more accurate, more efficient, and more powerful for the analysis of complex antennas and arrays, we have to address three major issues further. The first issue concerns truncation of the computational domain. In this chapter we employed a simple absorbing boundary condition for this truncation. Unfortunately, this absorbing boundary condition absorbs perfectly only the propagating field incident in the normal direction. For obliquely incident propagating fields, and for any incident evanescent fields, the absorbing boundary condition produces significant reflection back to the computational domain, which degrades the accuracy of the numerical solution. Although these undesirable reflections can be reduced a certain extent by placing the truncation boundary farther away from the antenna, doing so will greatly increase the size of the computational domain and thereby compromise the efficiency of the finite element solution. Therefore, more accurate and efficient mesh truncation techniques are needed to enhance the accuracy and efficiency of the finite element analysis, and this topic is addressed in Chapters 3 and 4.

The second issue is related to the modeling of antenna feeds and the calculation of antenna parameters, such as the input impedance. In this chapter the source region for the antenna was specified as a known electric or magnetic current. However, in practical applications, antennas are typically fed with some form of waveguide, which is often a coaxial transmission line. To predict antenna parameters accurately, especially those related to antenna feeds, such as the input and mutual impedances, more realistic models of the feed region are required. In addition, many advanced antennas are designed with small features such as thin wires and thin-material sheets. Accurate and efficient modeling of these small features is critical in the simulation of the antenna performance characteristics. The modeling of antenna feeds in both the frequency and time domains is discussed in Chapter 5, and accurate modeling of small details is addressed in Chapter 6.

As illustrated in this chapter, a finite element analysis requires solving a large, sparse matrix equation at each frequency or at each time step. For very large antenna applications, such as finite phased-array antennas and antennas placed on an electrically large platform, the finite element matrix equation can easily exceed tens or even hundreds of millions of unknowns. The solution of such a large matrix problem can be very time consuming, and perhaps even impractical, although the matrix is extremely sparse. Therefore, special techniques have to be developed to deal with such large applications. For example, a large phased-array antenna with a uniform excitation can be approximated as an infinitely periodic array, where the analysis can be confined to a unit cell. Furthermore, a large phased-array antenna is often comprised of many identical radiating elements, such that the finite element discretization of these radiating elements will yield identical matrix equations. The ability to utilize this geometrical repetition creates the possibility of designing extremely efficient algorithms to accommodate large finite phased-array antennas. For the case of an antenna on a large platform, a numerical simulation can be made much more efficient through a better understanding of the specific interaction mechanisms that exist between the platform and the antenna. These important topics are discussed in Chapters 8 through 11.

REFERENCES

1. R. L. Courant, "Variational methods for the solution of problems of equilibrium and vibration," *Bull. Am. Math. Soc.*, vol. 49, pp. 1–23, 1943.
2. J.-M. Jin, *The Finite Element Method in Electromagnetics*, 2nd ed. Hoboken, NJ: Wiley, 2002.
3. N. M. Newmark, "A method of computation for structural dynamics," *J. Eng. Mech. Div. Proc. Am. Soc. Civil Eng.*, vol. 85, pp. 67–94, July 1959.
4. J.-F. Lee and Z. Sacks, "Whitney elements time domain (WETD) methods," *IEEE Trans. Magn.*, vol. 31, no. 3, pp. 1325–1329, May 1995.
5. S. D. Gedney and U. Navsariwala, "Unconditionally stable finite element time-domain solution of the vector wave equation," *IEEE Microwave Guided Wave Lett.*, vol. 5, pp. 332–334, 1995.
6. D. Jiao and J. M. Jin, "Time-domain finite-element modeling of dispersive media," *IEEE Microwave Wireless Components Lett.*, vol. 11, no. 5, pp. 220–222, May 2001.
7. F. Maradei, "A frequency-dependent WETD formulation for dispersive materials," *IEEE Trans. Magn.*, vol. 37, no. 5, pt. 1, pp. 3303–3306, Sept. 2001.
8. F. Edelvik and B. Strand, "Frequency dispersive materials for 3-D hybrid solvers in time domain," *IEEE Trans. Antennas Propagat.*, vol. 51, pp. 1199–1205, June 2003.
9. D. Riley and J. M. Jin, "Modeling of magnetic loss in the finite-element time-domain method," *Microwave Opt. Tech. Lett.*, vol. 46, no. 2, pp. 165–168, July 2005.
10. J. M. Jin and D. Riley, "Modeling of anisotropic, dispersive, lossy, electric and magnetic materials in the finite-element time-domain solution of Maxwell's equations," Final Report on Contract 4500014491, Northrop Grumman Mission Systems, Reston, VA, Aug. 5, 2005.
11. J. M. Jin, Z. Lou, Y. J. Li, N. Riley, and D. Riley, "Finite element analysis of complex antennas and arrays," *IEEE Trans. Antennas Propagat.*, vol. 56, no. 8, pp. 2222–2240, Aug. 2008.
12. I.-T. Chiang and W. C. Chew, "Fast real-time convolution algorithm for microwave multi-port networks with nonlinear terminations," *IEEE Trans. Circuits Syst. II Express Briefs*, vol. 52, pp. 370–375, July 2005.

13. X. Li and J. M. Jin, "Time-domain finite-element modeling of electrically and magnetically dispersive medium via recursive FFT," *Microwave Opt. Tech. Lett.*, vol. 50, no. 6, pp. 1476–1481, June 2008.
14. B. M. Irons, "A frontal method solution program for finite element analysis," *Int. J. Numer. Methods Eng.*, vol. 2, pp. 5–32, 1970.
15. J. W. H. Liu, "The multifrontal method for sparse matrix solution: theory and practice," *SIAM Rev.*, vol. 34, pp. 82–109, Mar. 1992.
16. R. Barrett, M. Berry, T. F. Chan, J. Demmel, J. Donato, J. Dongarra, V. Eijkhout, R. Pozo, C. Romine, and H. V. der Vorst, *Templates for the Solution of Linear Systems: Building Blocks for Iterative Methods*, 2nd ed. Philadelphia, PA: SIAM, 1994.
17. Intel® Math Kernel Library (MKL), Reference Manual, 2007, http://www.intel.com/software/products/mkl/.
18. SGI Scientific Computing Software Library (SCSL) User's Guide, 2003, http://www.sgi.com/products/software/scsl.html.
19. A. Gupta, "WSMP: Watson Sparse Matrix Package Part I—Direct solution of symmetric sparse system," Tech. Rep. 21886 (98462), IBM T. J. Watson Research Center, Yorktown Heights, NY, 2000.
20. A. Gupta, "WSMP: Watson Sparse Matrix Package Part II—Direct solution of symmetric sparse system," Tech. Rep. 21888 (98472), IBM T. J. Watson Research Center, Yorktown Heights, NY, 2000.
21. T. A. Davis, "UMFPACK V3.2: an unsymmetric-pattern multifrontal method with a column pre-ordering strategy," Tech. Rep. TR-02-2002, Department of Computer Information Science, University of Florida, Gainesville, FL, 2002.
22. P. R. Amestoy, I. S. Duff, and J.-Y. L'Excellent, "Multifrontal parallel distributed symmetric and unsymmetric solvers," *Comput. Methods Appl. Mech. Eng.*, vol. 184, pp. 501–520, 2000.
23. X. Li and J. Demmel, "SuperLU DIST: a scalable distributed-memory sparse direct solver for unsymmetric linear systems," *ACM Trans. Math. Software*, vol. 29, no. 2, pp. 110–140, 2003.
24. S. Balay, K. Buschelman, V. Eijkhout, W. Gropp, D. Kaushik, M. Knepley, L. C. McInnes, B. Smith, and H. Zhang, *PETSc Users Manual*. Argonne National Laboratories, Argonne, IL, 2007. http://www.mcs.anl.gov/petsc.
25. Y. Saad, *SPARSKIT: A Basic Tool-Kit for Sparse Matrix Computations*, Version 2, Department of Computer Science and Engineering, University of Minnesota, Minneapolis, MN, 1994. http://www-users.cs.umn.edu/~saad/software/SPARSKIT/sparskit.html.
26. R. D. Graglia, D. R. Wilton, and A. F. Peterson, "Higher order interpolatory vector bases for computational electromagnetics," *IEEE Trans. Antennas Propagat.*, vol. 45, pp. 329–342, Mar. 1997.
27. J. P. Webb, "Hierarchal vector basis functions of arbitrary order for triangular and tetrahedral finite elements," *IEEE Trans. Antennas Propagat.*, vol. 47, pp. 1244–1253, Aug. 1999.

3 Finite Element Mesh Truncation

As illustrated in Chapter 2, one of the major challenges in the finite element analysis of antenna problems is the truncation of infinite space into a finite computational domain. This truncation can be accomplished by introducing an artificial surface to enclose the antenna. However, to emulate the original free-space environment, the artificial truncation surface should absorb as much of the radiated field as possible in order to reduce any artificially reflected fields back to the computational domain. This situation is similar to what occurs in antenna measurement in an anechoic chamber, where absorbers are used to cover the walls of the chamber so that reflections from the walls do not interfere with the measurement of antenna characteristics.

Although, in principle, we can use the finite element method to simulate an actual anechoic chamber for an antenna analysis, this approach is not practical because an anechoic chamber is electrically very large, and hence its numerical simulation is extremely time consuming. Fortunately, there are many other approaches to reducing the reflection of an artificial surface. These include the use of a mathematical boundary condition, the use of fictitious absorbing material layers, and the use of a surface integral equation. In this chapter we discuss these three approaches in detail.

3.1 ABSORBING BOUNDARY CONDITIONS

Among the three approaches for mesh truncation, the use of a mathematical boundary condition is simplest. In this section we describe the derivation and application of the first- and second-order absorbing boundary conditions.

3.1.1 First-Order Absorbing Boundary Condition

It is well known that when a plane wave is incident normally on a planar impedance surface, the impedance surface can absorb all the incident power without any reflection if the surface impedance matches the intrinsic impedance of free space. It can easily be verified that on such a surface the electric field $\mathbf{E}$ satisfies the relation

$$\hat{n} \times (\nabla \times \mathbf{E}) + jk_0\hat{n} \times (\hat{n} \times \mathbf{E}) = 0 \tag{3.1}$$

where $\hat{n}$ is the outward unit normal vector of the surface and k_0 is the wavenumber of the plane wave. Equation (3.1) can also be considered as a boundary condition

Finite Element Analysis of Antennas and Arrays, By Jian-Ming Jin and Douglas J. Riley

because it provides a relationship between the tangential components of the electric and magnetic fields. When this boundary condition is applied to the artificial truncation surface, the surface will absorb any normally incident plane wave. Hence, this boundary condition is often referred to as an *absorbing boundary condition* (ABC). Note that this condition is identical to the approximate Sommerfeld radiation condition (2.7). In reality, the radiated field at a point on the truncation surface is usually a summation of many plane waves propagating in different directions. For a plane wave incident at a large angle, a significant reflection may be caused by the boundary. However, when placed sufficiently far away from all sources of excitation, most waves would be incident on the truncation boundary only at small angles from normal and hence would mostly be absorbed. The truncation distance, measured from the antenna to the boundary, required for good absorptive performance depends on the nature of the radiation. Typically, a minimum distance of one-half wavelength (0.5λ) is necessary to obtain useful results.

3.1.2 Second-Order Absorbing Boundary Condition

It will be shown later that the boundary condition (3.1) is only the first-order approximation to the true Sommerfeld radiation condition (2.6). To derive more accurate ABCs, we can start with Wilcox's expansion,

$$\mathbf{E}(\mathbf{r}) = \frac{e^{-jk_0 r}}{r} \sum_{n=0}^{\infty} \frac{\mathbf{A}_n(\theta, \phi)}{r^n} \tag{3.2}$$

which converges uniformly for the field outside a spherical surface that encloses the source. By taking the curl of (3.2) and then the cross product with $\hat{r}$, we can derive a variety of approximate boundary conditions applicable at a spherical surface [1,2]. One set of such boundary conditions derived by Webb and Kanellopoulos [2] is given by

$$\mathcal{B}_m(\mathbf{E}) = O(r^{-2m-1}) \qquad m = 1, 2, \ldots \tag{3.3}$$

where

$$\mathcal{B}_m(\mathbf{E}) = (\mathcal{L}_{m-1})^m(\mathbf{E}_t) + s(\mathcal{L}_m)^{m-1}(\nabla_t E_r) \tag{3.4}$$

in which

$$\mathcal{L}_m(\mathbf{u}) = \hat{r} \times (\nabla \times \mathbf{u}) - \left(jk_0 + \frac{m}{r}\right)\mathbf{u} \qquad m = 0, 1, 2, \ldots \tag{3.5}$$

and s is a constant, which has no effect on the electric field transverse to r (TE_r field) and affects only the transverse magnetic field (TM_r field). In (3.4), the superscript on $\mathcal{L}_{m-1}$ denotes that the operator $\mathcal{L}_{m-1}$ is applied m times, and the superscript on

$\mathcal{L}_m$ denotes that the operator $\mathcal{L}_m$ is applied $m-1$ times. By letting $m=1$ and $s=1$, (3.3) becomes

$$\hat{r} \times (\nabla \times \mathbf{E}) + jk_0\hat{r} \times (\hat{r} \times \mathbf{E}) \approx 0 \tag{3.6}$$

which is identical to (3.1) and is referred to as the *first-order ABC*. It is clear now that this first-order ABC is accurate up to r^{-2} since it neglects all terms on the order of r^{-3} and higher. To derive a more accurate ABC, we can let $m=2$ in (3.3) to find that

$$\hat{r} \times (\nabla \times \mathbf{E}) \approx -jk_0\hat{r} \times (\hat{r} \times \mathbf{E}) + \beta\{\nabla \times [\hat{r}\hat{r} \cdot (\nabla \times \mathbf{E})] + \nabla_t(\nabla_t \cdot \mathbf{E})\} \tag{3.7}$$

where

$$\beta = \frac{1}{2}\left(jk_0 + \frac{1}{r}\right)^{-1}. \tag{3.8}$$

In deriving (3.7), we set $s=2$ in (3.4) to eliminate a term that would destroy the symmetry of the finite element matrix equation. Equation (3.7) is accurate up to r^{-4} since it neglects only terms on the order of r^{-5} and higher. It has much better accuracy than (3.6) and is referred to as the *second-order ABC*.

Although (3.6) and (3.7) are derived for a spherical surface, they can be applied to arbitrarily shaped convex surfaces with a certain loss of accuracy. This trade-off is worthwhile in many applications because one can often reduce the size of the computational domain significantly by using a surface other than a spherical shape. In this case, $\hat{r}$ is replaced by $\hat{n}$ and $1/r$ in (3.8) is replaced by the local curvature of the surface or is simply set to zero. The absorption performances of (3.6) and (3.7) can be examined and compared easily by applying them to a planar surface and evaluating their reflection coefficients [3], which are given by

$$R(\theta) = \frac{\cos\theta - 1}{\cos\theta + 1} \tag{3.9}$$

for the first-order ABC (3.6) and approximately by

$$R(\theta) = \frac{\cos\theta + \frac{1}{2}\sin^2\theta - 1}{\cos\theta - \frac{1}{2}\sin^2\theta + 1} \tag{3.10}$$

for the second-order ABC (3.7).† Note that even though higher-order ABCs provide improved absorbing performances, they still cannot absorb fields incident at

†To be more rigorous, (3.10) is valid only for the TE_r field. For the TM_r field, the reflection coefficient is slightly compromised by setting $s=2$ in (3.4) and hence a little bit larger than that predicted by (3.10). See Ref. 3 for detailed discussion on this issue.

near-grazing angles. Therefore, their application still requires a certain distance between the absorbing surface and the antenna structures.

The major advantage of ABCs, apart from their simplicity, is that they lead to a localized relation between boundary fields and consequently preserve the highly sparse and banded pattern of the finite element system matrices. In addition, the boundary condition (3.1) can easily be incorporated into a weak-form wave equation such as (2.13). Thus, implementation of the first-order ABC in the finite element method is straightforward, as demonstrated in Chapter 2. Implementation of the second-order ABC (3.7), however, is somewhat more involved. When we apply (3.7) to (2.12), we obtain

$$\begin{aligned}
&\iiint_V \left[(\nabla \times \mathbf{T}) \cdot \overleftrightarrow{\mu}_r^{-1} \cdot (\nabla \times \mathbf{E}) - k_0^2 \mathbf{T} \cdot \overleftrightarrow{\varepsilon}_r \cdot \mathbf{E}\right] dV \\
&= \oiint_{S_{\text{PEC}}} (\hat{n} \times \mathbf{T}) \cdot \overleftrightarrow{\mu}_r^{-1} \cdot (\nabla \times \mathbf{E})\, dS \\
&\quad + \oiint_{S_o} \mathbf{T} \cdot \{jk_0 \hat{r} \times (\hat{r} \times \mathbf{E}) - \beta \nabla \times [\hat{r}\hat{r} \cdot (\nabla \times \mathbf{E})] - \beta \nabla_t (\nabla_t \cdot \mathbf{E})\}\, dS \\
&\quad - \iiint_V \mathbf{T} \cdot \left[jk_0 Z_0 \mathbf{J}_{\text{imp}} + \nabla \times \left(\overleftrightarrow{\mu}_r^{-1} \cdot \mathbf{M}_{\text{imp}}\right)\right] dV. \qquad (3.11)
\end{aligned}$$

By some mathematical manipulations with the aid of vector identities and the surface divergence theorem [3], (3.11) can be written as

$$\begin{aligned}
&\iiint_V \left[(\nabla \times \mathbf{T}) \cdot \overleftrightarrow{\mu}_r^{-1} \cdot (\nabla \times \mathbf{E}) - k_0^2 \mathbf{T} \cdot \overleftrightarrow{\varepsilon}_r \cdot \mathbf{E}\right] dV \\
&= \oiint_{S_{\text{PEC}}} (\hat{n} \times \mathbf{T}) \cdot \overleftrightarrow{\mu}_r^{-1} \cdot (\nabla \times \mathbf{E})\, dS \\
&\quad - \oiint_{S_o} [jk_0 (\hat{r} \times \mathbf{T}) \cdot (\hat{r} \times \mathbf{E}) + \beta (\nabla \times \mathbf{T})_r \cdot (\nabla \times \mathbf{E})_r - \beta (\nabla_t \cdot \mathbf{T})(\nabla_t \cdot \mathbf{E})]\, dS \\
&\quad - \iiint_V \mathbf{T} \cdot \left[jk_0 Z_0 \mathbf{J}_{\text{imp}} + \nabla \times \left(\overleftrightarrow{\mu}_r^{-1} \cdot \mathbf{M}_{\text{imp}}\right)\right] dV. \qquad (3.12)
\end{aligned}$$

This equation is also applicable to an arbitrarily shaped *smooth* convex surface. The smoothness requirement is due to application of the surface divergence theorem. Therefore, one must always employ a smooth truncation surface for the second-order ABC, whereas there is no such requirement on use of the first-order ABC.

The weak-form equation in (3.12) has a desired symmetry, which can yield a symmetric finite element matrix equation. Unfortunately, the term $\beta(\nabla_t \cdot \mathbf{T})(\nabla_t \cdot \mathbf{E})$ cannot be implemented easily using the curl-conforming vector basis functions discussed in Chapter 2. As we pointed out, these basis functions are continuous in the tangential direction and discontinuous in the normal direction across interfaces between elements. As such, their surface divergence gives rise to a delta function, which cannot be evaluated. If we simply neglect this term [4], the absorption of the TM_r component of the field would be compromised, although the absorption of the TE_r component remains second order. The importance of the term $\beta(\nabla_t \cdot \mathbf{T})(\nabla_t \cdot \mathbf{E})$ to the second-order ABC has been studied together with a treatment using special basis functions [5]. Recently, a relatively simple remedy was proposed [6], which was shown to be highly effective and to retain the symmetry of the final finite element matrix equation. In this approach, an auxiliary scalar variable is defined on the surface S_o:

$$v = \nabla_t \cdot \mathbf{E} \qquad \mathbf{r} \in S_o. \tag{3.13}$$

With this, (3.12) can be written as

$$\begin{aligned}
&\iiint_V \left[(\nabla \times \mathbf{T}) \cdot \overleftrightarrow{\mu}_r^{-1} \cdot (\nabla \times \mathbf{E}) - k_0^2 \mathbf{T} \cdot \overleftrightarrow{\varepsilon}_r \cdot \mathbf{E}\right] dV \\
&= \oiint_{S_{\mathrm{PEC}}} (\hat{n} \times \mathbf{T}) \cdot \overleftrightarrow{\mu}_r^{-1} \cdot (\nabla \times \mathbf{E})\, dS \\
&\quad - \oiint_{S_o} [jk_0(\hat{r} \times \mathbf{T}) \cdot (\hat{r} \times \mathbf{E}) + \beta(\nabla \times \mathbf{T})_r \cdot (\nabla \times \mathbf{E})_r + \beta \mathbf{T} \cdot \nabla_t v]\, dS \\
&\quad - \iiint_V \mathbf{T} \cdot \left[jk_0 Z_0 \mathbf{J}_{\mathrm{imp}} + \nabla \times \left(\overleftrightarrow{\mu}_r^{-1} \cdot \mathbf{M}_{\mathrm{imp}}\right)\right] dV.
\end{aligned} \tag{3.14}$$

To discretize (3.14), the scalar auxiliary variable v can be expanded using scalar interpolatory basis functions defined on the surface S_o. Figure 3.1 shows a linear basis function for a node on a surface mesh. By using such basis functions, v can be expanded as

$$v = \sum_{i=1}^{N_{\mathrm{node}}} N_i v_i \tag{3.15}$$

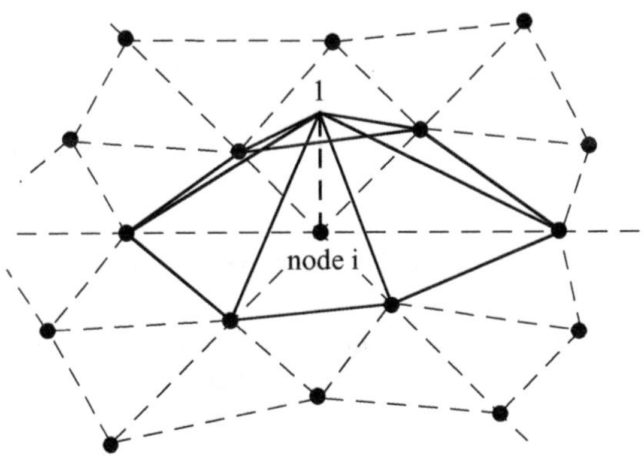

Figure 3.1 Scalar interpolatory basis function N_i for a node on a surface triangular mesh.

where N_{node} denotes the total number of nodes on S_o. By substituting (2.16) and (3.15) into (3.14) and using $\mathbf{N}_i$ as the weighting function $\mathbf{T}$, we obtain

$$[K \quad B]\begin{Bmatrix} E \\ v \end{Bmatrix} = \{b\} \tag{3.16}$$

where

$$K_{ij} = \iiint_V \left[(\nabla \times \mathbf{N}_i) \cdot \overleftrightarrow{\mu}_r^{-1} \cdot (\nabla \times \mathbf{N}_j) - k_0^2 \mathbf{N}_i \cdot \overleftrightarrow{\varepsilon}_r \cdot \mathbf{N}_j\right] dV$$
$$+ \oiint_{S_o} [jk_0(\hat{r} \times \mathbf{N}_i) \cdot (\hat{r} \times \mathbf{N}_j) + \beta(\nabla \times \mathbf{N}_i)_r \cdot (\nabla \times \mathbf{N}_j)_r]\, dS \tag{3.17}$$

$$B_{ij} = \oiint_{S_o} \beta \mathbf{N}_i \cdot \nabla_t N_j \, dS \tag{3.18}$$

and b_i is the same as given by (2.19). To complete (3.16), we multiply (3.13) by βN_i and integrate over S_o to find

$$\oiint_{S_o} \beta N_i v \, dS = \oiint_{S_o} \beta N_i \nabla_t \cdot \mathbf{E} \, dS. \tag{3.19}$$

By substituting (2.16) and (3.15) into (3.19) and applying the surface divergence theorem, we obtain

$$[B^{\text{T}} \quad C]\begin{Bmatrix} E \\ v \end{Bmatrix} = \{0\} \tag{3.20}$$

where

$$C_{ij} = \oiint_{S_o} \beta N_i N_j \, dS. \tag{3.21}$$

Combining (3.16) and (3.20), we obtain a symmetric matrix equation

$$\begin{bmatrix} K & B \\ B^{\text{T}} & C \end{bmatrix}\begin{Bmatrix} E \\ v \end{Bmatrix} = \begin{Bmatrix} b \\ 0 \end{Bmatrix} \tag{3.22}$$

for the solution of the electric field everywhere in the computational domain and the auxiliary variable on the absorbing surface.

It has been shown [6] that with this implementation, the second-order ABC consistently outperforms the first-order ABC. One example calculates the scattering of a plane wave by a conducting sphere having a radius of 0.1λ. A spherical truncation surface is placed 0.1λ away from the surface of the conducting sphere, and the first-order basis functions are used in the finite element simulation. The L2-norm

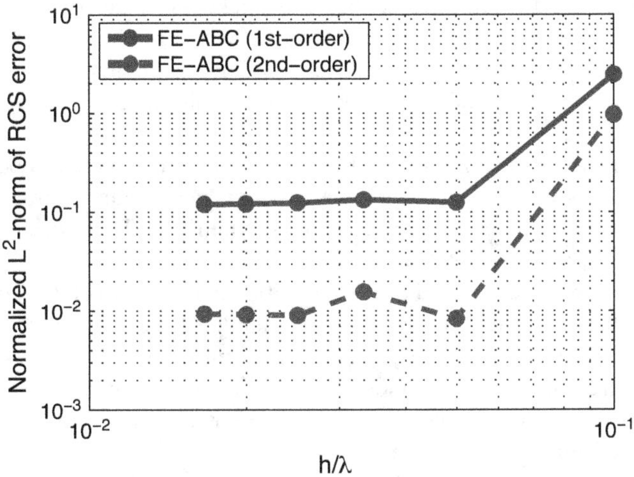

Figure 3.2 Normalized bistatic RCS error in the finite element solutions using first- and second-order ABCs versus the size of the finite elements in terms of wavelength.

error in the bistatic radar cross section (RCS) computed is plotted in Figure 3.2 as a function of the size of the finite elements used for the discretization. The L2-norm error shown is calculated by integrating the bistatic RCS error over all directions and then normalized by the integral of the analytical RCS. As can be seen, when the finite elements are sufficiently small, the error approaches a constant that depends on the specific ABC. The second-order ABC reduces the error by approximately one order of magnitude compared to the first-order ABC.

It should be emphasized once again that implementation of the second-order ABC, as described above, requires S_o to be a smooth surface, whereas there is no such requirement for implementation of the first-order ABC. Furthermore, the finite element matrix obtained using the first-order ABC is always better conditioned than the one using the second-order ABC. Therefore, when an iterative solver is employed to solve the finite element matrix equation, the first-order ABC is more efficient than the second-order ABC for a similarly discretized computational domain. These two distinctive advantages of the first-order ABC make it a preferred choice when one desires a quick and approximate solution for an initial antenna design.

3.2 PERFECTLY MATCHED LAYERS

Instead of using a mathematical boundary condition, an open computational domain may also be terminated by using absorptive materials. Whereas it is extremely costly to model the absorbers used in anechoic chambers due to their electrically large thickness and their wedged and pyramidal shapes, we can instead design thin layers of artificial absorbers solely for simulation purposes.

Specifically, by adjusting the number and thickness of layers, as well as their permittivity, permeability, and conductivity (all of which can be made tensors), these artificial absorbing layers can be designed to provide negligible reflection and sufficient attenuation to the field transmitted such that the field mostly vanishes as it propagates into the layers. An early attempt to use this method for finite element analysis was made in the frequency domain [7], which provided good absorbing performance at a specific frequency. An alternative and popular approach proposed by Berenger within the finite-difference time-domain (FDTD) setting [8] is known as *perfectly matched layers* (PMLs).

A PML is an artificial material that is theoretically defined to create no reflections regardless of the frequency, polarization, and angle of incidence of a plane wave. The frequency independence is especially important because it enables broadband simulation with a time-domain method. In its original form, the PML was formulated with the aid of nonphysical "split" fields [8,9]. Later, it was found that the PML could be derived alternatively from a modified form of Maxwell's equations based on stretched coordinates [10]. In addition, a PML can be derived based on an artificial anisotropic medium model [11,12].

3.2.1 PML in Terms of Stretched Coordinates

To introduce the basic idea of the PML, we consider the modified source-free Maxwell's equations [10]

$$\nabla_s \times \mathbf{E} = -j\omega\mu\mathbf{H} \tag{3.23}$$

$$\nabla_s \times \mathbf{H} = j\omega\varepsilon\mathbf{E} \tag{3.24}$$

$$\nabla_s \cdot (\varepsilon\mathbf{E}) = 0 \tag{3.25}$$

$$\nabla_s \cdot (\mu\mathbf{H}) = 0 \tag{3.26}$$

where ∇_s is defined by

$$\nabla_s = \hat{x}\frac{1}{s_x}\frac{\partial}{\partial x} + \hat{y}\frac{1}{s_y}\frac{\partial}{\partial y} + \hat{z}\frac{1}{s_z}\frac{\partial}{\partial z}. \tag{3.27}$$

Clearly, ∇_s can be considered as the standard ∇ operator in Cartesian coordinates, where the x-, y-, and z-axes are stretched by a factor of s_x, s_y, and s_z, respectively. Here, we assume that s_x, s_y, and s_z are either constants or functions of x, y, and z, respectively; that is, $s_x = s_x(x)$, $s_y = s_y(y)$, and $s_z = s_z(z)$.

For a plane wave governed by the modified Maxwell's equations, its propagation constant satisfies the dispersion relation

$$\left(\frac{k_x}{s_x}\right)^2 + \left(\frac{k_y}{s_y}\right)^2 + \left(\frac{k_z}{s_z}\right)^2 = \omega^2\mu\varepsilon = k^2 \tag{3.28}$$

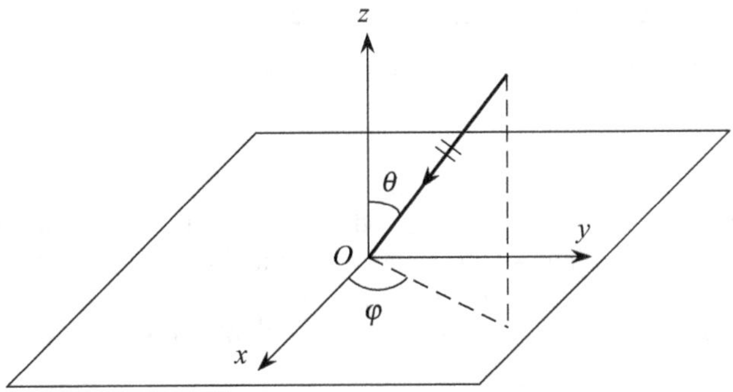

Figure 3.3 Plane wave incident on the interface between the upper and lower half-spaces.

where the solution is given by

$$k_x = ks_x \sin\theta \cos\phi, \quad k_y = ks_y \sin\theta \sin\phi, \quad k_z = ks_z \cos\theta. \tag{3.29}$$

This indicates that if s_x is a complex number with a negative imaginary part, the wave will be attenuated in the x-direction, and the same is true in the other two directions.

Consider further the reflection of a plane wave by an interface between two half-spaces in the stretched coordinate system. The interface is assumed to coincide with the xy-plane and the plane wave is incident upon the interface from the upper space (Figure 3.3). Using the phase-matching and tangential continuity conditions for **E** and **H**, we can find the reflection coefficient for the TE$_z$ case as

$$R_{\text{TE}} = \frac{k_{1z}s_{2z}\mu_2 - k_{2z}s_{1z}\mu_1}{k_{1z}s_{2z}\mu_2 + k_{2z}s_{1z}\mu_1} \tag{3.30}$$

where the subscript 1 is used to denote the parameters in the upper half-space and the subscript 2, to denote those in the lower half-space. Similarly, we can find the reflection coefficient for the TM$_z$ case as

$$R_{\text{TM}} = \frac{k_{1z}s_{2z}\varepsilon_2 - k_{2z}s_{1z}\varepsilon_1}{k_{1z}s_{2z}\varepsilon_2 + k_{2z}s_{1z}\varepsilon_1}. \tag{3.31}$$

If $\varepsilon_1 = \varepsilon_2 = \varepsilon$, $\mu_1 = \mu_2 = \mu$, $s_{1x} = s_{2x}$, and $s_{1y} = s_{2y}$, we find that $\theta_1 = \theta_2$ and $\phi_1 = \phi_2$, and consequently,

$$R_{\text{TE}} = 0, \qquad R_{\text{TM}} = 0 \tag{3.32}$$

which indicates that the interface is reflectionless. More important, this property remains true regardless of (1) the choice of s_{1z} and s_{2z}, (2) the angle of incidence, and (3) the frequency. Because of this, the interface is called a *perfectly matched interface* (PMI).

3.2.2 PML as an Anisotropic Material Absorber

The PML formulation based on stretched coordinates is inconvenient to implement in the finite element method because of the stretched operator in (3.27). A better approach is to consider the PML as an anisotropic absorbing medium [11,12] so that it can be implemented in the finite element method without much modification. This can be accomplished by converting the modified Maxwell's equations (3.23)–(3.26) based on stretched coordinates into a more traditional form using the standard ∇ operator. Let $\mathbf{E}^c$ and $\mathbf{H}^c$ denote the field quantities used in the modified Maxwell's equations (3.23)–(3.26). We can define the new field quantities $\mathbf{E}^a$ and $\mathbf{H}^a$ such that

$$\mathbf{E}^a = \begin{bmatrix} s_x & 0 & 0 \\ 0 & s_y & 0 \\ 0 & 0 & s_z \end{bmatrix} \cdot \mathbf{E}^c, \qquad \mathbf{H}^a = \begin{bmatrix} s_x & 0 & 0 \\ 0 & s_y & 0 \\ 0 & 0 & s_z \end{bmatrix} \cdot \mathbf{H}^c. \tag{3.33}$$

With this, we can easily find

$$\nabla_s \times \mathbf{E}^c = \begin{bmatrix} \dfrac{1}{s_y s_z} & 0 & 0 \\ 0 & \dfrac{1}{s_z s_x} & 0 \\ 0 & 0 & \dfrac{1}{s_x s_y} \end{bmatrix} \cdot \nabla \times \mathbf{E}^a, \qquad \mathbf{H}^c = \begin{bmatrix} \dfrac{1}{s_x} & 0 & 0 \\ 0 & \dfrac{1}{s_y} & 0 \\ 0 & 0 & \dfrac{1}{s_z} \end{bmatrix} \cdot \mathbf{H}^a. \tag{3.34}$$

Substituting these into (3.23), we obtain

$$\nabla \times \mathbf{E}^a = -j\omega\mu \overleftrightarrow{\mathbf{\Lambda}} \cdot \mathbf{H}^a \tag{3.35}$$

where

$$\overleftrightarrow{\mathbf{\Lambda}} = \begin{bmatrix} \dfrac{s_y s_z}{s_x} & 0 & 0 \\ 0 & \dfrac{s_z s_x}{s_y} & 0 \\ 0 & 0 & \dfrac{s_x s_y}{s_z} \end{bmatrix}. \tag{3.36}$$

Similarly, (3.24)–(3.26) can be converted into

$$\nabla \times \mathbf{H}^a = j\omega\varepsilon \overleftrightarrow{\mathbf{\Lambda}} \cdot \mathbf{E}^a \tag{3.37}$$

$$\nabla \cdot (\varepsilon \overleftrightarrow{\mathbf{\Lambda}} \cdot \mathbf{E}^a) = 0 \tag{3.38}$$

$$\nabla \cdot (\mu \overleftrightarrow{\mathbf{\Lambda}} \cdot \mathbf{H}^a) = 0. \tag{3.39}$$

These equations are recognized as the ordinary Maxwell's equations for an anisotropic medium with a permittivity tensor $\varepsilon\overleftrightarrow{\mathbf{\Lambda}}$ and a permeability tensor $\mu\overleftrightarrow{\mathbf{\Lambda}}$. Just like the modified Maxwell's equations (3.23)–(3.26), they reduce to the regular Maxwell's equations outside the PML where $s_x = s_y = s_z = 1$. However, inside the PML, the field quantities are different from those in (3.23)–(3.26), which nevertheless is not important because the fields inside the PML are of no interest. Since (3.35)–(3.39) are derived directly from (3.23)–(3.26), they define an anisotropic medium that has no reflection for an incident plane wave on its interface regardless of frequency, polarization, and angle of incidence.

The perfectly matched anisotropic absorbing medium described above is derived based on the property of the modified Maxwell's equations (3.23)–(3.26). It can, alternatively, be derived without such prior knowledge [11].

3.2.3 PML for Truncating the Computational Domain

Since the perfectly matched interface in the xy-plane is independent of s_{1z} and s_{2z}, we can choose any values for s_{1z} and s_{2z} without causing any reflection. If we choose $s_{2z} = s' - js''$, where s' and s'' are real numbers with $s' \geq 1$ and $s'' \geq 0$, then $k_{2z} = k_2(s' - js'')\cos\theta$. The wave transmitted will be attenuated exponentially by the factor $\exp(k_2 s'' z \cos\theta)$ in the $-\hat{z}$-direction. If we truncate medium 2 into a layer with a finite thickness L and place a conducting surface at its back (Figure 3.4), the magnitude of the reflection coefficient becomes

$$|R(\theta)| = e^{-2k_2 \int_0^L s''(z)\, dz \cos\theta}. \tag{3.40}$$

Clearly, this reflection has a minimum value for normal incidence and a maximum value at grazing incidence. This characteristic is very similar to that of the mathematical absorbing boundary conditions. However, for a PML, we can systematically reduce its reflection simply by increasing the value of $\int_0^L s''(z)\, dz$.

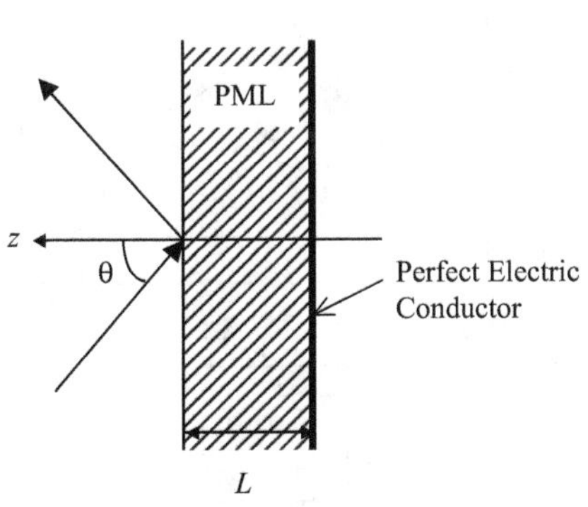

Figure 3.4 Perfectly matched layer backed by a conducting surface.

Because of the presence of k_2 in (3.40), the reflection coefficient is frequency dependent, which is not desirable for broadband simulations. This frequency dependence can be eliminated by choosing s'' as

$$s''(z) = \frac{\sigma(z)}{\omega\varepsilon} \tag{3.41}$$

which yields a frequency-independent reflection coefficient

$$|R(\theta)| = e^{-2\eta \int_0^L \sigma(z)\,dz \cos\theta} \tag{3.42}$$

where $\eta = \sqrt{\mu/\varepsilon}$. For a propagating wave such as that considered here, s' has no effect and can be set to 1. With this choice, s_{2z} can be written as

$$s_{2z} = s' - js'' = 1 + \frac{\sigma}{j\omega\varepsilon}. \tag{3.43}$$

The conductor-backed PML in Figure 3.4 can be used to truncate the computational domain for the finite element simulations. The basic scheme is sketched in Figure 3.5, where the problem of interest is boxed by a conductor-backed PML. The

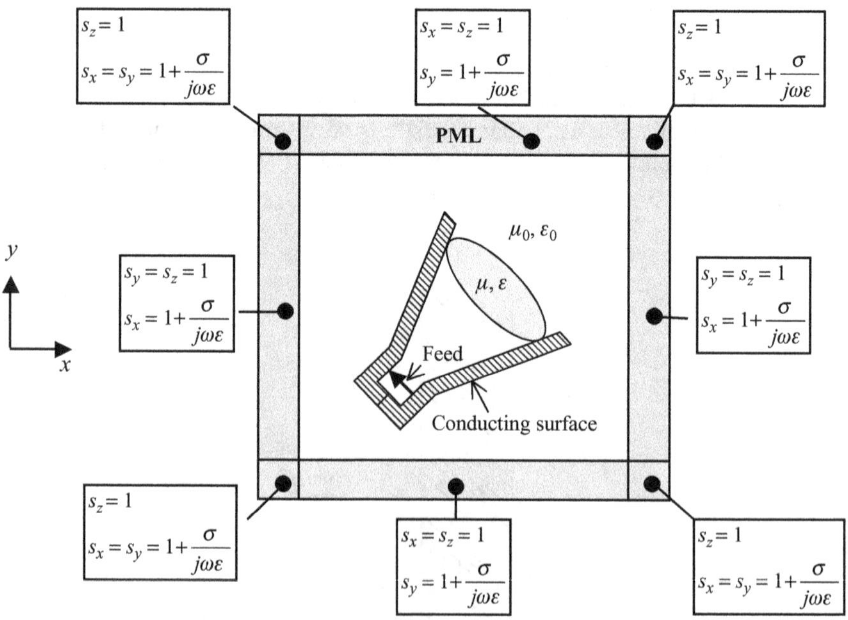

Figure 3.5 Computational domain truncated by perfectly matched layers.

choice of PML parameters depends on its position. For the PML perpendicular to the x-axis, we choose

$$s_x = s' - js'' = 1 + \frac{\sigma}{j\omega\varepsilon}, \qquad s_y = s_z = 1 \tag{3.44}$$

and for the PML perpendicular to the y-axis, we choose

$$s_y = s' - js'' = 1 + \frac{\sigma}{j\omega\varepsilon}, \qquad s_x = s_z = 1. \tag{3.45}$$

For the four corner regions, we choose

$$s_x = s_y = s' - js'' = 1 + \frac{\sigma}{j\omega\varepsilon}, \qquad s_z = 1 \tag{3.46}$$

to satisfy the conditions for a perfectly matched interface. The extension to the three-dimensional case is straightforward. Note that because of the truncation of the PML with a conducting surface, it is necessary to make sure that the truncated PML provides sufficient attenuation such that after the field is reflected by the truncation boundary, it becomes negligible when it reenters into the physical solution domain. This can be done relatively easily since the PML attenuation can be estimated with a simple formula given in (3.40). However, since the PML attenuates primarily waves propagating normal to the PML, the truncated PML will provide less attenuation for obliquely incident waves; consequently, a significant nonphysical reflection can occur for a wave incident at a near-grazing angle. For this reason, the PML has to be placed some distance (typically, a fourth to a half of a wavelength) away from all the sources of the field, similar to the case of a simple ABC termination.

3.2.4 Finite Element Implementation of PML

The PML modeled as an anisotropic medium can be implemented easily in the frequency-domain finite element method because the finite element formulation described in Chapter 2 can handle a general anisotropic medium. All that is required is to assign an appropriate value for $\overleftrightarrow{\varepsilon}$ and $\overleftrightarrow{\mu}$ for the elements in the PML based on their locations. The finite element implementation in the time domain is, however, more complicated because the PML represents an electrically and magnetically anisotropic, dispersive, and lossy medium. Toward this end, the general procedure for the time-domain finite element modeling of general dispersive media described in Section 2.3 can be applied. The first implementation of the PML in the time-domain finite element method was developed by Jiao et al. [13] and is summarized here.

By eliminating the magnetic field from Maxwell's equations in (3.35)–(3.39), the vector wave equation for the electric field in the PML becomes

$$\nabla \times \frac{1}{\mu}\left(\overleftrightarrow{\mathbf{\Lambda}}^{-1} \cdot \nabla \times \mathbf{E}\right) - \omega^2 \varepsilon \overleftrightarrow{\mathbf{\Lambda}} \cdot \mathbf{E} = 0. \tag{3.47}$$

To find the time-domain counterpart of (3.47), we first substitute (3.36) into (3.47) and then rearrange the terms according to their frequency dependency on $j\omega$. Doing so, we obtain

$$\nabla \times \frac{1}{\mu}\left[\overleftrightarrow{\mathbf{\Lambda}}^{-1}(s) \cdot \nabla \times \mathbf{E}\right] + s^2 \varepsilon \overleftrightarrow{\mathbf{\Lambda}}(s) \cdot \mathbf{E} = 0 \tag{3.48}$$

where $s = j\omega$ and

$$\overleftrightarrow{\mathbf{\Lambda}}(s) = \overleftrightarrow{I} + \frac{1}{s\varepsilon}\overleftrightarrow{J} + \frac{1}{s^2\varepsilon^2}\overleftrightarrow{K}_1 - \frac{1}{s^2\varepsilon^3}\overleftrightarrow{K}_2 \cdot \frac{\varepsilon}{s\varepsilon + \overleftrightarrow{\sigma}} \tag{3.49}$$

$$\overleftrightarrow{\mathbf{\Lambda}}^{-1}(s) = \overleftrightarrow{I} + \frac{1}{\varepsilon}\overleftrightarrow{L}_1 \cdot \frac{\varepsilon}{s\varepsilon + \overleftrightarrow{\sigma}^{+}} + \frac{1}{\varepsilon}\overleftrightarrow{L}_2 \cdot \frac{\varepsilon}{s\varepsilon + \overleftrightarrow{\sigma}^{++}}. \tag{3.50}$$

In the equations above, $\overleftrightarrow{I}$ denotes the identity tensor and the other tensors are defined by

$$\overleftrightarrow{J} = \begin{bmatrix} \sigma_y + \sigma_z - \sigma_x & 0 & 0 \\ 0 & \sigma_z + \sigma_x - \sigma_y & 0 \\ 0 & 0 & \sigma_x + \sigma_y - \sigma_z \end{bmatrix} \tag{3.51}$$

$$\overleftrightarrow{K}_1 = \begin{bmatrix} (\sigma_x - \sigma_y)(\sigma_x - \sigma_z) & 0 & 0 \\ 0 & (\sigma_y - \sigma_z)(\sigma_y - \sigma_x) & 0 \\ 0 & 0 & (\sigma_z - \sigma_x)(\sigma_z - \sigma_y) \end{bmatrix} \tag{3.52}$$

$$\overleftrightarrow{K}_2 = \begin{bmatrix} \sigma_x(\sigma_x - \sigma_y)(\sigma_x - \sigma_z) & 0 & 0 \\ 0 & \sigma_y(\sigma_y - \sigma_z)(\sigma_y - \sigma_x) & 0 \\ 0 & 0 & \sigma_z(\sigma_z - \sigma_x)(\sigma_z - \sigma_y) \end{bmatrix} \tag{3.53}$$

$$\overleftrightarrow{L}_1 = \begin{bmatrix} \frac{\sigma_x - \sigma_y}{\sigma_y - \sigma_z}\sigma_y & 0 & 0 \\ 0 & \frac{\sigma_y - \sigma_z}{\sigma_z - \sigma_x}\sigma_z & 0 \\ 0 & 0 & \frac{\sigma_z - \sigma_x}{\sigma_x - \sigma_y}\sigma_x \end{bmatrix} \tag{3.54}$$

$$\overleftrightarrow{L}_2 = \begin{bmatrix} \frac{\sigma_z - \sigma_x}{\sigma_y - \sigma_z}\sigma_z & 0 & 0 \\ 0 & \frac{\sigma_x - \sigma_y}{\sigma_z - \sigma_x}\sigma_x & 0 \\ 0 & 0 & \frac{\sigma_y - \sigma_z}{\sigma_x - \sigma_y}\sigma_y \end{bmatrix}. \tag{3.55}$$

The following shorthand notations are also employed in (3.49) and (3.50) to make them more compact:

$$\frac{\varepsilon}{s\varepsilon + \overleftrightarrow{\sigma}} = \begin{bmatrix} \frac{\varepsilon}{s\varepsilon + \sigma_x} & 0 & 0 \\ 0 & \frac{\varepsilon}{s\varepsilon + \sigma_y} & 0 \\ 0 & 0 & \frac{\varepsilon}{s\varepsilon + \sigma_z} \end{bmatrix} \tag{3.56}$$

$$\frac{\varepsilon}{s\varepsilon + \overleftrightarrow{\sigma}^{+}} = \begin{bmatrix} \frac{\varepsilon}{s\varepsilon + \sigma_y} & 0 & 0 \\ 0 & \frac{\varepsilon}{s\varepsilon + \sigma_z} & 0 \\ 0 & 0 & \frac{\varepsilon}{s\varepsilon + \sigma_x} \end{bmatrix} \tag{3.57}$$

$$\frac{\varepsilon}{s\varepsilon + \overleftrightarrow{\sigma}^{++}} = \begin{bmatrix} \frac{\varepsilon}{s\varepsilon + \sigma_z} & 0 & 0 \\ 0 & \frac{\varepsilon}{s\varepsilon + \sigma_x} & 0 \\ 0 & 0 & \frac{\varepsilon}{s\varepsilon + \sigma_y} \end{bmatrix}. \tag{3.58}$$

It is obvious here that the superscript + denotes one cyclic permutation and the superscript ++ denotes two cyclic permutations. Note that $\overleftrightarrow{J}$, $\overleftrightarrow{K}_1$, $\overleftrightarrow{K}_2$, $\overleftrightarrow{L}_1$, and $\overleftrightarrow{L}_2$ are all independent of frequency.

With the rearrangement above, (3.48) can now be transformed into the time domain using the inverse Laplace transform, yielding

$$\nabla \times \frac{1}{\mu}[\overleftrightarrow{\mathcal{L}}_2(t) \cdot \nabla \times \mathbf{E}(t)] + \varepsilon \overleftrightarrow{\mathcal{L}}_1(t) \cdot \mathbf{E}(t) = 0 \tag{3.59}$$

where $\overleftrightarrow{\mathcal{L}}_1(t)$ is the inverse Laplace transform of $s^2\overleftrightarrow{\mathbf{\Lambda}}(s)$ and is given by

$$\overleftrightarrow{\mathcal{L}}_1(t) = \overleftrightarrow{I}\frac{\partial^2}{\partial t^2} + \frac{1}{\varepsilon}\overleftrightarrow{J}\frac{\partial}{\partial t} + \frac{1}{\varepsilon^2}\overleftrightarrow{K}_1 - \frac{1}{\varepsilon^3}\overleftrightarrow{K}_2 \cdot e^{-\overleftrightarrow{\sigma} t/\varepsilon}u(t)* \tag{3.60}$$

and $\overleftrightarrow{\mathcal{L}}_2(t)$ is the inverse Laplace transform of $\overleftrightarrow{\mathbf{\Lambda}}^{-1}(s)$ and is given by

$$\overleftrightarrow{\mathcal{L}}_2(t) = \overleftrightarrow{I} + \frac{1}{\varepsilon}\overleftrightarrow{L}_1 \cdot e^{-\overleftrightarrow{\sigma}^{+} t/\varepsilon}u(t) * + \frac{1}{\varepsilon}\overleftrightarrow{L}_2 \cdot e^{-\overleftrightarrow{\sigma}^{++} t/\varepsilon}u(t)* \tag{3.61}$$

where $u(t)$ denotes the unit step function and $*$ denotes the convolution operator. The weak-form representation of (3.59) is given by

$$\iiint_V \left[\frac{1}{\mu}(\nabla \times \mathbf{T}) \cdot \overleftrightarrow{\mathcal{L}}_2(t) \cdot \nabla \times \mathbf{E}(t) + \varepsilon \mathbf{T} \cdot \overleftrightarrow{\mathcal{L}}_1(t) \cdot \mathbf{E}(t)\right] dV = 0 \tag{3.62}$$

and the corresponding finite element discretization via (2.26) yields

$$[T]\frac{d^2\{E\}}{dt^2} + [R]\frac{d\{E\}}{dt} + [S]\{E\} + \{h\} + \{g\} = \{0\} \tag{3.63}$$

where

$$T_{ij} = \iiint_V \varepsilon \mathbf{N}_i \cdot \mathbf{N}_j \, dV \tag{3.64}$$

$$R_{ij} = \iiint_V \mathbf{N}_i \cdot \overleftrightarrow{J} \cdot \mathbf{N}_j \, dV \tag{3.65}$$

$$S_{ij} = \iiint_V \frac{1}{\mu}(\nabla \times \mathbf{N}_i) \cdot (\nabla \times \mathbf{N}_j)\, dV + \iiint_V \frac{1}{\varepsilon}\mathbf{N}_i \cdot \overleftrightarrow{K}_1 \cdot \mathbf{N}_j \, dV \tag{3.66}$$

$$h_i = -\iiint_V \frac{1}{\varepsilon^2}\mathbf{N}_i \cdot \sum_j \left(\overleftrightarrow{K}_2 \cdot \overleftrightarrow{u}_j \cdot \mathbf{N}_j\right) dV \tag{3.67}$$

$$g_i = \iiint_V \frac{1}{\mu\varepsilon}(\nabla \times \mathbf{N}_i) \cdot \sum_j \left[\left(\overleftrightarrow{L}_1 \cdot \overleftrightarrow{u}_j^{+} + \overleftrightarrow{L}_2 \cdot \overleftrightarrow{u}_j^{++}\right) \cdot (\nabla \times \mathbf{N}_j)\right] dV. \tag{3.68}$$

In the equations above, the summation is carried out for all E_j inside and on the PML, and the tensors $\overleftrightarrow{u}_j$, $\overleftrightarrow{u}_j^{+}$, and $\overleftrightarrow{u}_j^{++}$ are given by

$$\overleftrightarrow{u}_j = \begin{bmatrix} e^{-\sigma_x t/\varepsilon}u(t) * E_j(t) & 0 & 0 \\ 0 & e^{-\sigma_y t/\varepsilon}u(t) * E_j(t) & 0 \\ 0 & 0 & e^{-\sigma_z t/\varepsilon}u(t) * E_j(t) \end{bmatrix} \tag{3.69}$$

$$\overleftrightarrow{u}_j^{+} = \begin{bmatrix} e^{-\sigma_y t/\varepsilon}u(t) * E_j(t) & 0 & 0 \\ 0 & e^{-\sigma_z t/\varepsilon}u(t) * E_j(t) & 0 \\ 0 & 0 & e^{-\sigma_x t/\varepsilon}u(t) * E_j(t) \end{bmatrix} \tag{3.70}$$

$$\overleftrightarrow{u}_j^{++} = \begin{bmatrix} e^{-\sigma_z t/\varepsilon}u(t) * E_j(t) & 0 & 0 \\ 0 & e^{-\sigma_x t/\varepsilon}u(t) * E_j(t) & 0 \\ 0 & 0 & e^{-\sigma_y t/\varepsilon}u(t) * E_j(t) \end{bmatrix}. \tag{3.71}$$

Next, we discretize the time domain to convert (3.63) into a time-marching equation. To this end, we can apply the Newmark-beta method, described in Chapter 2. However, we present a more general approach here, which employs the concept of basis and testing functions [14]. First, we expand the time-dependent variables $E_j(t)$ in terms of time-domain basis functions as

$$E_j(t) = \sum_n T_n(t) E_j^n \tag{3.72}$$

where $T_n(t)$ denotes a basis function defined on $(n-1)\,\Delta t \le t \le (n+1)\,\Delta t$. Substituting (3.72) into (3.63) and testing the resulting equation using testing functions $W_n(t)$ also defined on $(n-1)\,\Delta t \le t \le (n+1)\,\Delta t$, we obtain the discrete system

$$\begin{aligned}&[T](a_1\{E\}^{n+1} + a_0\{E\}^n + a_{-1}\{E\}^{n-1}) + [R](b_1\{E\}^{n+1} + b_0\{E\}^n + b_{-1}\{E\}^{n-1})\\ &+ [S](c_1\{E\}^{n+1} + c_0\{E\}^n + c_{-1}\{E\}^{n-1}) + c_1\{h\}^{n+1} + c_0\{h\}^n + c_{-1}\{h\}^{n-1}\\ &+ c_1\{g\}^{n+1} + c_0\{g\}^n + c_{-1}\{g\}^{n-1} = 0\end{aligned} \tag{3.73}$$

where

$$a_k = \int W_n(t)\frac{d^2}{dt^2}T_{n+k}(t)\,dt \qquad k = -1, 0, 1 \tag{3.74}$$

$$b_k = \int W_n(t)\frac{d}{dt}T_{n+k}(t)\,dt \qquad k = -1, 0, 1 \tag{3.75}$$

$$c_k = \int W_n(t)\,T_{n+k}(t)\,dt \qquad k = -1, 0, 1. \tag{3.76}$$

The expressions for $\{h\}^n$ and $\{g\}^n$ can be inferred from (3.67)–(3.71), and the required convolutions can be evaluated very efficiently in a recursive manner, as illustrated in Chapter 2. The values of the constants a_k, b_k, and c_k depend on the specific form of the basis and testing functions used in the temporal discretization. If linear functions are used for both $T_n(t)$ and $W_n(t)$, we obtain

$$a_{\pm 1} = \frac{1}{(\Delta t)^2}, \quad a_0 = -\frac{2}{(\Delta t)^2}, \quad b_{\pm 1} = \pm\frac{1}{2\Delta t}, \quad b_0 = 0, \quad c_{\pm 1} = \frac{1}{6}, \quad c_0 = \frac{2}{3}. \tag{3.77}$$

This result is equivalent to that of the Newmark-beta method with $\beta = 1/6$ for the non-PML region. The resulting time-marching equation is only conditionally stable. However, if we employ linear functions for $T_n(t)$ and a quadratic function for $W_n(t)$, which are illustrated in Figure 3.6, we obtain the same results as in (3.77) except

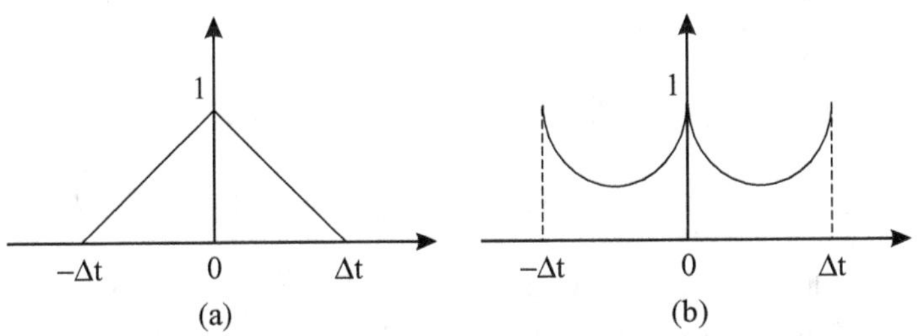

Figure 3.6 (a) Linear temporal basis function: $T_n(t) = 1 - |t - n\,\Delta t|/\Delta t$. (b) Quadratic temporal testing function: $W_n(t) = 1 - 3|t - n\,\Delta t|/\Delta t + 3(t - n\,\Delta t)^2/(\Delta t)^2$.

that $c_{\pm 1} = 1/4$ and $c_0 = 1/2$. This result is equivalent to that of the Newmark-beta method with $\beta = 1/4$ for the non-PML region, which is known to be unconditionally stable. Numerous tests have shown that this temporal discretization scheme is able to produce highly stable solutions through time marching [15].

Although the PML interface is reflectionless in the continuous space, this is not true in numerical simulations because of finite discretization. When an abrupt material change occurs and the finite element discretization is not sufficiently dense to resolve the change, undesirable numerical reflections may occur [16]. One approach to reducing these numerical reflections is to vary the material parameters smoothly within the PML. For example, for a PML perpendicular to the x-direction and interfacing with air, instead of using a constant σ, we can use a nonconstant σ expressed in terms of a pth-order polynomial:

$$\sigma = \sigma_{\max} \left| \frac{x - x_0}{L} \right|^p \qquad p = 1, 2, \ldots \tag{3.78}$$

where x_0 is the x-coordinate of the PML–air interface, L is the thickness of the PML, and $\sigma_{\max}$ is the maximum conductivity inside the PML. It is found that $p = 2$ or 3 is generally a good choice. For a given thickness L, the absorbing performance of the PML can be improved by increasing the maximum conductivity and the discretization level.

3.2.5 ABC-Backed, Complementary, CFS, and Second-Order PMLs

As noted earlier, when a PML has a finite thickness and is backed by a perfect conductor, it is no longer perfectly matched. Its absorption is controlled by its thickness and conductivity. To increase this absorption and reduce the reflection, one must increase either its thickness or its conductivity. In either case, one has to increase the number of finite elements across the thickness of the PML, resulting in a larger number of unknowns to deal with. There are a variety of simple tricks to improve the performance of a PML without increasing its finite element discretization density. One is to replace the perfect conductor backing of a PML with an impedance surface or an absorbing surface described by an absorbing boundary condition such as the simple first-order ABC [17]. When a PML is backed by an ABC, the field entering the PML is both attenuated inside the PML and partially absorbed by the ABC at the end. The total reflection then becomes

$$R_{\text{PML-ABC}}(\theta) = R_{\text{PML}}(\theta) \cdot R_{\text{ABC}}(\theta) \tag{3.79}$$

where $R_{\text{PML}}(\theta)$ is given by (3.42) and $R_{\text{ABC}}(\theta)$ is given by either (3.9) or (3.10), depending on which ABC is used.

Another simple approach to improving the performance of a PML is to employ a pair of PMLs that have exactly the same material composition or parameters, but one is backed by a perfect electric conductor (PEC) and the other by a perfect magnetic conductor (PMC) [18]. As a result, the PEC-backed PML has a reflection coefficient negative to that of the PMC-backed PML, or in other words, the PEC- and PMC-backed PMLs are complementary to each other. Assuming that the PML has a reflection coefficient R, the numerical solution obtained with the PEC-backed PML can be expressed as

$$E_{\text{PEC-PML}} = E_{\text{D}} - RE_{\text{D}} + R^2 E_{\text{D}} - \cdots \tag{3.80}$$

where E_{D} denotes the numerical solution without any reflection from the PML. The second term on the right-hand side is due to the first reflection by the PEC, and the third term is due to the second reflection by the PEC. When the same problem is solved using the PMC-backed PML, the numerical solution can be expressed as

$$E_{\text{PMC-PML}} = E_{\text{D}} + RE_{\text{D}} + R^2 E_{\text{D}} + \cdots. \tag{3.81}$$

By adding (3.80) and (3.81), we obtain

$$E_{\text{Ave}} = \tfrac{1}{2}(E_{\text{PEC-PML}} + E_{\text{PMC-PML}}) = E_{\text{D}} + R^2 E_{\text{D}} + \cdots \tag{3.82}$$

which eliminates the error proportional to R such that the remaining error is proportional to R^2, which is due to the double reflection of the fields. Therefore, by using complementary PMLs, the reflection error can be reduced to R^2 from R at the expense of doubling the computational time.

Apart from the errors due to numerical discretization, the PML is also known to have poor absorption for evanescent and near-grazing-propagating waves. Although the effect of such waves in free-space propagation can usually be neglected, this may not be the case in waveguide and antenna simulations. To demonstrate this, we consider the periodic dielectric slab illustrated in Figure 3.7(a), which is formed by alternating two different dielectric materials. A TM-polarized plane wave is incident upon this dielectric slab at an angle of $\theta^{\text{inc}} = 45°$. As a periodic structure, this dielectric slab can support higher-order Floquet modes. The first Floquet mode would start to appear at the cutoff frequency

$$f_c = \frac{c_0}{(1 + \sin\theta^{\text{inc}})T_x} \tag{3.83}$$

where c_0 denotes the speed of the incident wave and T_x denotes the periodic length ($T_x = 1.0$ cm). For $\theta^{\text{inc}} = 45°$, the cutoff frequency is 17.6 GHz. Below this frequency the first Floquet mode behaves as an evanescent wave, and above this frequency this mode becomes a propagating wave. The FDTD calculation using the regular PML [19] is shown in Figure 3.7(b), and the result is compared with the accurate modal solution. While the agreement between the two solutions is very good below 17.6 GHz, the FDTD calculation overpredicts the specular reflection by 20% around 17.6 GHz, due to ineffective absorption of evanescent and near-grazing-propagating waves by the regular PML.

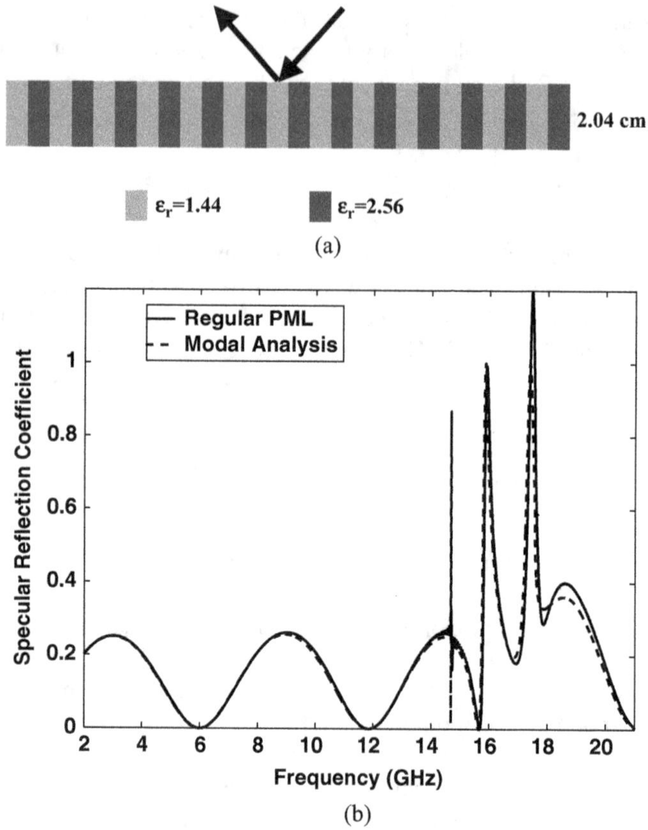

Figure 3.7 Plane-wave reflection by a one-dimensional periodic dielectric slab. (a) Geometry (each unit cell consists of two 0.5-cm-wide dielectrics). (b) Specular reflection coefficient calculated using the regular PML, which has a poor absorption around the cutoff frequency of the first Floquet mode (17.6 GHz). (After Correia and Jin [19], Copyright © IEEE 2005.)

To provide better absorption of evanescent and near-grazing-propagating waves, a modified version of the PML, referred to as the *complex-frequency shifted* (CFS) *PML*, has been proposed for the FDTD method [20,21]. For the CFS-PML, the PML parameters are modified as

$$s_\alpha = \kappa + \frac{\sigma}{a + j\omega\varepsilon} \qquad \alpha = x, y, z \tag{3.84}$$

where $\kappa \geq 1$ and $a \geq 0$ are introduced to better attenuate evanescent and near-grazing-propagating waves. The reflection coefficient associated with (3.84) is given by

$$R_{\text{CFS}} = e^{-2\eta\sigma L[\omega^2\varepsilon^2/(\omega^2\varepsilon^2+a^2)]\cos\theta} \tag{3.85}$$

where, for simplicity, the PML parameters are assumed to be constant in the PML. It is seen that a value of $a > 0$ compromises the absorption of low-frequency propagating waves since $R_{\mathrm{CFS}} \to 1$ as $\omega \to 0$. The performance of the CFS-PML is demonstrated clearly in Figure 3.8(a), which shows the FDTD calculation using the CFS-PML for the problem illustrated in Figure 3.7(a). Although an accurate solution is now obtained around 17.6 GHz, the solution at low frequencies deviates significantly from the reference solution, due to poor absorption of low-frequency propagating waves.

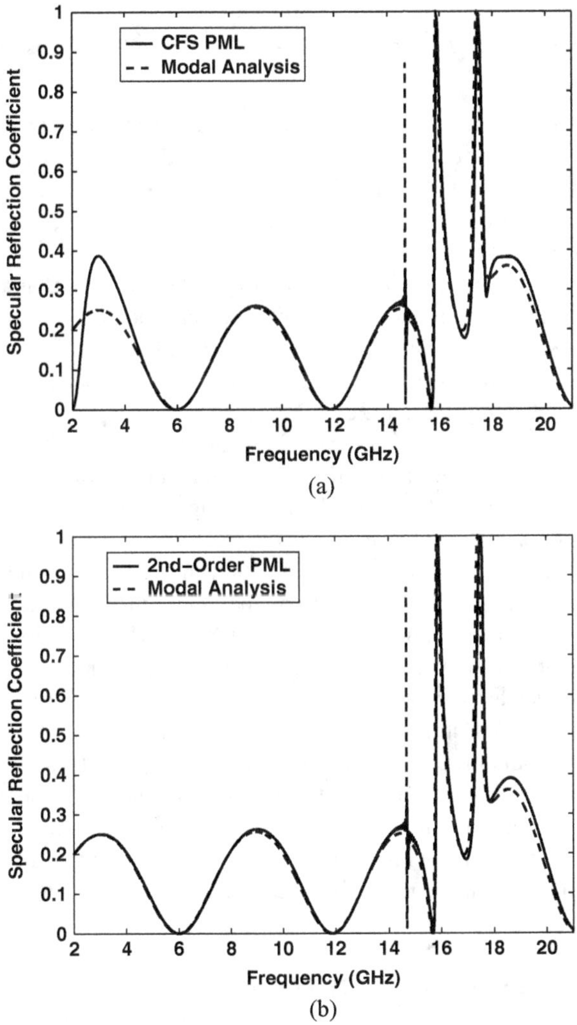

Figure 3.8 Specular reflection coefficient of a periodic dielectric slab. (a) Results calculated using the CFS-PML. (b) Results calculated using the second-order PML. (After Correia and Jin [19], Copyright © IEEE 2005.)

This shortcoming of the CFS-PML has recently been alleviated by using the second-order PML [19], whose parameters are defined as

$$s_\alpha = \left(1 + \frac{\sigma_1}{j\omega\varepsilon}\right)\left(\kappa + \frac{\sigma_2}{a + j\omega\varepsilon}\right). \tag{3.86}$$

This PML combines the excellent capability of the regular PML in absorbing propagating waves and the capability of the CFS-PML in absorbing evanescent and near-grazing-propagating waves. Its reflection coefficient is given by

$$R_{\text{2nd-order}} = \exp\left[-2\eta L\left(\kappa\sigma_1 + \frac{\sigma_2\omega^2\varepsilon^2}{\omega^2\varepsilon^2 + a^2} + \frac{\sigma_1\sigma_2 a}{\omega^2\varepsilon^2 + a^2}\right)\cos\theta\right] \tag{3.87}$$

which can be designed to absorb both the evanescent and propagating waves for all frequencies. The performance of the second-order PML is demonstrated in Figure 3.8(b), which shows the FDTD calculation using the second-order PML for the problem illustrated in Figure 3.7(a). In this case, an accurate solution is obtained in the entire frequency range.

It should be pointed out that implementation of the CFS-PML and second-order PML is trivial in the frequency-domain finite element method, and the overhead of using these PMLs is negligible. This is not, however, the case for the time-domain finite element analysis, which becomes much more complicated in the implementation and expensive in the computation [22].

A major advantage of the PML over the ABC approach is that the absorbing performance of the PML can be improved systematically simply by increasing the number of PML layers or, equivalently, by increasing the conductivity and discretization inside the PML. Despite the fact that the PML has been used extensively and successfully in the FDTD method, its application in the finite element method has not achieved similar popularity. In the frequency-domain finite element method, it has been observed that the system matrix becomes poorly conditioned when PML absorbing layers are present [23,24]. As a result, the number of iterations required by an iterative solver to converge increases substantially. In the time-domain finite element method, the PML implementation becomes much more complicated, due to the need to model anisotropic dispersive media. Moreover, it has been found that improper use of high PML conductivity values may cause undesirable instability problems. A better approach is to utilize the well-established implementation of the PML in the FDTD method by combining the time-domain finite element method and the FDTD method. This approach is discussed in Chapter 4.

3.3 BOUNDARY INTEGRAL EQUATIONS

Both the ABC and PML approaches have a distinct advantage in that they produce a highly sparse matrix. However, they share a disadvantage in that they are approximate, or in other words, they are not reflectionless for obliquely incident waves. Although the performance of the PML can be improved by increasing the number of layers, one still has to place it some distance (typically, 0.25λ to 0.5λ) away from the antenna, thereby increasing the computational volume. The third approach, discussed here, is the one based on boundary integral equations, which can provide a truly perfect boundary condition for mesh truncation at higher computational cost.

It is well known that integral equation–based methods are particularly suited for analyzing open-region radiation and scattering problems because they accurately model the wave propagation into free space via the use of appropriate Green's functions. However, these methods can encounter challenges when handling the complicated geometries and materials present in complex antenna and feed region designs. The finite element method, on the other hand, is most suitable for modeling complicated geometries and material compositions. The individual successes of the two methods in their own realms of applications have led to the development of a hybrid technique, known as the *finite element–boundary integral* (FE-BI) *method* [25–31], which combines the advantages of both methods and permits accurate analysis of highly complicated electromagnetic problems.

The FE-BI method employs an arbitrary boundary, tightly enclosing the object to be analyzed, to terminate the finite element mesh. Interior to the boundary, finite element discretization is applied. Exterior to the boundary, the fields are represented by boundary integral equations. The fields in the two regions are then coupled at the boundary via field continuity conditions, leading to a coupled system from which the interior and boundary fields can be determined. In the following we describe the basic formulations of the FE-BI method in the frequency and time domains.

3.3.1 Frequency-Domain Formulations

Consider the antenna radiation problem illustrated in Figure 2.1. To use the FE-BI method, we first introduce a surface S_o to truncate the computational domain (Figure 3.9). However, unlike what we usually do with an ABC or PML, this truncation

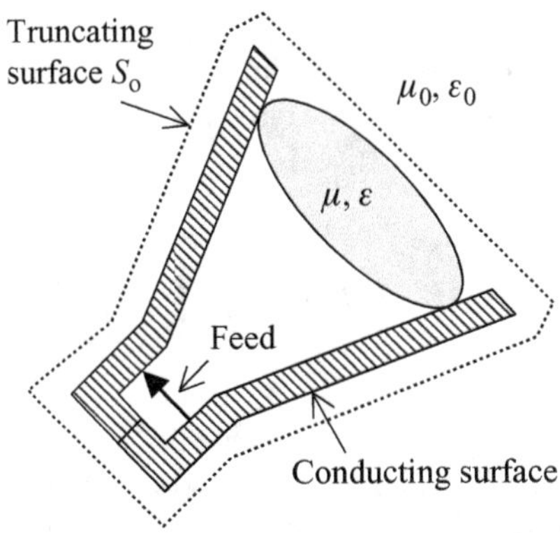

Figure 3.9 Computational domain truncated by a surface that tightly encloses the object to be analyzed.

surface can be placed very close to the antenna to be analyzed. In many problems it can be placed directly on the surface of the antenna. The only requirement is that it encloses the entire antenna such that there is nothing in the space exterior to this surface. With the introduction of this surface, the entire problem is decomposed into an interior and an exterior problem, where the fields can be formulated using the finite element and boundary integral methods, respectively.

In the interior region enclosed by S_o, the field satisfies the vector wave equation

$$\nabla \times \left(\overleftrightarrow{\mu}_r^{-1} \cdot \nabla \times \mathbf{E}\right) - k_0^2 \overleftrightarrow{\varepsilon}_r \cdot \mathbf{E} = -jk_0 Z_0 \mathbf{J}_{\text{imp}} - \nabla \times \left(\overleftrightarrow{\mu}_r^{-1} \cdot \mathbf{M}_{\text{imp}}\right) \qquad \mathbf{r} \in V \tag{3.88}$$

and certain boundary conditions for the specific problem. On the truncation surface, the boundary condition is unknown. However, since the field has to satisfy Maxwell's equations there, we can postulate a Neumann boundary condition with an unknown right-hand side:

$$\hat{n} \times \left(\overleftrightarrow{\mu}_r^{-1} \cdot \nabla \times \mathbf{E}\right) = -jk_0 Z_0 \hat{n} \times \mathbf{H} = -jk_0 \hat{n} \times \overline{\mathbf{H}} \qquad \mathbf{r} \in S_o \tag{3.89}$$

where $\overline{\mathbf{H}} = Z_0 \mathbf{H}$ and $\hat{n}$ denotes an outward normal to S_o. The boundary-value problem defined by (3.88) and (3.89) can be formulated into the weak-form wave equation as

$$\begin{aligned} &\iiint_V \left[(\nabla \times \mathbf{T}) \cdot \overleftrightarrow{\mu}_r^{-1} \cdot (\nabla \times \mathbf{E}) - k_0^2 \mathbf{T} \cdot \overleftrightarrow{\varepsilon}_r \cdot \mathbf{E}\right] dV + jk_0 \oiint_{S_o} \hat{n} \cdot (\mathbf{T} \times \overline{\mathbf{H}})\, dS \\ &= -\iiint_V \mathbf{T} \cdot \left[jk_0 Z_0 \mathbf{J}_{\text{imp}} + \nabla \times \left(\overleftrightarrow{\mu}_r^{-1} \cdot \mathbf{M}_{\text{imp}}\right)\right] dV. \end{aligned} \tag{3.90}$$

After expanding the electric field $\mathbf{E}$ according to (2.16) and a similar expansion of the surface magnetic field, (3.90) can be converted into the matrix equation

$$\begin{bmatrix} K_{II} & K_{IS} & 0 \\ K_{SI} & K_{SS} & B \end{bmatrix} \begin{Bmatrix} E_I \\ E_S \\ \overline{H}_S \end{Bmatrix} = \begin{Bmatrix} b_I \\ b_S \end{Bmatrix} \tag{3.91}$$

assuming that $\mathbf{J}_{\text{imp}}$ and $\mathbf{M}_{\text{imp}}$ are strictly inside S_o. In (3.91), $\{E_I\}$ represents the discrete electric field inside S_o, $\{E_S\}$ the discrete tangential electric field on S_o, and $\{\overline{H}_S\}$ the discrete tangential magnetic field on S_o. The matrix $[K]$ comprising four submatrices is a sparse symmetric matrix, and the matrix $[B]$ is also a sparse matrix. The expressions for the elements of $[K]$ and $[B]$ can easily be obtained from (3.90).

To be more specific, if the vector basis functions used to expand $\mathbf{E}$ and $\overline{\mathbf{H}}$ on S_o are denoted as $\mathbf{N}_i^S$, then

$$K_{ij} = \iiint_V \left[(\nabla \times \mathbf{N}_i) \cdot \overleftrightarrow{\mu}_r^{-1} \cdot (\nabla \times \mathbf{N}_j) - k_0^2 \mathbf{N}_i \cdot \overleftrightarrow{\varepsilon}_r \cdot \mathbf{N}_j\right] dV \tag{3.92}$$

$$B_{ij} = jk_0 \oiint_{S_o} \hat{n} \cdot \left(\mathbf{N}_i^S \times \mathbf{N}_j^S\right) dS. \tag{3.93}$$

The expression for the elements of $\{b\}$ is the same as (2.19).

Equation (3.91) is not sufficient to solve for $\{E_I\}$, $\{E_S\}$, and $\{\overline{H}_S\}$. It has to be complemented by an additional relationship between $\{E_S\}$ and $\{\overline{H}_S\}$, which comes from the formulation of the exterior field. In the exterior region occupied purely by free space, the $\mathbf{E}$ and $\mathbf{H}$ satisfy the vector wave equations

$$\nabla \times \nabla \times \mathbf{E}(\mathbf{r}) - k_0^2 \mathbf{E}(\mathbf{r}) = 0 \qquad \mathbf{r} \in V_\infty \tag{3.94}$$

$$\nabla \times \nabla \times \mathbf{H}(\mathbf{r}) - k_0^2 \mathbf{H}(\mathbf{r}) = 0 \qquad \mathbf{r} \in V_\infty \tag{3.95}$$

where V_∞ denotes the spatial volume external to V. To define the boundary-value problem properly, (3.94) and (3.95) are complemented by the Sommerfeld radiation condition (2.6). To formulate the exterior fields $\mathbf{E}$ and $\mathbf{H}$, we introduce the three-dimensional Green's function G_0 that satisfies the scalar Helmholtz equation

$$\nabla^2 G_0(\mathbf{r}, \mathbf{r}') + k_0^2 G_0(\mathbf{r}, \mathbf{r}') = -\delta(\mathbf{r} - \mathbf{r}') \tag{3.96}$$

and the radiation condition

$$r \left[\frac{\partial G_0(\mathbf{r}, \mathbf{r}')}{\partial r} + jk_0 G_0(\mathbf{r}, \mathbf{r}')\right] = 0 \qquad r \to \infty. \tag{3.97}$$

The solution to (3.96) and (3.97) has the well-known form

$$G_0(\mathbf{r}, \mathbf{r}') = \frac{e^{-jk_0|\mathbf{r}-\mathbf{r}'|}}{4\pi\, |\mathbf{r} - \mathbf{r}'|}. \tag{3.98}$$

Next we make use of the scalar–vector Green's theorem [32],

$$\iiint_V [b(\nabla \times \nabla \times \mathbf{a}) + \mathbf{a}\nabla^2 b + (\nabla \cdot \mathbf{a})\nabla b)]\, dV$$

$$= \oiint_S [(\hat{n} \cdot \mathbf{a})\nabla b + (\hat{n} \times \mathbf{a}) \times \nabla b + (\hat{n} \times \nabla \times \mathbf{a})b]\, dS. \tag{3.99}$$

By specially applying (3.99) to the external volume V_∞ and letting $\mathbf{a} = \mathbf{E}$ and $b = G_0$, we find that

$$\iiint_{V_\infty} [G_0(\nabla \times \nabla \times \mathbf{E}) + \mathbf{E}\,\nabla^2 G_0 + (\nabla \cdot \mathbf{E})\,\nabla G_0)]\,dV$$

$$= \oiint_{S_\infty} [(\hat{r} \cdot \mathbf{E})\,\nabla G_0 + (\hat{r} \times \mathbf{E}) \times \nabla G_0 + (\hat{r} \times \nabla \times \mathbf{E})G_0]\,dS$$

$$- \oiint_{S_o} [(\hat{n} \cdot \mathbf{E})\,\nabla G_0 + (\hat{n} \times \mathbf{E}) \times \nabla G_0 + (\hat{n} \times \nabla \times \mathbf{E})G_0]\,dS. \quad (3.100)$$

By making use of (3.94) and (3.96), the left-hand side of (3.100) becomes

$$\iiint_{V_\infty} [G_0(\nabla \times \nabla \times \mathbf{E}) + \mathbf{E}\,\nabla^2 G_0 + (\nabla \cdot \mathbf{E})\,\nabla G_0)]\,dV$$

$$= -\iiint_{V_\infty} \mathbf{E}\delta(\mathbf{r} - \mathbf{r}')\,dV = -\mathbf{E}(\mathbf{r}') \qquad \mathbf{r}' \in V_\infty. \quad (3.101)$$

Using the radiation conditions in (2.6) and (3.97), we can show that the integrand in the first integral on the right-hand side of (3.100) vanishes. As a result, (3.100) becomes

$$\oiint_{S_o} [(\hat{n}' \cdot \mathbf{E})\,\nabla' G_0 + (\hat{n}' \times \mathbf{E}) \times \nabla' G_0 - jk_0 Z_0 (\hat{n}' \times \mathbf{H}) G_0]\,dS' = \mathbf{E}(\mathbf{r}) \qquad \mathbf{r} \in V_\infty \quad (3.102)$$

where we have switched the primed and unprimed coordinates. Alternatively,

$$\mathbf{E}(\mathbf{r}) = -\oiint_{S_o} [(\hat{n}' \cdot \mathbf{E})\,\nabla G_0 + (\hat{n}' \times \mathbf{E}) \times \nabla G_0 + jk_0 Z_0 (\hat{n}' \times \mathbf{H}) G_0]\,dS' \qquad \mathbf{r} \in V_\infty \quad (3.103)$$

where we used the fact that $\nabla' G_0 = -\nabla G_0$. Similarly, if we let $\mathbf{a} = \mathbf{H}$, we obtain

$$\mathbf{H}(\mathbf{r}) = -\oiint_{S_o} [(\hat{n}' \cdot \mathbf{H})\,\nabla G_0 + (\hat{n}' \times \mathbf{H}) \times \nabla G_0 - jk_0 Y_0 (\hat{n}' \times \mathbf{E}) G_0]\,dS' \qquad \mathbf{r} \in V_\infty. \quad (3.104)$$

Both (3.103) and (3.104) contain the normal component of the field on surface S_o, which is not desirable. However, using the surface vector analysis, it can be shown that

$$\hat{n}' \cdot \mathbf{E} = \frac{jZ_0}{k_0} \nabla' \cdot (\hat{n}' \times \mathbf{H}) \quad (3.105)$$

$$\hat{n}' \cdot \mathbf{H} = \frac{Y_0}{jk_0} \nabla' \cdot (\hat{n}' \times \mathbf{E}). \quad (3.106)$$

Substituting these into (3.103) and (3.104) finally yields

$$\mathbf{E}(\mathbf{r}) = -\oiint_{S_o} \left[\frac{j}{k_0} \nabla' \cdot (\hat{n}' \times \overline{\mathbf{H}}) \nabla G_0 + (\hat{n}' \times \mathbf{E}) \times \nabla G_0 + j k_0 (\hat{n}' \times \overline{\mathbf{H}}) G_0 \right] dS' \qquad \mathbf{r} \in V_\infty \tag{3.107}$$

$$\overline{\mathbf{H}}(\mathbf{r}) = -\oiint_{S_o} \left[\frac{1}{j k_0} \nabla' \cdot (\hat{n}' \times \mathbf{E}) \nabla G_0 + (\hat{n}' \times \overline{\mathbf{H}}) \times \nabla G_0 - j k_0 (\hat{n}' \times \mathbf{E}) G_0 \right] dS' \qquad \mathbf{r} \in V_\infty. \tag{3.108}$$

To write these in a compact form, we define the operators

$$\mathcal{L}(\mathbf{X}) = j k_0 \oiint_{S_o} \left[\mathbf{X}(\mathbf{r}') G_0(\mathbf{r}, \mathbf{r}') + \frac{1}{k_0^2} \nabla' \cdot \mathbf{X}(\mathbf{r}') \nabla G_0(\mathbf{r}, \mathbf{r}') \right] dS' \tag{3.109}$$

$$\mathcal{K}(\mathbf{X}) = \oiint_{S_o} \mathbf{X}(\mathbf{r}') \times \nabla G_0(\mathbf{r}, \mathbf{r}')\, dS' \tag{3.110}$$

and introduce the equivalent surface currents

$$\overline{\mathbf{J}}_s(\mathbf{r}') = \hat{n}' \times \overline{\mathbf{H}}(\mathbf{r}'), \qquad \mathbf{M}_s(\mathbf{r}') = \mathbf{E}(\mathbf{r}') \times \hat{n}'. \tag{3.111}$$

As a result, (3.107) and (3.108) can be written as

$$\mathbf{E}(\mathbf{r}) = -\mathcal{L}(\overline{\mathbf{J}}_s) + \mathcal{K}(\mathbf{M}_s) \qquad \mathbf{r} \in V_\infty \tag{3.112}$$

$$\overline{\mathbf{H}}(\mathbf{r}) = -\mathcal{L}(\mathbf{M}_s) - \mathcal{K}(\overline{\mathbf{J}}_s) \qquad \mathbf{r} \in V_\infty. \tag{3.113}$$

The two equations above provide the foundation to derive integral equations for $\overline{\mathbf{J}}_s$ and $\mathbf{M}_s$, which can be obtained by taking the cross product with $\hat{n}$ and letting $\mathbf{r}$ approach S_o. However, when $\mathbf{r}$ approaches S_o, the integrals in (3.109) and (3.110) contain a singular point at $\mathbf{r} = \mathbf{r}'$. It can be shown that the singularity has no contribution to the evaluation of $\hat{n} \times \mathcal{L}(\mathbf{X})$. However, the singular term in $\hat{n} \times \mathcal{K}(\mathbf{X})$ does not vanish, and the result is

$$\hat{n} \times \mathcal{K}(\mathbf{X}) = \hat{n} \times \tilde{\mathcal{K}}(\mathbf{X}) - \tfrac{1}{2}\mathbf{X} \tag{3.114}$$

where $\tilde{\mathcal{K}}(\mathbf{X})$ is the same integral as in (3.110), except that the singular point $\mathbf{r} = \mathbf{r}'$ is now removed. By taking the cross product of (3.112) and (3.113) with $\hat{n}$, letting $\mathbf{r}$ approach S_o, and using (3.114), we obtain

$$\tfrac{1}{2}\mathbf{M}_s(\mathbf{r}) - \hat{n} \times \mathcal{L}(\overline{\mathbf{J}}_s) + \hat{n} \times \tilde{\mathcal{K}}(\mathbf{M}_s) = 0 \qquad \mathbf{r} \in S_o \tag{3.115}$$

$$\tfrac{1}{2}\overline{\mathbf{J}}_s(\mathbf{r}) + \hat{n} \times \mathcal{L}(\mathbf{M}_s) + \hat{n} \times \tilde{\mathcal{K}}(\overline{\mathbf{J}}_s) = 0 \qquad \mathbf{r} \in S_o. \tag{3.116}$$

Equation (3.115) is known as the *electric field integral equation* (EFIE) and (3.116) is called the *magnetic field integral equation* (MFIE).

Either (3.115) or (3.116) can be discretized into a desired matrix equation to be solved together with (3.91). However, when used independently for a closed surface S_o, the EFIE and MFIE can both support singular frequencies (nonphysical interior resonances), which can corrupt the numerical solutions. To eliminate this problem effectively, we may combine the EFIE and MFIE to form the *combined field integral equation* (CFIE)

$$\tfrac{1}{2}\bar{\mathbf{J}}_s(\mathbf{r}) + \hat{n} \times \mathcal{L}(\mathbf{M}_s) + \hat{n} \times \tilde{\mathcal{K}}(\bar{\mathbf{J}}_s) + \hat{n} \times \left[\tfrac{1}{2}\mathbf{M}_s(\mathbf{r}) - \hat{n} \times \mathcal{L}(\bar{\mathbf{J}}_s) + \hat{n} \times \tilde{\mathcal{K}}(\mathbf{M}_s)\right] = 0 \qquad \mathbf{r} \in S_o. \tag{3.117}$$

By adopting the same finite element expansions that were used for $\mathbf{E}$ and $\overline{\mathbf{H}}$, we can write the expansions for $\bar{\mathbf{J}}_s$ and $\mathbf{M}_s$ as

$$\bar{\mathbf{J}}_s(\mathbf{r}') = \hat{n}' \times \overline{\mathbf{H}}(\mathbf{r}') = \sum_{j=1}^{N_s} \mathbf{g}_j(\mathbf{r}')\overline{H}_j^S \tag{3.118}$$

$$\mathbf{M}_s(\mathbf{r}') = \mathbf{E}(\mathbf{r}') \times \hat{n}' = -\sum_{j=1}^{N_s} \mathbf{g}_j(\mathbf{r}')E_j^S \tag{3.119}$$

where $\mathbf{g}_j = \hat{n}' \times \mathbf{N}_j^S$, with $\mathbf{N}_j^S$ denoting the vector basis functions used to expand $\mathbf{E}$ and $\overline{\mathbf{H}}$ on the surface S_o, and N_s denoting the number of element edges on S_o. Substituting (3.118) and (3.119) into (3.117), we obtain

$$\sum_{j=1}^{N_s} \left\{\left[\tfrac{1}{2}\mathbf{g}_j - \hat{n} \times \hat{n} \times \mathcal{L}(\mathbf{g}_j) + \hat{n} \times \tilde{\mathcal{K}}(\mathbf{g}_j)\right]\overline{H}_j^S - \left[\tfrac{1}{2}\hat{n} \times \mathbf{g}_j + \hat{n} \times \mathcal{L}(\mathbf{g}_j) + \hat{n} \times \hat{n} \times \tilde{\mathcal{K}}(\mathbf{g}_j)\right] E_j^S\right\} = 0 \qquad \mathbf{r} \in S_o. \tag{3.120}$$

To convert (3.120) into a matrix equation, we need to choose a set of proper testing functions to take a dot product. A relatively simple analysis [28] shows that while $\mathbf{g}_i$ is a good testing function for the coefficient function of $\overline{H}_j^S$ in (3.120), it is a poor testing function for the coefficient function of E_j^S. On the other hand, although $\hat{n} \times \mathbf{g}_i$ is a good testing function for the coefficient function of E_j^S, it is a poor choice for the coefficient function of $\overline{H}_j^S$. Therefore, to test both coefficient functions well, we combine $\mathbf{g}_i$ and $\hat{n} \times \mathbf{g}_i$ to form a testing function

$$\mathbf{t}_i(\mathbf{r}) = \mathbf{g}_i(\mathbf{r}) + \hat{n} \times \mathbf{g}_i(\mathbf{r}). \tag{3.121}$$

Now, by taking the dot product of (3.120) by $\mathbf{t}_i(\mathbf{r})$ and integrating over S_o, we obtain the matrix equation

$$[P]\{E_S\} + [Q]\{\overline{H}_S\} = 0 \tag{3.122}$$

where $[P]$ and $[Q]$ represent two full matrices with elements that are given by

$$P_{ij} = \oiint_{S_o} [\hat{n} \times \mathbf{t}_i(\mathbf{r})] \cdot \left[\tfrac{1}{2}\mathbf{g}_j + \mathcal{L}(\mathbf{g}_j) + \hat{n} \times \tilde{\mathcal{K}}(\mathbf{g}_j)\right] dS \tag{3.123}$$

$$Q_{ij} = \oiint_{S_o} \mathbf{t}_i(\mathbf{r}) \cdot \left[\tfrac{1}{2}\mathbf{g}_j + \mathcal{L}(\mathbf{g}_j) + \hat{n} \times \tilde{\mathcal{K}}(\mathbf{g}_j)\right] dS. \tag{3.124}$$

In arriving at (3.123) and (3.124), we have applied vector identities and used the fact that $\mathbf{t}_i(\mathbf{r})$ is tangential to S_o. The evaluation of the double surface integrals in (3.123) and (3.124) can be carried out using Gaussian quadrature in conjunction with Duffy's transformation [3]. Equation (3.122) can now be combined with (3.91) to form the complete coupled system

$$\begin{bmatrix} K_{II} & K_{IS} & 0 \\ K_{SI} & K_{SS} & B \\ 0 & P & Q \end{bmatrix} \begin{Bmatrix} E_I \\ E_S \\ \overline{H}_S \end{Bmatrix} = \begin{Bmatrix} b_I \\ b_S \\ 0 \end{Bmatrix} \tag{3.125}$$

for the solution of $\{E_I\}$, $\{E_S\}$, and $\{\overline{H}_S\}$.

The coefficient matrix in (3.125) is a partly sparse and partly full matrix. It can be solved using a direct solver, where the efficiency can be improved by exploiting the sparsity of the matrix. For large problems involving many thousands of unknowns, an iterative solution is more efficient in both computational time and memory requirements. The iterative solution can be greatly accelerated by using a robust preconditioning technique. One such preconditioner can be obtained by neglecting the contributions of Green's function in the CFIE (3.117), which is then reduced to

$$\overline{\mathbf{J}}_s(\mathbf{r}) + \hat{n} \times \mathbf{M}_s(\mathbf{r}) \approx 0 \qquad \mathbf{r} \in S_o. \tag{3.126}$$

Note that (3.126) is simply the first-order absorbing boundary condition (2.7). With this ABC-based preconditioner [33], (3.125) can be written as

$$\begin{bmatrix} K_{II} & K_{IS} & 0 \\ K_{SI} & K_{SS} & B \\ 0 & \tilde{P} & \tilde{Q} \end{bmatrix}^{-1} \begin{bmatrix} K_{II} & K_{IS} & 0 \\ K_{SI} & K_{SS} & B \\ 0 & P & Q \end{bmatrix} \begin{Bmatrix} E_I \\ E_S \\ \overline{H}_S \end{Bmatrix} = \begin{bmatrix} K_{II} & K_{IS} & 0 \\ K_{SI} & K_{SS} & B \\ 0 & \tilde{P} & \tilde{Q} \end{bmatrix}^{-1} \begin{Bmatrix} b_I \\ b_S \\ 0 \end{Bmatrix} \tag{3.127}$$

where $[\tilde{P}]$ and $[\tilde{Q}]$ represent two purely sparse matrices given by

$$\tilde{P}_{ij} = \oiint_{S_o} [\hat{n} \times \mathbf{t}_i(\mathbf{r})] \cdot \mathbf{g}_j(\mathbf{r})\, dS \tag{3.128}$$

$$\tilde{Q}_{ij} = \oiint_{S_o} \mathbf{t}_i(\mathbf{r}) \cdot \mathbf{g}_j(\mathbf{r})\, dS. \tag{3.129}$$

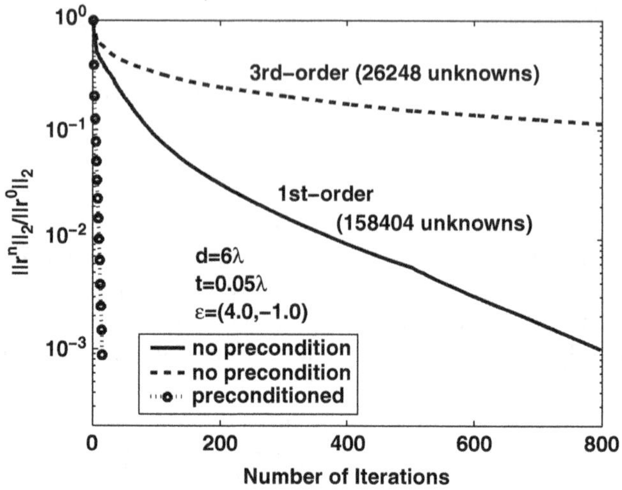

Figure 3.10 Convergence of the FE-BI iterative solution of scattering by a dielectric coated sphere. Comparison of the preconditioned solution and those without using a preconditioner. (After Liu and Jin [33], Copyright © IEEE 2002.)

Since the preconditioner is a purely sparse matrix, it can be constructed and factorized efficiently using a sparse direct solver. Alternatively, it can be solved iteratively with an incomplete *LU* preconditioner, resulting in a two-loop iteration algorithm. Extensive numerical studies showed that the ABC-based preconditioner is highly effective and in many applications, a converged solution can be obtained with a few tens of iterations [33]. An example is given in Figure 3.10, which shows the residual error of an iterative solution of the FE-BI equation for plane-wave scattering by a dielectric-coated conducting sphere. The diameter of the sphere is $6\lambda_0$ and the coating is $0.05\lambda_0$ thick and has a relative permittivity ε_r of $4.0 - j1.0$. Two finite element computations are performed: One uses first-order elements with 158,404 unknowns, and the other employs third-order elements with 26,248 unknowns. As can be seen, the iterative solutions without using any preconditioners converge very slowly, especially when higher-order elements are used. However, a rapid convergence is achieved using the preconditioner in (3.127), regardless of the order of finite elements.

It should be emphasized that the FE-BI method represents a general approach for solving open-region electromagnetic problems. The standard formulation described above is just one of many ways to formulate an FE-BI solution. For example, another fictitious surface can be introduced inside the truncation surface to define the equivalent surface currents $\bar{\mathbf{J}}_s$ and $\mathbf{M}_s$. When applying an integral equation to evaluate the field on the truncation surface by using these currents on the new shifted surface, we can avoid the singularity in the double surface integrals. In the following we describe briefly a new approach proposed recently to demonstrate a variety of ways to hybridize the finite element method and boundary integral equations. This approach, proposed in Refs. 30 and 31, makes use of the EFIE and MFIE separately on a single truncation surface. First, we substitute the MFIE (3.113) into (3.90) to obtain

$$\iiint_V \left[(\nabla \times \mathbf{T}) \cdot \overleftrightarrow{\mu}_r^{-1} \cdot (\nabla \times \mathbf{E}) - k_0^2 \mathbf{T} \cdot \overleftrightarrow{\varepsilon}_r \cdot \mathbf{E}\right] dV$$

$$- jk_0 \oiint_{S_o} (\hat{n} \times \mathbf{T}) \cdot [\mathcal{L}(\mathbf{M}_s) + \mathcal{K}(\bar{\mathbf{J}}_s)]\, dS$$

$$= - \iiint_V \mathbf{T} \cdot \left[jk_0 Z_0 \mathbf{J}_{\text{imp}} + \nabla \times \left(\overleftrightarrow{\mu}_r^{-1} \cdot \mathbf{M}_{\text{imp}}\right)\right] dV. \tag{3.130}$$

Next, we test the EFIE (3.112) with $\hat{n} \times \mathbf{T}$ and scale it with jk_0 to find another equation,

$$-jk_0 \oiint_{S_o} (\hat{n} \times \mathbf{T}) \cdot [\hat{n} \times \mathbf{M}_s + \mathcal{L}(\bar{\mathbf{J}}_s) - \mathcal{K}(\mathbf{M}_s)]\, dS = 0 \tag{3.131}$$

which complements (3.130) and completes the solution. The finite element discretization of these two equations yields

$$\begin{bmatrix} K_{II} & K_{IS} & 0 \\ K_{SI} & K_{SS} + U^M & V^M \\ 0 & V^E & U^E \end{bmatrix} \begin{Bmatrix} E_I \\ E_S \\ \bar{H}_S \end{Bmatrix} = \begin{Bmatrix} b_I \\ b_S \\ 0 \end{Bmatrix} \tag{3.132}$$

where $[K]$ is the same symmetric matrix as that defined in (3.92), and the elements of the other matrices are given by

$$U_{ij}^M = jk_0 \oiint_{S_o} \mathbf{g}_i \cdot \mathcal{L}(\mathbf{g}_j)\, dS \tag{3.133}$$

$$U_{ij}^E = -jk_0 \oiint_{S_o} \mathbf{g}_i \cdot \mathcal{L}(\mathbf{g}_j)\, dS \tag{3.134}$$

$$\begin{aligned} V_{ij}^M &= -jk_0 \oiint_{S_o} \mathbf{g}_i \cdot \mathcal{K}(\mathbf{g}_j)\, dS \\ &= \frac{jk_0}{2} \oiint_{S_o} (\hat{n} \times \mathbf{g}_i) \cdot \mathbf{g}_j\, dS - jk_0 \oiint_{S_o} \mathbf{g}_i \cdot \tilde{\mathcal{K}}(\mathbf{g}_j)\, dS \end{aligned} \tag{3.135}$$

$$\begin{aligned} V_{ij}^E &= jk_0 \oiint_{S_o} \mathbf{g}_i \cdot (\hat{n} \times \mathbf{g}_j)\, dS - jk_0 \oiint_{S_o} \mathbf{g}_i \cdot \mathcal{K}(\mathbf{g}_j)\, dS \\ &= \frac{jk_0}{2} \oiint_{S_o} \mathbf{g}_i \cdot (\hat{n} \times \mathbf{g}_j)\, dS - jk_0 \oiint_{S_o} \mathbf{g}_i \cdot \tilde{\mathcal{K}}(\mathbf{g}_j)\, dS. \end{aligned} \tag{3.136}$$

By using the definitions of the operators $\mathcal{L}$ and $\mathcal{K}$ in (3.109) and (3.110), it can be shown that both $[U^M]$ and $[U^E]$ are symmetric and $[V^M]$ and $[V^E]$ are transpose to each other. Hence, the coefficient matrix in (3.132) is completely symmetric, which can be solved using symmetric matrix solvers. Following the same approach, we can derive symmetric formulations for solving the interior magnetic field or both the interior electric and magnetic fields [31].

Unlike the ABC and PML, the boundary conditions derived from the boundary integral equations are exact. This allows the truncation boundary to be placed conformally to the antenna structure to be analyzed. However, this accuracy is achieved at the cost of an increased computational burden because the boundary integrals produce full matrices $[P]$ and $[Q]$ for the boundary unknowns, which are computationally much more expensive than those required for sparse matrices. Fortunately, fast algorithms such as the fast multipole method (FMM) [34,35] can be employed to speed up the computation, and one such formulation was described in Ref. 28.

3.3.2 Time-Domain Formulations

The hybridization of the finite element method with the boundary integral method can be achieved in the time domain as well. Recent developments in the time-domain finite element and time-domain integral equation methods have made such hybridization possible. The time-domain FE-BI method can be formulated in one of two different approaches, which have been fully studied and verified to be accurate and stable. We begin with a simpler approach and then proceed to a more complex one.

The first approach [36] employs two artificial surfaces: one, denoted as S_o, is to truncate the computational domain, and the other, denoted as S_{eq}, is to define a set of surface equivalent currents (Figure 3.11). Inside S_o, the electric field is governed by the vector wave equation

$$\begin{aligned}\nabla \times \left[\overleftrightarrow{\mu}^{-1} \cdot \nabla \times \mathbf{E}(\mathbf{r}, t)\right] + \overleftrightarrow{\varepsilon} \cdot \frac{\partial^2 \mathbf{E}(\mathbf{r}, t)}{\partial t^2} + \overleftrightarrow{\sigma}_e \cdot \frac{\partial \mathbf{E}(\mathbf{r}, t)}{\partial t} \\ = -\frac{\partial \mathbf{J}_{\text{imp}}(\mathbf{r}, t)}{\partial t} - \nabla \times \left[\overleftrightarrow{\mu}^{-1} \cdot \mathbf{M}_{\text{imp}}(\mathbf{r}, t)\right] \qquad \mathbf{r} \in V \qquad (3.137)\end{aligned}$$

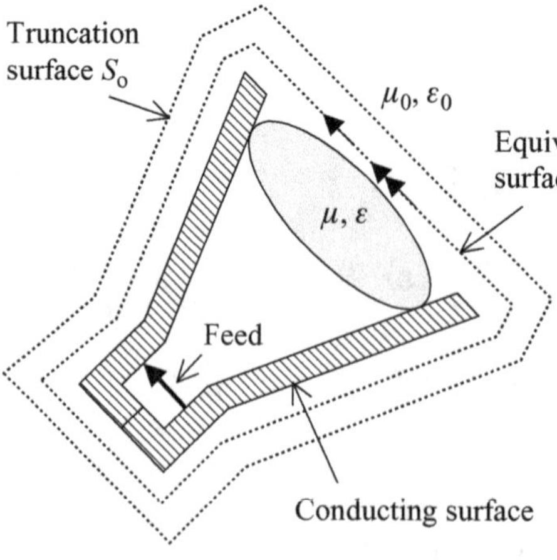

Figure 3.11 Equivalent source surface introduced to apply boundary integral equations.

where, for simplicity, we assumed that the medium involved is both nondispersive and magnetically lossless. The formulation to deal with general dispersive and lossy media was given in Section 2.3. To solve this equation, we need to define a boundary condition on S_o. Because this boundary condition is problem dependent, we postulate a mixed boundary condition with an unknown right-hand side,

$$\hat{n} \times \left[\frac{1}{\mu_0}\nabla \times \mathbf{E}(\mathbf{r}, t)\right] + Y_0\, \hat{n} \times \left[\hat{n} \times \frac{\partial}{\partial t}\mathbf{E}(\mathbf{r}, t)\right] = \mathbf{U}(\mathbf{r}, t) \qquad \mathbf{r} \in S_o \tag{3.138}$$

where S_o is assumed to reside in air. The weak-form representation for the interior electric field is then given by

$$\begin{aligned}
&\iiint_V \left[(\nabla \times \mathbf{T}) \cdot \overleftrightarrow{\mu}^{-1} \cdot (\nabla \times \mathbf{E}) + \mathbf{T} \cdot \overleftrightarrow{\varepsilon} \cdot \frac{\partial^2 \mathbf{E}}{\partial t^2} + \mathbf{T} \cdot \overleftrightarrow{\sigma}_e \cdot \frac{\partial \mathbf{E}}{\partial t}\right] dV \\
&\quad + \oiint_{S_o} \left[Y_0(\hat{n} \times \mathbf{T}) \cdot \left(\hat{n} \times \frac{\partial \mathbf{E}}{\partial t}\right) + \mathbf{T} \cdot \mathbf{U}\right] dS \\
&= -\iiint_V \mathbf{T} \cdot \left[\frac{\partial \mathbf{J}_{\text{imp}}}{\partial t} + \nabla \times (\overleftrightarrow{\mu}^{-1} \cdot \mathbf{M}_{\text{imp}})\right] dV
\end{aligned} \tag{3.139}$$

where the finite element discretization yields

$$[T]\frac{d^2\{E\}}{dt^2} + [R]\frac{d\{E\}}{dt} + [S]\{E\} = \{f\} + \{u\}. \tag{3.140}$$

The matrices $[T]$, $[R]$, and $[S]$ and the vectors $\{E\}$ and $\{f\}$ are the same as those defined in (2.28)–(2.31). The elements of the vector $\{u\}$ are given by

$$u_i(t) = -\oiint_{S_o} \mathbf{N}_i \cdot \mathbf{U}(\mathbf{r}, t)\, dS \tag{3.141}$$

which contains the unknown function $\mathbf{U}(\mathbf{r}, t)$. When central differencing is applied to (3.140), we obtain the time-marching equation

$$\begin{aligned}
\left\{\frac{1}{(\Delta t)^2}[T] + \frac{1}{2\Delta t}[R]\right\}\{E\}^{n+1} &= \left\{\frac{2}{(\Delta t)^2}[T] - [S]\right\}\{E\}^n \\
&\quad - \left\{\frac{1}{(\Delta t)^2}[T] - \frac{1}{2\Delta t}[R]\right\}\{E\}^{n-1} + \{f\}^n + \{u\}^n.
\end{aligned} \tag{3.142}$$

To calculate $\{E\}^{n+1}$, we require $\{u\}^n$, which can be computed using the boundary integral equations, and a technique to accomplish this is described below.

Given the electric field $\mathbf{E}(\mathbf{r}, t)$ on S_{eq}, we can define the surface equivalent electric and magnetic currents as

$$\mathbf{J}_s(\mathbf{r}, t) = \hat{n} \times \mathbf{H}(\mathbf{r}, t) = -\hat{n} \times \frac{1}{\mu_0} \int_0^t \nabla \times \mathbf{E}(\mathbf{r}, \tau)\, d\tau \qquad \mathbf{r} \in S_{\text{eq}} \tag{3.143}$$

and

$$\mathbf{M}_s(\mathbf{r}, t) = -\hat{n} \times \mathbf{E}(\mathbf{r}, t) \qquad \mathbf{r} \in S_{\text{eq}} \tag{3.144}$$

where S_{eq} is also assumed to reside in air. From these currents, the electric and magnetic vector potentials can be evaluated as

$$\mathbf{A}(\mathbf{r}, t) = \mu_0 \oiint_{S_{\text{eq}}} \frac{\mathbf{J}_s(\mathbf{r}', t - |\mathbf{r} - \mathbf{r}'|/c)}{4\pi |\mathbf{r} - \mathbf{r}'|}\, dS' \tag{3.145}$$

$$\mathbf{F}(\mathbf{r}, t) = \varepsilon_0 \oiint_{S_{\text{eq}}} \frac{\mathbf{M}_s(\mathbf{r}', t - |\mathbf{r} - \mathbf{r}'|/c)}{4\pi |\mathbf{r} - \mathbf{r}'|}\, dS' \tag{3.146}$$

where c is the speed of light. From the vector potentials $\mathbf{A}$ and $\mathbf{F}$, the fields on S_{o} can be computed as

$$\frac{\partial}{\partial t}\mathbf{E}(\mathbf{r}, t) = -\left(\frac{\partial^2}{\partial t^2} - c^2 \nabla\nabla\cdot\right)\mathbf{A}(\mathbf{r}, t) - \frac{1}{\varepsilon_0}\frac{\partial}{\partial t}\nabla \times \mathbf{F}(\mathbf{r}, t) \tag{3.147}$$

$$\frac{1}{\mu_0}\nabla \times \mathbf{E}(\mathbf{r}, t) = -\frac{1}{\mu_0}\frac{\partial}{\partial t}\nabla \times \mathbf{A}(\mathbf{r}, t) + \left(\frac{\partial^2}{\partial t^2} - c^2 \nabla\nabla\cdot\right)\mathbf{F}(\mathbf{r}, t) \tag{3.148}$$

and when these are substituted into (3.138), we obtain $\mathbf{U}(\mathbf{r}, t)$. Hence, (3.141) can be written as

$$\begin{aligned} u_i(t) = &\oiint_{S_{\text{o}}} \mathbf{N}_i \cdot \left\{\hat{n} \times \left[\frac{1}{\mu_0}\frac{\partial}{\partial t}\nabla \times \mathbf{A}(\mathbf{r}, t) - \left(\frac{\partial^2}{\partial t^2} - c^2 \nabla\nabla\cdot\right)\mathbf{F}(\mathbf{r}, t)\right]\right\} dS \\ &- Y_0 \oiint_{S_{\text{o}}} (\hat{n} \times \mathbf{N}_i) \cdot \left\{\hat{n} \times \left[\frac{1}{\varepsilon_0}\frac{\partial}{\partial t}\nabla \times \mathbf{F}(\mathbf{r}, t) + \left(\frac{\partial^2}{\partial t^2} - c^2 \nabla\nabla\cdot\right)\mathbf{A}(\mathbf{r}, t)\right]\right\} dS. \end{aligned} \tag{3.149}$$

Using this equation in conjunction with a temporal expansion for the electric field, we can compute $\{u\}^n$ based on $\{E\}^n$ and the values calculated previously, which then can be used in (3.142) to calculate $\{E\}^{n+1}$.

The approach described above has an advantage in that the integrals in (3.149) are nonsingular, due to the offset between S_{eq} and S_{o}; hence, they can be evaluated relatively easily. The solution is also free of spurious modes where these frequencies correspond to the resonant frequencies of a cavity that is formed by covering S_{o} with either a perfect electric or a perfect magnetic conductor and is filled with the materials in the exterior medium, which is air in the current formulation. The major disadvantage of this formulation is that the time-marching scheme is only conditionally stable because of the use of the central-difference method for temporal discretization. The Newmark-beta method cannot be applied directly in this approach because its discretization of (3.140) would require $\{u\}^{n+1}$, which cannot be calculated prior to the calculation of $\{E\}^{n+1}$.

The second approach, proposed in Ref. 37, transforms the standard FE-BI formulation directly from the frequency domain into the time domain using the Laplace transform. For example, (3.90) is transformed into the time domain as

$$\iiint_V \left[(\nabla \times \mathbf{T}) \cdot \overleftrightarrow{\mu}^{-1} \cdot (\nabla \times \mathbf{E}) + \mathbf{T} \cdot \overleftrightarrow{\varepsilon} \cdot \frac{\partial^2 \mathbf{E}}{\partial t^2} + \mathbf{T} \cdot \overleftrightarrow{\sigma}_e \cdot \frac{\partial \mathbf{E}}{\partial t} \right] dV$$
$$- Y_0 \oiint_{S_o} \mathbf{T} \cdot \left(\hat{n} \times \frac{\partial \overline{\mathbf{H}}}{\partial t} \right) dS = - \iiint_V \mathbf{T} \cdot \left[\frac{\partial \mathbf{J}_{\text{imp}}}{\partial t} + \nabla \times (\overleftrightarrow{\mu}^{-1} \cdot \mathbf{M}_{\text{imp}}) \right] dV \tag{3.150}$$

and the corresponding finite element matrix equation becomes

$$[T] \frac{d^2\{E\}}{dt^2} + [R] \frac{d\{E\}}{dt} + [S]\{E\} + [B] \frac{d\{\overline{H}_S\}}{dt} = \{f\}. \tag{3.151}$$

The matrices $[T]$ and $[S]$ and the vectors $\{E\}$ and $\{f\}$ are the same as those defined in (2.28), (2.30), and (2.31). The elements of the matrices $[R]$ and $[B]$ are given by

$$R_{ij} = \iiint_V \mathbf{N}_i \cdot \overleftrightarrow{\sigma}_e \cdot \mathbf{N}_j \, dV \tag{3.152}$$

$$B_{ij} = Y_0 \oiint_{S_o} \hat{n} \cdot \left(\mathbf{N}_i \times \mathbf{N}_j^S \right) dS. \tag{3.153}$$

The EFIE (3.112) and MFIE (3.113) can also be transformed into the time domain as

$$\mathbf{E}(\mathbf{r}, t) = -\mathcal{L}(\overline{\mathbf{J}}_s, \mathbf{r}, t) + \mathcal{K}(\mathbf{M}_s, \mathbf{r}, t) \qquad \mathbf{r} \in V_\infty \tag{3.154}$$

$$\overline{\mathbf{H}}(\mathbf{r}, t) = -\mathcal{L}(\mathbf{M}_s, \mathbf{r}, t) - \mathcal{K}(\overline{\mathbf{J}}_s, \mathbf{r}, t) \qquad \mathbf{r} \in V_\infty \tag{3.155}$$

where

$$\mathcal{L}(\mathbf{X}, \mathbf{r}, t) = \frac{1}{c_0} \oiint_{S_o} \frac{\partial_t \mathbf{X}(\mathbf{r}', t - |\mathbf{r} - \mathbf{r}'|/c_0)}{4\pi |\mathbf{r} - \mathbf{r}'|} dS'$$
$$- c_0 \nabla \oiint_{S_o} \int_0^{t - |\mathbf{r} - \mathbf{r}'|/c_0} \frac{\nabla' \cdot \mathbf{X}(\mathbf{r}', \tau)}{4\pi |\mathbf{r} - \mathbf{r}'|} d\tau \, dS' \tag{3.156}$$

$$\mathcal{K}(\mathbf{X}, \mathbf{r}, t) = -\nabla \times \oiint_{S_o} \frac{\mathbf{X}(\mathbf{r}', t - |\mathbf{r} - \mathbf{r}'|/c_0)}{4\pi |\mathbf{r} - \mathbf{r}'|} dS'. \tag{3.157}$$

These can be used to construct the CFIE, which after taking a time derivative and testing with $\mathbf{t}_i(\mathbf{r})$ becomes

$$\oiint_{S_o} \mathbf{t}_i(\mathbf{r})\cdot \partial_t \left\{[\bar{\mathbf{J}}_s(\mathbf{r},t)+\hat{n}\times\mathcal{L}(\mathbf{M}_s,\mathbf{r},t)+\hat{n}\times\mathcal{K}(\bar{\mathbf{J}}_s,\mathbf{r},t)]\right.$$
$$\left.+\hat{n}\times[\mathbf{M}_s(\mathbf{r},t)-\hat{n}\times\mathcal{L}(\bar{\mathbf{J}}_s,\mathbf{r},t)+\hat{n}\times\mathcal{K}(\mathbf{M}_s,\mathbf{r},t)]\right\} dS=0 \qquad \mathbf{r}\in S_o. \tag{3.158}$$

The discretization of (3.158) yields the matrix equation

$$[P]*\{E_S\}+[Q]*\{\overline{H}_S\}=0 \tag{3.159}$$

where $[P]$ and $[Q]$ represent two full matrices with elements that are given by

$$P_{ij}(t)*E_j^S(t)=\partial_t \oiint_{S_o}[\hat{n}\times\mathbf{t}_i(\mathbf{r})]\cdot\left[\mathbf{g}_j E_j^S(t)+\mathcal{L}\left(\mathbf{g}_j E_j^S,\mathbf{r},t\right)\right.$$
$$\left.+\hat{n}\times\mathcal{K}\left(\mathbf{g}_j E_j^S,\mathbf{r},t\right)\right]dS \tag{3.160}$$

$$Q_{ij}(t)*\overline{H}_j^S(t)=\partial_t \oiint_{S_o}\mathbf{t}_i(\mathbf{r})\cdot\left[\mathbf{g}_j \overline{H}_j^S(t)+\mathcal{L}\left(\mathbf{g}_j \overline{H}_j^S,\mathbf{r},t\right)\right.$$
$$\left.+\hat{n}\times\mathcal{K}\left(\mathbf{g}_j \overline{H}_j^S,\mathbf{r},t\right)\right]dS. \tag{3.161}$$

By applying the Newmark-beta method to discretize (3.151) and converting (3.159) into a discrete convolution, we obtain the time-marching equation to update $\{E\}$ simultaneously, including $\{E_S\}$, and $\{\overline{H}_S\}$:

$$\begin{bmatrix} K_{II}^0 & K_{IS}^0 & 0 \\ K_{SI}^0 & K_{SS}^0 & B^0 \\ 0 & P^0 & Q^0 \end{bmatrix} \begin{Bmatrix} E_I^{n+1} \\ E_S^{n+1} \\ \overline{H}_S^{n+1} \end{Bmatrix} = \begin{Bmatrix} b_I^{n+1} \\ b_S^{n+1} \\ \overline{b}_S^{n+1} \end{Bmatrix} \tag{3.162}$$

where for the sake of clarity, $\{E\}$ has been explicitly split into $\{E_I\}$ and $\{E_S\}$. In (3.162),

$$\begin{Bmatrix} b_I^{n+1} \\ b_S^{n+1} \\ \overline{b}_S^{n+1} \end{Bmatrix} = \beta \begin{Bmatrix} f_I^{n+1} \\ f_S^{n+1} \\ 0 \end{Bmatrix} + (1-2\beta)\begin{Bmatrix} f_I^{n} \\ f_S^{n} \\ 0 \end{Bmatrix} + \beta \begin{Bmatrix} f_I^{n-1} \\ f_S^{n-1} \\ 0 \end{Bmatrix}$$
$$-\begin{bmatrix} K_{II}^1 & K_{IS}^1 & 0 \\ K_{SI}^1 & K_{SS}^1 & 0 \\ 0 & P^1 & Q^1 \end{bmatrix} \begin{Bmatrix} E_I^{n} \\ E_S^{n} \\ \overline{H}_S^{n} \end{Bmatrix} - \begin{bmatrix} K_{II}^2 & K_{IS}^2 & 0 \\ K_{SI}^2 & K_{SS}^2 & B^2 \\ 0 & P^2 & Q^2 \end{bmatrix} \begin{Bmatrix} E_I^{n-1} \\ E_S^{n-1} \\ \overline{H}_S^{n-1} \end{Bmatrix}$$
$$-\sum_{k=0}^{n-2} \begin{bmatrix} 0 & 0 & 0 \\ 0 & 0 & 0 \\ 0 & P^{n-k+1} & Q^{n-k+1} \end{bmatrix} \begin{Bmatrix} 0 \\ E_S^{k} \\ \overline{H}_S^{k} \end{Bmatrix} \tag{3.163}$$

where

$$[K^0] = \frac{1}{(\Delta t)^2}[T] + \frac{1}{2\Delta t}[R] + \beta[S] \tag{3.164}$$

$$[K^1] = -\frac{2}{(\Delta t)^2}[T] + (1 - 2\beta)[S] \tag{3.165}$$

$$[K^2] = \frac{1}{(\Delta t)^2}[T] - \frac{1}{2\Delta t}[R] + \beta[S] \tag{3.166}$$

$$[B^0] = \frac{1}{2\Delta t}[B], \qquad [B^2] = -\frac{1}{2\Delta t}[B] \tag{3.167}$$

$$P_{ij}^{n-k+1} = P_{ij}(t) * T(t - k\Delta t)\big|_{t=(n+1)\Delta t} \tag{3.168}$$

$$Q_{ij}^{n-k+1} = Q_{ij}(t) * T(t - k\Delta t)\big|_{t=(n+1)\Delta t}. \tag{3.169}$$

In (3.168) and (3.169), $T(t)$ denotes temporal basis functions used to approximate the convolution integrals [37]. Equation (3.162) is highly stable, due to use of the Newmark-beta method. Of course, care must be exercised to ensure accurate expansion of the unknown functions and accurate evaluation of the temporal derivatives and computation of the singular integrals in the discretization of the boundary integral equation (3.158). More detailed discussion on this topic can be found in Ref. 37.

Similar to the frequency-domain FE-BI method, the most computationally time-consuming part in the time-domain FE-BI solution is calculation of the time-domain boundary integrals. This calculation can be accelerated using fast algorithms based either on the plane-wave time-domain method or the time-domain adaptive integral method, as demonstrated in Refs. 36 and 37. As an example to demonstrate application of the time-domain FE-BI to antenna analysis, we consider a Vivaldi antenna residing on a finite ground plane (Figure 3.12). The antenna is excited by a TEM pulse with

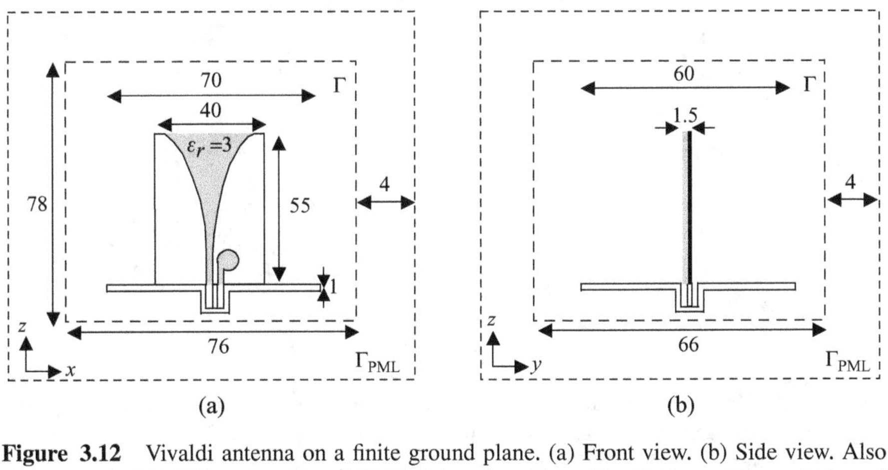

Figure 3.12 Vivaldi antenna on a finite ground plane. (a) Front view. (b) Side view. Also shown are the FE-BI interface Γ and FE-PML interface Γ_{PML}. The PML are 16 mm thick, and all dimensions are in millimeters. (After Yilmaz et al. [37], Copyright © IEEE 2007.)

$f_0 = 3.2$ GHz and $f_{bw} = 2.8$ GHz. The inner and outer coaxial cable radii are 1 and 3 mm, respectively. The FE-BI interface Γ (which is S_o) and the FE-PML interface Γ_{PML} are chosen as shown in Figure 3.12. Figure 3.13 plots the reflection coefficient of the antenna as a function of frequency and the radiation pattern in the E-plane at 3.0 GHz. It can be seen that the time-domain FE-BI solution agrees very well

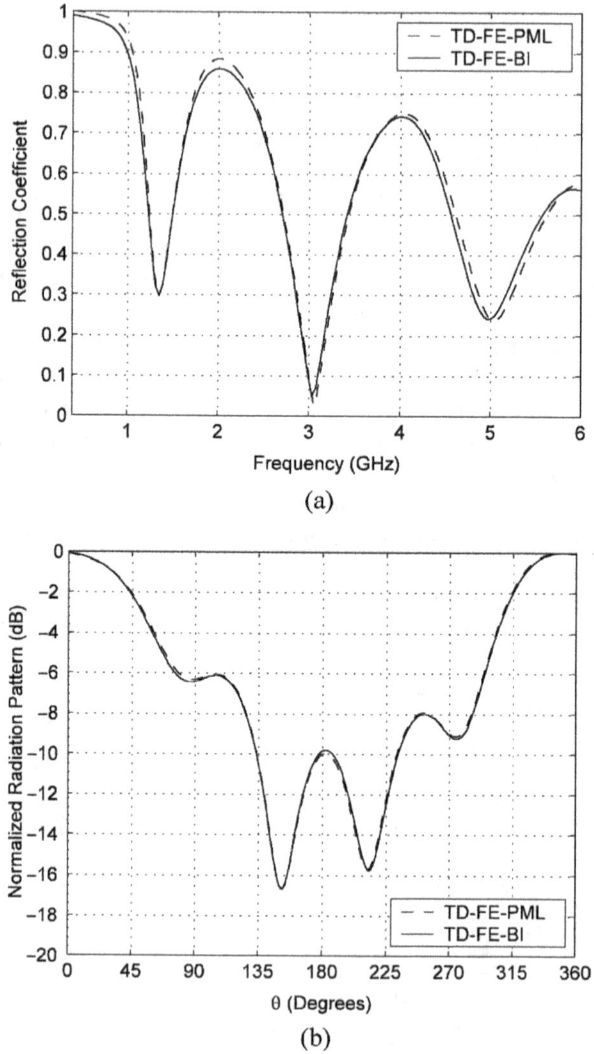

Figure 3.13 Radiation from a Vivaldi antenna on a finite ground plane. (a) Reflection coefficients versus frequency observed at the feed. (b) Radiation patterns at 3 GHz. (After Yilmaz et al. [37], Copyright © IEEE 2007.)

with those obtained by the time-domain finite element method using PML for mesh truncation.

3.3.3 Treatment of the Infinite Ground Plane

In practical applications, some antennas, such as aperture and conformal antennas, are designed to work in the presence of a ground plane. To analyze this type of antenna, the boundary integral equations have to be modified to take into account the effect of the ground plane. A simple approach is to employ the image theory and make use of the equations derived above for the free-space case.

Consider the problem illustrated in Figure 3.14(a), where an antenna structure is situated partially in a cavity below a ground plane and partially protruded above the ground plane, which is assumed to coincide with the xy-plane. To analyze this problem using the FE-BI method, we introduce a mathematical surface S_o to tightly cover the antenna structure above the ground plane. In contrast to the free-space case, S_o is no longer a closed surface. To formulate the field exterior to S_o using the equivalent surface currents on S_o, we can first remove the ground plane and replace its effect with the image currents on the image surface S_{im}, as shown in Figure 3.14(b). In accordance with the image theory, the image currents are given by

$$\mathbf{J}_s^{im}(x, y, z) = 2\hat{z}\hat{z} \cdot \mathbf{J}_s(x, y, -z) - \mathbf{J}_s(x, y, -z) \tag{3.170}$$

$$\mathbf{M}_s^{im}(x, y, z) = -2\hat{z}\hat{z} \cdot \mathbf{M}_s(x, y, -z) + \mathbf{M}_s(x, y, -z) \tag{3.171}$$

where $\mathbf{J}_s$ and $\mathbf{M}_s$ represent the original currents on S_o. With the currents on a closed surface formed by S_o and S_{im} radiating in free space, we can use the integral equations derived earlier to formulate the exterior fields. For example, we can use the EFIE (3.112) to find

$$\mathbf{E}(\mathbf{r}) = -\mathcal{L}(\bar{\mathbf{J}}_s) + \mathcal{K}(\mathbf{M}_s) \qquad \mathbf{r} \in V_\infty \tag{3.172}$$

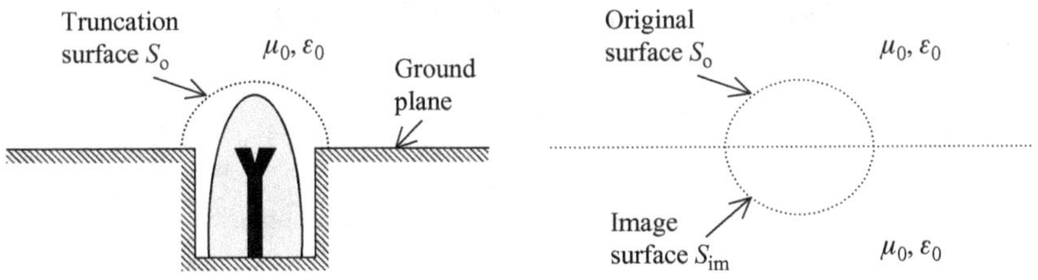

(a) (b)

Figure 3.14 Antenna structure partially residing in a cavity and protruding above a ground plane. (a) Original problem. (b) Equivalent image problem for the truncation surface.

where

$$\mathcal{L}(\bar{\mathbf{J}}_s) = jk_0 \iint_{S_o} \left[\bar{\mathbf{J}}_s(\mathbf{r}')G_0(\mathbf{r}, \mathbf{r}') + \frac{1}{k_0^2}\nabla' \cdot \bar{\mathbf{J}}_s(\mathbf{r}')\nabla G_0(\mathbf{r}, \mathbf{r}') \right] dS'$$
$$+ jk_0 \iint_{S_{im}} \left[\bar{\mathbf{J}}_s^{im}(\mathbf{r}')G_0(\mathbf{r}, \mathbf{r}') + \frac{1}{k_0^2}\nabla' \cdot \bar{\mathbf{J}}_s^{im}(\mathbf{r}')\nabla G_0(\mathbf{r}, \mathbf{r}') \right] dS' \quad (3.173)$$

$$\mathcal{K}(\mathbf{M}_s) = \iint_{S_o} \mathbf{M}_s(\mathbf{r}') \times \nabla G_0(\mathbf{r}, \mathbf{r}')\,dS' + \iint_{S_{im}} \mathbf{M}_s^{im}(\mathbf{r}') \times \nabla G_0(\mathbf{r}, \mathbf{r}')\,dS'. \quad (3.174)$$

Substituting (3.170) and (3.171) into the equations above, we have

$$\mathcal{L}(\bar{\mathbf{J}}_s) = jk_0 \iint_{S_o} \Big\{ \bar{\mathbf{J}}_s(\mathbf{r}')[G_0(\mathbf{r}, \mathbf{r}') - G_0(\mathbf{r}, \mathbf{r}_i')] + 2\hat{z}\hat{z} \cdot \bar{\mathbf{J}}_s(\mathbf{r}')G_0(\mathbf{r}, \mathbf{r}_i')$$
$$+ \frac{1}{k_0^2}\nabla' \cdot \bar{\mathbf{J}}_s(\mathbf{r}')\nabla[G_0(\mathbf{r}, \mathbf{r}') - G_0(\mathbf{r}, \mathbf{r}_i')] \Big\}\, dS' \quad (3.175)$$

$$\mathcal{K}(\mathbf{M}_s) = \iint_{S_o} \{ \mathbf{M}_s(\mathbf{r}') \times \nabla[G_0(\mathbf{r}, \mathbf{r}') + G_0(\mathbf{r}, \mathbf{r}_i')] - 2\hat{z}\hat{z} \cdot \mathbf{M}_s(\mathbf{r}') \times \nabla G_0(\mathbf{r}, \mathbf{r}_i') \}\, dS' \quad (3.176)$$

where $\mathbf{r}_i = x\hat{x} + y\hat{y} - z\hat{z}$ denotes the image position. Equation (3.172) can now be used in the FE-BI solution of the problem in the same manner as for the free-space case. The singularity treatment is also the same. Since S_o now is not a closed surface, there is no internal resonance problem and hence no need to use the CFIE.

If the entire antenna structure is confined in the cavity and there is no protrusion above the ground plane, the cavity's aperture can then be used as S_o (Figure 3.15). In this case, $\mathcal{L}(\bar{\mathbf{J}}_s) = 0$ and

$$\mathcal{K}(\mathbf{M}_s) = 2 \iint_{S_o} \mathbf{M}_s(\mathbf{r}') \times \nabla G_0(\mathbf{r}, \mathbf{r}')\,dS'. \quad (3.177)$$

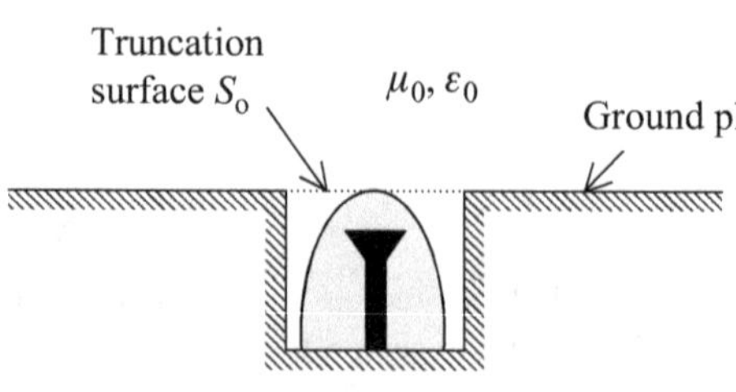

Figure 3.15 Antenna structure completely residing in a cavity recessed in a ground plane.

Equation (3.172) then becomes

$$\mathbf{E}(\mathbf{r}) = 2 \iint_{S_o} \mathbf{M}_s(\mathbf{r}') \times \nabla G_0(\mathbf{r}, \mathbf{r}')\, dS' \qquad \mathbf{r} \in V_\infty. \tag{3.178}$$

From the MFIE (3.113), we find that

$$\overline{\mathbf{H}}(\mathbf{r}) = \frac{2}{jk_0} \iint_{S_o} \left[k_0^2 \mathbf{M}_s(\mathbf{r}') G_0(\mathbf{r}, \mathbf{r}') + \nabla' \cdot \mathbf{M}_s(\mathbf{r}') \nabla G_0(\mathbf{r}, \mathbf{r}') \right] dS' \qquad \mathbf{r} \in V_\infty \tag{3.179}$$

which can also be obtained from (3.178) using Maxwell's equations. Substituting (3.179) into the weak-form representation (3.90) for the field inside the cavity, we obtain

$$\begin{aligned} &\iiint_V \left[(\nabla \times \mathbf{T}) \cdot \overleftrightarrow{\mu}_r^{-1} \cdot (\nabla \times \mathbf{E}) - k_0^2 \mathbf{T} \cdot \overleftrightarrow{\varepsilon}_r \cdot \mathbf{E} \right] dV + 2 \iint_{S_o} (\hat{z} \times \mathbf{T}) \\ &\qquad \cdot \iint_{S_o} \left[k_0^2 \mathbf{M}_s(\mathbf{r}') G_0(\mathbf{r}, \mathbf{r}') + \nabla' \cdot \mathbf{M}_s(\mathbf{r}') \nabla G_0(\mathbf{r}, \mathbf{r}') \right] dS'\, dS \\ &\quad = - \iiint_V \mathbf{T} \cdot \left[jk_0 Z_0 \mathbf{J}_{\text{imp}} + \nabla \times \left(\overleftrightarrow{\mu}_r^{-1} \cdot \mathbf{M}_{\text{imp}} \right) \right] dV. \end{aligned} \tag{3.180}$$

By using vector identities and Gauss's theorem, (3.180) can be written as

$$\begin{aligned} &\iiint_V \left[(\nabla \times \mathbf{T}) \cdot \overleftrightarrow{\mu}_r^{-1} \cdot (\nabla \times \mathbf{E}) - k_0^2 \mathbf{T} \cdot \overleftrightarrow{\varepsilon}_r \cdot \mathbf{E} \right] dV + 2k_0^2 \iint_{S_o} (\hat{z} \times \mathbf{T}) \\ &\qquad \cdot \iint_{S_o} \mathbf{M}_s(\mathbf{r}') G_0(\mathbf{r}, \mathbf{r}')\, dS'\, dS - 2 \iint_{S_o} \nabla \cdot (\hat{z} \times \mathbf{T}) \iint_{S_o} \nabla' \cdot \mathbf{M}_s(\mathbf{r}') G_0(\mathbf{r}, \mathbf{r}')\, dS'\, dS \\ &= - \iiint_V \mathbf{T} \cdot \left[jk_0 Z_0 \mathbf{J}_{\text{imp}} + \nabla \times (\overleftrightarrow{\mu}_r^{-1} \cdot \mathbf{M}_{\text{imp}}) \right] dV. \end{aligned} \tag{3.181}$$

The finite element discretization of (3.181) yields the matrix equation

$$\begin{bmatrix} K_{II} & K_{IS} \\ K_{SI} & K_{SS} + P \end{bmatrix} \begin{Bmatrix} E_I \\ E_S \end{Bmatrix} = \begin{Bmatrix} b_I \\ b_S \end{Bmatrix} \tag{3.182}$$

where $[K]$ is defined by (3.92) and the elements of $[P]$ are given by

$$\begin{aligned} P_{ij} = 2 \iint_{S_o} \nabla \cdot \mathbf{g}_i(\mathbf{r}) \iint_{S_o} \nabla' \cdot \mathbf{g}_j(\mathbf{r}') G_0(\mathbf{r}, \mathbf{r}')\, dS'\, dS - 2k_0^2 \iint_{S_o} \mathbf{g}_i(\mathbf{r}) \\ \cdot \iint_{S_o} \mathbf{g}_j(\mathbf{r}') G_0(\mathbf{r}, \mathbf{r}')\, dS'\, dS. \end{aligned} \tag{3.183}$$

Since both $[K]$ and $[P]$ are symmetric, the coefficient matrix in (3.182) is also symmetric. Once the aperture field is solved for, the fields in the exterior region can be computed using (3.178) and (3.179).

3.4 SUMMARY

In this chapter we discussed three approaches to truncating an infinite free-space environment surrounding an antenna into a finite computational domain that is suitable for a finite element formulation. Of the three approaches, the one based on absorbing boundary conditions is the most efficient and also the easiest to implement. In fact, the implementation of the first-order absorbing boundary condition was covered in Chapter 2 in both the frequency and time domains. The second-order absorbing boundary condition can be implemented with the aid of an auxiliary variable for a smooth surface, which maintains the symmetry of the finite element matrix.

The second approach, based on perfectly matched layers, has gained a great deal of popularity in recent years mainly because of its excellent performance in the FDTD method. This approach is also simple to implement in the frequency-domain finite element method, although its implementation in the time domain becomes much more complicated and hence was described in detail. The performance of perfectly matched layers can be improved by several relatively simple modifications, which include using an absorbing boundary condition to terminate the PML region, as well as alternative formulations such as the complementary, complex-frequency shifted, and second-order perfectly matched layers.

The third approach employs a boundary integral equation to truncate the computational domain, and the resulting numerical technique is often referred to as the hybrid finite element–boundary integral (FE-BI) method. This approach is the most accurate, but it is also the most difficult to implement. We described two formulations for the frequency-domain implementation: One is the standard formulation and the other yields a symmetric final system of equations, together with the derivation of the boundary integral equations. We also described two formulations to implement the FE-BI method in the time domain. One uses two mathematical surfaces, yielding a conditionally stable time-marching process. The other employs a single surface and results in a more stable time-marching scheme when implemented with an accurate discretization of the boundary integral equation involved. The second formulation can be regarded as the Laplace transform of the standard FE-BI formulation in the frequency domain. The efficiency of the FE-BI solution as compared with the other two approaches depends on individual problems. In general, the FE-BI solution is more costly because of the necessity to compute and solve the full submatrices due to the boundary integral equations. However, the cost can be reduced significantly by using fast integral equation solvers.

At the end of the chapter we discussed the modeling of antennas in the presence of an infinite ground plane and derived the corresponding integral equations that can be used in the FE-BI solution of such problems. This treatment is important because many antennas, such as aperture and conformal antennas or any antennas placed near a large smooth conducting surface, can be approximated as having an infinite ground plane. For antennas designed with a layered medium, we can either use the finite element method combined with an absorbing boundary condition or perfectly matched layers to discretize the layered medium directly or incorporate the effect of the layered medium into the boundary integral equation in the FE-BI formulation. The first approach is straightforward, and the second is rather involved because of the complexity of the Green's function for a layered medium [38].

REFERENCES

1. A. F. Peterson, "Absorbing boundary conditions for the vector wave equation," *Microwave Opt. Tech. Lett.*, vol. 1, pp. 62–64, Apr. 1988.
2. J. P. Webb and V. N. Kanellopoulos, "Absorbing boundary conditions for the finite element solution of the vector wave equation," *Microwave Opt. Tech. Lett.*, vol. 2, pp. 370–372, Oct. 1989.
3. J.-M. Jin, *The Finite Element Method in Electromagnetics*, 2nd ed. Hoboken, NJ: Wiley, 2002.
4. A. Chatterjee, J. M. Jin, and J. L. Volakis, "Edge-based finite elements and vector ABCs applied to 3D scattering," *IEEE Trans. Antennas Propagat.*, vol. 41, no. 2, pp. 221–226, Feb. 1993.
5. V. N. Kanellopoulos and J. P. Webb, "The importance of the surface divergence term in the finite element–vector absorbing boundary condition method," *IEEE Trans. Microwave Theory Tech.*, vol. 43, no. 9, pp. 2168–2170, Sept. 1995.
6. M. M. Botha and D. B. Davidson, "Rigorous, auxiliary variable–based implementation of a second-order ABC for the vector FEM," *IEEE Trans. Antennas Propagat.*, vol. 54, no. 11, pp. 3499–3504, Nov. 2006.
7. J. M. Jin, J. L. Volakis, and V. V. Liepa, "A fictitious absorber for truncating finite element meshes in scattering," *Proc. Inst. Elec. Eng. H*, vol. 139, no. 5, pp. 472–476, Oct. 1992.
8. J.-P. Berenger, "A perfectly matched layer for the absorption of electromagnetic waves," *J. Comput. Phys.*, vol. 114, no. 2, pp. 185–200, 1994.
9. D. S. Katz, E. T. Thiele, and A. Taflove, "Validation and extension to three dimensions of the Berenger PML absorbing boundary condition for FD-TD meshes," *IEEE Microwave Guided Wave Lett.*, vol. 4, pp. 268–270, Aug. 1994.
10. W. C. Chew and W. H. Weedon, "A 3D perfectly matched medium from modified Maxwell's equations with stretched coordinates," *Microwave Opt. Tech. Lett.*, vol. 7, pp. 599–604, Sept. 1994.
11. Z. S. Sacks, D. M. Kingsland, R. Lee, and J.-F. Lee, "A perfectly matched anisotropic absorber for use as an absorbing boundary condition," *IEEE Trans. Antennas Propagat.*, vol. 43, pp. 1460–1463, Dec. 1995.
12. S. D. Gedney, "An anisotropic perfectly matched layer absorbing medium for the truncation of FDTD lattices," *IEEE Trans. Antennas Propagat.*, vol. 44, pp. 1630–1639, Dec. 1996.
13. D. Jiao, J. M. Jin, E. Michielssen, and D. Riley, "Time-domain finite-element simulation of three-dimensional scattering and radiation problems using perfectly matched layers," *IEEE Trans. Antennas Propagat.*, vol. 51, no. 2, pp. 296–305, Feb. 2003.

14. T. Rylander and J. M. Jin, "Perfectly matched layers in three dimensions for the time-domain finite element method applied to radiation problems," *IEEE Trans. Antennas Propagat.*, vol. 53, no. 4, pp. 1489–1499, Apr. 2005.
15. Z. Lou and J. M. Jin, "Modeling and simulation of broadband antennas using the time-domain finite element method," *IEEE Trans. Antennas Propagat.*, vol. 53, no. 12, pp. 4099–4110, Dec. 2005.
16. W. C. Chew and J. M. Jin, "Perfectly matched layers in the discretized space: an analysis and optimization," *Electromagnetics*, vol. 16, no. 4, pp. 325–340, July 1996.
17. J. M. Jin and W. C. Chew, "Combining PML and ABC for finite element analysis of scattering problems," *Microwave Opt. Tech. Lett.*, vol. 12, no. 4, pp. 192–197, July 1996.
18. J. M. Jin, X. Q. Sheng, and W. C. Chew, "Complementary perfectly matched layers to reduce reflection errors," *Microwave Opt. Tech. Lett.*, vol. 14, no. 5, pp. 284–287, Apr. 1997.
19. D. Correia and J. M. Jin, "On the development of a higher-order PML," *IEEE Trans. Antennas Propagat.*, vol. 53, no. 12, pp. 4157–4163, Dec. 2005.
20. M. Kuzuoglu and R. Mittra, "Frequency dependence of the constitutive parameters of causal perfectly anisotropic absorbers," *IEEE Microwave Guided Wave Lett.*, vol. 6, no. 12, pp. 447–449, Dec. 1996.
21. J. A. Roden and S. D. Gedney, "Convolution PML (CPML): an efficient FDTD implementation of the CFS-PML for arbitrary media," *Microwave Opt. Tech. Lett.*, vol. 27, no. 5, pp. 334–339, Dec. 2000.
22. Z. Lou, D. Correia, and J. M. Jin, "Second-order perfectly matched layers for the time-domain finite element method," *IEEE Trans. Antennas Propagat.*, vol. 55, no. 3, pp. 1000–1004, Mar. 2007.
23. J.-Y. Wu, D. M. Kingsland, J.-F. Lee, and R. Lee, "A comparison of anisotropic PML to Berenger's PML and its application to the finite-element method for EM scattering," *IEEE Trans. Antennas Propagat.*, vol. 45, no. 1, pp. 40–50, Jan. 1997.
24. C. T. Wolfe, U. Navsariwala, and S. D. Gedney, "A parallel finite-element tearing and interconnecting algorithm for solution of the vector wave equation with PML absorbing medium," *IEEE Trans. Antennas Propagat.*, vol. 48, no. 2, pp. 278–284, Feb. 2000.
25. J. M. Jin and V. V. Liepa, "Application of a hybrid finite element method to electromagnetic scattering from coated cylinders," *IEEE Trans. Antennas Propagat.*, vol. 36, no. 1, pp. 50–54, Jan. 1988.
26. X. Yuan, "Three-dimensional electromagnetic scattering from inhomogeneous objects by the hybrid moment and finite element method," *IEEE Trans. Microwave Theory Tech.*, vol. 38, pp. 1053–1058, Aug. 1990.
27. T. Eibert and V. Hansen, "Calculation of unbounded field problems in free space by a 3D FEM/BEM-hybrid approach," *J. Electromagn. Waves Appl.*, vol. 10, no. 1, pp. 61–78, 1996.
28. X. Q. Sheng, J. M. Jin, J. M. Song, C. C. Lu, and W. C. Chew, "On the formulation of hybrid finite-element and boundary-integral method for 3D scattering," *IEEE Trans. Antennas Propagat.*, vol. 46, no. 3, pp. 303–311, Mar. 1998.
29. J. Liu and J. M. Jin, "A novel hybridization of higher order finite element and boundary integral methods for electromagnetic scattering and radiation problems," *IEEE Trans. Antennas Propagat.*, vol. 49, no. 12, pp. 1794–1806, Dec. 2001.

30. M. N. Vouvakis, S.-C. Lee, K. Zhao, and J.-F. Lee, "A symmetric FEM–IE formulation with a single-level IE–QR algorithm for solving electromagnetic radiation and scattering problems," *IEEE Trans. Antennas Propagat.*, vol. 52, no. 11, pp. 3060–3070, Nov. 2004.
31. M. M. Botha and J. M. Jin, "On the variational formulation of hybrid finite element–boundary integral techniques for electromagnetic analysis," *IEEE Trans. Antennas Propagat.*, vol. 52, no. 11, pp. 3037–3047, Nov. 2004.
32. C. T. Tai, "Direct integration of field equations," *IEEE AP-S Int. Symp. Dig.*, vol. 2, p. 884, Atlanta, GA, June 1998.
33. J. Liu and J. M. Jin, "A highly effective preconditioner for solving the finite element–boundary integral matrix equation of 3-D scattering," *IEEE Trans. Antennas Propagat.*, vol. 50, no. 9, pp. 1212–1221, Sept. 2002.
34. R. Coifman, V. Rokhlin, and S. Wandzura, "The fast multipole method for the wave equation: a pedestrian prescription," *IEEE Antennas Propagat. Mag.*, vol. 35, pp. 7–12, June 1993.
35. W. C. Chew, J. M. Jin, E. Michielssen, and J. M. Song, Eds., *Fast and Efficient Algorithms in Computational Electromagnetics*. Norwood, MA: Artech House, 2001.
36. D. Jiao, A. Ergin, B. Shanker, E. Michielssen, and J. M. Jin, "A fast time-domain higher-order finite element–boundary integral method for 3-D electromagnetic scattering analysis," *IEEE Trans. Antennas Propagat.*, vol. 50, no. 9, pp. 1192–1202, Sept. 2002.
37. A. E. Yilmaz, Z. Lou, E. Michielssen, and J. M. Jin, "A single-boundary, implicit, and FFT-accelerated time-domain finite element–boundary integral solver," *IEEE Trans. Antennas Propagat.*, vol. 55, no. 5, pp. 1382–1397, May 2007.
38. K. A. Michalski and J. R. Mosig, "Multilayered media Green's functions in integral equation formulations," *IEEE Trans. Antennas Propagat.*, vol. 45, pp. 508–519, Mar. 1997.

4 Hybrid FETD–FDTD Technique

The finite-difference time-domain (FDTD) method has been widely used for the analysis and design of antennas [1]. The basic FDTD technique is highly efficient because it utilizes both a structured grid and explicit time marching. A three-dimensional structured grid is defined by three integer indices, typically *i–j–k*, and as such, storage of the full mesh is often not required. In contrast, finite element grids based on tetrahedral elements are typically unstructured, and in this case the entire mesh generally needs to be stored and a simple integer cell-referencing scheme is generally not possible. In addition, because the standard FDTD algorithm gives rise to a diagonal system matrix, it yields an explicit time-marching algorithm. Explicit time-marching algorithms are conditionally stable, meaning that the time step must be proportionally reduced with finer levels of spatial discretization. The FDTD method can also be parallelized in a distributed sense in a very straightforward manner [2], and the grid-truncation technology based on the perfectly matched layer (PML) is mature and numerically stable [3]. These features make the FDTD method a powerful and popular tool for many applications. However, because the traditional FDTD technique is based primarily on a rectilinear grid, the accurate modeling of complex antenna feed regions and/or complex antenna geometry may require an exceedingly fine level of spatial discretization along with extensive computational resources. In addition, because the formulation is conditionally stable, which is again a requirement for explicit time marching, the Courant limit dictates that a fine spatial discretization also requires a corresponding decrease in the time-step size and thus an increase in the number of time steps to obtain an accurate solution.

An alternative approach to the global application of the traditional FDTD algorithm is to utilize a hybridization of locally conformal solution techniques with the standard FDTD method. The techniques that have been proposed for the conformal portion include (1) the use of local modifications to the basic FDTD method to accommodate surface curvature [4,5], (2) finite-volume techniques [6–9], and (3) finite element methods [10–20]. The first technique is known as the conformal FDTD method and is based on a weighting procedure that analyzes how an arbitrary surface intersects the traditional FDTD rectangular cells. This has been found to be a useful and efficient approach, although there are limits to the geometrical complexity that can be accommodated and the technique often requires using locally refined rectangular grids, which then reduce the global time step.

In this chapter we address primarily the local application of finite element methods; specifically, a finite element time-domain (FETD) hybridization with FDTD. By utilizing this technique, the antenna details can be characterized by the unconditionally stable and conformal aspects of the FETD, while the efficiency of the rectangular-grid FDTD method can be utilized in the external regions of the problem domain. The advantage of this approach is that fine discretization, common in antenna feed regions such as spirals and other antennas, does not require an increase in the number of time steps. In addition, the efficiency of FDTD is advantageous in the regions surrounding the antenna that typically contain only air or simple materials out to the termination of the grid boundary, at which point the well-established FDTD PML [3] can easily be applied. This hybrid FETD-FDTD concept is provably numerically stable up to the Courant limit of the FDTD region that supports the finite element domains [16]. In other words, the grid associated with the internal FETD regions can expand and contract as needed to characterize the local geometry without imposing further restrictions on the global time step with regard to numerical stability.

The success of this hybridization depends fundamentally on an underlying equivalence between the FDTD and the FETD formulations, which are based on vector edge basis functions. Specifically, when trapezoidal integration is applied to rectangular hexahedral cells used in the FETD method, the FDTD algorithm can be recovered [13,21,22]. Consequently, a symmetric interface between these two methods can be constructed, which is typically a necessary condition for proving numerical stability [16,20].

The topics discussed in this chapter include the following. First, we provide an overview of the FDTD method along with the well-established FDTD PML formulations based on the stretched-coordinate and anisotropic-medium techniques. Because of the importance in obtaining far-field simulation data in antenna applications, we also discuss an effective near-to-far-field transformation technique for the FDTD method. We then examine an FETD formulation that is based fundamentally on Maxwell's two first-order curl equations and note the relationship of this FETD formulation as a generalization of FDTD to unstructured grids. We show further that FDTD can be constructed as a special case of the vector edge FETD method, which will ultimately enable construction of a numerically stable FETD–FDTD interface. Because the construction of the hybridized mesh imposes certain requirements on grid generation, we subsequently discuss techniques to facilitate this process. Also, the wave equation is known to support solutions that can drift in the late time for some applications, and therefore a simple suppression technique is discussed. Finally, we present a few applications that demonstrate this powerful hybrid technique, which synergistically leverages many of the best features of the FETD and FDTD methods.

4.1 FDTD METHOD

The finite-difference time-domain (FDTD) method represents an intuitive and powerful solution technique for the transient differential Maxwell's equations. Originally developed by Yee [23], the method has been applied extensively and enhanced algorithmically over its long history [24]. Some of the desirable features of the FDTD method are the following: (1) natural representation of Ampère's and Faraday's laws, (2) second-order convergence in both space and time, (3) divergence free in the absence of electric and magnetic charges, (4) continuity of the tangential components of the electric and magnetic fields, (5) nondissipative and hence energy conserving, and (6) fully explicit in time. Although the most common implementation of the FDTD method conforms to a Cartesian grid, spherical [25] and cylindrical [26] coordinate implementations have also been reported. In the following we restrict the discussion to the Cartesian grid, where this grid consists of a lattice of rectangular, or brick, cells with indexing defined by $(i, j, k) = (i\,\Delta x, j\,\Delta y, k\,\Delta z)$. As noted previously, this indexing is associated with a structured mesh, or alternatively, a mapped mesh. For simplicity, the discussion here is limited to the uniform spatial discretization $(\Delta x, \Delta y, \Delta z)$ in each of the three Cartesian axes (x, y, z), although more general implementations can incorporate a nonuniform bias along any or all of the axes.

The traditional Yee cell with the electric and magnetic field components is shown in Figure 4.1. Based on this orientation, it is easy to see how the field circulations correspond directly to both Faraday's and Ampère's laws. Mathematically, Faraday's and Ampère's laws in linear, lossless, and isotropic media are given by (2.21) and (2.22), respectively, and are repeated here for convenience:

$$\mu \frac{\partial \mathbf{H}}{\partial t} = -\nabla \times \mathbf{E} - \sigma_m \mathbf{H} - \mathbf{M}_{\text{imp}} \tag{4.1}$$

$$\varepsilon \frac{\partial \mathbf{E}}{\partial t} = \nabla \times \mathbf{H} - \sigma_e \mathbf{E} - \mathbf{J}_{\text{imp}}. \tag{4.2}$$

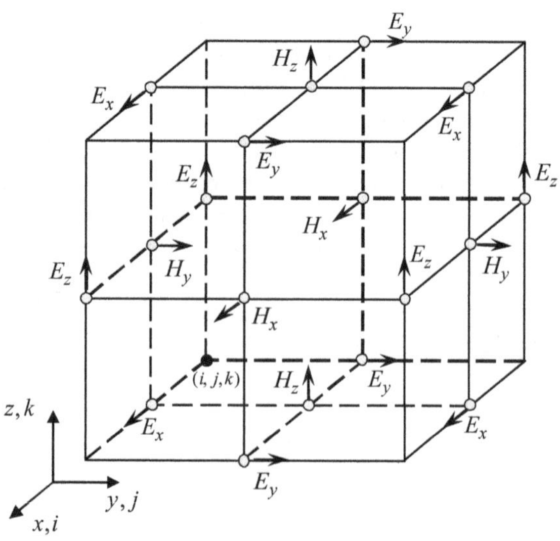

Figure 4.1 FDTD "Yee" basic cell showing the offset nature of electric and magnetic fields.

In the Cartesian coordinate system, the specific components of (4.1) and (4.2) are given by

$$\frac{\partial H_x}{\partial t} = \frac{1}{\mu}\left(\frac{\partial E_y}{\partial z} - \frac{\partial E_z}{\partial y} - \sigma_m H_x - M_{\mathrm{imp},x}\right) \tag{4.3}$$

$$\frac{\partial H_y}{\partial t} = \frac{1}{\mu}\left(\frac{\partial E_z}{\partial x} - \frac{\partial E_x}{\partial z} - \sigma_m H_y - M_{\mathrm{imp},y}\right) \tag{4.4}$$

$$\frac{\partial H_z}{\partial t} = \frac{1}{\mu}\left(\frac{\partial E_x}{\partial y} - \frac{\partial E_y}{\partial x} - \sigma_m H_z - M_{\mathrm{imp},z}\right) \tag{4.5}$$

$$\frac{\partial E_x}{\partial t} = \frac{1}{\varepsilon}\left(\frac{\partial H_z}{\partial y} - \frac{\partial H_y}{\partial z} - \sigma_e E_x - J_{\mathrm{imp},x}\right) \tag{4.6}$$

$$\frac{\partial E_y}{\partial t} = \frac{1}{\varepsilon}\left(\frac{\partial H_x}{\partial z} - \frac{\partial H_z}{\partial x} - \sigma_e E_y - J_{\mathrm{imp},y}\right) \tag{4.7}$$

$$\frac{\partial E_z}{\partial t} = \frac{1}{\varepsilon}\left(\frac{\partial H_y}{\partial x} - \frac{\partial H_x}{\partial y} - \sigma_e E_z - J_{\mathrm{imp},z}\right). \tag{4.8}$$

By adopting the spatial grid shown in Figure 4.1 along with a centered finite difference for the time derivatives, the fully discrete FDTD form for (4.3)–(4.8) is easily obtained. The resulting time advancement of the electric and magnetic fields is known as *leapfrogging*. The truncation error is $O(\Delta t^2)$ in time and $O(\Delta^2)$ in space, where Δ corresponds to Δx, Δy, or Δz. In a fully discrete form, the traditional FDTD equations are given by

$$\begin{aligned}
H_x^{n+1/2}\left(i, j+\tfrac{1}{2}, k+\tfrac{1}{2}\right)\left(1+\frac{\sigma_m \Delta t}{2\mu}\right) &= \left(1-\frac{\sigma_m \Delta t}{2\mu}\right) H_x^{n-1/2}\left(i, j+\tfrac{1}{2}, k+\tfrac{1}{2}\right)\\
&+ \frac{\Delta t}{\mu \Delta z}\left[E_y^n\left(i, j+\tfrac{1}{2}, k+1\right) - E_y^n\left(i, j+\tfrac{1}{2}, k\right)\right]\\
&- \frac{\Delta t}{\mu \Delta y}\left[E_z^n\left(i, j+1, k+\tfrac{1}{2}\right) - E_z^n\left(i, j, k+\tfrac{1}{2}\right)\right] - \frac{\Delta t}{\mu} M_{\mathrm{imp},x}^n\left(i, j+\tfrac{1}{2}, k+\tfrac{1}{2}\right)
\end{aligned} \tag{4.9}$$

$$\begin{aligned}
H_y^{n+1/2}\left(i+\tfrac{1}{2}, j, k+\tfrac{1}{2}\right)\left(1+\frac{\sigma_m \Delta t}{2\mu}\right) &= \left(1-\frac{\sigma_m \Delta t}{2\mu}\right) H_y^{n-1/2}\left(i+\tfrac{1}{2}, j, k+\tfrac{1}{2}\right)\\
&+ \frac{\Delta t}{\mu \Delta x}\left[E_z^n\left(i+1, j, k+\tfrac{1}{2}\right) - E_z^n\left(i, j, k+\tfrac{1}{2}\right)\right]\\
&- \frac{\Delta t}{\mu \Delta z}\left[E_x^n\left(i+\tfrac{1}{2}, j, k+1\right) - E_x^n\left(i+\tfrac{1}{2}, j, k\right)\right] - \frac{\Delta t}{\mu} M_{\mathrm{imp},y}^n\left(i+\tfrac{1}{2}, j, k+\tfrac{1}{2}\right)
\end{aligned} \tag{4.10}$$

$$H_z^{n+1/2}(i+\tfrac{1}{2},j+\tfrac{1}{2},k)\left(1+\frac{\sigma_m\Delta t}{2\mu}\right)=\left(1-\frac{\sigma_m\Delta t}{2\mu}\right)H_z^{n-1/2}(i+\tfrac{1}{2},j+\tfrac{1}{2},k)$$
$$+\frac{\Delta t}{\mu\Delta y}\left[E_x^n(i+\tfrac{1}{2},j+1,k)-E_x^n(i+\tfrac{1}{2},j,k)\right]$$
$$-\frac{\Delta t}{\mu\Delta x}\left[E_y^n(i+1,j+\tfrac{1}{2},k)-E_y^n(i,j+\tfrac{1}{2},k)\right]-\frac{\Delta t}{\mu}M_{\text{imp},z}^n(i+\tfrac{1}{2},j+\tfrac{1}{2},k) \tag{4.11}$$

$$E_x^{n+1}(i+\tfrac{1}{2},j,k)\left(1+\frac{\sigma_e\Delta t}{2\varepsilon}\right)=\left(1-\frac{\sigma_e\Delta t}{2\varepsilon}\right)E_x^n(i+\tfrac{1}{2},j,k)$$
$$+\frac{\Delta t}{\varepsilon\Delta y}\left[H_z^{n+1/2}(i+\tfrac{1}{2},j+\tfrac{1}{2},k)-H_z^{n+1/2}(i+\tfrac{1}{2},j-\tfrac{1}{2},k)\right]$$
$$-\frac{\Delta t}{\varepsilon\Delta z}\left[H_y^{n+1/2}(i+\tfrac{1}{2},j,k+\tfrac{1}{2})-H_y^{n+1/2}(i+\tfrac{1}{2},j,k-\tfrac{1}{2})\right]$$
$$-\frac{\Delta t}{\varepsilon}J_{\text{imp},x}^{n+1/2}(i+\tfrac{1}{2},j,k) \tag{4.12}$$

$$E_y^{n+1}(i,j+\tfrac{1}{2},k)\left(1+\frac{\sigma_e\Delta t}{2\varepsilon}\right)=\left(1-\frac{\sigma_e\Delta t}{2\varepsilon}\right)E_y^n(i,j+\tfrac{1}{2},k)$$
$$+\frac{\Delta t}{\varepsilon\Delta z}\left[H_x^{n+1/2}(i,j+\tfrac{1}{2},k+\tfrac{1}{2})-H_x^{n+1/2}(i,j+\tfrac{1}{2},k-\tfrac{1}{2})\right]$$
$$-\frac{\Delta t}{\varepsilon\Delta x}\left[H_z^{n+1/2}(i+\tfrac{1}{2},j+\tfrac{1}{2},k)-H_z^{n+1/2}(i-\tfrac{1}{2},j+\tfrac{1}{2},k)\right]$$
$$-\frac{\Delta t}{\varepsilon}J_{\text{imp},y}^{n+1/2}(i,j+\tfrac{1}{2},k) \tag{4.13}$$

$$E_z^{n+1}(i,j,k+\tfrac{1}{2})\left(1+\frac{\sigma_e\Delta t}{2\varepsilon}\right)=\left(1-\frac{\sigma_e\Delta t}{2\varepsilon}\right)E_z^n(i,j,k+\tfrac{1}{2})$$
$$+\frac{\Delta t}{\varepsilon\Delta x}\left[H_y^{n+1/2}(i+\tfrac{1}{2},j,k+\tfrac{1}{2})-H_y^{n+1/2}(i-\tfrac{1}{2},j,k+\tfrac{1}{2})\right]$$
$$-\frac{\Delta t}{\varepsilon\Delta y}\left[H_x^{n+1/2}(i,j+\tfrac{1}{2},k+\tfrac{1}{2})-H_x^{n+1/2}(i,j-\tfrac{1}{2},k+\tfrac{1}{2})\right]$$
$$-\frac{\Delta t}{\varepsilon}J_{\text{imp},z}^{n+1/2}(i,j,k+\tfrac{1}{2}). \tag{4.14}$$

When constructing finite-difference representations such as those used in (4.9)–(4.14), it is important to keep track of where the particular differences are centered. If a particular formulation requires a field value at a time or spatial location that is not directly available, an average value of the known fields is adopted. For example, one possible time-averaging form is $J_{\text{imp},x}^{n+1/2}=(J_{\text{imp},x}^{n+1}+J_{\text{imp},x}^{n})/2$, and a spatial average will involve a stencil across two or more cells, where the term *stencil* is used to denote the field locations involved in the construction process. Averages of these types do not compromise the second-order accuracy of the FDTD solution. Note that finite element formulations permit the interpolation to arbitrary spatial locations within the finite element cell, due to the use of basis functions.

Many of the numerical properties of the coupled system of equations described by (4.9)–(4.14) can be established by using a technique known as *Fourier stability analysis* [27]. Properties of interest include the stability criterion, which relates the time step to the spatial cell size, as well as the dispersion and dissipation aspects of the numerical scheme. In the Fourier method, the various discrete field components are assumed to have a sinusoidal representation, such as

$$U^n(i, j, k) = U_0 e^{j\omega n\Delta t} e^{-j(\tilde{k}_x i\Delta x + \tilde{k}_y j\Delta y + \tilde{k}_z k\Delta z)} \tag{4.15}$$

where $\tilde{k}_{x,y,z}$ denote the numerical wavenumbers in the directions x, y, and z. By applying (4.15) to the lossless Yee algorithm with $\sigma_e = \sigma_m = 0$, the following dispersion relationship is obtained:

$$\left(\frac{1}{c\Delta t}\sin\frac{\omega\Delta t}{2}\right)^2 = \left(\frac{1}{\Delta x}\sin\frac{\tilde{k}_x\Delta x}{2}\right)^2 + \left(\frac{1}{\Delta y}\sin\frac{\tilde{k}_y\Delta y}{2}\right)^2 + \left(\frac{1}{\Delta z}\sin\frac{\tilde{k}_z\Delta z}{2}\right)^2. \tag{4.16}$$

To derive (4.16), Maxwell's equations are first written in a normalized form as $j\nabla \times \mathbf{U} = \partial\mathbf{U}/\partial t$, where $\mathbf{U} \equiv (\mathbf{H} + j\mathbf{E})$ and $j = \sqrt{-1}$ [28]. The representation (4.15) is then substituted into a Yee discretization of $j\nabla \times \mathbf{U} = \partial\mathbf{U}/\partial t$, which is similar to (4.9)–(4.14). A 3×3 system of equations for the components of $\mathbf{U}$ will result, and by setting the determinant equal to zero, (4.16) is obtained. Note that (4.16) is for a grid of infinite extent; in other words, the influence of boundary conditions is not included in the analysis.

The numerical dispersion relationship (4.16) can be compared against the ideal theoretical dispersion relationship for a plane wave propagating in an infinite free-space medium:

$$\left(\frac{\omega}{c}\right)^2 = k_x^2 + k_y^2 + k_z^2. \tag{4.17}$$

Note that by applying a small argument expansion for the sine functions in (4.16)—namely, $\sin x \approx x$ for $x \ll 1$—(4.16) reduces to (4.17). However, for more practical choices of the time step and the cell size, (4.16) reveals that the dispersion characteristics of the Yee algorithm are anisotropic over a three-dimensional space. The largest dispersion error occurs along the principal axes, whereas the smallest errors occur for propagation in directions diagonal to the grid. In fact, for a plane wave propagating along the diagonal of a three-dimensional Yee lattice under the conditions of cubical cells with edge length Δ and $c\Delta t/\Delta = 1/\sqrt{3}$, (4.16) reduces to (4.17), which indicates that ideal propagation is obtained. It is also noted that in a three-dimensional space with propagation along a principal axis, the three-dimensional dispersion relationship is equivalent to that associated with the one-dimensional Maxwell's equations discretized with the Yee algorithm. Detailed studies and additional observations for the dispersion relationship (4.16) as compared to ideal propagation can be found in Ref. 29.

Equation (4.16) is also used to determine the numerical stability condition, or alternatively, the *Courant criterion*, of equations (4.9)–(4.14). This is accomplished initially by substituting $\tilde{\omega}$ for ω in (4.15) and (4.16), where $\tilde{\omega} = \tilde{\omega}_r + j\tilde{\omega}_i$ represents a complex angular frequency, and then solving (4.16) for $\tilde{\omega}$. By restricting the solution to only real values for $\tilde{\omega}$ [since complex values would lead to exponential growth in time as seen in (4.15)], the following relationship between the time and spatial steps is required:

$$c\Delta t \le \frac{1}{\sqrt{1/\Delta x^2 + 1/\Delta y^2 + 1/\Delta z^2}}. \tag{4.18}$$

In the case of cubical elements such that $\Delta = \Delta x = \Delta y = \Delta z$, (4.18) simplifies to $c\,\Delta t \le \Delta/\sqrt{3}$. Also note that because (4.18) is associated only with real values for $\tilde{\omega}$, the solution (4.15) does not decay in time, which is a necessary condition for the algorithm to be nondissipative and hence energy conserving.

For typical applications to which the FDTD method is applied, 15 to 20 cells per wavelength is the traditional spatial discretization for the definition of Δx, Δy, and Δz. However, for electrically large applications involving propagation over, say, tens of wavelengths, much finer spatial discretization may be necessary to obtain an acceptable dispersion error. Equation (4.16) can be used to provide an estimate of the phase-velocity error to be expected for a given spatial resolution. For example, under the conditions of propagation along a principal axis, for cubical cells with an edge length Δ, where Δ is chosen to provide 30 points per free-space wavelength, and for a time step defined by $c\Delta t/\Delta = 0.5$, the phase-velocity error of the Yee algorithm relative to ideal propagation is approximately 0.13% per wavelength of travel. Thus, through the use of (4.16) an estimate of the cell size required for a given application is readily obtained. In addition, most practical implementations of the three-dimensional Yee algorithm choose the time step Δt to be slightly smaller than the theoretical limit defined by (4.18), where a typical value for cubical elements would be $c\Delta t = 0.99\,\Delta/\sqrt{3}$.

4.2 PML IMPLEMENTATION IN FDTD

A beneficial feature of adopting an FETD–FDTD hybridized formulation is to leverage the advances in the mature grid-truncation techniques that are widely used in the FDTD method. One of the more robust and accurate methods for truncation of the FDTD grid is the PML concept originally developed by Berenger [30]. By splitting the fields to add additional degrees of freedom, Berenger found that a theoretically reflectionless interface could be constructed for any frequency and angle of incidence. This split-field PML was later found to be represented equivalently by a modified form for Maxwell's equations based on stretched coordinates. The stretched-coordinate PML was described in Section 3.2.1, and it was shown to provide a reflectionless interface independent of frequency and the angle of incidence. In an FDTD setting, the split-field PML in the three-dimensional space revealed reflections from the outer grid boundary more than 40 dB lower than were obtained previously with analytical absorbing boundary conditions that were similar to those described in Section 3.1 [31]. In Sections 4.2.1 and 4.2.2, we apply the stretched-coordinate PML and anisotropic-medium PML, respectively, to the FDTD setting.

4.2.1 FDTD Stretched-Coordinate PML

The source-free Maxwell's equations in a stretched-coordinate space are given by (3.23)–(3.26). By expanding the stretched-coordinate curl operator in (3.23), we have

$$-j\omega\mu\mathbf{H} = \nabla_s \times \mathbf{E} = \hat{x}\left(\frac{1}{s_y}\frac{\partial E_z}{\partial y} - \frac{1}{s_z}\frac{\partial E_y}{\partial z}\right) + \hat{y}\left(\frac{1}{s_z}\frac{\partial E_x}{\partial z} - \frac{1}{s_x}\frac{\partial E_z}{\partial x}\right) + \hat{z}\left(\frac{1}{s_x}\frac{\partial E_y}{\partial x} - \frac{1}{s_y}\frac{\partial E_x}{\partial y}\right) \tag{4.19}$$

and similarly for (3.24),

$$j\omega\varepsilon\mathbf{E} = \nabla_s \times \mathbf{H} = \hat{x}\left(\frac{1}{s_y}\frac{\partial H_z}{\partial y} - \frac{1}{s_z}\frac{\partial H_y}{\partial z}\right) + \hat{y}\left(\frac{1}{s_z}\frac{\partial H_x}{\partial z} - \frac{1}{s_x}\frac{\partial H_z}{\partial x}\right) + \hat{z}\left(\frac{1}{s_x}\frac{\partial H_y}{\partial x} - \frac{1}{s_y}\frac{\partial H_x}{\partial y}\right). \tag{4.20}$$

By further splitting the electric and magnetic fields in (4.19) and (4.20), it is apparent that s_x, s_y, and s_z can be isolated and removed from the denominators. This will be useful for obtaining a simple time-domain PML implementation. Specifically, the field components can be split as

$$\hat{x}H_x = \hat{x}(H_{xy} + H_{xz}); \quad \hat{y}H_y = \hat{y}(H_{yx} + H_{yz}); \quad \hat{z}H_z = \hat{z}(H_{zx} + H_{zy}) \tag{4.21}$$

$$\hat{x}E_x = \hat{x}(E_{xy} + E_{xz}); \quad \hat{y}E_y = \hat{y}(E_{yx} + E_{yz}); \quad \hat{z}E_z = \hat{z}(E_{zx} + E_{zy}). \tag{4.22}$$

With (4.21) and (4.22), the components of (4.19) and (4.20) become

$$j\omega\mu s_y H_{xy}\hat{x} = -\frac{\partial(E_{zx} + E_{zy})}{\partial y}\hat{x} \tag{4.23}$$

$$j\omega\mu s_z H_{xz}\hat{x} = \frac{\partial(E_{yx} + E_{yz})}{\partial z}\hat{x} \tag{4.24}$$

$$j\omega\mu s_z H_{yz}\hat{y} = -\frac{\partial(E_{xy}+E_{xz})}{\partial z}\hat{y} \tag{4.25}$$

$$j\omega\mu s_x H_{yx}\hat{y} = \frac{\partial(E_{zx}+E_{zy})}{\partial x}\hat{y} \tag{4.26}$$

$$j\omega\mu s_x H_{zx}\hat{z} = -\frac{\partial(E_{yx}+E_{yz})}{\partial x}\hat{z} \tag{4.27}$$

$$j\omega\mu s_y H_{zy}\hat{z} = \frac{\partial(E_{xy}+E_{xz})}{\partial y}\hat{z} \tag{4.28}$$

$$j\omega\varepsilon s_y E_{xy}\hat{x} = \frac{\partial(H_{zx}+H_{zy})}{\partial y}\hat{x} \tag{4.29}$$

$$j\omega\varepsilon s_z E_{xz}\hat{x} = -\frac{\partial(H_{yx}+H_{yz})}{\partial z}\hat{x} \tag{4.30}$$

$$j\omega\varepsilon s_z E_{yz}\hat{y} = \frac{\partial(H_{xy}+H_{xz})}{\partial z}\hat{y} \tag{4.31}$$

$$j\omega\varepsilon s_x E_{yx}\hat{y} = -\frac{\partial(H_{zx}+H_{zy})}{\partial x}\hat{y} \tag{4.32}$$

$$j\omega\varepsilon s_x E_{zx}\hat{z} = \frac{\partial(H_{yx}+H_{yz})}{\partial x}\hat{z} \tag{4.33}$$

$$j\omega\varepsilon s_y E_{zy}\hat{z} = -\frac{\partial(H_{xy}+H_{xz})}{\partial y}\hat{z}. \tag{4.34}$$

Although s_x, s_y, and s_z can have a general form, and the particular form selected gives rise to different PML behavior, the following simple definitions are made to facilitate transformation into the time domain:

$$s_x = 1 - j\frac{\sigma_x}{\omega\varepsilon}; \quad s_y = 1 - j\frac{\sigma_y}{\omega\varepsilon}; \quad s_z = 1 - j\frac{\sigma_z}{\omega\varepsilon}. \tag{4.35}$$

With (4.35), (4.23)–(4.34) are easily written in time-dependent form as

$$\mu\frac{\partial H_{xy}}{\partial t}\hat{x} + \sigma_y\frac{\mu}{\varepsilon}H_{xy}\hat{x} = -\frac{\partial(E_{zx}+E_{zy})}{\partial y}\hat{x} \tag{4.36}$$

$$\mu\frac{\partial H_{xz}}{\partial t}\hat{x} + \sigma_z\frac{\mu}{\varepsilon}H_{xz}\hat{x} = \frac{\partial(E_{yx}+E_{yz})}{\partial z}\hat{x} \tag{4.37}$$

$$\mu\frac{\partial H_{yz}}{\partial t}\hat{y} + \sigma_z\frac{\mu}{\varepsilon}H_{yz}\hat{y} = -\frac{\partial(E_{xy}+E_{xz})}{\partial z}\hat{y} \tag{4.38}$$

$$\mu\frac{\partial H_{yx}}{\partial t}\hat{y} + \sigma_x\frac{\mu}{\varepsilon}H_{yx}\hat{y} = \frac{\partial(E_{zx}+E_{zy})}{\partial x}\hat{y} \tag{4.39}$$

$$\mu\frac{\partial H_{zx}}{\partial t}\hat{z} + \sigma_x\frac{\mu}{\varepsilon}H_{zx}\hat{z} = -\frac{\partial(E_{yx}+E_{yz})}{\partial x}\hat{z} \tag{4.40}$$

$$\mu \frac{\partial H_{zy}}{\partial t}\hat{z} + \sigma_y \frac{\mu}{\varepsilon} H_{zy}\hat{z} = \frac{\partial (E_{xy} + E_{xz})}{\partial y}\hat{z} \tag{4.41}$$

$$\varepsilon \frac{\partial E_{xy}}{\partial t}\hat{x} + \sigma_y E_{xy}\hat{x} = \frac{\partial (H_{zx} + H_{zy})}{\partial y}\hat{x} \tag{4.42}$$

$$\varepsilon \frac{\partial E_{xz}}{\partial t}\hat{x} + \sigma_z E_{xz}\hat{x} = -\frac{\partial (H_{yx} + H_{yz})}{\partial z}\hat{x} \tag{4.43}$$

$$\varepsilon \frac{\partial E_{yz}}{\partial t}\hat{y} + \sigma_z E_{yz}\hat{y} = \frac{\partial (H_{xy} + H_{xz})}{\partial z}\hat{y} \tag{4.44}$$

$$\varepsilon \frac{\partial E_{yx}}{\partial t}\hat{y} + \sigma_x E_{yx}\hat{y} = -\frac{\partial (H_{zx} + H_{zy})}{\partial x}\hat{y} \tag{4.45}$$

$$\varepsilon \frac{\partial E_{zx}}{\partial t}\hat{z} + \sigma_x E_{zx}\hat{z} = \frac{\partial (H_{yx} + H_{yz})}{\partial x}\hat{z} \tag{4.46}$$

$$\varepsilon \frac{\partial E_{zy}}{\partial t}\hat{z} + \sigma_y E_{zy}\hat{z} = -\frac{\partial (H_{xy} + H_{xz})}{\partial y}\hat{z}. \tag{4.47}$$

By applying the Yee algorithm to (4.36)–(4.47), we obtain the following time-advancement equations in the PML region:

$$\begin{aligned} H_{xy}^{n+1/2}\left(i, j+\tfrac{1}{2}, k+\tfrac{1}{2}\right)\left(1+\frac{\sigma_y \Delta t}{2\varepsilon}\right) &= \left(1-\frac{\sigma_y \Delta t}{2\varepsilon}\right) H_{xy}^{n-1/2}\left(i, j+\tfrac{1}{2}, k+\tfrac{1}{2}\right) \\ &- \frac{\Delta t}{\mu \Delta y}\left[E_{zx}^{n}\left(i, j+1, k+\tfrac{1}{2}\right) - E_{zx}^{n}\left(i, j, k+\tfrac{1}{2}\right)\right] \\ &- \frac{\Delta t}{\mu \Delta y}\left[E_{zy}^{n}\left(i, j+1, k+\tfrac{1}{2}\right) - E_{zy}^{n}\left(i, j, k+\tfrac{1}{2}\right)\right] \end{aligned} \tag{4.48}$$

$$\begin{aligned} H_{xz}^{n+1/2}\left(i, j+\tfrac{1}{2}, k+\tfrac{1}{2}\right)\left(1+\frac{\sigma_z \Delta t}{2\varepsilon}\right) &= \left(1-\frac{\sigma_z \Delta t}{2\varepsilon}\right) H_{xz}^{n-1/2}\left(i, j+\tfrac{1}{2}, k+\tfrac{1}{2}\right) \\ &+ \frac{\Delta t}{\mu \Delta z}\left[E_{yx}^{n}\left(i, j+\tfrac{1}{2}, k+1\right) - E_{yx}^{n}\left(i, j+\tfrac{1}{2}, k\right)\right] \\ &+ \frac{\Delta t}{\mu \Delta z}\left[E_{yz}^{n}\left(i, j+\tfrac{1}{2}, k+1\right) - E_{yz}^{n}\left(i, j+\tfrac{1}{2}, k\right)\right] \end{aligned} \tag{4.49}$$

$$\begin{aligned} H_{yz}^{n+1/2}\left(i+\tfrac{1}{2}, j, k+\tfrac{1}{2}\right)\left(1+\frac{\sigma_z \Delta t}{2\varepsilon}\right) &= \left(1-\frac{\sigma_z \Delta t}{2\varepsilon}\right) H_{yz}^{n-1/2}\left(i+\tfrac{1}{2}, j, k+\tfrac{1}{2}\right) \\ &- \frac{\Delta t}{\mu \Delta z}\left[E_{xy}^{n}\left(i+\tfrac{1}{2}, j, k+1\right) - E_{xy}^{n}\left(i+\tfrac{1}{2}, j, k\right)\right] \\ &- \frac{\Delta t}{\mu \Delta z}\left[E_{xz}^{n}\left(i+\tfrac{1}{2}, j, k+1\right) - E_{xz}^{n}\left(i+\tfrac{1}{2}, j, k\right)\right] \end{aligned} \tag{4.50}$$

$$H_{yx}^{n+1/2}(i+\tfrac{1}{2},j,k+\tfrac{1}{2})\left(1+\frac{\sigma_x\Delta t}{2\varepsilon}\right)=\left(1-\frac{\sigma_x\Delta t}{2\varepsilon}\right)H_{yx}^{n-1/2}(i+\tfrac{1}{2},j,k+\tfrac{1}{2})$$
$$+\frac{\Delta t}{\mu\Delta x}\left[E_{zx}^{n}(i+1,j,k+\tfrac{1}{2})-E_{zx}^{n}(i,j,k+\tfrac{1}{2})\right]$$
$$+\frac{\Delta t}{\mu\Delta x}\left[E_{zy}^{n}(i+1,j,k+\tfrac{1}{2})-E_{zy}^{n}(i,j,k+\tfrac{1}{2})\right] \tag{4.51}$$

$$H_{zx}^{n+1/2}(i+\tfrac{1}{2},j+\tfrac{1}{2},k)\left(1+\frac{\sigma_x\Delta t}{2\varepsilon}\right)=\left(1-\frac{\sigma_x\Delta t}{2\varepsilon}\right)H_{zx}^{n-1/2}(i+\tfrac{1}{2},j+\tfrac{1}{2},k)$$
$$-\frac{\Delta t}{\mu\Delta x}\left[E_{yz}^{n}(i+1,j+\tfrac{1}{2},k)-E_{yz}^{n}(i,j+\tfrac{1}{2},k)\right]$$
$$-\frac{\Delta t}{\mu\Delta x}\left[E_{yx}^{n}(i+1,j+\tfrac{1}{2},k)-E_{yx}^{n}(i,j+\tfrac{1}{2},k)\right] \tag{4.52}$$

$$H_{zy}^{n+1/2}(i+\tfrac{1}{2},j+\tfrac{1}{2},k)\left(1+\frac{\sigma_y\Delta t}{2\varepsilon}\right)=\left(1-\frac{\sigma_y\Delta t}{2\varepsilon}\right)H_{zy}^{n-1/2}(i+\tfrac{1}{2},j+\tfrac{1}{2},k)$$
$$+\frac{\Delta t}{\mu\Delta y}\left[E_{xy}^{n}(i+\tfrac{1}{2},j+1,k)-E_{xy}^{n}(i+\tfrac{1}{2},j,k)\right]$$
$$+\frac{\Delta t}{\mu\Delta y}\left[E_{xz}^{n}(i+\tfrac{1}{2},j+1,k)-E_{xz}^{n}(i+\tfrac{1}{2},j,k)\right] \tag{4.53}$$

$$E_{xy}^{n+1}(i+\tfrac{1}{2},j,k)\left(1+\frac{\sigma_y\Delta t}{2\varepsilon}\right)=\left(1-\frac{\sigma_y\Delta t}{2\varepsilon}\right)E_{xy}^{n}(i+\tfrac{1}{2},j,k)$$
$$+\frac{\Delta t}{\varepsilon\Delta y}\left[H_{zx}^{n+1/2}(i+\tfrac{1}{2},j+\tfrac{1}{2},k)-H_{zx}^{n+1/2}(i+\tfrac{1}{2},j-\tfrac{1}{2},k)\right]$$
$$+\frac{\Delta t}{\varepsilon\Delta y}\left[H_{zy}^{n+1/2}(i+\tfrac{1}{2},j+\tfrac{1}{2},k)-H_{zy}^{n+1/2}(i+\tfrac{1}{2},j-\tfrac{1}{2},k)\right] \tag{4.54}$$

$$E_{xz}^{n+1}(i+\tfrac{1}{2},j,k)\left(1+\frac{\sigma_z\Delta t}{2\varepsilon}\right)=\left(1-\frac{\sigma_z\Delta t}{2\varepsilon}\right)E_{xz}^{n}(i+\tfrac{1}{2},j,k)$$
$$-\frac{\Delta t}{\varepsilon\Delta z}\left[H_{yx}^{n+1/2}(i+\tfrac{1}{2},j,k+\tfrac{1}{2})-H_{yx}^{n+1/2}(i+\tfrac{1}{2},j,k-\tfrac{1}{2})\right]$$
$$-\frac{\Delta t}{\varepsilon\Delta z}\left[H_{yz}^{n+1/2}(i+\tfrac{1}{2},j,k+\tfrac{1}{2})-H_{yz}^{n+1/2}(i+\tfrac{1}{2},j,k-\tfrac{1}{2})\right] \tag{4.55}$$

$$E_{yz}^{n+1}(i,j+\tfrac{1}{2},k)\left(1+\frac{\sigma_z\Delta t}{2\varepsilon}\right)=\left(1-\frac{\sigma_z\Delta t}{2\varepsilon}\right)E_{yz}^{n}(i,j+\tfrac{1}{2},k)$$
$$+\frac{\Delta t}{\varepsilon\Delta z}\left[H_{xy}^{n+1/2}(i,j+\tfrac{1}{2},k+\tfrac{1}{2})-H_{xy}^{n+1/2}(i,j+\tfrac{1}{2},k-\tfrac{1}{2})\right]$$
$$+\frac{\Delta t}{\varepsilon\Delta z}\left[H_{xz}^{n+1/2}(i,j+\tfrac{1}{2},k+\tfrac{1}{2})-H_{xz}^{n+1/2}(i,j+\tfrac{1}{2},k-\tfrac{1}{2})\right] \tag{4.56}$$

$$E_{yx}^{n+1}(i, j+\tfrac{1}{2}, k)\left(1+\frac{\sigma_x \Delta t}{2\varepsilon}\right) = \left(1-\frac{\sigma_x \Delta t}{2\varepsilon}\right) E_{yx}^{n}(i, j+\tfrac{1}{2}, k)$$
$$-\frac{\Delta t}{\varepsilon \Delta x}\left[H_{zx}^{n+1/2}(i+\tfrac{1}{2}, j+\tfrac{1}{2}, k) - H_{zx}^{n+1/2}(i-\tfrac{1}{2}, j+\tfrac{1}{2}, k)\right]$$
$$-\frac{\Delta t}{\varepsilon \Delta x}\left[H_{zy}^{n+1/2}(i+\tfrac{1}{2}, j+\tfrac{1}{2}, k) - H_{zy}^{n+1/2}(i-\tfrac{1}{2}, j+\tfrac{1}{2}, k)\right] \tag{4.57}$$

$$E_{zx}^{n+1}(i, j, k+\tfrac{1}{2})\left(1+\frac{\sigma_x \Delta t}{2\varepsilon}\right) = \left(1-\frac{\sigma_x \Delta t}{2\varepsilon}\right) E_{zx}^{n}(i, j, k+\tfrac{1}{2})$$
$$+\frac{\Delta t}{\varepsilon \Delta x}\left[H_{yx}^{n+1/2}(i+\tfrac{1}{2}, j, k+\tfrac{1}{2}) - H_{yx}^{n+1/2}(i-\tfrac{1}{2}, j, k+\tfrac{1}{2})\right]$$
$$+\frac{\Delta t}{\varepsilon \Delta x}\left[H_{yz}^{n+1/2}(i+\tfrac{1}{2}, j, k+\tfrac{1}{2}) - H_{yz}^{n+1/2}(i-\tfrac{1}{2}, j, k+\tfrac{1}{2})\right] \tag{4.58}$$

$$E_{zy}^{n+1}(i, j, k+\tfrac{1}{2})\left(1+\frac{\sigma_y \Delta t}{2\varepsilon}\right) = \left(1-\frac{\sigma_y \Delta t}{2\varepsilon}\right) E_{zy}^{n}(i, j, k+\tfrac{1}{2})$$
$$-\frac{\Delta t}{\varepsilon \Delta y}\left[H_{xy}^{n+1/2}(i, j+\tfrac{1}{2}, k+\tfrac{1}{2}) - H_{xy}^{n+1/2}(i, j-\tfrac{1}{2}, k+\tfrac{1}{2})\right]$$
$$-\frac{\Delta t}{\varepsilon \Delta y}\left[H_{xz}^{n+1/2}(i, j+\tfrac{1}{2}, k+\tfrac{1}{2}) - H_{xz}^{n+1/2}(i, j-\tfrac{1}{2}, k+\tfrac{1}{2})\right]. \tag{4.59}$$

4.2.2 FDTD Anisotropic-Medium PML

The split-field or stretched-coordinate PML is based on a mathematical as opposed to a physical medium description. However, a PML based on a physical model can be formulated in terms of a dispersive and anisotropic medium, with a performance that is similar to that of the mathematical model of PMLs. This approach was discussed in a finite element setting in Section 3.2.2, and applies similarly using the FDTD formulation. The uniaxial PML (UPML [32]) is based on an anisotropic medium formulation and is widely used in FDTD applications and commercial software. For the FDTD implementation of the UPML, we begin with Faraday's and Ampère's laws in time-harmonic form,

$$\nabla \times \mathbf{E} = -j\omega\mu \overleftrightarrow{\mathbf{\Lambda}} \cdot \mathbf{H} \tag{4.60}$$

$$\nabla \times \mathbf{H} = j\omega\varepsilon \overleftrightarrow{\mathbf{\Lambda}} \cdot \mathbf{E} \tag{4.61}$$

where $\overleftrightarrow{\mathbf{\Lambda}}$ is given by (3.36). For an efficient FDTD implementation, we use the constitutive relationships

$$B_x = \mu \frac{s_z}{s_x} H_x; \quad B_y = \mu \frac{s_x}{s_y} H_y; \quad B_z = \mu \frac{s_y}{s_z} H_z \tag{4.62}$$

$$D_x = \varepsilon \frac{s_z}{s_x} E_x; \quad D_y = \varepsilon \frac{s_x}{s_y} E_y; \quad D_z = \varepsilon \frac{s_y}{s_z} E_z \tag{4.63}$$

and further generalize s_x, s_y, and s_z such that

$$s_x = \kappa_x - j\frac{\sigma_x}{\omega\,\varepsilon}; \quad s_y = \kappa_y - j\frac{\sigma_y}{\omega\,\varepsilon}; \quad s_z = \kappa_z - j\frac{\sigma_z}{\omega\,\varepsilon} \tag{4.64}$$

where $\kappa_x = \kappa_x(x)$, $\kappa_y = \kappa_y(y)$, and $\kappa_z = \kappa_z(z)$. A representative spatial profile for κ_ξ is $\kappa_\xi = 1 + (\kappa_{\max,\xi} - 1)(\xi/L)^p$, where $\xi = x, y, z$, and a typical spatial profile for σ_ξ is $\sigma_\xi(\xi) = \sigma_{\max}(\xi/L)^p$, where $1 \le p \le 4$ and L denotes the total thickness of the PML. It has been found that $p = 3$ is a good overall choice for the FDTD uniaxial PML.

By using (4.62) in (4.60) and (3.36), and (4.63) in (4.61) and (3.36), and then inverse Fourier transforming to the time domain, we obtain

$$\overleftrightarrow{\mathbf{\Lambda}}_\kappa \cdot \frac{\partial \mathbf{B}}{\partial t} = -\nabla \times \mathbf{E} - \frac{1}{\varepsilon}\overleftrightarrow{\mathbf{\Lambda}}_\sigma \cdot \mathbf{B} \tag{4.65}$$

$$\overleftrightarrow{\mathbf{\Lambda}}_\kappa \cdot \frac{\partial \mathbf{D}}{\partial t} = \nabla \times \mathbf{H} - \frac{1}{\varepsilon}\overleftrightarrow{\mathbf{\Lambda}}_\sigma \cdot \mathbf{D} \tag{4.66}$$

where

$$\overleftrightarrow{\mathbf{\Lambda}}_\kappa \equiv \begin{bmatrix} \kappa_y & 0 & 0 \\ 0 & \kappa_z & 0 \\ 0 & 0 & \kappa_x \end{bmatrix}; \quad \overleftrightarrow{\mathbf{\Lambda}}_\sigma \equiv \begin{bmatrix} \sigma_y & 0 & 0 \\ 0 & \sigma_z & 0 \\ 0 & 0 & \sigma_x \end{bmatrix}. \tag{4.67}$$

The Yee algorithm can be applied to (4.65) and (4.66), and advancement equations for **B** and **D** similar to (4.9)–(4.14) are obtained. However, the constitutive relationships (4.62) and (4.63) are additionally required to properly relate **B** with **H** and **D** with **E**. This is easily accomplished by using (4.64) in (4.62) and (4.63) and then inverse Fourier transforming the results into the time domain. For example, $s_x B_x = \mu s_z H_x$ has the time-dependent form

$$\mu\kappa_z \frac{\partial H_x}{\partial t} + \mu\frac{\sigma_z}{\varepsilon}H_x = \kappa_x\frac{\partial B_x}{\partial t} + \frac{\sigma_x}{\varepsilon}B_x \tag{4.68}$$

and the corresponding finite-difference relationship on a Yee lattice is

$$\begin{aligned} H_x^{n+1/2}(i, j+\tfrac{1}{2}, k+\tfrac{1}{2})\left(\kappa_z + \frac{\sigma_z\Delta t}{2\varepsilon}\right) &= \left(\kappa_z - \frac{\sigma_z\Delta t}{2\varepsilon}\right) H_x^{n-1/2}(i, j+\tfrac{1}{2}, k+\tfrac{1}{2}) \\ &+ B_x^{n+1/2}(i, j+\tfrac{1}{2}, k+\tfrac{1}{2})\frac{1}{\mu}\left(\kappa_x + \frac{\sigma_x\Delta t}{2\varepsilon}\right) \\ &- B_x^{n-1/2}(i, j+\tfrac{1}{2}, k+\tfrac{1}{2})\frac{1}{\mu}\left(\kappa_x - \frac{\sigma_x\Delta t}{2\varepsilon}\right). \end{aligned} \tag{4.69}$$

Finite-difference equations for the other components in (4.62) and (4.63) are obtained similarly. The discrete constitutive relationships in conjunction with the finite-difference representation of (4.65) and (4.66) form a self-consistent system of equations to advance the fields within the PML region of the FDTD grid. In fact, for an FDTD computational volume with lossless materials outside the PML region, this system of equations can be used for the entire volume simply by defining $\overleftrightarrow{\Lambda} = \overleftrightarrow{I}$ outside the PML region, where $\overleftrightarrow{I}$ denotes the identity dyadic.

Within the FDTD setting, the split-field PML, as well as the UPML, are known to suffer from additional inaccuracies that are associated with low frequency as well as evanescent fields. This has led to the investigation of alternative PML representations, many of which were described in Section 3.3.5. An additional PML formulation that also shows promise for improving evanescent-field performance is the convolutional PML (CPML) [33]. The CPML is related to the CFS PML discussed in Section 3.3.5 and is based on the use of the recursive convolution technique to accelerate the convolution integrals inherent to this method, where the convolutions result from the inverse Fourier transformation of (4.19) and (4.20), specifically the product of the reciprocals of s_x, s_y, and s_z with the fields. The CPML has been found to provide on the order of 20-dB lower reflections on the FDTD grid compared to the split-field and UPML techniques traditionally used for FDTD applications for problems involving evanescent fields.

Although there are currently a variety of formulations and implementations of the PML available to FDTD codes, the PML concept has proven to be a robust and accurate grid termination for open-region scattering and radiation applications. Because certain formulations may permit the PML to be placed closer to a scatterer than is possible with other approximate formulations, examining the convergence of the solution with different placements of the PML, as well as different PML thicknesses, is recommended.

4.3 NEAR-TO-FAR-FIELD TRANSFORMATION IN FDTD

The finite element and finite-difference formulations for Maxwell's equations typically provide near-field data. Although it is possible to extend the computational volume into the far-field region, the computational resources required will often be prohibitive. However, far-field data are extremely important for the analysis and design of antennas and arrays since fundamental quantities, such as gain, require the knowledge of far-field performance.

The FDTD near-to-far-field (NTF) techniques are related directly to the similar NTF techniques for the finite element method described in Section 5.3. However, there are some differences due primarily to the offset nature of the electric and magnetic fields in the Yee stencil, as well as the use of finite differencing instead of basis functions. Consequently, an accurate FDTD NTF formulation based on the technique described in Ref. 34 is reviewed in this section. The basic strategy of this method is to apply the NTF to two distinct equivalent surfaces that naturally follow the offset of the Yee discretization.

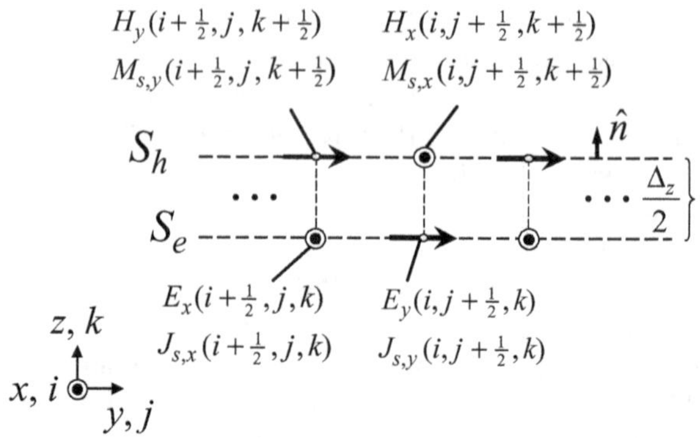

Figure 4.2 FDTD near-to-far-field transformation technique based on offset surfaces for the electric and magnetic fields [34].

The two surfaces S_h and S_e along with their tangential field components are shown in Figure 4.2. On the surface S_e the equivalent electric current is given by

$$\mathbf{J}_s|_{S_e} = \hat{n} \times \mathbf{H}|_{S_h} \tag{4.70}$$

and on the surface S_h the equivalent magnetic current is given by

$$\mathbf{M}_s|_{S_h} = -\,\hat{n} \times \mathbf{E}|_{S_e}. \tag{4.71}$$

Following Ref. 34, the motivation for using two offset equivalent surfaces is rooted in the Huygens' surface concept for launching plane waves into the total-field region of the FDTD computational volume. This can be seen by rewriting (4.12) for the x-component of the electric field in a lossless medium:

$$\begin{aligned} &E_x^{n+1}\left(i+\tfrac{1}{2},j,k\right) \\ &\quad= E_x^{n}\left(i+\tfrac{1}{2},j,k\right) + \frac{\Delta t}{\varepsilon\Delta y}\left[H_z^{n+1/2}\left(i+\tfrac{1}{2},j+\tfrac{1}{2},k\right) - H_z^{n+1/2}\left(i+\tfrac{1}{2},j-\tfrac{1}{2},k\right)\right] \\ &\quad\quad- \frac{\Delta t}{\varepsilon\Delta z}\left[H_y^{n+1/2}\left(i+\tfrac{1}{2},j,k+\tfrac{1}{2}\right) - H_y^{n+1/2}\left(i+\tfrac{1}{2},j,k-\tfrac{1}{2}\right)\right] \\ &\quad\quad- \frac{\Delta t}{\varepsilon} J_{\text{imp},x}^{n+1/2}\left(i+\tfrac{1}{2},j,k\right) \end{aligned} \tag{4.72}$$

where the incident magnetic field can be written in terms of the surface electric current as

$$J_{\text{imp},x}^{n+1/2}\left(i+\tfrac{1}{2},j,k\right) = H_y^{\text{inc},n+1/2}\left(i+\tfrac{1}{2},j,k+\tfrac{1}{2}\right). \tag{4.73}$$

The important observation is that although the impressed surface electric current is collocated with the electric field $E_x(i+\frac{1}{2}, j, k)$, it is dependent on the specified incident magnetic field shifted a half-cell in the z-direction. Similarly, the incident electric field is shifted a half-cell relative to the impressed surface magnetic current. The placement of the current sources is an important aspect in an FDTD total- and scattered-field decomposition technique so that only scattered fields exist outside the equivalent source surfaces.

This same concept is applied in the construction of an NTF transformation. In the time-harmonic case, the far electric and magnetic fields are given in terms of the electric and magnetic vector potentials by

$$E_\theta(\mathbf{r}, \omega) = -\frac{j\omega}{4\pi r}\left(A_\theta + \sqrt{\frac{\mu_0}{\varepsilon_0}}F_\phi\right) \tag{4.74}$$

$$E_\phi(\mathbf{r}, \omega) = -\frac{j\omega}{4\pi r}\left(A_\phi - \sqrt{\frac{\mu_0}{\varepsilon_0}}F_\theta\right) \tag{4.75}$$

and in time-dependent form by

$$E_\theta(\mathbf{r}, t) = -\frac{1}{4\pi r}\left(\frac{\partial A_\theta}{\partial t} + \sqrt{\frac{\mu_0}{\varepsilon_0}}\frac{\partial F_\phi}{\partial t}\right) \tag{4.76}$$

$$E_\phi(\mathbf{r}, t) = -\frac{1}{4\pi r}\left(\frac{\partial A_\phi}{\partial t} - \sqrt{\frac{\mu_0}{\varepsilon_0}}\frac{\partial F_\theta}{\partial t}\right). \tag{4.77}$$

The vector potentials specialized to the offset equivalent surfaces S_h and S_e are given in time-harmonic form by

$$\mathbf{A}(\mathbf{r}, \omega) = \mu_0 e^{\;jk_0 r} \oiint_{S_e} \hat{n} \times \mathbf{H}|_{S_h}\, e^{jk_0\hat{\mathbf{r}}\;\mathbf{r}'_e}\, dS' \tag{4.78}$$

$$\mathbf{F}(\mathbf{r}, \omega) = -\varepsilon_0 e^{-jk_0 r} \oiint_{S_h} \hat{n} \times \mathbf{E}|_{S_e}\, e^{jk_0\hat{\mathbf{r}}\cdot\mathbf{r}'_h}\, dS' \tag{4.79}$$

and in time-dependent form by

$$\mathbf{A}(\mathbf{r}, t) = \mu_0 \oiint_{S_e} \hat{n} \times \mathbf{H}|_{S_h}\left(t - \left\{\frac{r - \hat{\mathbf{r}}\cdot\mathbf{r}'_e}{c}\right\}\right) dS' \tag{4.80}$$

$$\mathbf{F}(\mathbf{r}, t) = -\varepsilon_0 \oiint_{S_h} \hat{n} \times \mathbf{E}|_{S_e}\left(t - \left\{\frac{r - \hat{\mathbf{r}}\cdot\mathbf{r}'_h}{c}\right\}\right) dS'. \tag{4.81}$$

Computing the far fields directly in the time domain is useful for obtaining broadband information, such as frequency-swept antenna gain, at a few observation angles. An efficient time-domain implementation procedure is based on shifting the retarded time indices off the fields in (4.80) and (4.81) and onto the potentials [35], or in other words, from the terms on the right-hand sides of (4.80) and (4.81) to the potentials on the left-hand sides. For example, with reference to (4.81), consider the contribution to the F_z component by the tangential field E_x^n over one discrete face of the surface S_e, where the surface area of this face is defined by $\Delta x\,\Delta z$. Specifically, at time step n and assuming that the fields are constant over this cell face, we can write

$$\Delta F_z(\mathbf{r}, n\Delta t + (r - \hat{\mathbf{r}}\cdot\mathbf{r}'_h)/c) = \varepsilon_0\,\Delta x\,\Delta z\,E_x^n. \tag{4.82}$$

The fields from the other discrete elemental faces of S_e contribute similarly to the total vector potential. For an efficient computer implementation, it is convenient to store the potentials in the form of arrays with an integer indexing scheme. However, two issues need to be addressed to accomplish this. First, for a given observation angle and source location (i.e., $\hat{\mathbf{r}}\cdot\mathbf{r}'_h$), the argument of the electric potential will typically not be an integral multiple of Δt, and second, $\hat{\mathbf{r}}\cdot\mathbf{r}'_h$ will generally lead to multiple contributions with the same retarded time index. The first issue can be resolved by taking the integer part of $n\Delta t + (r - \hat{\mathbf{r}}\cdot\mathbf{r}'_h)/c$ such that

$$m = \text{int}\left\{t_c/\Delta t + \tfrac{1}{2}\right\} \tag{4.83}$$

where $t_c = n\Delta t + (r - \hat{\mathbf{r}}\cdot\mathbf{r}'_h)/c$. The second issue is resolved by interpolating and apportioning the field contributions across several integral time indices associated with the vector potentials. For example, on a surface with a $\hat{y}$ outward normal and using a linear interpolation, E_x^n will contribute across three time indices for the vector potential such that

$$\begin{aligned}
\Delta F_z^{m-1} &= \left(\frac{1}{2} - \frac{t_c}{\Delta t} + m\right)\varepsilon_0\,\Delta x\,\Delta z\,E_x^n\\
\Delta F_z^{m} &= 2\left(\frac{t_c}{\Delta t} - m\right)\varepsilon_0\,\Delta x\,\Delta z\,E_x^n\\
\Delta F_z^{m+1} &= -\left(\frac{1}{2} + \frac{t_c}{\Delta t} - m\right)\varepsilon_0\,\Delta x\,\Delta z\,E_x^n.
\end{aligned} \tag{4.84}$$

Note that the contributions will accumulate in the various array registers depending on the source and observation locations as well as the particular time step. Higher-order interpolations can be used, although this is not necessary. The additional components of the vector potentials (4.81) and (4.80) are evaluated similarly, noting that an additional factor of 1/2 is included in (4.83) for the magnetic field evaluation in (4.80) due to the time offset between the electric and magnetic fields. In this way the vector potentials are built during the course of the time-marching simulation, and the far fields are then constructed with a simple postprocessing procedure following (4.76) and (4.77).

The time-dependent NTF described above can be used to compute frequency-dependent patterns from an antenna; however, to do so requires generating the full time history at numerous observation angles and then extracting the desired frequencies from the Fourier transform for each angle of interest. A much more efficient procedure for computing the patterns at a few frequencies, yet with numerous observation angles, is to alternatively build up the Fourier transformation at specific frequencies inline with the time-marching process. For example, the discrete Fourier transform (DFT) for the tangential electric and magnetic fields on the equivalent surfaces S_e and S_h is efficiently computed directly from the calculated transient fields by

$$\hat{n} \times \mathbf{E}(\mathbf{r}, \omega)|_{S_e} = \sum_{n=1}^{N} \hat{n} \times \mathbf{E}(\mathbf{r}, n\Delta t)|_{S_e} e^{j\omega n \Delta t}$$

$$\hat{n} \times \mathbf{H}(\mathbf{r}, \omega)|_{S_h} = \sum_{n=1}^{N} \hat{n} \times \mathbf{H}\left(\mathbf{r}, \left(n + \tfrac{1}{2}\right)\Delta t\right)\Big|_{S_h} e^{j\omega\left(n+\frac{1}{2}\right)\Delta t} \tag{4.85}$$

where ω defines the angular frequency for the pattern computation desired, and the summations are simply accumulated with each time iteration over the duration of the simulation, which is taken to be a total of N time steps in (4.85). The vector potentials are then computed directly in the frequency domain from (4.78) and (4.79), and the patterns from (4.74) and (4.75).

4.4 ALTERNATIVE FETD FORMULATION

The FETD method described in Chapter 2 was formulated based on the second-order wave equation for the electric field, where the electric field was represented by vector edge basis functions. When formulated in this way, use of the Newmark-beta time-integration technique leads to an unconditionally stable time-marching scheme. An alternative FETD formulation [21] is based directly on the two first-order Maxwell's equations: Ampère's and Faraday's laws. The resulting time-marching algorithm can be considered as a generalization of the FDTD method to unstructured grids. Because this formulation treats both the electric and magnetic fields within a finite element setting, vector basis functions are now required for each type of field. The location of the vector fields will be defined so that the electric field resides on the edges of the finite elements, whereas the magnetic field will be defined on the faces, or facets, of the finite elements. Hence, the traditional vector edge basis functions described in Chapter 2 are again used for the electric fields, whereas a new vector basis function is introduced for the magnetic fields. The new vector basis functions will be defined so that the normal component of the magnetic (**B**) field will be continuous across neighboring finite elements, and such basis functions are known as *vector facet basis functions*.

When the traditional FDTD leapfrogging time-marching technique is applied to this alternative FETD representation, a direct analogy between this FETD formulation and the traditional FDTD method will be apparent. Further, and as shown in Section 4.5, when the cells of the computational mesh are defined to be rectangular brick elements and trapezoidal integration is used in the construction of the elemental matrices, the FETD algorithm can reduce *identically* to the FDTD algorithm. This holds whether the FETD is formulated using the traditional wave-equation representation based only on vector edge basis functions, or using the one based on the first-order Maxwell's equations with the vector edge and vector facet basis functions to be described in this section. We will see that the possible equivalence of all these methods will be critical to the construction of a numerically stable hybridization scheme that combines FETD and FDTD.

To formulate an FETD algorithm that is based on the first-order Maxwell's equations, we begin with Maxwell's equations in the time domain that are defined throughout a linear, isotropic, lossless ($\sigma_e = \sigma_m = 0$) volume V with boundary S_o by

$$\frac{\partial}{\partial t}\mathbf{B} = -\nabla \times \mathbf{E} - \mathbf{M}_{\text{imp}} \tag{4.86}$$

$$\varepsilon\frac{\partial}{\partial t}\mathbf{E} = \nabla \times \frac{1}{\mu}\mathbf{B} - \mathbf{J}_{\text{imp}} \tag{4.87}$$

where $\mathbf{E}$ and $\mathbf{B}$ denote the time-dependent electric and magnetic fields, respectively. For simplicity, a Dirichlet boundary condition on S_o is assumed:

$$\hat{n} \times \mathbf{E}(\mathbf{r}, t) = \gamma(\mathbf{r}, t) \qquad \mathbf{r} \in S_o \tag{4.88}$$

along with the initial conditions

$$\mathbf{B}(\mathbf{r}, 0) = \mathbf{B}_0(\mathbf{r}) \qquad \mathbf{r} \in V \tag{4.89}$$

$$\mathbf{E}(\mathbf{r}, 0) = \mathbf{E}_0(\mathbf{r}) \qquad \mathbf{r} \in V. \tag{4.90}$$

To construct a finite element representation of (4.86)–(4.90), we seek a weak solution such that $\mathbf{E} \in W$, $\mathbf{B} \in W_f$, where W denotes the space of curl-conforming functions defined by $W = H(\text{curl};V) = \{\mathbf{N} \in (L^2(V))^3, \nabla \times \mathbf{N} \in (L^2(V))^3\}$, and W_f denotes the space of divergence-conforming functions defined by $W_f = H(\text{div};V) = \{\mathbf{N}^f \in (L^2(V))^3, \nabla \cdot \mathbf{N}^f \in (L^2(V))^3\}$ [36,37]. The edge (curl-conforming) basis functions provide for tangential continuity between neighboring finite elements, whereas facet (divergence-conforming) basis functions provide for normal continuity. Note that the edge basis functions defined here are identical to those used in previous chapters. The weak-form representations for (4.86) and (4.87) are given by

$$\iiint_V \left\{(\nabla \times \mathbf{E}) \cdot \left(\frac{1}{\mu}\mathbf{N}^f\right) + \mathbf{N}^f \cdot \frac{\partial}{\partial t}\left(\frac{1}{\mu}\mathbf{B}\right)\right\} dV + \iiint_V \mathbf{N}^f \cdot \left(\frac{1}{\mu}\mathbf{M}_{\text{imp}}\right) dV = 0 \qquad \forall \mathbf{N}^f \in W_f \tag{4.91}$$

$$\iiint_V \left\{-(\nabla \times \mathbf{N}) \cdot \left(\frac{1}{\mu}\mathbf{B}\right) + \varepsilon\mathbf{N} \cdot \frac{\partial}{\partial t}\mathbf{E}\right\} dV + \iiint_V \mathbf{N} \cdot \mathbf{J}_{\text{imp}}\, dV = 0 \qquad \forall \mathbf{N} \in W \tag{4.92}$$

along with (4.88)–(4.90) to complete the boundary-value problem. The system of equations defined by (4.91) and (4.92) can be viewed as a variational analog of the FDTD method.

The semidiscrete representation for (4.91) and (4.92) is obtained by expanding $\mathbf{B}$ in terms of time-dependent coefficients B_i and spatial facet basis functions $\mathbf{N}_i^f$ as

$$\mathbf{B}(\mathbf{r},t) = \sum_i B_i(t)\mathbf{N}_i^f(\mathbf{r}) \tag{4.93}$$

where $i = 1, 2, \ldots, N_f$. Similarly, the expansion for $\mathbf{E}$ in terms of spatial edge basis functions $\mathbf{N}_i$ is given by

$$\mathbf{E}(\mathbf{r},t) = \sum_i E_i(t)\,\mathbf{N}_i(\mathbf{r}) \tag{4.94}$$

where $i = 1, 2, \ldots, N$. Note that these expansions correspond to a Galerkin formulation for (4.91) and (4.92) due to the equivalence of the basis and testing functions.

Substituting (4.93) and (4.94) into (4.91) and (4.92) leads to the semidiscrete system of equations

$$[T^f]\frac{d\{B\}}{dt} + [A]\{E\} = -\{g\} \tag{4.95}$$

$$[T]\frac{d\{E\}}{dt} - [A]^{\mathrm{T}}\{B\} = \{f\} \tag{4.96}$$

with $\{E\} = [E_1(t), E_2(t), \ldots, E_N(t)]^{\mathrm{T}}$ and $\{B\} = \left[B_1(t), B_2(t), \ldots, B_{N_f}(t)\right]^{\mathrm{T}}$, where the superscript T denotes the transpose, and

$$T_{ij} = \varepsilon \iiint_V \mathbf{N}_i \cdot \mathbf{N}_j \, dV \tag{4.97}$$

$$T_{ij}^f = \iiint_V \frac{1}{\mu}\mathbf{N}_i^f \cdot \mathbf{N}_j^f \, dV \tag{4.98}$$

$$A_{ij} = \iiint_V \frac{1}{\mu}\mathbf{N}_i^f \cdot (\nabla \times \mathbf{N}_j)\, dV \tag{4.99}$$

$$f_i = -\iiint_V \mathbf{N}_i \cdot \mathbf{J}_{\mathrm{imp}} \, dV \tag{4.100}$$

$$g_i = \iiint_V \mathbf{N}_i^f \cdot \left(\frac{1}{\mu}\mathbf{M}_{\mathrm{imp}}\right) dV. \tag{4.101}$$

Note that the matrix $[T]$ is of dimension $N \times N$, $[T^f]$ is of dimension $N_f \times N_f$, and $[A]$ is of dimension $N_f \times N$. Also, $\{H\}$ and $\{B\}$ are related by

$$\{H\} = [T^f]\{B\}. \tag{4.102}$$

Well-defined relationships exist between the vector edge and vector facet basis functions [37]. Because of these fundamental properties, (4.95) can be rewritten conveniently as

$$\frac{d}{dt}\{B\} = -[C]\{E\} - [T^f]^{-1}\{g\} \tag{4.103}$$

where $[C] \equiv [T^f]^{-1}[A]$ is a sparse rectangular (incidence) matrix with nonzero entries consisting simply of ± 1 [38]. Consequently, the magnetic field becomes related to the circulation of the electric field around the cell faces, and a direct analogy is found with the FDTD representation of Faraday's law.

The fully discrete representation of (4.95) and (4.96) is obtained through the definition of a suitable time-integration scheme. If a traditional leapfrog method is adopted through the use of central differences, a conditionally stable scheme results. In this case the fully discrete system is given by

$$\{B\}^{n+1/2} = \{B\}^{n-1/2} - \Delta t\left([C]\{E\}^n + [T^f]^{-1}\,\{g\}|_{t=n\Delta t}\right) \tag{4.104}$$

$$\{E\}^{n+1} = \{E\}^n + \Delta t[T]^{-1}\left([A]^T\{B\}^{n+1/2} + \{f\}|_{t=(n+1/2)\Delta t}\right). \tag{4.105}$$

Equations (4.104) and (4.105) represent a compact description of a finite element formulation to time advance both the electric and magnetic fields on a three-dimensional mesh.

In the next section we show that the elemental matrices associated with the finite element formulation for rectangular brick elements can be reduced to a representation that is equivalent to a finite-difference formulation. In this case, (4.104) and (4.105) are equivalent to a wave-equation formulation based on central time differencing, which in turn is equivalent to the FDTD method.

4.5 EQUIVALENCE BETWEEN FETD AND FDTD

Explicit time-stepping algorithms, such as the traditional FDTD method, are appealing because the field advancement to the next time step requires only the simple inversion of a diagonal matrix. On the other hand, implicit methods require the inversion of matrices that are nondiagonal. Although this adds to the computational complexity, implicit methods can give rise to an unconditionally stable time-stepping algorithm, which is something that is not theoretically possible [39] with explicit schemes. This is easily seen since a large time step could permit a wavefront to interact simultaneously with numerous spatial cells, and hence this potential multi-cell interaction must be accounted for within the governing matrices.

The elemental matrices associated with a finite element analysis are of small dimension and are constructed by forming the reaction integrals of the vector basis functions within a given discrete cell. In the standard finite element method, these integrals are evaluated either by exact analytical integration or by Gaussian quadrature, which usually gives rise to densely populated elemental matrices [40]. The assembly of these local elemental matrices into the global finite element matrices will lead to global matrices that are highly sparse, although not diagonal. However, it is well known that by alternatively using a trapezoidal integration process in the construction of the elemental matrices instead of an exact integration process, an inherently implicit finite element algorithm can be reduced to an explicit finite-difference scheme [39]. Indeed, in the field of acoustics, this concept was applied over 30 years ago to construct a provably stable explicit–implicit transient numerical algorithm [41].

In a broader sense, the process of diagonalizing the inherently nondiagonal finite element system matrix is known as *lumping* [42,43], with the goal of being able to obtain an explicit algorithm. Here, the off-diagonal matrix entries are either combined or discarded, with the final result being included, or lumped, into the diagonal entries. Depending on the technique used and the elements it is applied to, this process can have undesirable consequences, such as the creation of singular matrices or the corruption of the eigenvalue spectrum [22]. Although the undesirable aspects of lumping often occur when applied to the more general elemental shapes, such as tetrahedral elements, good results have been obtained when lumping is applied to rectangular hexahedral, or brick, elements [43]. In fact, lumping techniques have been used to establish an equivalence between vector edge–based finite element formulations on brick elements and the traditional FDTD algorithm [21,38]. Trapezoidal integration can, in a sense, be considered to be a lumping technique; however, it is more general and systematic in nature and provides a mechanism to recover the FDTD method readily from the edge-based FETD method when applied to rectangular brick-shaped finite elements. We now examine in more detail the use of trapezoidal integration in establishing an equivalence between these two techniques.

In Chapter 2 it was shown that the traditional finite element formulation gives rise to matrices with elements given by (2.28)–(2.30). For the purposes of this section, only the so-called "mass" matrix $[T]$ and "stiffness" matrix $[S]$ will be considered, and in the case of a free-space environment, the matrix elements are described by

$$T_{ij} = \varepsilon_0 \iiint_V \mathbf{N}_i \cdot \mathbf{N}_j \, dV \tag{4.106}$$

$$S_{ij} = \frac{1}{\mu_0} \iiint_V (\nabla \times \mathbf{N}_i) \cdot (\nabla \times \mathbf{N}_j) \, dV. \tag{4.107}$$

For definiteness, we consider the brick-shaped element shown in Figure 4.3. There are 12 first-order vector edge basis functions $\mathbf{N}_i$, $i = 1, \ldots, 12$ for this element, and

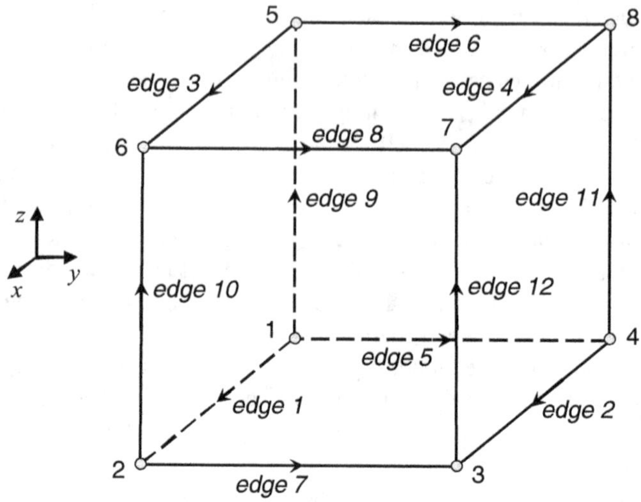

Figure 4.3 Reference rectangular hexahedral (brick) element used to construct elemental matrices [40]. Edge and node numbers are shown in the figure.

the resulting elemental matrices are of dimension 12 × 12 [40]. For a cubical brick element centered at the origin of a Cartesian space and with edge length Δ, the four basis functions associated with the x-directed edges are given by

$$
\begin{aligned}
N_{x1}^{e} &= \frac{1}{\Delta^2}\left(\frac{\Delta}{2} - y\right)\left(\frac{\Delta}{2} - z\right) \\
N_{x2}^{e} &= \frac{1}{\Delta^2}\left(y + \frac{\Delta}{2}\right)\left(\frac{\Delta}{2} - z\right) \\
N_{x3}^{e} &= \frac{1}{\Delta^2}\left(\frac{\Delta}{2} - y\right)\left(z + \frac{\Delta}{2}\right) \\
N_{x4}^{e} &= \frac{1}{\Delta^2}\left(y + \frac{\Delta}{2}\right)\left(z + \frac{\Delta}{2}\right)
\end{aligned}
\tag{4.108}
$$

with similar definitions for the y- and z-directed edges. By substituting (4.108) into (4.106) and evaluating the integrals exactly, we obtain the following matrix for the xx-portion of the full elemental matrix:

$$
[T_{xx}^{e}] = \varepsilon_0 \frac{\Delta^3}{36}\begin{bmatrix} 4 & 2 & 2 & 1 \\ 2 & 4 & 1 & 2 \\ 2 & 1 & 4 & 2 \\ 1 & 2 & 2 & 4 \end{bmatrix} \tag{4.109}
$$

where the notation $[T^e_{xx}]$ corresponds to the xx block of a single element's contribution to $[T]$. Clearly, an elemental matrix with fully populated subblocks will lead to a nondiagonal mass matrix $[T]$ and consequently, an implicit time-stepping algorithm.

We now consider, alternatively, the evaluation of (4.106) by trapezoidal integration. In one dimension, integrating over a line segment of length Δ gives rise to the integral evaluation

$$\int_0^{\Delta} v(z)\,dz = \frac{\Delta}{2}[v(\Delta) + v(0)]. \tag{4.110}$$

Applying this technique to (4.106) now leads to a block diagonal form for (4.109):

$$\left[T^e_{xx}\right] = \varepsilon_0 \frac{\Delta^3}{4}[I] \tag{4.111}$$

where $[I]$ denotes the identity matrix. The resulting mass matrix $[T]$ will now be fully diagonalized.

In the absence of sources and loss, the finite element problem (2.27) can be written as

$$[T]\frac{d^2\{E\}}{dt^2} + [S]\{E\} = 0. \tag{4.112}$$

By applying central time differencing to (4.112), specifically the case $\beta = 0$ in (2.36), we obtain

$$[T]\{E\}^{n+1} = 2[T]\{E\}^n - [T]\{E\}^{n-1} - (\Delta t)^2[S]\{E\}^n \tag{4.113}$$

which is clearly an explicit time-advancement scheme when trapezoidal spatial integration is used because of the resulting diagonalization of $[T]$. The trapezoidal integration technique is applied similarly in the evaluation of the stiffness matrix $[S]$; however, to preserve the coupling between the various components of the electric field, this matrix will obviously not be diagonal.

It is straightforward to establish the equivalence of (4.113) and the traditional FDTD method in one-dimensional space [22]. To accomplish this, we proceed similarly to Section 4.1 and write the standard one-dimensional FDTD update equations for the electric and magnetic fields as

$$\begin{aligned} E_x^{n+1}(i) &= E_x^n(i) - \frac{\Delta t}{\varepsilon_0 \Delta}\left[H_y^{n+1/2}\left(i+\tfrac{1}{2}\right) - H_y^{n+1/2}\left(i-\tfrac{1}{2}\right)\right] \\ H_y^{n+1/2}\left(i+\tfrac{1}{2}\right) &= H_y^{n-1/2}\left(i+\tfrac{1}{2}\right) - \frac{\Delta t}{\mu_0 \Delta}\left[E_x^n(i+1) - E_x^n(i)\right]. \end{aligned} \tag{4.114}$$

We then combine the coupled equations to obtain a single equation for the electric field:

$$E_x^{n+1}(i) = 2E_x^n(i) - E_x^{n-1}(i) + \left(\frac{c\Delta t}{\Delta}\right)^2 \left[E_x^n(i+1) - 2E_x^n(i) + E_x^n(i-1)\right] \tag{4.115}$$

which corresponds to a central-difference representation for the wave equation $\partial^2 E_x/\partial t^2 = c^2\, \partial^2 E_x/\partial x^2$. We now consider (4.113) in a one-dimensional infinite space. Using linear (rooftop) basis functions [40] along with trapezoidal integration, the stiffness matrix $[S]$ in (4.113) is tridiagonal, with the elements on a typical row given by $[0\ldots,(-1/\mu_0\Delta),(2/\mu_0\Delta),(-1/\mu_0\Delta),\ldots 0]$. Also, the mass matrix becomes $[T] = \varepsilon_0\Delta[I]$. By substituting these matrices into (4.113) and examining the central row entries, the resulting equivalence of (4.113) and (4.115) is apparent. Thus, the traditional FDTD leapfrog algorithm (4.114) for the two coupled curl equations has an equivalent wave-equation representation based on central differencing in both space and time, and the use of trapezoidal integration in the finite element formulation (4.113), along with central time differencing, is also equivalent to the discrete wave equation (4.115) based directly on central time and spatial differencing.

A similar equivalence between finite elements and the FDTD method exists in two- and three-dimensional spaces; however, this is awkward to show analytically in three dimensions, so a numerical example will instead be used. The example is a closed, conducting rectangular resonator that is meshed with nonuniform brick elements, as shown in Figure 4.4(a). We selected interior element edges randomly to impress a Gaussian-pulse voltage source and to monitor the transient response induced. The voltages observed for solutions based on the traditional FDTD, the FETD using the wave equation based exclusively on vector edge basis functions, and the FETD using Maxwell's two first-order curl equations based on vector edge and vector facet basis functions are shown in Figure 4.4(b). All solutions differ only by the machine precision of the computational hardware.

4.6 STABLE FETD–FDTD INTERFACE

The development of a stable finite element and FDTD interface evolved from a similar hybridization concept that was based initially on finite-volume techniques. In this section we review briefly the initial approaches to developing the hybrid-grid interface and then describe the current finite element–based methodologies. The grid-generation requirements for constructing a conformal-grid interface that is suitable for a direct connection to FDTD are largely independent of the particular conformal method used, and practical methodologies to realize this interface construction are described in Section 4.7.

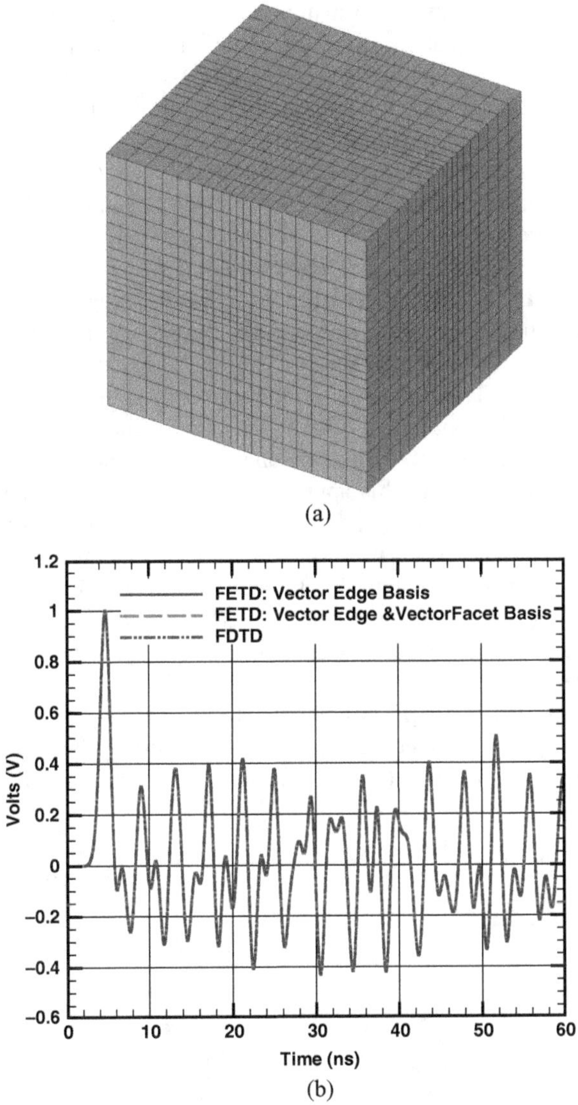

Figure 4.4 Equivalence of three-dimensional FETD and FDTD using brick elements with trapezoidal integration. (a) Nonuniform mesh for a rectangular resonator. (b) Internal voltage response based on three solution methods. All solutions differ only by the machine precision of the computational hardware.

4.6.1 Initial Approaches

Over the past 15 years there have been a variety of techniques proposed to interface conformal grids with the rectilinear FDTD method [6–20]. Indeed, hybrid techniques leverage the beneficial features of two or more methods with the goal of producing a more powerful solution technique, so an interface of this type represents a natural progression for differential equation–based computational electromagnetics. The original hybrid methods in this area were based on finite-volume time-domain (FVTD) methods interfaced to the FDTD method. In their simplest form, finite-volume methods assume constant fields throughout the volumetric integration domain of the small cells used to partition a spatial domain. Although a variety of FVTD methods are used in computational electromagnetics [44–47], the techniques that are naturally suited to hybridization with FDTD are those based on the offset mesh concept that employ primal and dual grids for the electric and magnetic fields, respectively [44].

An FVTD–FDTD hybrid that was based on overlapping grids, similar to the Chimara grids that have been widely used in the computational fluid dynamics discipline [48], was first described by Yee et al. [6,7]. An example of this type of grid is shown in Figure 4.5. The surface conformal mesh used in this approach is based on extruding surface elements, such as triangles or quadrilaterals, normal to the surface of the geometry so that three-dimensional prisms or hexahedral elements are created. A beneficial feature of this approach is that the resulting conformal mesh can be represented using an *i–j–k* indexing system, that is, a structured mesh, just as in the case of the standard FDTD grid. However, this method has two disadvantages. First, structured conformal meshes typically lack the geometrical flexibility of fully unstructured grids, and second, complicated interpolations are required at the boundary of the conformal grid and the background FDTD mesh. In addition, a more serious consequence of the latter is that these types of interpolations often give rise to numerical instabilities that appear in the late time.

An alternative approach to interfacing fully unstructured grids directly to the rectilinear FDTD grid was described in Refs. 8 and 9. A two-dimensional cross section

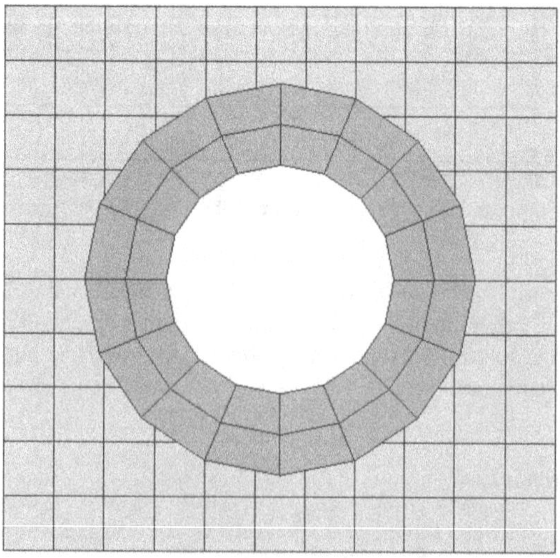

Figure 4.5 Overlapping conformal mesh with a background rectangular mesh.

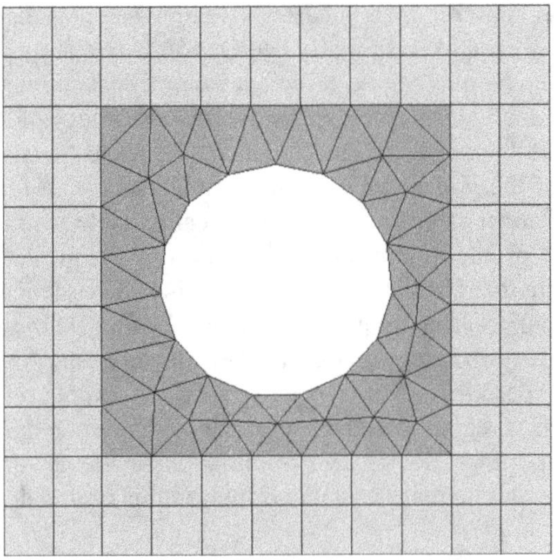

Figure 4.6 Directly interfacing conformal unstructured mesh with a uniform rectangular mesh.

of this type of hybridization is shown in Figure 4.6, where in three dimensions the triangular elements typically correspond to tetrahedra and the quadrilaterals correspond to the FDTD brick elements. The original approach to the numerical solution in the unstructured (tetrahedral) portion of this mixed-element grid was also based on FVTD concepts. Similar to the equivalence between FETD and FDTD shown in the preceding section, an important feature of the particular FVTD method used was that the technique could be shown to reduce identically to the standard FDTD algorithm when the cells took on the shape of the traditional FDTD rectangular bricks [9].

In three dimensions, the triangular faces of tetrahedra do not fully edge-align with the quadrilateral faces associated with the rectangular brick elements, so special care is required in this interface region. Because finite-volume techniques simply integrate over cell volumes, this apparent misalignment challenge is readily resolved by introducing a new interfacing hexahedral element that had one of its traditional quadrilateral faces subdivided into two triangles [8]. In this way, the direct interface of tetrahedral and hexahedral elements can be accomplished. Alternatively, a single layer of pyramidal elements can also be used to accomplish this transition without the creation of nonstandard elemental cells.

Although finite-volume techniques that are based on offset primal and dual grids facilitate the realization of an interface of unstructured grids with the FDTD method, all known offset-grid FVTD methods suffer from late-time instabilities on nonrectangular cell shapes. Consequently, all known hybridizations involving offset-grid FVTD methods with the FDTD method similarly experience these types of instabilities in the late time. Lowpass filtering schemes have been found to be somewhat effective at suppressing or delaying the onset of instabilities [9]; however, this is accomplished by introducing dissipative loss, and consequently, the resulting algorithms no longer conserve energy. Alternative time-stepping algorithms have also been investigated [18] and have similarly been reported to simply delay the onset of instability, which may or may not be sufficient for practical applications.

Although conceptually the same with regard to the topology of the FVTD–FDTD hybrid-grid interface, the use of finite element techniques to advance the fields in the conformal region of the grid provides significant advantages. Most notably, FETD methods are well known to be provably numerically stable, and hence there is no issue with the onset of late-time instabilities caused by the presence of eigenvalues that fall outside the unit circle in the complex plane [49]. Consequently, early work to construct a conformal mesh technique that interfaced with FDTD also utilized FETD methods in the unstructured mesh region [10–12]. However, the possible equivalence of FETD and FDTD described in the preceding section was not utilized in these early studies, so the interface region lacked the necessary symmetry requirements to obtain a globally stable algorithm. Subsequent formulations made use of this equivalence and a stable interface was obtained. An algorithm of this type is described next.

4.6.2 Stable Formulation

The electric-field time advancement in a finite element domain was given previously by (2.36) and is repeated here for convenience:

$$\left(\frac{1}{(\Delta t)^2}[T] + \frac{1}{2\Delta t}[R] + \beta[S]\right)\{E\}^{n+1} = \left(\frac{2}{(\Delta t)^2}[T] - (1-2\beta)[S]\right)\{E\}^n - \left(\frac{1}{(\Delta t)^2}[T] - \frac{1}{2\Delta t}[R] + \beta[S]\right)\{E\}^{n-1} + \beta\{f\}^{n+1} + (1-2\beta)\{f\}^n + \beta\{f\}^{n-1}. \tag{4.116}$$

Two cases of this equation will be of interest. By defining $\beta = 0$, we obtain the conditionally stable central-difference field advancement:

$$\left([T] + \frac{\Delta t}{2}[R]\right)\{E\}^{n+1} = \left(2[T] - (\Delta t)^2[S]\right)\{E\}^n - \left([T] - \frac{\Delta t}{2}[R]\right)\{E\}^{n-1} + (\Delta t)^2\{f\}^n \tag{4.117}$$

and by defining $\beta = 1/4$, we obtain the unconditionally stable field advancement

$$\left([T] + \frac{\Delta t}{2}[R] + \frac{(\Delta t)^2}{4}[S]\right)\{E\}^{n+1} = 2\left([T] - \frac{(\Delta t)^2}{4}[S]\right)\{E\}^n - \left([T] - \frac{\Delta t}{2}[R] + \frac{(\Delta t)^2}{4}[S]\right)\{E\}^{n-1} + \frac{(\Delta t)^2}{4}\left(\{f\}^{n+1} + 2\{f\}^n + \{f\}^{n-1}\right). \tag{4.118}$$

As described in the preceding section, when trapezoidal integration is used in the construction of the finite element matrices for the case of brick elements, (4.117) can be viewed as an *explicit* FDTD field advancement for the electric field. However, for the case of more general elemental shapes, such as tetrahedra, skewed hexahedra, prisms, and/or pyramidal elements, we will use (4.118) and construction of the elemental matrices will be based on exact integration techniques, which will give rise to a traditional *implicit* finite element field advancement for these cells. Thus, the possible realization of a global implicit–explicit algorithm is apparent. Similar to what is described in Refs. 16 and 20, a succinct global algorithm that combines (4.117) and (4.118) can be written in terms of the "implicitness" factor β_{el} as well as the form of the spatial integration taken over the ensemble of elements in the computational domain: namely,

$$\begin{aligned}
&\sum_{el=1}^{N_{el}} \left([T_{el}] + \frac{\Delta t}{2}[R_{el}] + \beta_{el}(\Delta t)^2[S_{el}] \right) \{E\}^{n+1} \\
&\quad = \sum_{el=1}^{N_{el}} \left(2[T_{el}] - (1 - 2\beta_{el})(\Delta t)^2[S_{el}] \right) \{E\}^{n} \\
&\quad\quad - \sum_{el=1}^{N_{el}} \left([T_{el}] - \frac{\Delta t}{2}[R_{el}] + \beta_{el}(\Delta t)^2[S_{el}] \right) \{E\}^{n-1} \\
&\quad\quad + (\Delta t)^2 \sum_{el=1}^{N_{el}} \left(\beta_{el}\{f_{el}\}^{n+1} + (1 - 2\beta_{el})\{f_{el}\}^{n} + \beta_{el}\{f_{el}\}^{n-1} \right).
\end{aligned} \tag{4.119}$$

In (4.119), $[T] = [T_1] + [T_2] + \cdots + [T_{N_{el}}]$, where $[T]$ has simply been decomposed into the contributing elements $[T_{el}]$, $el = 1, \ldots, N_{el}$. Decomposing the finite element matrices into their regional elemental contributions highlights the two different integration schemes that are used: trapezoidal and exact spatial integration. Similar interpretations are used for the matrices $[R]$ and $[S]$ as well as for the forcing function. The definition of the implicitness parameter β_{el} has also been extended to accommodate the type of time differencing to be used on an elemental level; specifically, $\beta_{el} = 0$ in the case of the rectangular bricks that are to be associated with the FDTD region of the hybrid grid (the explicit time-stepping region), and $\beta_{el} = 1/4$ elsewhere (the implicit time-stepping region). An analytical proof of the algorithmic stability of (4.119) is available [16,20], and the specific form for the vector edge basis functions for the brick, tetrahedral, and pyramidal elements used in the FETD portion of the hybrid grid can be found in Ref. 20.

Note that by writing the time-advancement scheme in the form of (4.119), it is seen that the entire computational domain is formally considered to be based on a global finite element discretization. The distinction between finite element and FDTD regions of the hybridized grid is largely one of preference and coding. Specifically, the FDTD algorithm based on the first-order coupled-curl Maxwell's equations is traditionally used for structured grids and the resulting coding can be written in

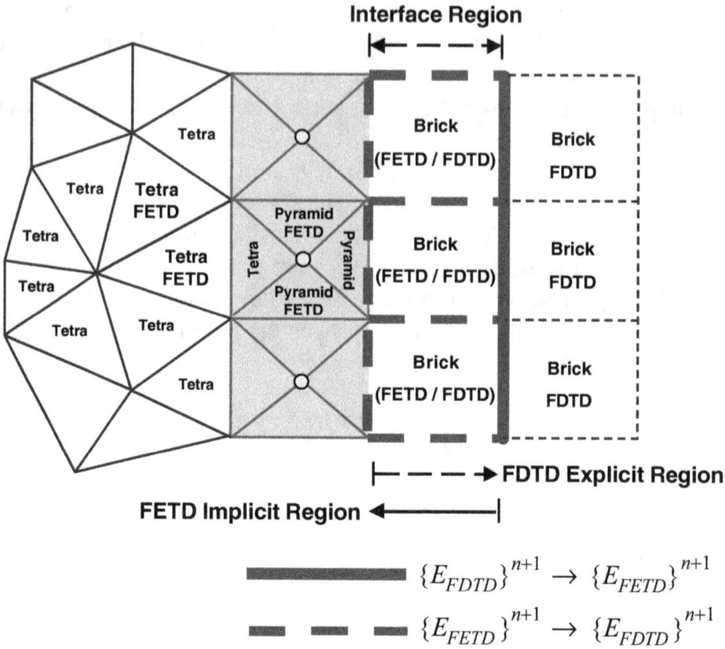

Figure 4.7 Portion of the interface between an explicit and structured FDTD grid and an implicit and unstructured FETD grid. Data exchange from the FDTD grid to the FETD grid occurs on the thick solid line, whereas data exchange from the FETD grid to the FDTD grid occurs on the thick dashed lines. Trapezoidal integration is used for the brick elements in the interface region that are associated with the FETD region.

a highly efficient manner. Consequently, to maximize computational efficiency and minimize the use of computer memory, practical implementations of (4.119) typically utilize the traditional FDTD algorithms for the uniform portions of the hybrid grid and finite element algorithms elsewhere. In this way, the matrices $[T]$, $[R]$, and $[S]$ do not need to be stored for the FDTD region of the grid.

At the interface between the FETD and FDTD algorithms, a simple data exchange is required. More specifically, and with reference to Figure 4.7, the global time advancement can be accomplished by the following algorithm:

> The electric fields in the FDTD region are updated explicitly to time step $n + 1$: namely, $\{E_{\mathrm{FDTD}}\}^{n+1}$. The portion of these electric fields that resides on the thick solid line in Figure 4.7 defines a Dirichlet boundary condition for the finite element domain such that $\{E_{\mathrm{FDTD}}\}^{n+1} \rightarrow \{E\}^{n+1}$. Subject to this interface boundary condition, the remaining electric fields in the finite element region are advanced implicitly to time step $n + 1$. The resulting solution for the finite element–based electric fields that reside on the thick dashed line shown in Figure 4.7 are then transferred to the FDTD region according to $\{E\}^{n+1} \rightarrow \{E_{\mathrm{FDTD}}\}^{n+1}$. With these electric fields in place, the magnetic fields on the FDTD portion of the grid can be advanced to time step $n + 3/2$, and the time step is completed. This procedure continues for the duration of the simulation.

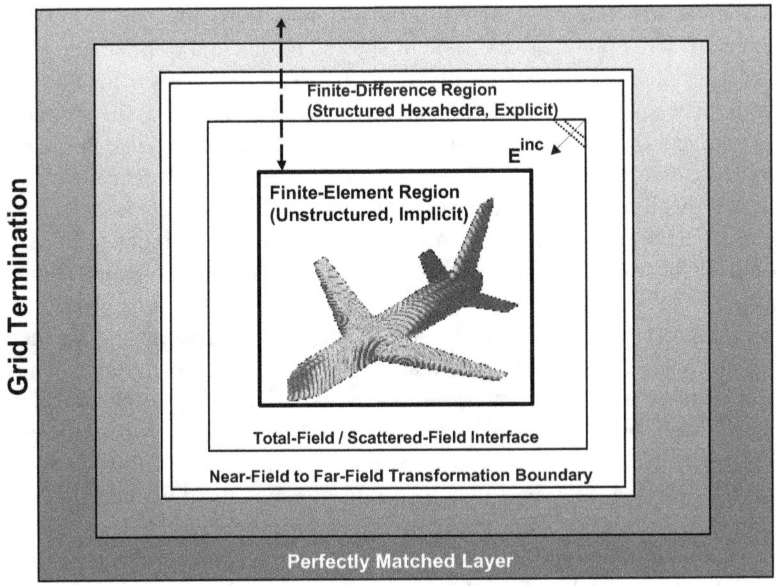

Figure 4.8 Locations of the FETD, FDTD, and PML regions, as well as the near-to-far-field transformation surface used for a complete FETD–FDTD implementation. (*See insert for color representation of figure.*)

Note that this basic data-exchange procedure is applicable to many hybridized-grid formulations. For example, for a finite volume–based algorithm in the unstructured-grid region, $\{E\}^{n+1}$ would simply be computed by the finite-volume technique. Thus, the adaptation to a variety of new hybrid-grid algorithms is apparent and will probably be a topic of future research. As noted previously, the key to the success of the FETD–FDTD scheme is the potential equivalence of FETD to FDTD on brick elements, which leads to a symmetric interface condition. To obtain a globally stable algorithm, a similar symmetry condition would need to be established for alternative hybridized algorithms. More specifically, the stability of each constituent algorithm is not sufficient to ensure the stability of a hybridized formulation.

The locations of the FETD, FDTD, PML, and near-to-far-field transformation surface for a complete FETD–FDTD implementation are shown in Figure 4.8. Although only a single FETD region is shown, multiple FETD instances may be present.

4.7 BUILDING HYBRID MESHES

The realization of a global mesh that permits interfacing unstructured finite elements with a structured uniform grid is independent of the particular regional algorithm to be used to solve for the fields. In the specific case of the FETD–FDTD hybrid technique, unstructured tetrahedral elements are typically used in the finite element region, whereas structured rectangular brick elements are used in the FDTD region. Alternative element combinations, such as skewed hexahedral or prismatic elements are, of course, possible within the FETD spatial region; however, for purposes of this section we consider the tetrahedral element case. Obviously, a direct connection between tetrahedral and brick elements is not possible, due to the edge misalignment between these two elements, and consequently, a transitional element such as a pyramid is typically required. However, because pyramidal elements are not widely used in practical applications across a variety of scientific disciplines, this element type is not widely supported by commercial mesh generation software, although a few exceptions exist [50,51].

In the event that pyramidal elements are not supported by specific mesh generation software, they can still be constructed readily in a straightforward manner. The basic procedure is either to merge the interior faces of two neighboring tetrahedral elements that connect to the grid interface, or to add an additional node at the barycenter of a brick element to which new edges are constructed. For example, in the latter case the brick element may be subdivided into five pyramidal elements along with two tetrahedra. In this way, the triangular faces of the constructed tetrahedra will naturally match to the tetrahedral elements within the unstructured grid region, and the quadrilateral faces of the bases of the pyramids will edge-align with the structured-grid brick elements used throughout the FDTD region. Both approaches have been adopted in the development of effective FETD–FDTD solutions.

Even though the construction of pyramidal elements is straightforward, certain constraints are required for the termination of the tetrahedral-grid region if a direct connection to an FDTD grid is to be accomplished. Specifically, the nodes on the tetrahedral-grid outer boundary ultimately need to align with those of the FDTD brick elements. A simple way to accomplish this is to place a uniform, or mapped, triangular mesh on this boundary as first described in Ref. 8. More specifically, Figure 4.9 shows a possible path to constructing a tetrahedral grid that will interface directly to a structured FDTD grid. In Figure 4.9(a) we have defined the solid model of a scattering geometry of interest. The desired geometry is then placed into a rectangular container as shown in Figure 4.9(b). On the boundary of the container we place a mapped triangular mesh as shown in Figure 4.9(c), where the specified size of the triangles is based on the desired Δx, Δy, and Δz edge lengths associated with the FDTD cells to which the tetrahedra will interface. The scatterer within the container is then meshed as desired, and finally, the region exterior to the scatterer is meshed out volumetrically to the mapped triangular mesh that was defined on the boundary of the container. This unstructured tetrahedral mesh is now suitable for placement within an FDTD background mesh, and the pyramidal transitioning elements can be constructed following one of the two procedures described above. The simulation on the hybrid mesh can then be realized, as illustrated in Figure 4.9(d).

A useful aspect to this procedure is that multiple instances of the original geometry can easily be placed within the FDTD background grid through a simple translation and/or rotation process. Thus, geometry can be relocated without recreating the tetrahedral mesh. In addition, materials can extend through the interface regions between the solution methods. For example, complex antennas and feed regions can

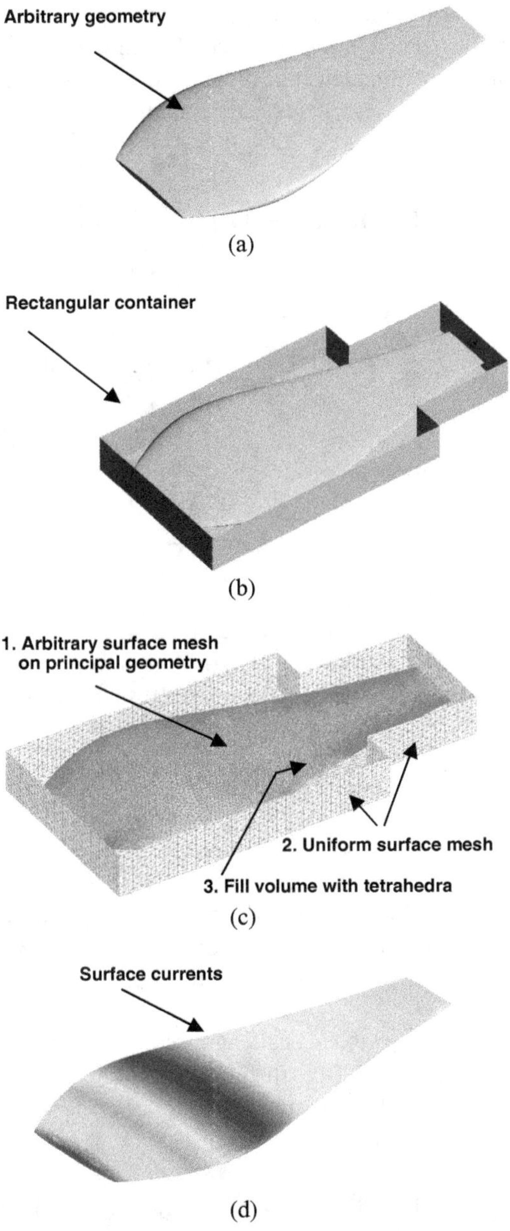

Figure 4.9 Procedure to construct a simple unstructured tetrahedral grid interfaced to a structured brick-element grid. (a) Solid geometry desired. (b) Suitably sized rectangular box to hold the geometry. (c) Uniform triangular surface mesh on the box and arbitrary triangular mesh on the geometry (filled with a tetrahedral mesh in between). (d) Unstructured mesh embedded in a background FDTD mesh (not shown) and the solution. (*See insert for color representation of figure.*)

be modeled accurately using finite element technology, and then their ground planes and/or substrates can be carried into and solved efficiently within the FDTD region. The FDTD region also accommodates the global grid termination, where the FDTD PML techniques described in Section 4.2 can be applied.

4.8 WAVE-EQUATION STABILIZATION

The wave equation

$$\nabla \times \left[\frac{1}{\mu} \nabla \times \mathbf{E}(t) \right] + \varepsilon \frac{\partial^2 \mathbf{E}(t)}{\partial t^2} + \sigma_e \frac{\partial \mathbf{E}(t)}{\partial t} = -\frac{\partial \mathbf{J}_{\text{imp}}(t)}{\partial t} \tag{4.120}$$

supports the nontrivial solution

$$\mathbf{E}(t) = -(at + b)\,\nabla\varphi \tag{4.121}$$

with $\sigma_e = 0$ and $\mathbf{J}_{\text{imp}} = 0$, where φ denotes a scalar potential and a and b are constants. Because a is not necessarily zero in (4.121), a solution to (4.120) can theoretically drift linearly in time. However, by integrating (4.120) such that

$$\nabla \times \left[\frac{1}{\mu} \nabla \times \int_0^t \mathbf{E}(\tau)\, d\tau \right] + \varepsilon \frac{\partial \mathbf{E}(t)}{\partial t} + \sigma_e\, \mathbf{E}(t) = -\mathbf{J}_{\text{imp}}(t) \tag{4.122}$$

the value of a must now be exactly zero in (4.121). It is noted that formulations based on the two first-order Maxwell's equations instead of the second-order wave equation, such as the traditional FDTD algorithm and the alternative FETD algorithm defined by (4.104) and (4.105), will not typically support a solution that will drift in time.

To remove this potential theoretical challenge associated with the direct numerical solution of the wave equation (4.120), Artuzi [52] proposed solving (4.122) using the alternative discrete finite element equation

$$\left([T] + \frac{\Delta t}{2}[R] + \frac{(\Delta t)^2}{4}[S] \right) \{v\}^n = \{f\}^n - [R]\{E\}^{n-1/2} - [S]\{w\}^n \tag{4.123}$$

with

$$\{E\}^{n+1/2} = \{E\}^{n-1/2} + \Delta t\{v\}^n \tag{4.124}$$

$$\{w\}^{n+1} = \{w\}^n + \Delta t\{E\}^{n+1/2} \tag{4.125}$$

where the desired solution is given by $\{E\}^{n+1/2}$. The elements of the matrices $[T]$, $[R]$, and $[S]$ are the same as those defined previously by (2.28)–(2.30), respectively; however, the elements of the excitation vector $\{f\}$ are given by

$$f_i(t) = -\iiint_V \mathbf{N}_i \cdot \mathbf{J}_{\text{imp}}\, dV \tag{4.126}$$

which corresponds to the time integral of (2.31) in the absence of an impressed magnetic current source. The system (4.123)–(4.125) has been found to be unconditionally stable and free of a low-frequency or late-time drift.

Practical antenna applications with realistic loss mechanisms, which include radiation loss, are typically well characterized by the traditional Newmark-beta formulation for the wave equation defined by (4.118). However, lossless internal resonator simulations perpetuating for hundreds of thousands of time iterations and low-frequency applications utilizing a time step that is hundreds or perhaps thousands of times larger than a Courant-limited algorithm can both benefit from the alternative formulation (4.123)–(4.125).

To demonstrate the utility of the finite element formulation defined by (4.123)–(4.125), we consider a lossless rectangular resonator example. This example is solved by both the traditional Newmark-beta discretization given by (4.118) and the alternative formulation (4.123)–(4.125), with the results presented in Figures 4.10 and 4.11. The nonphysical drift in the traditional solution of this type of application is shown in Figure 4.10(a), where an iterative matrix solution with a 10^{-4} residual error was used. When using iterative techniques to solve the governing system of equations associated with the traditional Newmark-beta formulation, a simple reduction in the residual error can be effective in delaying the onset of the linear growth term in (4.121) beyond a steady-state solution for most practical applications. For example, in the case of the lossless resonator, iteratively solving (4.118) with a residual error of 10^{-8} instead of 10^{-4} leads to the interior solution shown in Figure 4.10(b), where the drift seen in Figure 4.10(a) has been largely eliminated over the 100,000 time iterations used for this simulation. Direct solution methods, wherever feasible based on the number of unknowns, can provide further benefits. The reason for this improvement is that either a smaller residual error in the case of iterative solutions or the use of direct solution methods will lead to globally reduced numerical error on the grid and thereby further suppress the emergence of solutions such as (4.121) in the traditional Newmark-beta formulation. Alternatively, representation (4.123) provides a simple approach to fully remove the linear transient growth term in the finite element solution of the wave equation even with larger definitions for the residual error, and this is shown in Figure 4.11(a) for the lossless resonator. Finally, since the precise temporal behavior predicted by the two solution procedures is difficult to see when visualized over numerous time steps, an early-time comparison of both (4.118) and (4.123) is shown in Figure 4.11(b), and the results are seen to overlay. In addition to the approach described by (4.123)–(4.125), alternative techniques to stabilize the wave equation have been proposed [53,54].

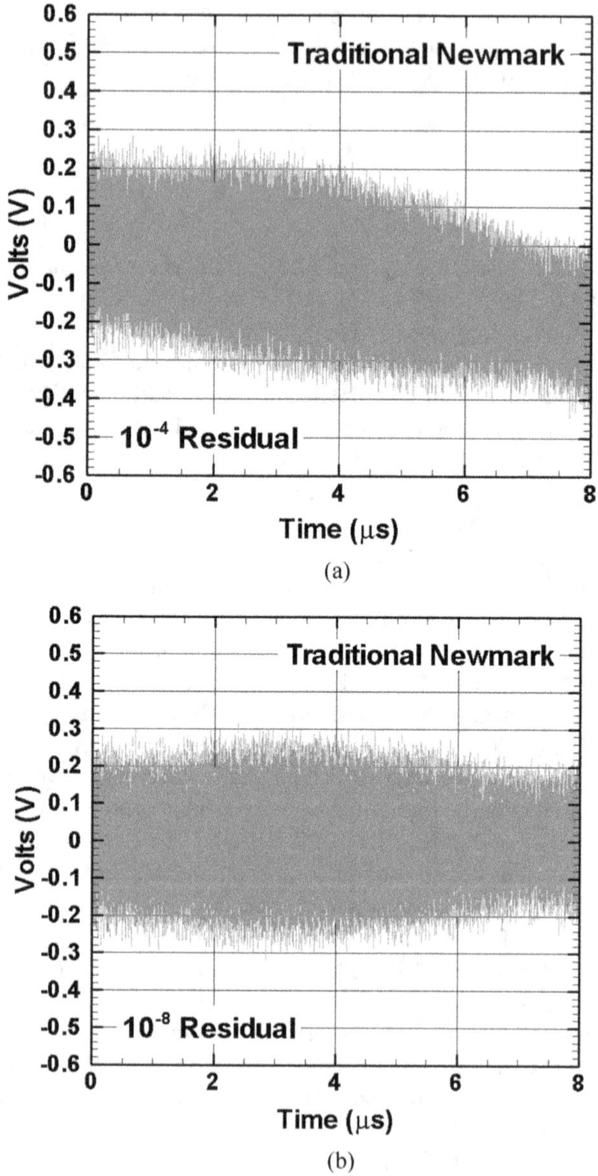

Figure 4.10 Traditional Newmark solution based on (4.118) for a lossless rectangular resonator. (a) Traditional Newmark method with 10^{-4} residual. (b) Traditional Newmark method with 10^{-8} residual. The source term was a differentiated Gaussian pulse.

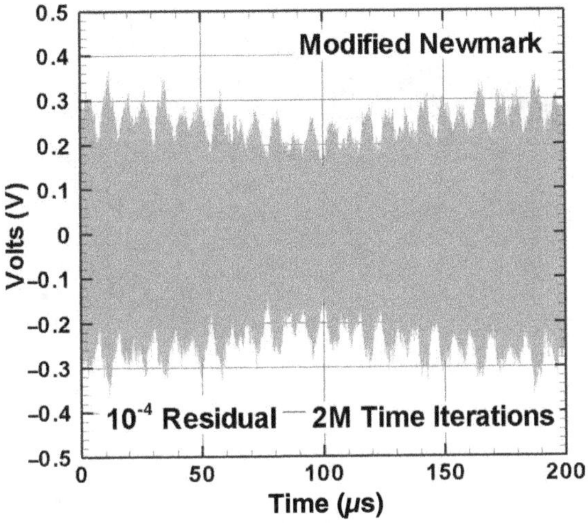

(a)

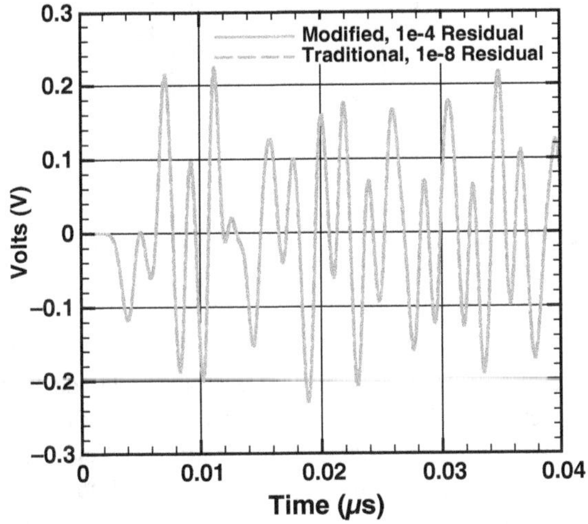

(b)

Figure 4.11 Modified Newmark solution based on (4.123) for a lossless rectangular resonator. (a) Modified Newmark method with 10^{-4} residual and 2 million time iterations. (b) Early time comparison of the traditional Newmark method with 10^{-8} residual and the modified Newmark method with 10^{-4} residual. The source term was a differentiated Gaussian pulse.

4.9 VALIDATION EXAMPLES

The first example of the FETD–FDTD hybrid technique is a coaxial-fed recessed patch antenna with an infinite flange, as shown in Figure 4.12. The patch antenna resides on a substrate with a thickness of 0.08779 cm and a relative permittivity of 2.17.

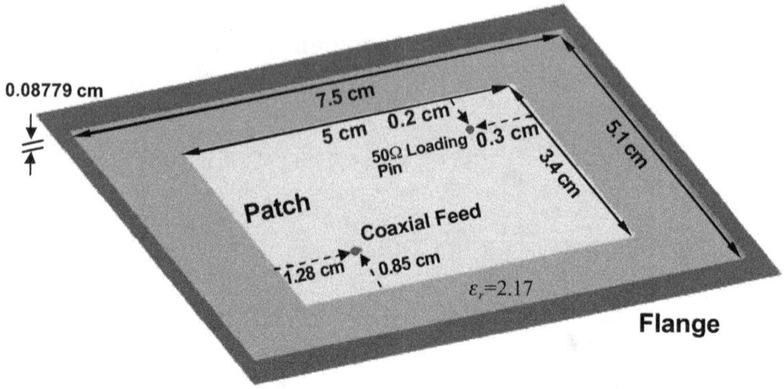

(a)

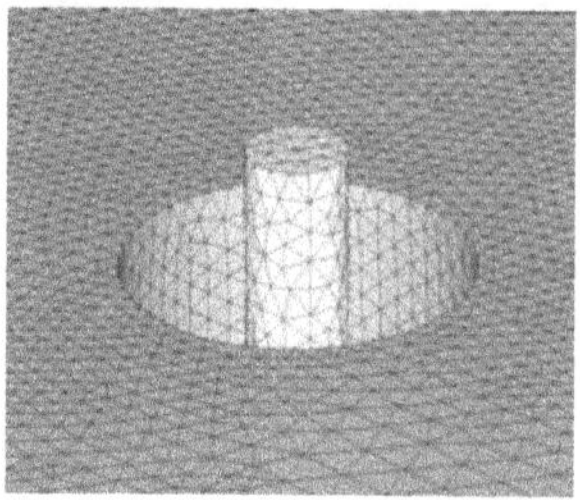

(b)

Figure 4.12 Coaxial-fed patch antenna recessed in a ground plane. (a) Geometry. (b) Local finite element surface mesh of the coaxial feed with an inner radius of 0.045 cm, an outer radius of 0.155 cm, and a relative permittivity of 2.2.

The patch was driven by a 50-Ω coaxial feed with an inner radius of 0.045 cm, an outer radius of 0.155 cm, and a relative permittivity of 2.2. A 50-Ω resistor was used to load the antenna by using a technique described in Chapter 6. The antenna and a portion of the conducting flange were included within the conformal FETD mesh, and the continuation of the flange was accomplished by pushing it initially through the FETD–FDTD interface region and onto the FDTD grid. The FDTD grid was terminated with a PML. Predicted and measured [55] data for the input resistance and reactance are shown in Figure 4.13(a) and (b), and the data correlation is generally very good.

The second example is plane-wave scattering by a metallic double ogive, where the geometry is shown in Figure 4.14(a). The double ogive is formed by combining two half-ogives. The first half-ogive has a length of 6.36 cm, a maximum radius of 2.54 cm, and a half-angle at the tip of 46.4°. The second half-ogive has a length of 12.72 cm, a maximum radius of 2.54 cm, and a half-angle of 22.62° at the tip. The double ogive was modeled by first-order tetrahedral elements within the conformal FETD region of the hybrid FETD–FDTD grid, whereas the surrounding space was modeled by the FDTD technique with a PML grid termination. The surface

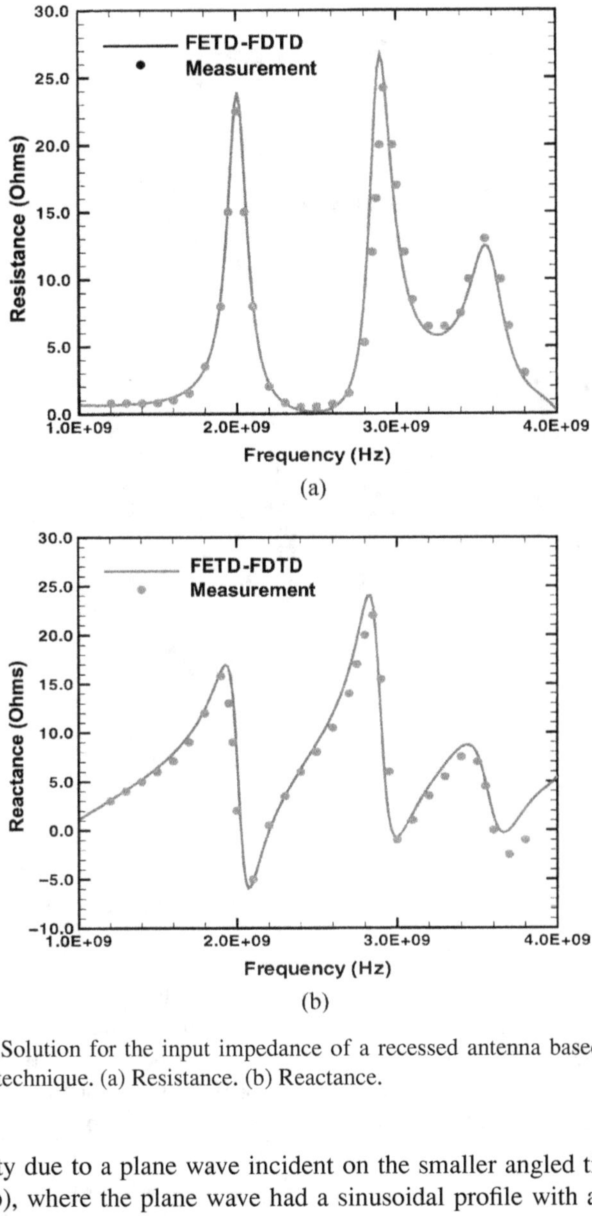

Figure 4.13 Solution for the input impedance of a recessed antenna based on the hybrid FETD–FDTD technique. (a) Resistance. (b) Reactance.

current density due to a plane wave incident on the smaller angled tip is shown in Figure 4.14(b), where the plane wave had a sinusoidal profile with a frequency of 30 GHz. At 30 GHz, the double ogive has an electrical length of 19.05 wavelengths. Measurements [56] and predictions for the monostatic RCS of the double ogive in the xz-plane at 9 GHz are shown in Figure 4.15, where the data correlation is generally very good.

Because the FETD–FDTD hybrid formulation is a transient solution method, the technique computes the monostatic RCS directly at a single incident angle over a wide bandwidth. For example, for a unit amplitude plane wave with a Gaussian

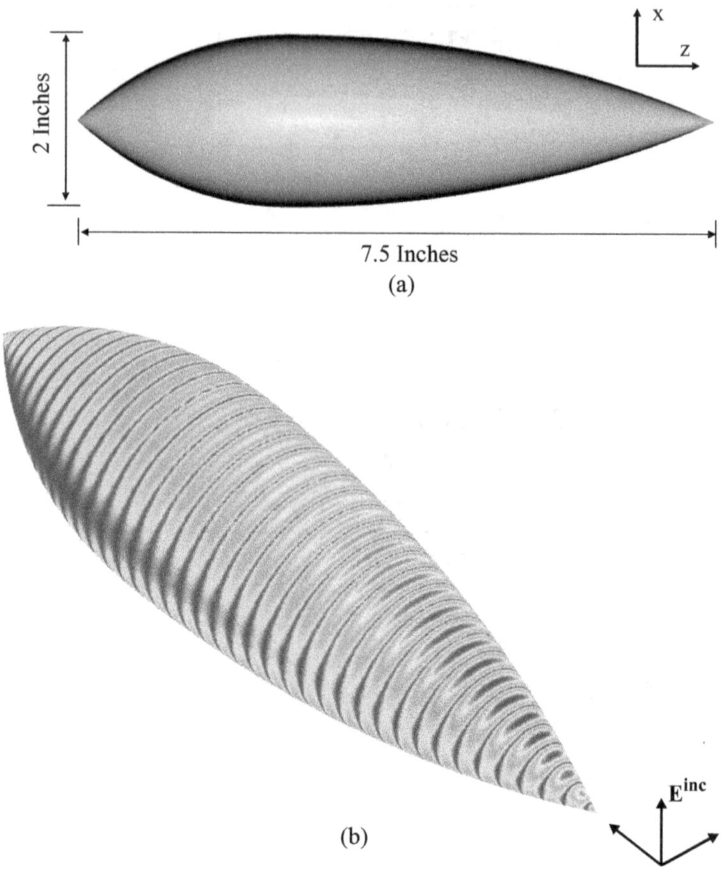

Figure 4.14 Metallic double ogive. (a) Geometry. (b) Surface current density for a sinusoidal plane-wave excitation at 30 GHz. (*See insert for color representation of figure.*)

time profile incident on the double ogive shown in Figure 4.14(a), the transient VV-polarized monostatic electric field in the elevation direction $\theta = 60^\circ$ is shown in Figure 4.16(a). Using the Fourier transform, the broadband, monostatic VV-polarized RCS in this direction is shown in Figure 4.16(b), and it is noted that the 9-GHz numerical value of -52 dB corresponds to the $\theta = 60^\circ$ monostatic data plotted in Figure 4.15(a).

4.10 SUMMARY

The FETD method represents a powerful numerical method for the broadband solution of practical antenna, array, and scattering applications. This is because an FETD implementation will conform precisely to a complex geometry of arbitrary shape and can be formulated to be unconditionally stable so that the time step does not

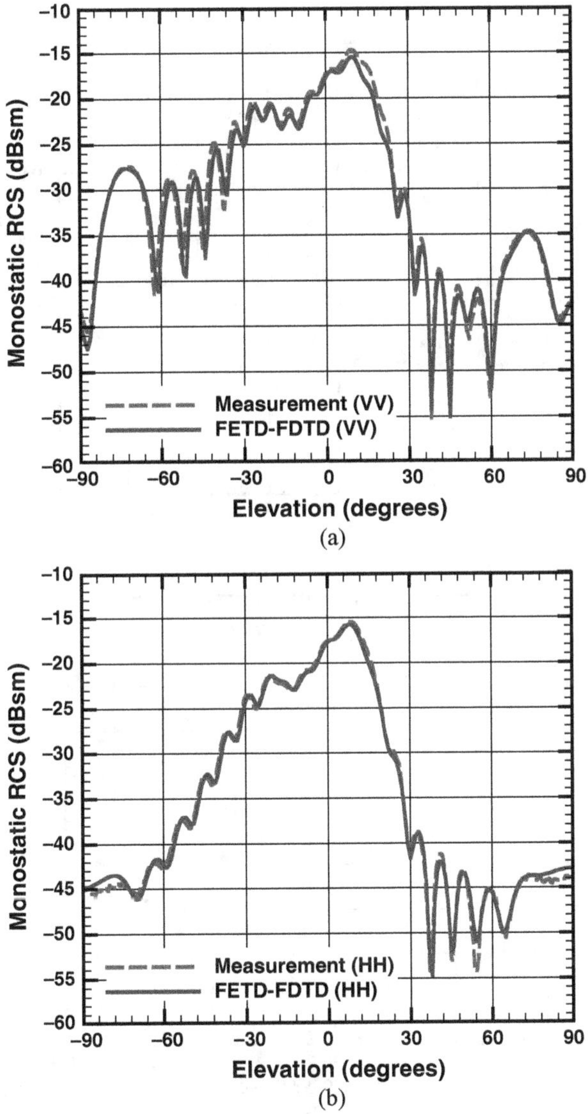

Figure 4.15 Monostatic RCS of the metallic double ogive at 9 GHz. (a) VV polarization. (b) HH polarization.

have to be reduced when spatially resolving small features, such as the antenna feed region. However, due to the typical use of unstructured grids, these methods are more computationally expensive on a per-cell basis than approaches based exclusively on structured grids, such as the traditional FDTD method. Consequently, it is of interest to formulate techniques that can utilize the best features of both methodologies, and an effective hybridization of FETD and FDTD was the topic of this chapter.

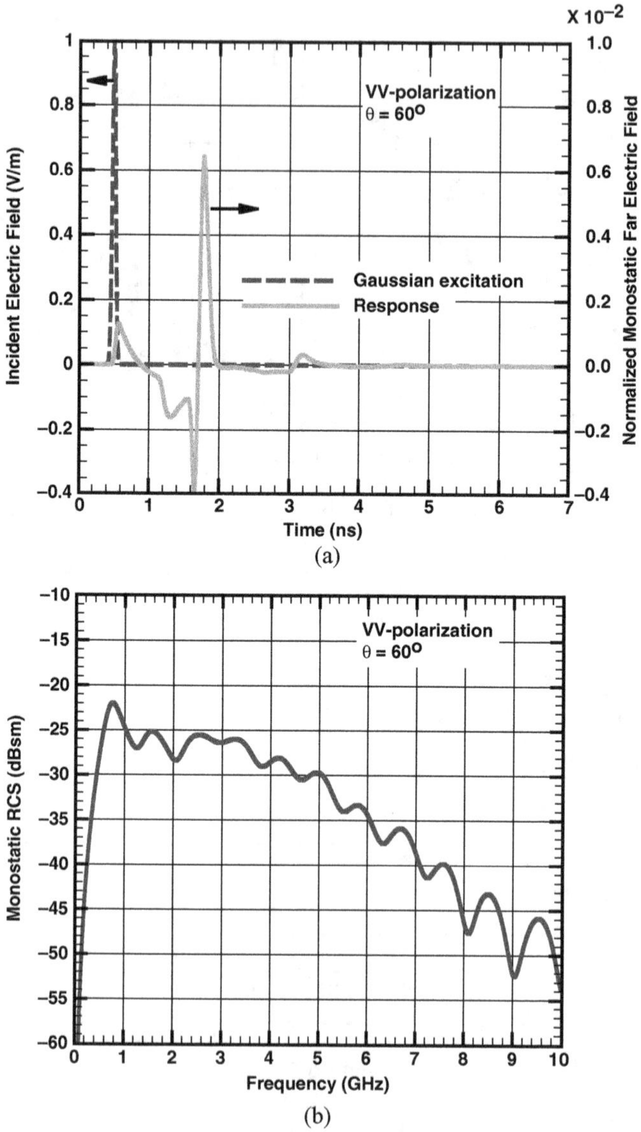

Figure 4.16 Monostatic RCS of the metallic double ogive at elevation angle $\theta = 60°$. (a) Transient excitation and response. (b) Monostatic RCS as a function of frequency.

By demonstrating that the FDTD method is a special case of the vector edge–based FETD technique, it was shown that a globally stable FETD–FDTD hydridization can be constructed. The resulting method is an implicit–explicit time-stepping algorithm where the unconditionally stable implicit FETD method is used to accommodate the complex surface, material, and feeding regions of antennas and arrays, whereas the structured-grid and explicit FDTD technique is used for efficient accommodation to the geometrically less complex regions of the global grid.

Because of the importance and maturity of both the FDTD PML and FDTD near-to-far-field transformation techniques in the context of the FETD–FDTD hybrid concept, these techniques were also reviewed in this chapter. In addition, an alternative FETD technique based on the two first-order Maxwell's curl equations, as well as grid generation approaches and a technique to stabilize the transient solution for the wave equation, were discussed further in this chapter. The alternative FETD formulation based on the two first-order equations is also suitable for interfacing to the FDTD method.

The FETD–FDTD technique is a mature technology that has been applied successfully to diverse electromagnetic applications, ranging from electromagnetic compatibility and interference (EMC and EMI) of circuit and system components to large-scale, ultrawideband finite phased arrays, as well as complex antennas having small details located on electrically large platforms. Many of these applications can be found throughout the book, as well as in Refs. 13–20.

REFERENCES

1. J. Maloney, G. Smith, E. Thiele, O. Gandhi, N. Chavannes, and S. Hagness, "Modeling of antennas," in *Computational Electrodynamics: The Finite-Difference Time-Domain Method*, 3rd ed., A. Taflove and S. C. Hagness, Eds. Norwood, MA: Artech House, pp. 607–676, 2005.
2. S. K. Maxumdar, J. E. Lumpp, and S. D. Gedney, "Performance modeling of the finite-difference time-domain method on high performance parallel systems," *Appl. Comput. Electromagn. J.*, vol. 13, no. 2, pp. 147–159, 1998.
3. S. D. Gedney, "Perfectly matched layer absorbing boundary conditions," in *Computational Electrodynamics: The Finite-Difference Time-Domain Method*, 3rd ed., A. Taflove and S. C. Hagness, Eds. Norwood, MA: Artech House, pp. 273–328, 2005.
4. S. Dey and R. Mittra, "A locally conformal finite-difference time-domain algorithm for modeling three-dimensional perfectly conducting objects," *IEEE Microwave Guided Wave Lett.*, vol. 7, pp. 273–275, Sept. 1997.
5. B. Krietenstein, R. Schuhmann, P. Thoma, and T. Weiland, "The perfect boundary approximation technique facing the big challenge of high precision field computation," Presented at the 19th International Linear Accelerator Conference (LINAC 98), Chicago, Aug. 1998.
6. K. S. Yee, J. S. Chen, and A. H. Chang, "Conformal finite-difference time-domain (FDTD) with overlapping grids," *IEEE Trans. Antennas Propagat.*, vol. 40, pp. 1068–1075, Sept. 1992.
7. K. S. Yee and J. S. Chen, "Conformal hybrid finite difference time domain and finite volume time domain," *IEEE Trans. Antennas Propagat.*, vol. 42, pp. 1450–1455, Oct. 1994.
8. D. J. Riley and C. D. Turner, "Interfacing unstructured tetrahedron grids to structured-grid FDTD," *IEEE Microwave Guided Wave Lett.*, vol. 5, no. 9, pp. 284–286, Sept. 1995.

9. D. J. Riley and C. D. Turner, "VOLMAX: A solid-model-based, transient, volumetric Maxwell solver using hybrid grids," *IEEE Antennas Propagat. Mag.*, vol. 39, no. 1, pp. 20–33, Feb. 1997.
10. R. B. Wu and T. Itoh, "Hybridizing FDTD analysis with unconditionally stable FEM for objects of curved boundary," *IEEE MTT-S Int. Microwave Symp. Dig.*, vol. 2, pp. 833–836, May 1995.
11. R. B. Wu and T. Itoh, "Hybrid finite-difference time-domain modeling of curved surfaces using tetrahedral edge elements," *IEEE Trans. Antennas Propagat.*, vol. 45, pp. 1302–1309, Aug. 1997.
12. M. Feliziani and F. Maradei, "Mixed finite-difference/Whitney-elements time domain (FD/WE-TD) method," *IEEE Trans. Magn.*, vol. 34, no. 5, p. 1, pp. 3222–3227, Sept. 1998.
13. T. Rylander and A. Bondeson, "Stable FDTD–FEM hybrid method for Maxwell's equations," *Comput. Phys. Commun.*, vol. 125, pp. 75–82, Mar. 2000.
14. D. Riley, "Transient finite elements for computational electromagnetics: hybridization with finite differences, modeling thin wire and thin slots, and parallel processing," *Proc. 17th Annu. Rev. Prog. Appl. Comput. Electromagn. (ACES)*, Monterey, CA, pp. 128–138, Mar. 2001.
15. N. Montgomery, R. Hutchins, and D. J. Riley, "Thin wire hybrid FETD/FDTD broadband antenna predictions," *USNC/URSI Natl. Radio Sci. Mtg. Dig.*, p. 194, July 2001.
16. T. Rylander and A. Bondeson, "Stability of explicit–implicit hybrid time-stepping schemes for Maxwell's equations," *J. Comput. Phys.*, vol. 179, pp. 426–438, July 2002.
17. T. Rylander, "Stable FDTD–FEM hybrid method for Maxwell's equations," Ph.D. thesis, Department of Electromagnetics, Chalmers University of Technology, Gothenburg, Sweden, 2002.
18. F. Edelvik, "Hybrid solvers for the Maxwell equations in time-domain," Ph.D. thesis, Department of Information Technology, Scientific Computing, Uppsala University, Uppsala, Sweden, 2002.
19. D. Riley, N. Riley, M. Pasik, J. Kotulski, and C. D. Turner, "Analysis of airframe-mounted antennas using parallel and hybridized finite element time-domain methods," *IEEE AP-S Int. Symp. Dig.*, vol. 3, pp. 168–171, San Antonio, TX, June 2002.
20. T. Rylander, F. Edelvik, A. Bondeson, and D. Riley, "Advances in hybrid FDTD-FE techniques," in *Computational Electrodynamics: The Finite-Difference Time-Domain Method*, 3rd ed., A. Taflove and S. C. Hagness, Eds. Norwood, MA: Artech House, pp. 907–953, 2005.
21. G. Cohen and P. Monk. "Gauss point mass lumping schemes for Maxwell's equations," *Numer. Methods Part. Diff. Eq.*, vol. 14, pp. 63–88, Jan. 1998.
22. R. Lee, "A note on mass lumping in the finite element time domain method," *IEEE Trans. Antennas Propagat.*, vol. 54, 2, pp. 760–762, Feb. 2006.
23. K. S. Yee, "Numerical solution of initial boundary value problems involving Maxwell's equations in isotropic media," *IEEE Trans. Antennas Propagat.*, vol. 14, pp. 302–307, May 1966.
24. K. L. Shlager and J. B. Schneider, "A survey of the finite-difference time-domain literature," in *Advances in Computational Electrodynamics: The Finite-Difference Time-Domain Method*, A. Taflove, Ed. Norwood, MA: Artech House, 1998.

25. R. Holland, "THREDS: A finite-difference time-domain EMP code in 3D spherical coordinates," *IEEE Trans. Nucl. Sci.*, vol. 30, pp. 4592–4595, Dec. 1983.
26. J. Van Hese and D. De Zutter, "Modeling of discontinuities in general coaxial waveguide structures by the FDTD method," *IEEE Trans. Microwave Theory Tech.*, vol. 40, pp. 547–556, Mar. 1992.
27. W. F. Ames, *Numerical Methods for Partial Differential Equations*. New York: Academic Press, 1977.
28. A. Taflove and M. E. Brodwin, "Numerical solution of steady-state electromagnetic scattering problems using the time-dependent Maxwell's equations," *IEEE Trans. Microwave Theory Tech.*, vol. 23, pp. 623–630, Aug. 1975.
29. A. Taflove and S. C. Hagness, "Numerical dispersion and stability," in *Computational Electrodynamics: The Finite-Difference Time-Domain Method*, 3rd ed., A. Taflove and S. C. Hagness, Eds. Norwood, MA: Artech House, pp. 107–167, 2005.
30. J. P. Berenger, "A perfectly matched layer for the absorption of electromagnetic waves," *J. Comput. Phys.*, vol. 114, pp. 185–200, Oct. 1994.
31. D. S. Katz, E. T. Thiele, and A. Taflove, "Validation and extension to three-dimensions of the Berenger PML absorbing boundary condition for FDTD meshes," *IEEE Microwave Guided Wave Lett.*, vol. 4, pp. 268–270, Aug. 1994.
32. S. D. Gedney, "An anisotropic PML absorbing media for FDTD simulation of fields in lossy dispersive media," *Electromagnetics*, vol. 16, pp. 399–415, July–Aug. 1996.
33. J. A. Roden and S. D. Gedney, "Convolutional PML (CPML): An efficient FDTD implementation of the CFS-PML for arbitrary media," *Microwave Opt. Tech. Lett.*, vol. 27, pp. 334–339, Dec. 5, 2000.
34. T. Martin, "An improved near- to far-zone transformation for the finite-difference time-domain method," *IEEE Trans. Antennas Propagat.*, vol. 46, pp. 1263–1291, Sept. 1998.
35. R. J. Lubbers, K. S. Kunz, M. Schneider, and F. Hunsberger, "A finite-difference time-domain near zone to far zone transformation," *IEEE Trans. Antennas Propagat.*, vol. 39, pp. 429–433, Apr. 1991.
36. J. C. Nedelec, "Mixed finite elements in R^3," *Numer. Math.*, vol. 35, pp. 315–341, Sept. 1980.
37. A. Bossavit, "Whitney forms: A class of finite elements for three-dimensional computations in electromagnetism," *IEE Proc. H*, vol. 135, pp. 493–500, Nov. 1988.
38. M. Wong, O. Picon, and V. Hanna, "A finite element method based on Whitney forms to solve Maxwell equations in the time domain," *IEEE Trans. Magn.*, vol. 31, pp. 1618–1621, May 1995.
39. T. Hughes, *The Finite Element Method: Linear Static and Dynamic Finite Element Analysis*. Englewood Cliffs, NJ: Prentice-Hall, 1987.
40. J.-M. Jin, *The Finite Element Method in Electromagnetics*, 2nd ed. Hoboken, NJ: Wiley, 2002.
41. T. Belytschko and R. Mullen, "Stability of explicit–implicit mesh partitions in time integration," *Int. J. Numer. Methods Eng.*, vol. 12, pp. 1575–1586, 1978.
42. O. C. Zienkiewicz and R. L. Taylor, *The Finite Element Method*, 5th ed. Oxford, UK: Butterworth-Heinemann, 2000.
43. G. C. Cohen, *Higher Order Numerical Methods for Transient Wave Equations*. Berlin: Springer-Verlag, 2002.

44. N. Madsen and R. Ziolkowski, "A three-dimensional modified finite volume technique for Maxwell's equations," *Electromagnetics*, vol. 10, pp. 147–161, Jan. 1990.

45. V. Shankar, A. H. Mohammadian, and W. F. Hall, "A time-domain finite-volume treatment for the Maxwell's equations," *Electromagnetics*, vol. 10, pp. 127–145, Jan. 1990.

46. J. S. Shang, "Characteristic-based algorithms for solving the Maxwell equations in the time-domain," *IEEE Antennas Propagat. Mag.*, vol. 37, no. 3, pp. 15–25, June 1995.

47. N. Madsen, "Divergence preserving discrete surface integral methods for Maxwell's equations using nonorthogonal unstructured grids," *J. Comput. Phys.*, vol. 119, pp. 34–45, June 1995.

48. J. H. Ferziger and M. Peric, *Computational Methods for Fluid Dynamics*. Berlin: Springer-Verlag, 1996.

49. J. F. Lee and Z. Sacks, "Whitney elements time domain (WETD) methods," *IEEE Trans. Magn.*, vol. 31, pp. 1325–1329, May 1995.

50. ANSYS ICEM-CFD, ANSYS Corporation, Canonsburg, PA.

51. GRIDGEN, Pointwise Inc., Fort Worth, TX.

52. W. Artuzi, "Improving the Newmark time integration scheme in finite element time domain method," *IEEE Microwave Wireless Components Lett.*, vol. 15, no. 12, pp. 898–900, Dec. 2005.

53. N. Venkatarayalu, M. Vouvakis, Y.-G. Gan, and J.-F. Lee, "Suppressing linear time growth in edge element based finite element time domain solution using divergence free constraint equation," *IEEE AP-S Int. Symp. Dig.*, vol. 4B, pp. 193–196, July 2005.

54. R. A. Chilton and R. Lee, "The discrete origin of FETD–Newmark late time instability, and a correction scheme," *J. Comput. Phys.*, vol. 224, pp. 1293–1306, June 2007.

55. D. P. Forrai and E. H. Newman, "Radiation and scattering from loaded microstrip antennas over a wide bandwidth," Tech. Rep. 719493-1, Ohio State University Columbus, OH, Sept. 1988.

56. A. C. Woo, H. T. G. Wang, M. J. Schuh, and M. L. Sanders, "Benchmark radar targets for the validation of computational electromagnetics programs," *IEEE Antennas Propagat. Mag.*, vol. 35, no. 1, pp. 84–89, Feb. 1993.

5 Antenna Source Modeling and Parameter Calculation

To characterize an antenna, the principal measured parameters of interest are the input impedance and radiation patterns, from which other parameters can be derived. For example, the reflection coefficient and the voltage standing-wave ratio (VSWR) observed at the feed can be calculated from the input impedance, and the antenna directivity and gain can be calculated from the radiation pattern. Calculation of the antenna input impedance depends on the feed model employed in the numerical simulation; hence, an appropriate feed model is often essential to obtaining correlation of measured and predicted input impedances. Simplified feed models, such as voltage gap sources, are frequently used to predict accurate far-field patterns and in initial antenna design studies. By using Huygens' principle, the antenna radiation pattern can be computed from a near-field surface integral based on equivalent currents located on the surfaces within the finite element mesh. Because integration tends to average out small errors in the currents, antenna radiation patterns are much less sensitive to the fidelity of the feed model than are impedance predictions.

In this chapter, the numerical modeling of antenna feeds and plane-wave excitations and the calculation of the input impedance and far-field patterns are discussed in detail. The antenna feed models discussed include current probes, voltage gaps, and accurate models for coaxial and other types of waveguides. In addition to the calculation of the input impedance and far-field patterns, the near-field distribution may also reveal useful information about the underlying physics of the antenna performance. Therefore, visualization for the near field is also described briefly at the end of the chapter.

5.1 ANTENNA FEED MODELING

A proper antenna feed model provides not only the desired excitation to the antenna but also a means by which the input impedance and other parameters can be calculated. Despite the various types of feeds that an antenna may actually employ, the numerical models for an antenna feed can be classified into two categories. The first type of feed model incorporates the precise feed geometry into the numerical simulation. Although numerically accurate, these models increase the burden of mesh

Finite Element Analysis of Antennas and Arrays, By Jian-Ming Jin and Douglas J. Riley

generation. More important, challenges may arise when the antenna feed is small in dimensions compared to the wavelength. In these situations, a much denser spatial discretization has to be employed in the vicinity of the feed to resolve the feed geometry, resulting in a highly unbalanced mesh over the entire computational domain. Handled inappropriately, this may cause a significant increase in run time, due to an increase in the condition number of the system matrix.

The second type of feed model is based on a simplified approach that uses a current probe or a voltage gap (usually the low-frequency approximation) to represent the actual antenna feed. Such models are numerically convenient and efficient to implement and do not require especially dense spatial discretization in the vicinity of the feed. Moreover, they enable the separation of antenna properties from the effect of feeding networks. This may be helpful during the design process, in which the objective is to understand the behavior of the antenna itself. However, the accuracy of these models may depend on the frequency, the antenna type, and the parameters of interest.

5.1.1 Current Probe

Simplified feed models are widely used in the numerical simulation of antennas. Although a voltage source such as a delta-gap source is commonly used in moment-method simulations, it is more convenient to model a current source in finite element simulations. The simplest current source model is a short and infinitesimally thin current probe inserted at the feed point of the antenna, as illustrated in Figure 5.1(a). This model was first proposed to excite a microstrip patch antenna in a finite element simulation [1]. In this simple model, the excitation current density $\mathbf{J}_{\text{imp}}(\mathbf{r})$ can be modeled as a delta function in three-dimensional space; for example, a short current probe oriented in the z-direction and located at (x_f, y_f) can be modeled as

$$\mathbf{J}_{\text{imp}}(x, y, z) = \hat{z} I_o \delta(x - x_f,\ y - y_f) \qquad z \in L \tag{5.1}$$

where I_o is the lumped current flowing into the antenna and L is the length of the probe. Such a current source can also be considered as an infinitesimal dipole with a

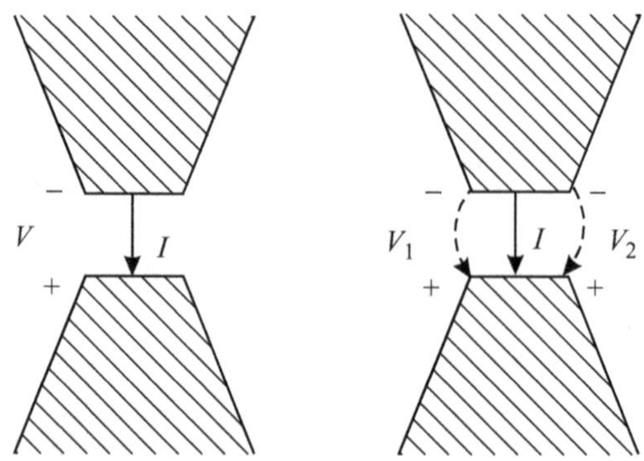

Figure 5.1 Electric current probe excitation. (a) Simplified model. (b) Improved model that calculates the voltage along offset lines. (After Lou and Jin [4], Copyright © IEEE 2005.)

constant current distribution. Using (5.1) as excitation, the electric field everywhere inside the computational domain can be determined uniquely. Once the electric field is determined, the voltage across the probe can be computed according to

$$V = \int_L \hat{z} \cdot \mathbf{E}(x_f, y_f)\, dz \tag{5.2}$$

and the input impedance of the antenna, by its definition, can be calculated as

$$Z_{\text{in}} = V/I. \tag{5.3}$$

The use of an electrically small input current source corresponds to a Norton equivalent network for the source. As such, if one desires a voltage source with amplitude V_o and internal resistance R_o, the definition of I_o in (5.1) would be V_o/R_o. In addition, the input current I in (5.3) can be represented by $I = I_o - V/R_o$. This strategy can be extended to model commonly used circuit elements in the finite element method [2], and this topic is discussed in more detail in Section 6.3.

The current probe model can be used to approximate a variety of actual antenna feeds, such as coaxial cable and microstrip lines. Typically, the length of the probe has to be small compared to the shortest wavelength in the frequency band of interest in order for the distributed effect to be negligible. To illustrate the performance of the current probe model, we consider a microstrip patch antenna geometry consisting of a 5.0-cm × 3.4-cm rectangular conducting patch residing on a dielectric substrate of thickness 0.0877 cm, relative permittivity 2.17, and conductivity 0.362 mS/m. The substrate is housed in a 7.5-cm × 5.1-cm rectangular cavity recessed in a ground plane (Figure 5.2). The patch is excited by a current probe applied at $x_f = 1.22$ cm and $y_f = 0.85$ cm. A 50-Ω impedance load is placed at $x_L = -2.2$ cm and $y_L = -1.5$ cm.

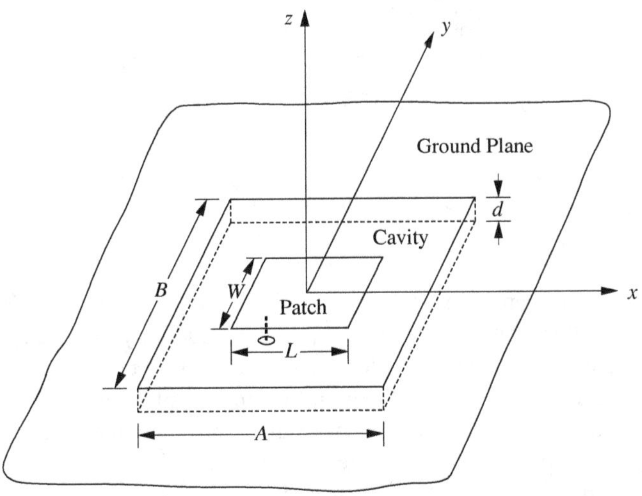

Figure 5.2 Microstrip patch antenna recessed in a ground plane ($A = 7.5$ cm, $B = 5.1$ cm, $L = 5.0$ cm, $W = 3.4$ cm, $d = 0.0877$ cm). The antenna is fed by a coaxial line, which is modeled as an electric current probe.

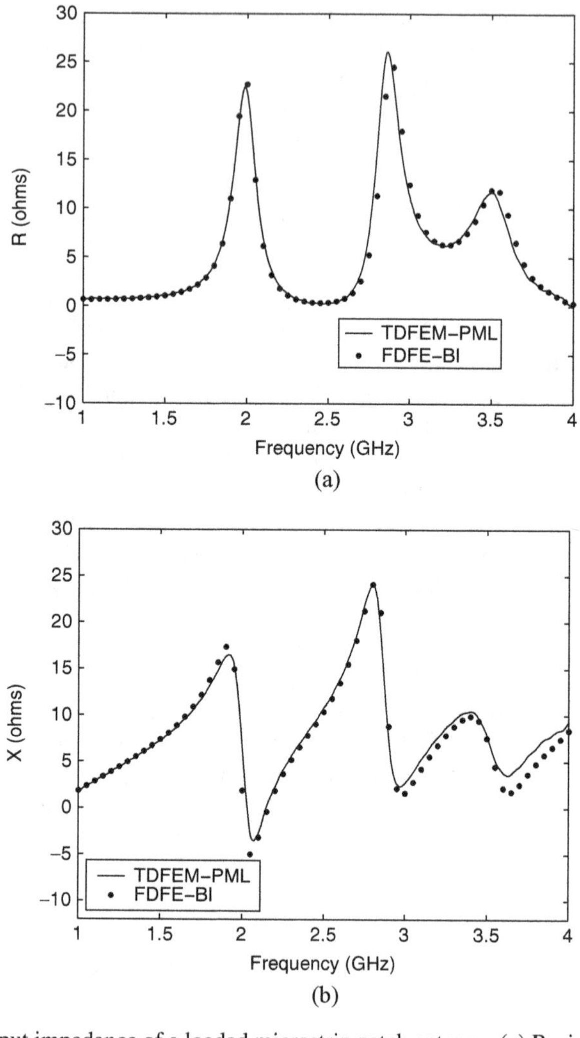

Figure 5.3 Input impedance of a loaded microstrip patch antenna. (a) Resistance. (b) Reactance. (After Jiao and Jin [3], Copyright © Wiley 2002.)

This impedance load can be modeled as a post of finite conductivity between the patch antenna and the ground plane. The required conductivity of the post is given by $\sigma = L/Z_L s$, where L and s denote the length and the cross section of the post and Z_L is the impedance. Figure 5.3 shows the input impedance of the antenna as a function of frequency from 1 to 4 GHz. Two sets of results are shown in the figure. One set is calculated by the frequency-domain finite element method using a boundary integral equation for mesh truncation [1]. The other set is calculated using the time-domain finite element method using perfectly matched layers for mesh truncation [3]. Since both calculations employ the same probe model, their results agree well. Although not included in Figure 5.3, the first set of results has also been compared to experimental data, and the agreement is quite good [1]. This example is the same as the one treated in Figures 4.12 and 4.13 using the FETD–FDTD hybrid technique, where the coaxial feed was modeled physically. It can be seen that the results shown in Figure 5.3 correlate very well with those in Figure 4.13.

In the example above, a good agreement with measurement and the solution of a precise feed modeling is obtained because the dielectric substrate is very thin; hence, the current probe is very short. If we increase the substrate thickness—and, thus, the length of the current probe—the calculation of the input impedance becomes less accurate, although the prediction of far-field behaviors such as radiation patterns and polarization characteristics is not affected. One reason is that for a long probe, the current is no longer constant along the probe. In this case, one can break up the feed line into several segments and replace one segment with a current probe so that the probe can be made short. The fundamental reason for the inaccuracy of the current probe model is that the simplified numerical model does not correspond exactly to the original geometry of the antenna feed. For example, since a radius is not specified for the probe and is usually assumed to be infinitesimally thin, the electromagnetic field is singular in the vicinity of the probe, even though this singularity is dampened in the numerical solution. Numerical errors due to this imprecise feed model affect the calculation of the input impedance much more so than far-field quantities because this calculation depends directly on the field solution local to the feed.

To demonstrate the problem of field singularity, we calculate the input admittance of a monopole antenna over a ground plane, as illustrated in Figure 5.4(a), using the simplified probe model shown in Figure 5.4(b). In the simplified probe model, a 1.6-mm-long electric probe is placed between the ground plane and the monopole as an excitation. The calculated input conductance and susceptance are shown in Figure 5.5 and are compared with the calculation based on the accurate coaxial model to be discussed in Section 5.1.3 [4]. It is clearly evident that there is a significant difference between the probe-model (solid line) and coaxial-model (dotted line) results, despite the fact that both employ an overly dense mesh (to ensure that the numerical discretization error does not affect the comparison). The difference persists when an even denser mesh is employed, which indicates that the difference is indeed due to the aforementioned intrinsic drawback in the simplified probe model. A remedy to avoid the problem caused by the singular field at the probe is to calculate the voltage using the field on a cylindrical surface centered at the probe and having a radius equal to that of the physical feed (the inner conductor in the case of a coaxial feed). Since the field on this cylindrical surface is quite axisymmetric, it is sufficient to calculate the field along one or two lines (called observation probes here) on the surface, as illustrated in Figure 5.1(b). When two observation probes are used, the terminal voltage is then calculated as the average voltage on the two probes. The input admittance calculated by using this improved model is also shown (dashed line) in Figure 5.5. It is noted that the result agrees with the coaxial model much better than does the original probe model. In contrast with the delta-gap model (discussed next), where the susceptance is slightly smaller compared to the accurate measurement [5], it is observed that the input susceptance calculated by the probe model is always slightly

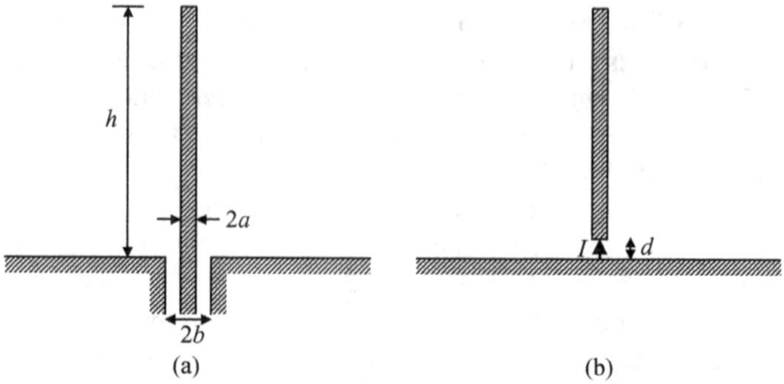

Figure 5.4 Cylindrical monopole antenna on a ground plane. (a) Geometry ($a = 1.52$ mm, $b = 3.5$ mm, $h = 5$ cm). (b) Simplified probe feed model ($d = 1.6$ mm).

larger than the corresponding coaxial-model result, as can be seen in Figure 5.5(b). The difference can be explained by the additional susceptance associated with the artificial gap between the conductors. Figure 5.6 shows an equivalent circuit for the feed, where the gap is modeled as a capacitor in parallel with the antenna. In the case of the monopole antenna, the gap is modeled conveniently as a parallel-plate capacitor, and its capacitance is determined by

$$C = \frac{\varepsilon_0 A}{d} \tag{5.4}$$

where A is the cross-sectional area of the monopole and d is the length of the gap. After C is determined, the antenna admittance Y_a can be corrected as

$$Y_a = Y_p - j\omega C \tag{5.5}$$

where Y_p denotes the antenna admittance calculated using the improved probe model. With this correction, the input susceptance result becomes more accurate, as shown in Figure 5.5(b), especially at higher frequencies. It should be noted that the good agreement shown in Figure 5.5 is obtained specific to the coaxial-fed monopole antenna problem. The accuracy of the simplified source model with respect to the original feeding structure will, of course, depend on the specific geometries and configurations. The example above is intended to demonstrate that the simplified probe feed model can still be used to predict the antenna input impedance or admittance if the electromagnetic field in the feed region is modeled properly.

5.1.2 Voltage Gap Generator

As noted previously, the concept of the delta-gap voltage generator is used widely for the analysis of linear antennas based on the moment method [6] and the finite-difference time-domain (FDTD) method [5]. Although the current-probe model

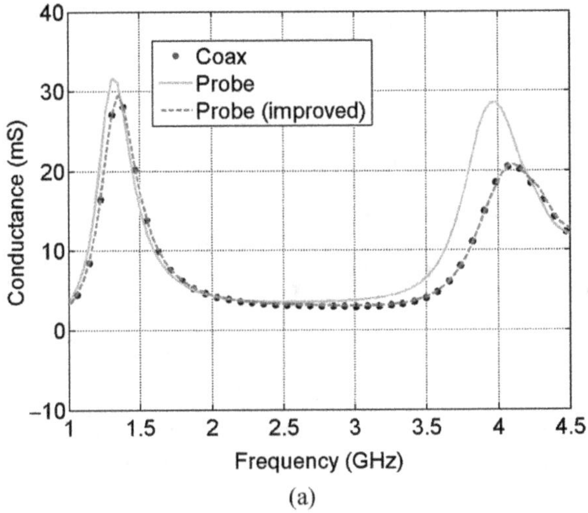

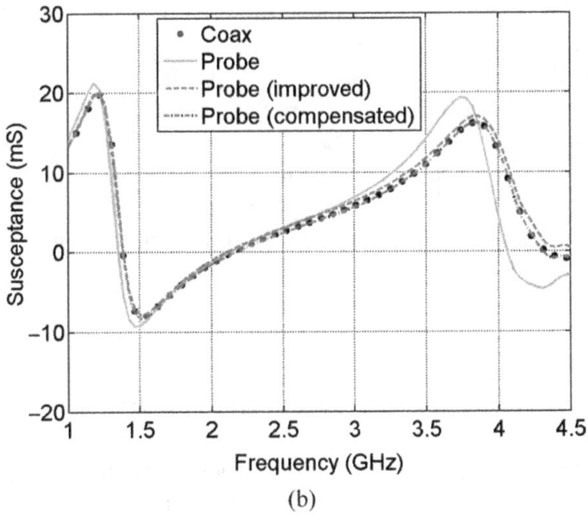

Figure 5.5 Cylindrical monopole antenna. (a) Input conductance. (b) Input susceptance. (After Lou and Jin [4], Copyright © IEEE 2005.)

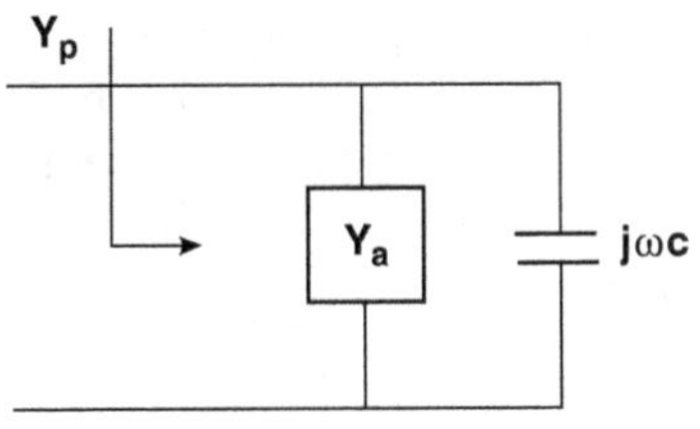

Figure 5.6 Equivalent circuit of the probe feed.

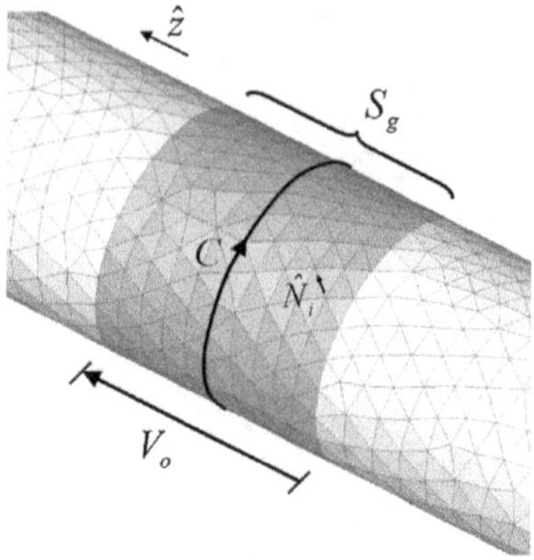

Figure 5.7 Linear dipole antenna showing local spatial grid region for impressing a distributed voltage source for modeling a delta-gap voltage generator.

discussed in the preceding section is commonly used in finite element analyses, the delta-gap voltage generator is also straightforward to apply within the finite element method [1]. The delta-gap source assumes that the impressed voltage induces an incident electric field that exists only within the region of the gap. A simple dipole antenna is shown in Figure 5.7, and the feed region is represented by the surface S_g, which corresponds to a "belt" of width d ($\ll \lambda$) around the circumference of the $\hat{z}$-directed linear antenna with radius a. The potential difference across the belt is defined to be V_o. Consequently, the electric field on S_g can be defined as

$$\mathbf{E}(\mathbf{r}) = -\frac{V_o}{d}\,\hat{z} \qquad \mathbf{r} \in S_g. \tag{5.6}$$

In the finite element setting, this expression can be incorporated conveniently as a Dirichlet boundary condition by projecting the defined electric field onto the finite element edges located on S_g. Specifically, the coefficient for the gap electric field associated with the ith finite element edge on S_g is defined by

$$E_i = -\frac{V_o}{d}\,\frac{\hat{z}\cdot\mathbf{N}_i}{\mathbf{N}_i\cdot\mathbf{N}_i}. \tag{5.7}$$

Once the electric field is computed globally, we can calculate the magnetic field and integrate along a closed contour encircling the feed to find the current flowing along the feed, from which the input impedance can subsequently be evaluated as

$$Z_{\text{in}} = \frac{V_o}{\oint_C \mathbf{H}\cdot d\mathbf{l}} \tag{5.8}$$

where the integration contour C is illustrated in Figure 5.7.

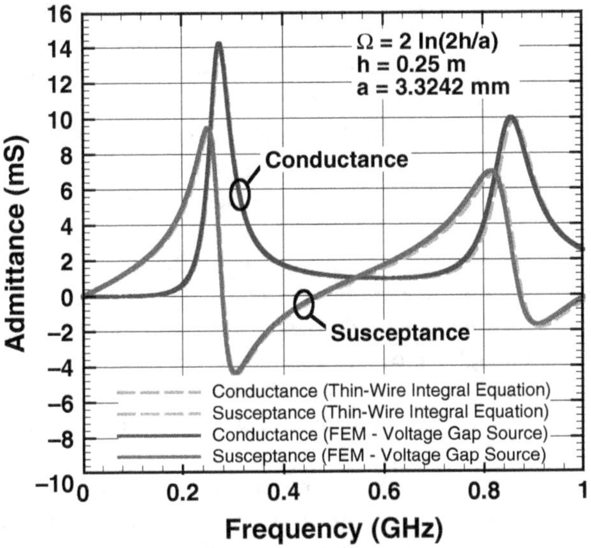

Figure 5.8 Input admittance of a straight wire antenna of half-length 0.25 m and radius 3.32 mm. The finite element solution is compared with the moment-method solution of the thin-wire integral equation.

As an example to demonstrate the application of the voltage gap generator, we simulate a straight wire antenna having a half-length $h = 0.25$ m and a radius $a = 3.32$ mm using the hybrid FETD–FDTD algorithm described in Chapter 4. The antenna thickness factor is $2\ \ln(2h/a) = 10$, which corresponds to a moderately thin wire antenna. A voltage gap generator is created by impressing (5.7) over the entire circumference of the 5-mm-long gap region created at the middle of the antenna. The input admittance calculated using (5.8) is plotted in Figure 5.8 and compared with the well-established solution obtained by solving a thin-wire integral equation using the method of moments. Even though the two solutions are obtained using two very different approaches, the agreement is remarkably good over the frequency band from 0 to 1 GHz. It is observed that the moment-method solution appears to model a very slightly shorter antenna because of the thin-wire approximation. As pointed out earlier, the voltage gap model may underestimate the susceptance slightly and has to be used carefully to obtain converged results [5].

5.1.3 Waveguide Feed Model

Although simplified feed models are numerically attractive, their use is limited in certain applications. For example, in the simulation of a horn antenna fed by a rectangular waveguide, it is impossible to employ a simplified feed model to account for the modal distribution inside the waveguide and to absorb the field reflected back into the waveguide. In these situations it is necessary to model the antenna feed in a precise manner. This modeling requires the truncation of the feeding waveguide to create a waveguide port and the formulation of an appropriate boundary condition to be applied at the waveguide port. The desired boundary condition should be able to launch an incident wave into the waveguide and, at the same time, absorb the reflected wave from the antenna without any spurious reflection. Moreover, it is desirable to place the port boundary as close to the antenna as possible to reduce the size of the computational domain. One approach to terminating a waveguide port is to use perfectly matched layers. However, the absorption of perfectly matched layers becomes less effective as the frequency approaches the cutoff frequency of the operating mode. Furthermore, the original perfectly matched layers were not designed to absorb the evanescent waves that are present in the waveguide for modes below their cutoff frequencies, although a modified formulation of the perfectly matched layers has been proposed recently to improve its ability to absorb evanescent waves without compromising its ability to absorb propagating waves, as discussed in Section 3.2.3. Nevertheless, the modified formulation is computationally intensive, thus making perfectly matched layers less attractive for terminating waveguides.

A more accurate and efficient approach is to impose a mixed boundary condition, specifically designed for the waveguide, on the waveguide port [7]. The exact boundary condition can be derived based on waveguide modal expansions. All waveguide modes, including both propagation and evanescent modes, can be perfectly absorbed by the boundary condition. This allows the waveguide to be truncated very close to the antenna. To derive this boundary condition, consider a waveguide port with a given incident electric field $\mathbf{E}^{\text{inc}}$ and an unknown reflected field $\mathbf{E}^{\text{ref}}$ (Figure 5.9). If this waveguide is homogeneous and isotropic, the reflected field can be expressed as the superposition of orthogonal TEM (if it exists), TE, and TM waveguide modes as

$$\begin{aligned}\mathbf{E} &= \mathbf{E}^{\text{inc}} + \mathbf{E}^{\text{ref}} \\ &= \mathbf{E}^{\text{inc}} + a_0 \mathbf{e}_0^{\text{TEM}} e^{\gamma_0^{\text{TEM}} z} + \sum_{m=1}^{\infty} a_m \mathbf{e}_m^{\text{TE}} e^{\gamma_m^{\text{TE}} z} + \sum_{m=1}^{\infty} b_m \mathbf{e}_m^{\text{TM}} e^{\gamma_m^{\text{TM}} z}\end{aligned} \tag{5.9}$$

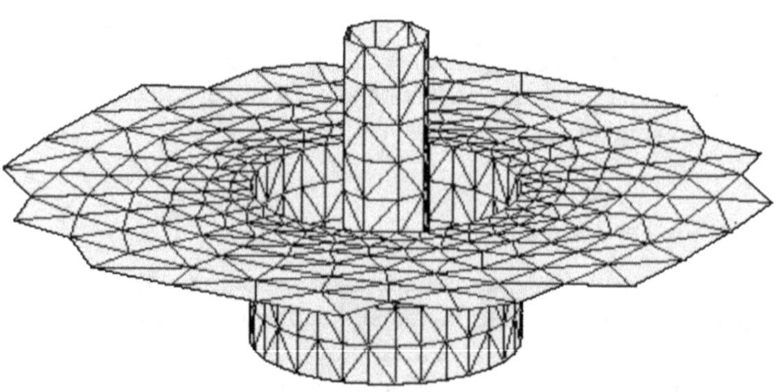

Figure 5.9 Coaxial feed model.

where m is the modal index; a_0, a_m, and b_m are the modal amplitudes; γ_0^{TEM}, γ_m^{TE}, and γ_m^{TM} are the modal propagation constants; and $\mathbf{e}_0^{\mathrm{TEM}}$, $\mathbf{e}_m^{\mathrm{TE}}$, and $\mathbf{e}_m^{\mathrm{TM}}$ are the modal functions for the TEM, TE, and TM modes, respectively. In (5.9), z is a local coordinate, which is defined to be perpendicular to the port surface and point in the direction of the propagation of the incident field. Applying a curl operator to both sides of (5.9) yields

$$\hat{n} \times (\nabla \times \mathbf{E}) = \hat{n} \times (\nabla \times \mathbf{E}^{\mathrm{inc}}) + \gamma_0^{\mathrm{TEM}} a_0 \mathbf{e}_0^{\mathrm{TEM}} e^{\gamma_0^{\mathrm{TEM}} z} + \sum_{m=1}^{\infty} \gamma_m^{\mathrm{TE}} a_m \mathbf{e}_m^{\mathrm{TE}} e^{\gamma_m^{\mathrm{TE}} z} + \sum_{m=1}^{\infty} \frac{-k^2}{\gamma_m^{\mathrm{TM}}} b_m \mathbf{e}_{tm}^{\mathrm{TM}} e^{\gamma_m^{\mathrm{TM}} z} \tag{5.10}$$

where $\mathbf{e}_{tm}^{\mathrm{TM}}$ denotes the transverse component of $\mathbf{e}_m^{\mathrm{TM}}$. By modal orthogonality, the modal amplitudes a_0, a_m, and b_m can be found as

$$a_0 = \iint_{S_p} \mathbf{e}_0^{\mathrm{TEM}} \cdot [\mathbf{E} - \mathbf{E}^{\mathrm{inc}}]\, dS \tag{5.11}$$

$$a_m = \iint_{S_p} \mathbf{e}_m^{\mathrm{TE}} \cdot [\mathbf{E} - \mathbf{E}^{\mathrm{inc}}]\, dS \qquad m = 1, 2, \ldots \tag{5.12}$$

$$b_m = \iint_{S_p} \mathbf{e}_{tm}^{\mathrm{TM}} \cdot [\mathbf{E} - \mathbf{E}^{\mathrm{inc}}]\, dS \qquad m = 1, 2, \ldots \tag{5.13}$$

where S_p denotes the waveguide port cross section. By substituting (5.11)–(5.13) into (5.10), (5.10) can be written in compact form as

$$\hat{n} \times (\nabla \times \mathbf{E}) + P(\mathbf{E}) = \mathbf{U}^{\mathrm{inc}} \tag{5.14}$$

where

$$P(\mathbf{E}) = -\gamma_0^{\mathrm{TEM}} \mathbf{e}_0^{\mathrm{TEM}} \iint_{S_p} \mathbf{e}_0^{\mathrm{TEM}} \cdot \mathbf{E}\, dS - \sum_{m=1}^{\infty} \gamma_m^{\mathrm{TE}} \mathbf{e}_m^{\mathrm{TE}} \iint_{S_p} \mathbf{e}_m^{\mathrm{TE}} \cdot \mathbf{E}\, dS - \sum_{m=1}^{\infty} \frac{-k^2}{\gamma_m^{\mathrm{TM}}} \mathbf{e}_{tm}^{\mathrm{TM}} \iint_{S_p} \mathbf{e}_{tm}^{\mathrm{TM}} \cdot \mathbf{E}\, dS \tag{5.15}$$

$$\mathbf{U}^{\mathrm{inc}} = \hat{n} \times (\nabla \times \mathbf{E}^{\mathrm{inc}}) - \gamma_0^{\mathrm{TEM}} \mathbf{e}_0^{\mathrm{TEM}} \iint_{S_p} \mathbf{e}_0^{\mathrm{TEM}} \cdot \mathbf{E}^{\mathrm{inc}}\, dS - \sum_{m=1}^{\infty} \gamma_m^{\mathrm{TE}} \mathbf{e}_m^{\mathrm{TE}} \iint_{S_p} \mathbf{e}_m^{\mathrm{TE}} \cdot \mathbf{E}^{\mathrm{inc}}\, dS - \sum_{m=1}^{\infty} \frac{-k^2}{\gamma_m^{\mathrm{TM}}} \mathbf{e}_{tm}^{\mathrm{TM}} \iint_{S_p} \mathbf{e}_{tm}^{\mathrm{TM}} \cdot \mathbf{E}^{\mathrm{inc}}\, dS. \tag{5.16}$$

Equation (5.14) was first derived in the first edition of Ref. 7, applied to the analysis of a variety of waveguide devices by Liu et al. [8], and was later referred to as the *waveguide port boundary condition* (WPBC) [9]. Note that similar formulations based on the modal expansion and mode matching have also been developed by others [10]. The analytical modal functions and their propagation constants ($\gamma_m = \sqrt{k_{cm}^2 - k^2}$) can be found for certain common waveguide geometries, such as coaxial, rectangular, and circular waveguides (Tables 5.1 to 5.3). For waveguides with an irregular cross section, the modal functions and their propagation constants can be computed using a two-dimensional finite element method [7]. In fact, even for regular coaxial, rectangular, and circular waveguide ports, it is often more efficient to employ the numerical modal functions than the analytical ones in Tables 5.1 to 5.3.

The waveguide port boundary condition in (5.14) can readily be incorporated into the frequency-domain weak-form wave equation (2.12), yielding

$$\iiint_V \left[(\nabla \times \mathbf{T}) \cdot \overleftrightarrow{\mu}_r^{-1} \cdot (\nabla \times \mathbf{E}) - k_0^2 \mathbf{T} \cdot \overleftrightarrow{\varepsilon}_r \cdot \mathbf{E}\right] dV + jk_0 \oiint_{S_o} (\hat{n} \times \mathbf{T}) \cdot (\hat{n} \times \mathbf{E})\, dS$$
$$- \iint_{S_p} \frac{1}{\mu_r} \mathbf{T} \cdot P(\mathbf{E})\, dS = - \iint_{S_p} \frac{1}{\mu_r} \mathbf{T} \cdot \mathbf{U}^{\text{inc}}\, dS \tag{5.17}$$

where μ_r denotes the relative permeability of the material filling at the waveguide port, which is assumed to be homogeneous and isotropic. Following the spatial discretization procedure outlined in Chapter 2, a linear system of equations can be derived from (5.17) for a unique determination of the electric field everywhere inside the antenna and at the waveguide port.

The procedure described above can be extended to inhomogeneously filled waveguides [11]. For such a waveguide, the modes can no longer be expressed as TEM,

TABLE 5.1 Modal Functions and Their Cutoff Wavenumbers for a Rectangular Waveguide with a Cross Section of $a \times b$

Modal Functions $\mathbf{e}_0^{\text{TEM}}$, $\mathbf{e}_m^{\text{TE}}$, and $\mathbf{e}_{tm}^{\text{TM}}$	Cutoff Wavenumbers k_{cm}
$\mathbf{e}_0^{\text{TEM}} = 0$	
$\mathbf{e}_{pq}^{\text{TE}} = \sqrt{\frac{v_p v_q}{ab}} \frac{1}{k_{cpq}^{\text{TE}}} \left(\hat{x} \frac{q\pi}{b} \cos \frac{p\pi x}{a} \sin \frac{q\pi y}{b} - \hat{y} \frac{p\pi}{a} \sin \frac{p\pi x}{a} \cos \frac{q\pi y}{b}\right)$	$k_{cpq}^{\text{TE}} = \sqrt{\left(\frac{p\pi}{a}\right)^2 + \left(\frac{q\pi}{b}\right)^2}$
$\mathbf{e}_{tpq}^{\text{TM}} = \sqrt{\frac{v_p v_q}{ab}} \frac{1}{k_{cpq}^{\text{TM}}} \left(\hat{x} \frac{p\pi}{a} \cos \frac{p\pi x}{a} \sin \frac{q\pi y}{b} + \hat{y} \frac{q\pi}{b} \sin \frac{p\pi x}{a} \cos \frac{q\pi y}{b}\right)$	$k_{cpq}^{\text{TM}} = \sqrt{\left(\frac{p\pi}{a}\right)^2 + \left(\frac{q\pi}{b}\right)^2}$
$v_p = \begin{cases} 1 & p = 0 \\ 2 & p \neq 0 \end{cases} \qquad v_q = \begin{cases} 1 & q = 0 \\ 2 & q \neq 0 \end{cases}$	

TABLE 5.2 Modal Functions and Their Cutoff Wavenumbers for a Circular Cylindrical Waveguide with a Radius of *a*

Modal Functions $\mathbf{e}_0^{\text{TEM}}$, $\mathbf{e}_m^{\text{TE}}$, and $\mathbf{e}_{tm}^{\text{TM}}$	Cutoff Wavenumbers k_{cm}
$\mathbf{e}_0^{\text{TEM}} = 0$	
$\mathbf{e}_{pq}^{\text{TE}} = \dfrac{\sqrt{\nu_p/2}}{\pi\sqrt{\chi_{pq}'^2 - p^2}J_p(\chi'_{pq})}\left[\hat{\rho}\dfrac{p}{\rho}J_p(\chi'_{pq}\rho/a)\begin{Bmatrix}-\sin p\phi\\ \cos p\phi\end{Bmatrix} - \hat{\phi}\dfrac{\chi'_{pq}}{a}J'_p(\chi'_{pq}\rho/a)\begin{Bmatrix}\cos p\phi\\ \sin p\phi\end{Bmatrix}\right]$	$k_{cpq}^{\text{TE}} = \dfrac{\chi'_{pq}}{a}, \quad J'_p(\chi'_{pq}) = 0$
$\mathbf{e}_{tpq}^{\text{TM}} = \dfrac{\sqrt{\nu_p/2}}{\pi\chi_{pq}J'_p(\chi_{pq})}\left[\hat{\rho}\dfrac{\chi_{pq}}{a}J'_p(\chi_{pq}\rho/a)\begin{Bmatrix}\cos p\phi\\ \sin p\phi\end{Bmatrix} + \hat{\phi}\dfrac{p}{\rho}J_p(\chi_{pq}\rho/a)\begin{Bmatrix}-\sin p\phi\\ \cos p\phi\end{Bmatrix}\right]$	$k_{cpq}^{\text{TM}} = \dfrac{\chi_{pq}}{a}, \quad J_p(\chi_{pq}) = 0$

TABLE 5.3 Modal Functions and Their Cutoff Wavenumbers for a Coaxial Waveguide with an Inner Radius *a* and an Outer Radius *b*

Modal Functions $\mathbf{e}_0^{\text{TEM}}$, $\mathbf{e}_m^{\text{TE}}$, and $\mathbf{e}_{tm}^{\text{TM}}$	Cutoff Wavenumbers k_{cm}
$\mathbf{e}_0^{\text{TEM}} = \hat{\rho}\dfrac{1}{\sqrt{2\pi\ln(b/a)}}\dfrac{1}{\rho}$	$k_{c0}^{\text{TEM}} = 0$
$\mathbf{e}_{pq}^{\text{TE}} = \dfrac{b\sqrt{\nu_p}}{2\pi\chi'_{pq}I_T(\chi'_{pq})}\left[\hat{\rho}\dfrac{p}{\rho}T_p(\chi'_{pq}\rho/b)\begin{Bmatrix}-\sin p\phi\\ \cos p\phi\end{Bmatrix} - \hat{\phi}\dfrac{\chi'_{pq}}{b}T'_p(\chi'_{pq}\rho/b)\begin{Bmatrix}\cos p\phi\\ \sin p\phi\end{Bmatrix}\right]$	$k_{cpq}^{\text{TE}} = \dfrac{\chi'_{pq}}{b}, \quad T'_p(\chi'_{pq}) = 0$
$I_T(\chi'_{pq}) = \left[\displaystyle\int_a^b T_p^2(\chi'_{pq}\rho/b)d\rho\right]^{1/2}$	$T_p(\chi'_{pq}\rho/b) = Y'_p(\chi'_{pq}a/b)J_p(\chi'_{pq}\rho/b) - J'_p(\chi'_{pq}a/b)Y_p(\chi'_{pq}\rho/b)$
$\mathbf{e}_{tpq}^{\text{TM}} = \dfrac{b\sqrt{\nu_p}}{2\pi\chi_{pq}I_S(\chi_{pq})}\left[\hat{\rho}\dfrac{\chi_{pq}}{b}S'_p(\chi_{pq}\rho/b)\begin{Bmatrix}\cos p\phi\\ \sin p\phi\end{Bmatrix} + \hat{\phi}\dfrac{p}{\rho}S_p(\chi_{pq}\rho/b)\begin{Bmatrix}-\sin p\phi\\ \cos p\phi\end{Bmatrix}\right]$	$k_{cpq}^{\text{TM}} = \dfrac{\chi_{pq}}{b}, \quad S_p(\chi_{pq}) = 0$
$I_S(\chi_{pq}) = \left[\displaystyle\int_a^b S_p^2(\chi_{pq}\rho/b)d\rho\right]^{1/2}$	$S_p(\chi_{pq}\rho/b) = Y_p(\chi_{pq}a/b)J_p(\chi_{pq}\rho/b) - J_p(\chi_{pq}a/b)Y_p(\chi_{pq}\rho/b)$

TE, and TM modes. Furthermore, except for a few special cases, the modal functions cannot be derived analytically. Instead, they have to be calculated numerically using a two-dimensional finite element method [12,13]. If we denote the electric and magnetic modal functions as $\mathbf{e}_m$ and $\mathbf{h}_m$ for mode m, which satisfy the modal orthogonal relation

$$\iint_{S_p} (\mathbf{h}_m \times \mathbf{e}_n) \cdot \hat{z} \, dS = \begin{cases} \kappa_m & m = n \\ 0 & m \neq n \end{cases} \tag{5.18}$$

the total field at the waveguide port can be expanded as

$$\mathbf{E} = \mathbf{E}^{\text{inc}} + \mathbf{E}^{\text{ref}} = \mathbf{E}^{\text{inc}} + \sum_{m=1}^{\infty} a_m \mathbf{e}_m e^{\gamma_m z}. \tag{5.19}$$

The expansion coefficient can be determined using (5.18) as

$$a_m = \frac{1}{\kappa_m} \iint_{S_p} [\mathbf{h}_m \times (\mathbf{E} - \mathbf{E}^{\text{inc}})] \cdot \hat{z} \, dS. \tag{5.20}$$

Applying a curl operator to both sides of (5.19) and using Maxwell's equations for the modal fields, we obtain

$$\hat{n} \times (\nabla \times \mathbf{E}) = \hat{n} \times (\nabla \times \mathbf{E}^{\text{inc}}) - \frac{j\omega\mu}{\kappa_m} \sum_{m=1}^{\infty} a_m (\hat{n} \times \mathbf{h}_m) e^{\gamma_m z}. \tag{5.21}$$

Since

$$\mathbf{h}_m e^{\gamma_m z} = \frac{j}{\omega\mu} \nabla \times (\mathbf{e}_m e^{\gamma_m z}) = \frac{j}{\omega\mu} (\nabla_t + \gamma_m \hat{z}) \times \mathbf{e}_m e^{\gamma_m z} \tag{5.22}$$

we have

$$\hat{z} \times \mathbf{h}_m = \frac{1}{j\omega\mu} (\gamma_m \mathbf{e}_{tm} - \nabla_t e_{zm}). \tag{5.23}$$

Substituting this into (5.20) and (5.21), we obtain the waveguide port boundary condition for an inhomogeneously filled waveguide:

$$\hat{n} \times (\nabla \times \mathbf{E}) + P(\mathbf{E}) = \mathbf{U}^{\text{inc}} \tag{5.24}$$

with

$$P(\mathbf{E}) = \sum_{m=1}^{\infty} \frac{1}{j\omega\mu\kappa_m}(\gamma_m \mathbf{e}_{tm} - \nabla_t e_{zm}) \iint_{S_p} (\gamma_m \mathbf{e}_{tm} - \nabla_t e_{zm}) \cdot \mathbf{E}\ dS \tag{5.25}$$

$$\mathbf{U}^{\text{inc}} = \hat{n} \times (\nabla \times \mathbf{E}^{\text{inc}}) + \sum_{m=1}^{\infty} \frac{1}{j\omega\mu\kappa_m}(\gamma_m \mathbf{e}_{tm} - \nabla_t e_{zm}) \iint_{S_p} (\gamma_m \mathbf{e}_{tm} - \nabla_t e_{zm}) \cdot \mathbf{E}^{\text{inc}}\ dS. \tag{5.26}$$

Note that for an inhomogeneously filled waveguide, the modal fields are functions of frequency, whereas for a homogeneously filled waveguide, they are independent of frequency.

The frequency-domain waveguide port boundary condition for a homogeneous waveguide derived above can be converted to the time domain in a straightforward manner. Using the Laplace transform pair, which is denoted by $\leftrightarrow$, we find that

$$\gamma_0^{\text{TEM}} = jk = \frac{j\omega}{c} \leftrightarrow \frac{1}{c}\frac{\partial}{\partial t} \tag{5.27}$$

$$\gamma_m^{\text{TE}} = \sqrt{k_{cm}^2 - k^2} = \sqrt{k_{cm}^2 + (j\omega/c)^2} \leftrightarrow \frac{1}{c}\frac{\partial}{\partial t} + h_m(t) * \tag{5.28}$$

$$\frac{-k^2}{\gamma_m^{\text{TM}}} = \frac{-k^2}{\sqrt{k_{cm}^2 - k^2}} = \frac{(j\omega/c)^2}{\sqrt{k_{cm}^2 + (j\omega/c)^2}} \leftrightarrow \frac{1}{c}\frac{\partial}{\partial t} + g_m(t) * \tag{5.29}$$

where k_{cm} is the cutoff wavenumber of the mth mode, $c = 1/\sqrt{\mu\varepsilon}$, $*$ again stands for the time-domain convolution, and h_m and g_m are time-domain functions with expressions that can be found analytically as

$$h_m(t) = \frac{k_{cm}}{t} J_1(k_{cm}ct)u(t) \tag{5.30}$$

$$g_m(t) = \frac{k_{cm}}{t} J_1(k_{cm}ct)u(t) - k_{cm}^2 c J_0(k_{cm}ct)u(t) \tag{5.31}$$

in which $u(t)$ denotes a step function and J_0 and J_1 denote the first- and second-order Bessel functions, respectively. Using these results, we obtain the waveguide port boundary condition in the time domain as

$$\hat{n} \times (\nabla \times \mathbf{E}) + P(\mathbf{E}) = \mathbf{U}^{\text{inc}} \tag{5.32}$$

where

$$P(\mathbf{E}) = -\mathbf{e}_0^{\mathrm{TEM}} \iint_{S_p} \mathbf{e}_0^{\mathrm{TEM}} \cdot \frac{1}{c}\frac{\partial}{\partial t}\mathbf{E}\, dS - \sum_{m=1}^{\infty} \mathbf{e}_m^{\mathrm{TE}} \iint_{S_p} \mathbf{e}_m^{\mathrm{TE}} \cdot \left(\frac{1}{c}\frac{\partial}{\partial t}\mathbf{E} + h_m * \mathbf{E}\right) dS$$
$$- \sum_{m=1}^{\infty} \mathbf{e}_{tm}^{\mathrm{TM}} \iint_{S_p} \mathbf{e}_{tm}^{\mathrm{TM}} \cdot \left(\frac{1}{c}\frac{\partial}{\partial t}\mathbf{E} + g_m * \mathbf{E}\right) dS \tag{5.33}$$

$$\mathbf{U}^{\mathrm{inc}} = \hat{n} \times (\nabla \times \mathbf{E}^{\mathrm{inc}}) - \mathbf{e}_0^{\mathrm{TEM}} \iint_{S_p} \mathbf{e}_0^{\mathrm{TEM}} \cdot \frac{1}{c}\frac{\partial}{\partial t}\mathbf{E}^{\mathrm{inc}}\, dS$$
$$- \sum_{m=1}^{\infty} \mathbf{e}_m^{\mathrm{TE}} \iint_{S_p} \mathbf{e}_m^{\mathrm{TE}} \cdot \left(\frac{1}{c}\frac{\partial}{\partial t}\mathbf{E}^{\mathrm{inc}} + h_m * \mathbf{E}^{\mathrm{inc}}\right) dS$$
$$- \sum_{m=1}^{\infty} \mathbf{e}_{tm}^{\mathrm{TM}} \iint_{S_p} \mathbf{e}_{tm}^{\mathrm{TM}} \cdot \left(\frac{1}{c}\frac{\partial}{\partial t}\mathbf{E}^{\mathrm{inc}} + g_m * \mathbf{E}^{\mathrm{inc}}\right) dS. \tag{5.34}$$

Equations (5.32), (5.33), and (5.34) can easily be incorporated into the time-domain weak-form wave equation (2.25), which after spatial and temporal discretization can be converted into a time-marching system [9].

In both the frequency- and time-domain versions, the waveguide port boundary condition is expressed as a summation of an infinite number of waveguide modes. When the series are not truncated, the expression is the exact boundary condition satisfied on the waveguide port. In a numerical implementation, however, the infinite summation is always truncated, and hence the boundary condition is only approximate. Nevertheless, since the magnitudes of the higher-order evanescent modes decay quickly, it is usually sufficient to include only a small number of evanescent modes in the boundary condition in addition to all the propagating modes, or alternatively, increase the separation distance to the waveguide port such that most evanescent modes are attenuated before reaching the waveguide port. Once the field at the waveguide port is computed, the reflection coefficient (or S_{11}) at the waveguide port can be extracted according to

$$R = \frac{\iint_{S_p} (\mathbf{E} - \mathbf{E}^{\mathrm{inc}}) \cdot \mathbf{E}^{\mathrm{inc}}\, dS}{\iint_{S_p} |\mathbf{E}^{\mathrm{inc}}|^2\, dS}. \tag{5.35}$$

Then the input impedance can be calculated from the reflection coefficient R according to

$$Z_{\mathrm{in}} = \frac{1+R}{1-R} Z_0 \tag{5.36}$$

where Z_0 is the characteristic impedance of the waveguide.

The waveguide port boundary condition has been implemented and studied extensively for the finite element analysis of waveguide devices and antennas in both the frequency domain [8,11] and the time domain [4,9]. Its accuracy has been demonstrated by many successful examples. Here, we consider three examples related to antenna simulation. The first example is the simulation of a cylindrical monopole antenna, illustrated in Figure 5.10(a). The antenna is fed by a coaxial line with an incident TEM mode. The coaxial line is terminated by the waveguide port boundary condition a small distance below the ground plane. The open computational domain above the ground plane is truncated by a hemispherical surface on which the first-order absorbing boundary condition is imposed to absorb the radiated fields. Accurate analysis and measurement of the same antenna have been reported in the past [14] and the results are used here as reference data. The reflection coefficients calculated

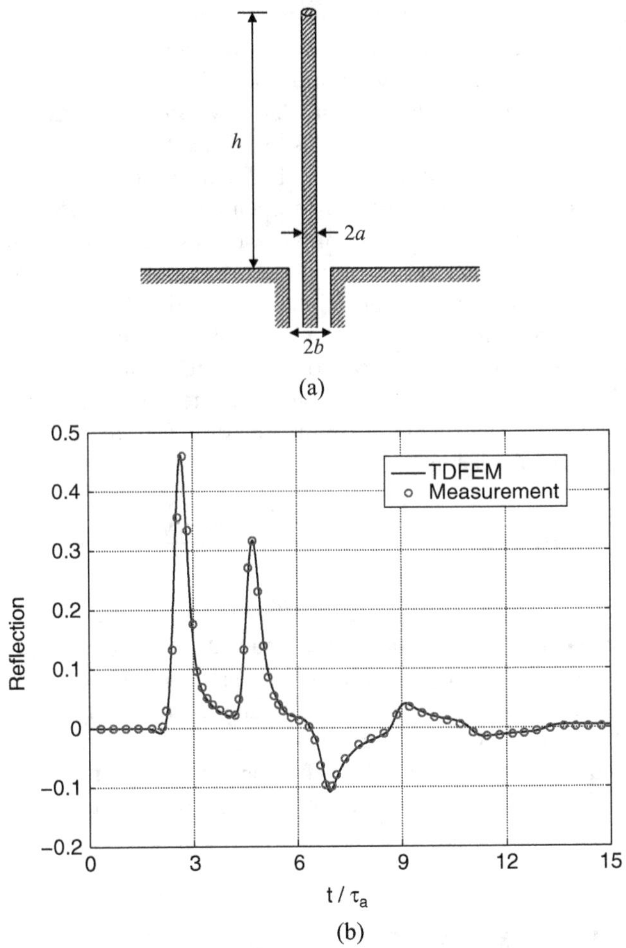

Figure 5.10 Cylindrical monopole antenna. (a) Geometry (a = 1 mm, b = 2.3 mm, h = 32.8 mm). (b) Time-domain reflection coefficient at the coaxial port ($\tau_a = h/c$). (After Lou and Jin [9], Copyright © IEEE 2005.)

on the coaxial port are shown in Figure 5.10(b). As can be seen, the simulated results are in excellent agreement with the measured data.

The second example is a 2 × 2 microstrip patch antenna array with the specific geometry described in Figure 5.11(a). The coaxial line feed has inner and outer radii $r_i = 0.48$ mm and $r_o = 1.5$ mm and is filled with dielectric material with an ε_r value of 1.86. The entire array is printed on a substrate with $\varepsilon_r = 2.67$ and $\mu_r = 1.0$ and is housed in a 127.2-mm × 116-mm × 7-mm cavity in a ground plane. Figure 5.11(b) shows the calculation of the mutual coupling (S-parameters) among its four antenna elements [15]. More specifically, it gives the plots of the 4 × 4 scattering matrix from 1 to 3 GHz. For the calculation, one antenna is excited and the other three are terminated with a matched load, all using the waveguide port boundary condition. The results obtained by the frequency-domain finite element method using a boundary integral equation for mesh truncation (circles) compare very well with the results (lines) of another calculation by the time-domain finite element method using perfectly matched layers for mesh truncation [16].

The third example deals with a rectangular waveguide port, as opposed to the coaxial ports in the two preceding examples. Specifically, we consider a pyramidal horn antenna, which is fed by a 2.29-cm × 1.02-cm rectangular waveguide with a TE_{10} incident mode. The size of the horn's open aperture is 6.75 cm × 4.95 cm, and the length of the horn measured from the waveguide/horn junction to the open aperture is 13.87 cm. The inclusion of higher-order modes in the waveguide port boundary condition allows truncation of the waveguide placed very close to the horn (1.3 cm away from the junction). The horn antenna is enclosed by 1-cm-thick perfectly matched layers for mesh truncation. The simulation is carried out using the time-domain finite element method. The far field radiated by the antenna is computed by a time-domain near-to-far-field transformation (discussed in Section 5.3), and the antenna gain is calculated using

$$G(\theta, \phi) = 4\pi \frac{U(\theta, \phi)}{P^{\text{inc}}} = 4\pi r^2 \left| \frac{\mathbf{E}(\theta, \phi)}{\mathbf{E}_0} \right|^2 \frac{Z_{10}}{Z_0} \tag{5.37}$$

where Z_0 and Z_{10} are the wave impedance of free space and the TE_{10} mode, respectively. The antenna gain versus frequency at broadside is plotted in Figure 5.12 for 8 to 12 GHz, which is in good agreement with the predication given by an analytical model [17]. The E- and H-plane antenna gain patterns at 10 GHz are shown in Figure 5.13. The finite element results are found to agree well with the moment-method results published by Liu et al. [18]. The difference occurring at larger angles in the H-plane pattern is probably due to different modeling of the exterior surfaces of the antenna.

5.2 PLANE-WAVE EXCITATION

For many practical antenna applications, the radar cross section (RCS) of the antenna is of interest. For the analysis of the scattering characteristics of an antenna, the source

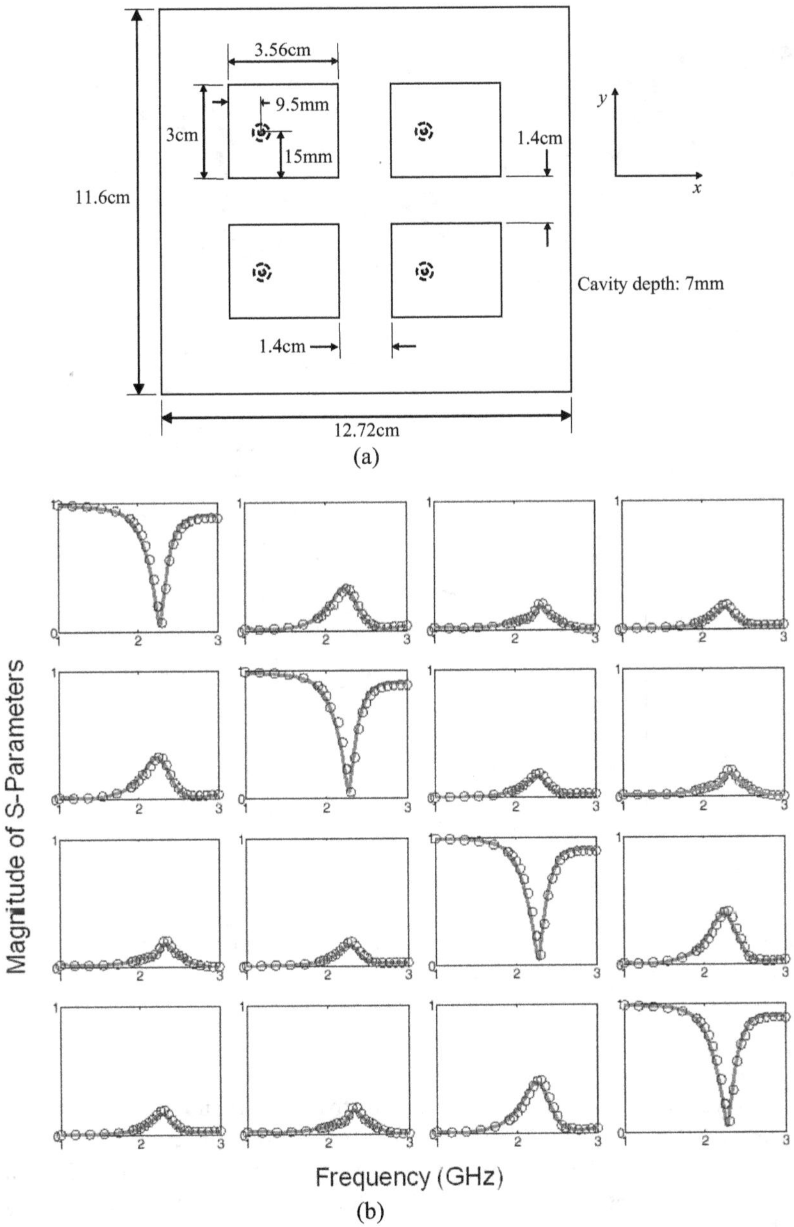

Figure 5.11 (a) Layout of a 2 × 2 microstrip patch antenna array. (b) Scattering matrix as a function of frequency from 1 to 3 GHz. (Circles: The frequency-domain FE-BI solution. Lines: The time-domain finite element solution). The graph on the ith row and jth column shows $|S_{ij}(f)|$. (After Mao et al. [15], Copyright © Taylor & Francis 2006.)

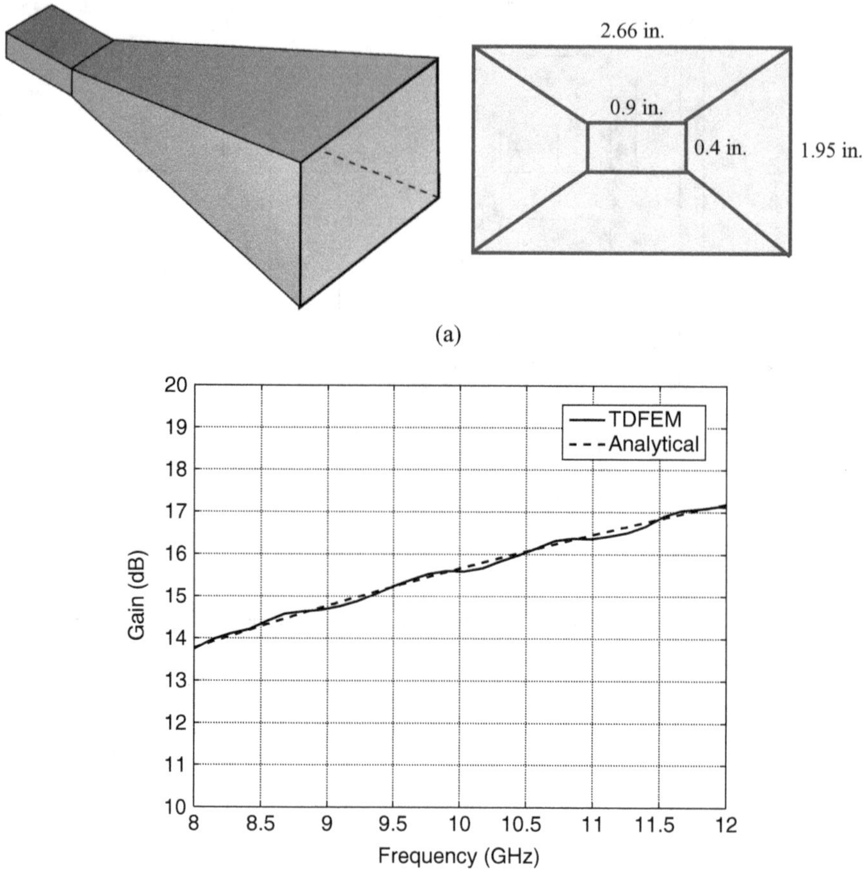

(b)

Figure 5.12 Pyramidal horn antenna. (a) Geometry. (b) Gain versus frequency at broadside. (After Lou and Jin [4], Copyright © IEEE 2005.)

of excitation is usually a plane wave incident from a certain direction. For such a scattering problem, the finite element solution can be formulated in terms of either the total or the scattered field. In this section we discuss these two formulations in the context of different mesh truncation techniques using absorbing boundary conditions, perfectly matched layers, and boundary integral equations. It will be shown that the two formulations excite the problem in a different manner, and consequently, they have different limitations and accuracy for scattering applications. We then introduce a third formulation that employs a decomposition of the total and scattered fields to retain the benefits of both the total- and scattered-field formulations while avoiding their limitations.

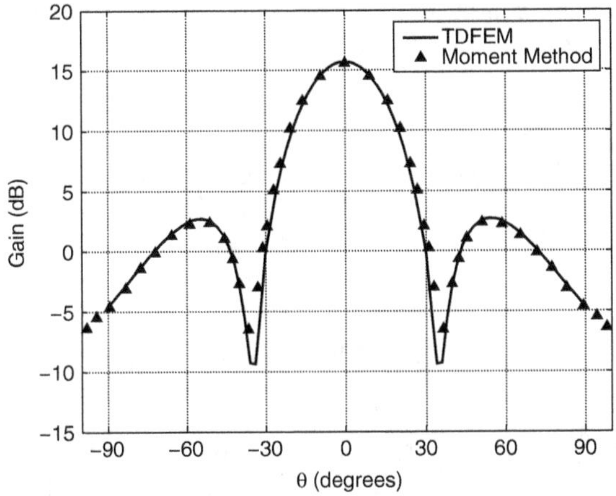

(a)

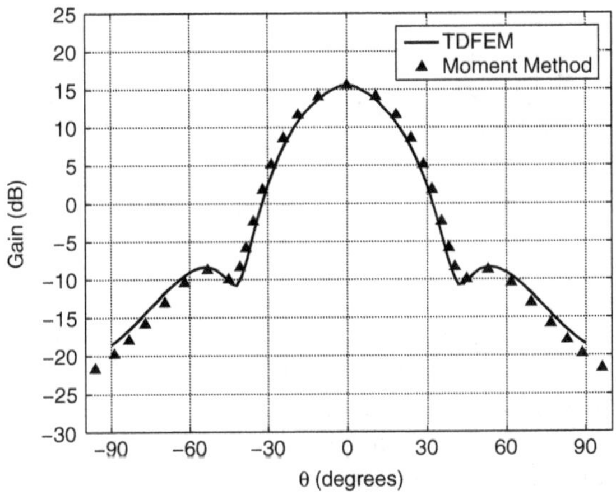

(b)

Figure 5.13 Radiation pattern for a 15-dB standard pyramidal gain horn at 10 GHz. (a) E-plane pattern. (b) H-plane pattern. (After Lou and Jin [4], Copyright © IEEE 2005.)

5.2.1 Total-Field Formulation

When the total field is employed to formulate the finite element solution, the boundary condition on a perfectly conducting surface remains the same as (2.5) and the vector wave equation remains the same as (2.8), except that now $\mathbf{J}_{\text{imp}} = \mathbf{M}_{\text{imp}} = 0$ since there are no internal sources in the computational domain. The only difference is the formulation of the field at the truncation surface. For the radiation case, the radiated field has to satisfy the absorbing boundary condition (2.7) so that the radiated field can leave the computational domain without any significant nonphysical reflection. For the scattering case, it is the scattered field, instead of the total field, that should satisfy the absorbing boundary condition (2.7). To convert (2.7) into a boundary condition applicable to the total field, we substitute the scattered field $\mathbf{E}^{\rm sc} = \mathbf{E} - \mathbf{E}^{\rm inc}$ for $\mathbf{E}$ in (2.7) to obtain

$$\hat{n} \times (\nabla \times \mathbf{E}) + jk_0\hat{n} \times (\hat{n} \times \mathbf{E}) \approx \mathbf{U}^{\rm inc} \qquad \mathbf{r} \in S_{\rm o} \tag{5.38}$$

where

$$\mathbf{U}^{\rm inc} = \hat{n} \times (\nabla \times \mathbf{E}^{\rm inc}) + jk_0\hat{n} \times (\hat{n} \times \mathbf{E}^{\rm inc}) \tag{5.39}$$

and $\mathbf{E}^{\rm inc}$ denotes the incident electric field. Equation (5.38) represents the absorbing boundary condition for the total field, and its right-hand side denotes a boundary excitation due to the incident field. The weak-form solution to the electric field becomes

$$\iiint_V \left[(\nabla \times \mathbf{T}) \cdot \overleftrightarrow{\mu}_r^{-1} \cdot (\nabla \times \mathbf{E}) - k_0^2 \mathbf{T} \cdot \overleftrightarrow{\varepsilon}_r \cdot \mathbf{E}\right] dV + jk_0 \oiint_{S_{\rm o}} (\hat{n} \times \mathbf{T}) \cdot (\hat{n} \times \mathbf{E})\, dS$$
$$= -\oiint_{S_{\rm o}} \mathbf{T} \cdot \mathbf{U}^{\rm inc}\, dS \tag{5.40}$$

together with the explicit enforcement of (2.5).

It is evident from (5.40) that in the total-field formulation of the finite element solution, the excitation is effected by generating the incident field on the truncation surface, as illustrated in Figure 5.14(a). The advantage of this formulation is that it requires little modification of the finite element solution developed for the radiation analysis to deal with scattering problems. For example, the treatment of the boundary condition on perfectly conducting surfaces would remain the same. The disadvantage is that when used together with approximate absorbing boundary conditions such as the first-order absorbing boundary condition, the errors associated with these absorbing boundary conditions can often require that the truncation surface be placed a wavelength or more away from the scattering object of interest. Consequently, the resulting impressed wave may undergo a considerable dispersion error as it propagates across the large computational domain between the truncation surface and the scattering object. This would result in a loss of accuracy in the final field computation. This disadvantage is mitigated when a boundary integral equation is employed for mesh truncation because in this case the truncation surface $S_{\rm o}$ can be placed very close to the surface of the scatterer.

When the boundary integral equations are used for mesh truncation, as discussed in Section 3.3, the electric field integral equation (EFIE) becomes

$$\tfrac{1}{2}\mathbf{M}_s(\mathbf{r}) - \hat{n} \times \mathcal{L}(\bar{\mathbf{J}}_s) + \hat{n} \times \tilde{\mathcal{K}}(\mathbf{M}_s) = -\hat{n} \times \mathbf{E}^{\rm inc}(\mathbf{r}) \qquad \mathbf{r} \in S_{\rm o} \tag{5.41}$$

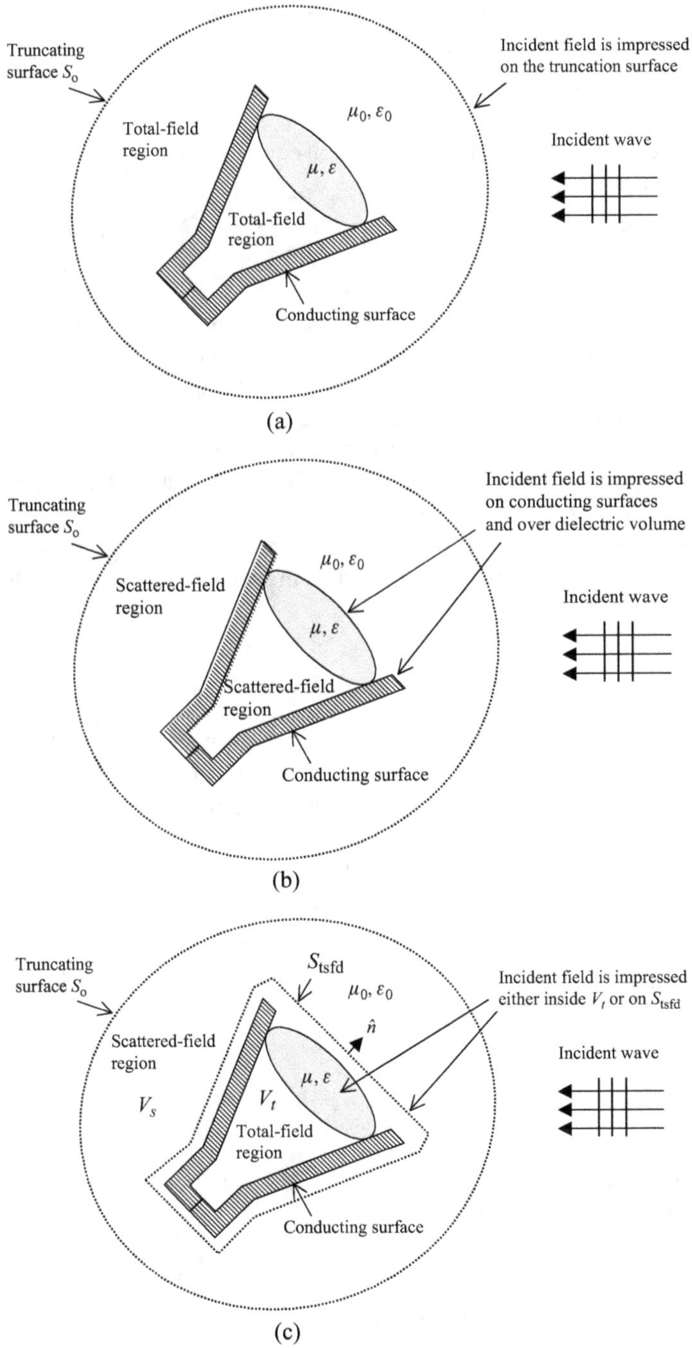

Figure 5.14 Incident-field excitation in three scattering formulations. (a) Total-field formulation. (b) Scattered-field formulation. (c) Total- and scattered-field decomposition.

and the magnetic field integral equation (MFIE) becomes

$$\tfrac{1}{2}\bar{\mathbf{J}}_s + \hat{n} \times \mathcal{L}(\mathbf{M}_s) + \hat{n} \times \tilde{\mathcal{K}}(\bar{\mathbf{J}}_s) = \hat{n} \times \overline{\mathbf{H}}^{\text{inc}}(\mathbf{r}) \qquad \mathbf{r} \in S_o. \tag{5.42}$$

The discretization of either of these two integral equations or their combination yields the boundary matrix equation

$$[P]\{E_S\} + [Q]\{H_S\} = \{b^{\text{inc}}\} \tag{5.43}$$

where the right-hand-side vector is due entirely to the incident field. Equation (5.43) can be solved together with the finite element matrix equation, which defines a finite element–boundary integral formulation for the scattering case.

Note that the implementation of the total-field formulation for the finite element solution using perfectly matched layers as mesh truncation can be quite cumbersome because perfectly matched layers should absorb only the scattered field, not the total field, and without the explicit separation of the scattered field from the total field, this selective absorption is difficult to implement. Therefore, the total-field formulation is often best used in the finite element solution using a boundary integral equation for mesh truncation.

5.2.2 Scattered-Field Formulation

In addition to the total-field formulation, the finite element solution of a scattering problem can also be formulated in terms of the scattered field. With the scattered field, the absorbing boundary condition remains the same as (2.7); that is,

$$\hat{n} \times (\nabla \times \mathbf{E}^{\text{sc}}) + jk_0\hat{n} \times (\hat{n} \times \mathbf{E}^{\text{sc}}) \approx 0 \qquad \mathbf{r} \in S_o. \tag{5.44}$$

However, the usual boundary condition and vector wave equation become different. By substituting the relation $\mathbf{E} = \mathbf{E}^{\text{inc}} + \mathbf{E}^{\text{sc}}$ into (2.5) and (2.8), we obtain

$$\hat{n} \times \mathbf{E}^{\text{sc}} = -\hat{n} \times \mathbf{E}^{\text{inc}} \qquad \mathbf{r} \in S_{\text{PEC}} \tag{5.45}$$

and

$$\nabla \times \left(\overleftrightarrow{\mu}_r^{-1} \cdot \nabla \times \mathbf{E}^{\text{sc}}\right) - k_0^2 \overleftrightarrow{\varepsilon}_r \cdot \mathbf{E}^{\text{sc}} = \mathbf{F}^{\text{inc}} \qquad \mathbf{r} \in V \tag{5.46}$$

where

$$\mathbf{F}^{\text{inc}} = -\nabla \times \left(\overleftrightarrow{\mu}_r^{-1} \cdot \nabla \times \mathbf{E}^{\text{inc}}\right) + k_0^2 \overleftrightarrow{\varepsilon}_r \cdot \mathbf{E}^{\text{inc}} \qquad \mathbf{r} \in V. \tag{5.47}$$

Note that $\mathbf{F}^{\text{inc}} = 0$ in the free-space region since the incident field satisfies the vector wave equation with $\mu_r = \varepsilon_r = 1$. The weak-form solution to the boundary-value problem defined above is given by

$$\iiint_V \left[(\nabla \times \mathbf{T}) \cdot \overleftrightarrow{\mu}_r^{-1} \cdot (\nabla \times \mathbf{E}^{\mathrm{sc}}) - k_0^2 \mathbf{T} \cdot \overleftrightarrow{\varepsilon}_r \cdot \mathbf{E}^{\mathrm{sc}} \right] dV$$

$$+ jk_0 \oiint_{S_o} (\hat{n} \times \mathbf{T}) \cdot (\hat{n} \times \mathbf{E}^{\mathrm{sc}})\, dS = \iiint_{V_{\mathrm{sc}}} \mathbf{T} \cdot \mathbf{F}^{\mathrm{inc}}\, dV \qquad (5.48)$$

together with the explicit enforcement of (5.45). In (5.48), V_{sc} denotes the penetrable volume of the scatterer, which vanishes if the scatterer contains only perfect conductors. This scattered-field formulation can be applied directly to the case where perfectly matched layers are employed for mesh truncation. In this case, one need only place perfectly matched layers on the inside of S_o.

It is clear that in the scattered-field formulation, incident fields are only impressed on the conducting surface of the scatterer using (5.45) and/or within the penetrable volume of the scatterer throughout the computational domain, as illustrated in Figure 5.14(b). When formulated in this manner, dispersion errors associated with an incident field are eliminated because the incident field does not have to propagate numerically across the computational domain between the truncation surface and the scattering object. Because of this, the scattered-field approach is generally believed to provide better accuracy than the total-field formulation. However, it offers no significant advantage when a boundary integral equation is used for mesh truncation since the truncation surface can be placed very close to the scatterer in this case.

5.2.3 Total- and Scattered-Field Decomposition Approach

As discussed above, the scattered-field formulation is advantageous because it can eliminate the dispersion error associated with the propagation of the incident field from the mesh truncation surface to the scatterer, and it can facilitate the use of perfectly matched layers for the finite element solution of wave scattering. However, there are applications where it is desirable to formulate the problem directly with the total-field variables. One such example is interior coupling through small apertures, where the incident and scattered fields may be nearly equal in amplitude and opposite in polarization such that the internal total field is small.

To retain the advantages of the scattered-field formulation while preserving a total-field region local to the geometry of interest, it is desirable to split the finite element computational domain into a scattered-field and a total-field region so that the incident field can be impressed closer to the scatterer. The mathematical foundation of this total- and scattered-field decomposition (TSFD) concept has been addressed for the frequency-domain Helmholtz equation [19]. In addition, the total- and scattered-field decomposition concept has been widely applied and developed for the FDTD simulations [20–23]. It has also been developed recently for the time-domain finite element method [24–27]. In particular, a volumetric TSFD scheme has been proposed [26,27] as an alternative to the more traditional Huygens' surface formulations

[25,26]. This volumetric scheme permits an essentially dispersion-free plane wave to be impressed within the total-field region while providing nearly zero leakage of the incident wave into the scattered-field region, independent of the mesh density and propagation angle. Consequently, the volumetric TSFD method provides a solution accuracy that is very similar to a formulation based globally on the scattered-field variables. This approach is particularly useful when it is desirable to solve the total-field equations local to a scatterer along with a requirement of high phase accuracy in the incident wave over electrically long propagation distances.

To describe the total- and scattered-field decomposition scheme, consider the problem illustrated in Figure 5.14(c). The surface, S_{tsfd}, splits the entire computational domain into total- and scattered-field volumes, V_t and V_s, respectively, while the surface, S_{o}, defines the outer surface of the overall finite element region. For convenience, the scattered-field region, V_s, is assumed to consist of only free space, and the decomposition surface, S_{tsfd}, is surrounded by free space. In this region, the boundary-value problem in terms of the scattered electric field, $\mathbf{E}^{\text{sc}}$, is then given by

$$\nabla \times (\nabla \times \mathbf{E}^{\text{sc}}) - k_0^2 \mathbf{E}^{\text{sc}} = 0 \qquad \mathbf{r} \in V_s \tag{5.49}$$

$$\hat{n} \times (\nabla \times \mathbf{E}^{\text{sc}}) + jk_0 \hat{n} \times (\hat{n} \times \mathbf{E}^{\text{sc}}) \approx 0 \qquad \mathbf{r} \in S_{\text{o}}. \tag{5.50}$$

The vector wave equation in the total-field region is

$$\nabla \times (\overleftrightarrow{\mu}_r^{-1} \cdot \nabla \times \mathbf{E}) - k_0^2 \overleftrightarrow{\varepsilon}_r \cdot \mathbf{E} = 0 \qquad \mathbf{r} \in V_t. \tag{5.51}$$

The weak-form equation for the scattered-field region is given by

$$\iiint_{V_s} \left[(\nabla \times \mathbf{T}) \cdot (\nabla \times \mathbf{E}^{\text{sc}}) - k_0^2 \mathbf{T} \cdot \mathbf{E}^{\text{sc}} \right] dV + jk_0 \oiint_{S_{\text{o}}} (\hat{n} \times \mathbf{T}) \cdot (\hat{n} \times \mathbf{E}^{\text{sc}})\, dS$$
$$+ \oiint_{S_{\text{tsfd}}} (\hat{n} \times \mathbf{T}) \cdot (\nabla \times \mathbf{E}^{\text{sc}})\, dS = 0 \tag{5.52}$$

where the unit normal on S_{tsfd} points from inside to outside. The weak-form equation for the total-field region is given by

$$\iiint_{V_t} \left[(\nabla \times \mathbf{T}) \cdot \overleftrightarrow{\mu}_r^{-1} \cdot (\nabla \times \mathbf{E}) - k_0^2 \mathbf{T} \cdot \overleftrightarrow{\varepsilon}_r \cdot \mathbf{E} \right] dV$$
$$- \oiint_{S_{\text{tsfd}}} (\hat{n} \times \mathbf{T}) \cdot (\nabla \times \mathbf{E})\, dS = 0. \tag{5.53}$$

Combining the weak-form equations for the two regions yields

$$\iiint_{V_t} \left[(\nabla \times \mathbf{T}) \cdot \overleftrightarrow{\mu}_r^{-1} \cdot (\nabla \times \mathbf{E}) - k_0^2 \mathbf{T} \cdot \overleftrightarrow{\varepsilon}_r \cdot \mathbf{E}\right] dV + \iiint_{V_s} \left[(\nabla \times \mathbf{T}) \cdot (\nabla \times \mathbf{E}^{\mathrm{sc}})\right.$$
$$\left. - k_0^2 \mathbf{T} \cdot \mathbf{E}^{\mathrm{sc}}\right] dV + jk_0 \oiint_{S_o} (\hat{n} \times \mathbf{T}) \cdot (\hat{n} \times \mathbf{E}^{\mathrm{sc}})\, dS = \oiint_{S_{\mathrm{tsfd}}} (\hat{n} \times \mathbf{T}) \cdot (\nabla \times \mathbf{E}^{\mathrm{inc}})\, dS \tag{5.54}$$

where $\mathbf{E} = \mathbf{E}^{\mathrm{inc}} + \mathbf{E}^{\mathrm{sc}}$ has been used to yield the right-hand side.

Since the incident field satisfies the free-space wave equation, the surface integral on S_{tsfd} can be written as

$$\oiint_{S_{\mathrm{tsfd}}} (\hat{n} \times \mathbf{T}) \cdot (\nabla \times \mathbf{E}^{\mathrm{inc}})\, dS = \iiint_{V_t} \left[(\nabla \times \mathbf{T}) \cdot (\nabla \times \mathbf{E}^{\mathrm{inc}}) - k_0^2 \mathbf{T} \cdot \mathbf{E}^{\mathrm{inc}}\right] dV. \tag{5.55}$$

Alternatively, we can use Maxwell's equations for the incident field to write it as

$$\oiint_{S_{\mathrm{tsfd}}} (\hat{n} \times \mathbf{T}) \cdot (\nabla \times \mathbf{E}^{\mathrm{inc}})\, dS = -jk_0 Z_0 \oiint_{S_{\mathrm{tsfd}}} (\hat{n} \times \mathbf{T}) \cdot \mathbf{H}^{\mathrm{inc}}\, dS. \tag{5.56}$$

Equation (5.55) yields a volumetric representation for the impressed wave throughout the total-field region, and in this case (5.54) becomes

$$\iiint_{V_t} \left[(\nabla \times \mathbf{T}) \cdot \overleftrightarrow{\mu}_r^{-1} \cdot (\nabla \times \mathbf{E}) - k_0^2 \mathbf{T} \cdot \overleftrightarrow{\varepsilon}_r \cdot \mathbf{E}\right] dV$$
$$+ \iiint_{V_s} \left[(\nabla \times \mathbf{T}) \cdot (\nabla \times \mathbf{E}^{\mathrm{sc}}) - k_0^2 \mathbf{T} \cdot \mathbf{E}^{\mathrm{sc}}\right] dV$$
$$+ jk_0 \oiint_{S_o} (\hat{n} \times \mathbf{T}) \cdot (\hat{n} \times \mathbf{E}^{\mathrm{sc}})\, dS = \iiint_{V_t} \left[(\nabla \times \mathbf{T}) \cdot (\nabla \times \mathbf{E}^{\mathrm{inc}}) - k_0^2 \mathbf{T} \cdot \mathbf{E}^{\mathrm{inc}}\right] dV. \tag{5.57}$$

Equation (5.56), on the other hand, leads to a surface representation for the impressed wave, which is also known as *Huygens' surface representation.* In this case, (5.54) can be written as

$$\iiint_{V_t} \left[(\nabla \times \mathbf{T}) \cdot \overleftrightarrow{\mu}_r^{-1} \cdot (\nabla \times \mathbf{E}) - k_0^2 \mathbf{T} \cdot \overleftrightarrow{\varepsilon}_r \cdot \mathbf{E}\right] dV$$
$$+ \iiint_{V_s} \left[(\nabla \times \mathbf{T}) \cdot (\nabla \times \mathbf{E}^{\mathrm{sc}}) - k_0^2 \mathbf{T} \cdot \mathbf{E}^{\mathrm{sc}}\right] dV$$
$$+ jk_0 \oiint_{S_o} (\hat{n} \times \mathbf{T}) \cdot (\hat{n} \times \mathbf{E}^{\mathrm{sc}})\, dS = jk_0 Z_0 \oiint_{S_{\mathrm{tsfd}}} \mathbf{T} \cdot (\hat{n} \times \mathbf{H}^{\mathrm{inc}})\, dS \tag{5.58}$$

where $\hat{n} \times \mathbf{H}^{\text{inc}}$ can be interpreted as the surface equivalent electric current due to the incident field.

In the numerical implementation, the unknown expansion coefficients on the interface surface S_{tsfd} can be associated with either the total field or the scattered field. If they are associated with the total field, the incident field should be subtracted from these expansion coefficients in the evaluation of the corresponding volume integral over V_s in (5.57) or (5.58). On the other hand, if they are associated with the scattered field, the incident field should be added to these expansion coefficients in the evaluation of the corresponding volume integral over V_t in (5.57) or (5.58). Note again that this is done only for the basis functions with their unknown expansion coefficients residing on S_{tsfd}. To facilitate either the subtraction or addition of the incident field, the incident field should be expanded in a similar fashion to the scattered and total fields. Theoretically, the volumetric approach gives rise to zero leakage from the total-field region into the scattered-field region. In practice, the undesired leakage is limited by the machine precision and the residual of the solution method. In the Huygens' surface approach, the electric current is entering through the surface integral in (5.58), whereas the magnetic current is entering through volume integrals; consequently, there will be a slight spatial misalignment in the implementation of this scheme within the finite element method. Because of this, there will not be perfect cancellation of these sources in the scattered-field region.

The total- and scattered-field decomposition scheme described above has been extended to the finite element analysis of scattering in free space in the time domain [26]. It has also been extended to the analysis of scattering in a layered medium [27]. Extensive numerical tests have shown that by using the volumetric approach, the undesired leakage in the scattered-field region could easily be reduced to −120 dB, and the field computation has accuracy comparable to that of the scattered-field approach. This is illustrated by considering plane-wave propagation through a cubical computational domain having a side length of 130 cm. Since there are no scattering objects in this computational domain, the scattered field should be identically zero. To test the effectiveness of the volumetric and surface TSFD schemes, we place a spherical surface with a radius of 60 cm to impress the incident field inside the surface. The entire computational domain is discretized with first-order tetrahedral and hexahedral elements with an average edge length of 4 cm and is terminated with perfectly matched layers, as illustrated in Figure 5.15(a).

Figure 5.16 shows forward- and backward-scattered far-field results versus frequency for the case of plane-wave incidence propagating broadside to the left face of the cubical domain shown in Figure 5.15(a). Three sets of results are shown in the figure. The first two sets are computed by the time-domain finite element method using the volumetric and Huygens' surface TSFD schemes, respectively. The third set is computed by the FDTD method using a Huygens' surface technique [23–25] implemented on a pure FDTD grid having a grid size of 5 cm, which is shown in Figure 5.15(b). As can be seen in Figure 5.16, the Huygens' surface excitation employed in the finite element simulation provides results in the backscatter direction comparable to those obtained using the Huygens' surface scheme on a pure FDTD grid (note that no dispersion-compensation methods were used for the

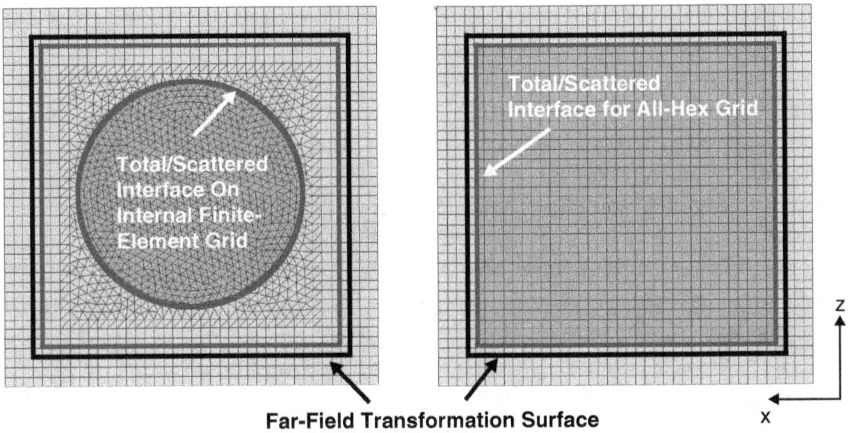

Figure 5.15 Two-dimensional cross section of finite element and FDTD grids, showing location of total- and scattered-field interface surfaces as well as near-to-far-field transformation surfaces. (a) Mixed tetrahedral and hexahedral element grid. (b) Traditional all-hexahedral-element (FDTD) grid.

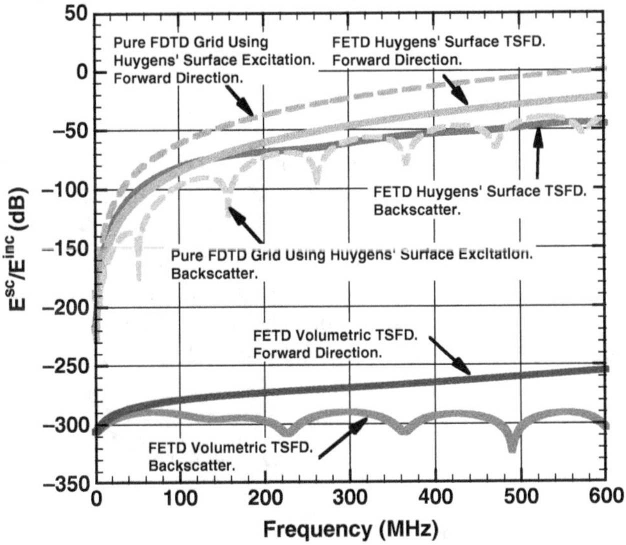

Figure 5.16 Forward and backward scattering versus frequency for plane-wave incidence using the proposed TSFD techniques on the finite element grid as well as a traditional Huygens' surface technique on a pure FDTD grid. Ideally, there should be zero leakage of the incident field into the scattered-field region. The volumetric TSFD excitation proposed is nearly ideal by suppressing leakage by over 12 orders of magnitude. All Huygens' surface–based methods exhibit much higher levels of leakage. (After Riley et al. [26], Copyright © IEEE 2006.)

FDTD results). However, the volumetric excitation technique provides significantly improved results. The dispersion error associated with the basis functions, as well as the misalignment between the electric and magnetic current sources on the interface, affects all results based on the Huygens' surface excitation, whereas there is essentially no dispersion error or source misalignment introduced into the impressed wave with the volumetric approach.

To demonstrate the scattering application, consider a simple case of plane-wave scattering by a perfectly conducting sphere computed by the time-domain finite element method using the volumetric TSFD scheme. The sphere has a radius of 50 cm and is enclosed by a cubic computational domain having a side length of 130 cm. The total- and scattered-field regions are split by a spherical surface having a radius of 60 cm. The entire computational domain is discretized with first-order tetrahedral elements with an average edge length of 2.5 cm and is terminated with perfectly matched layers. The monostatic RCS as a function of frequency is shown in Figure 5.17(a). The solutions shown are based on the Mie series (exact), the volumetric TSFD excitation method, and a pure scattered-field formulation described in Section 5.2.2. The absolute difference in the RCS as calculated by the volumetric TSFD method and the pure scattered-field formulation is shown in Figure 5.17(b). Note that the RCS computed using these two very different formulations are in agreement to at least five decimal places. This level of agreement with a pure scattered-field formulation is generally not possible with traditional Huygens' surface excitation methods.

5.3 FAR-FIELD PATTERN COMPUTATION

The far field radiated or scattered by an antenna can be calculated numerically from the near field on an arbitrary surface completely enclosing the antenna structure, denoted here as S_{NTF}. This procedure is referred to as the *near-to-far-field* (NTF) *transformation*. In Section 4.3 we described an accurate NTF technique for the FDTD method. In this section we reconsider it from the finite element perspective, where surfaces of more general shape could be used. Given the near-zone electric and magnetic fields on S_{NTF}, it can be shown that the electric field in the far zone can be expressed as

$$E_\theta(\mathbf{r}) = -\frac{jk_0 e^{-jk_0 r}}{4\pi r}(L_\phi + Z_0 N_\theta) \tag{5.59}$$

$$E_\phi(\mathbf{r}) = \frac{jk_0 e^{-jk_0 r}}{4\pi r}(L_\theta - Z_0 N_\phi) \tag{5.60}$$

where Z_0 is the free-space impedance and

$$\mathbf{N}(\hat{\mathbf{r}}) = \iint_{S_{\text{NTF}}} \mathbf{J}(\mathbf{r}')e^{jk_0 \mathbf{r}'\cdot\hat{\mathbf{r}}}\, dS' \tag{5.61}$$

$$\mathbf{L}(\hat{\mathbf{r}}) = \iint_{S_{\text{NTF}}} \mathbf{M}(\mathbf{r}')e^{jk_0 \mathbf{r}'\cdot\hat{\mathbf{r}}}\, dS' \tag{5.62}$$

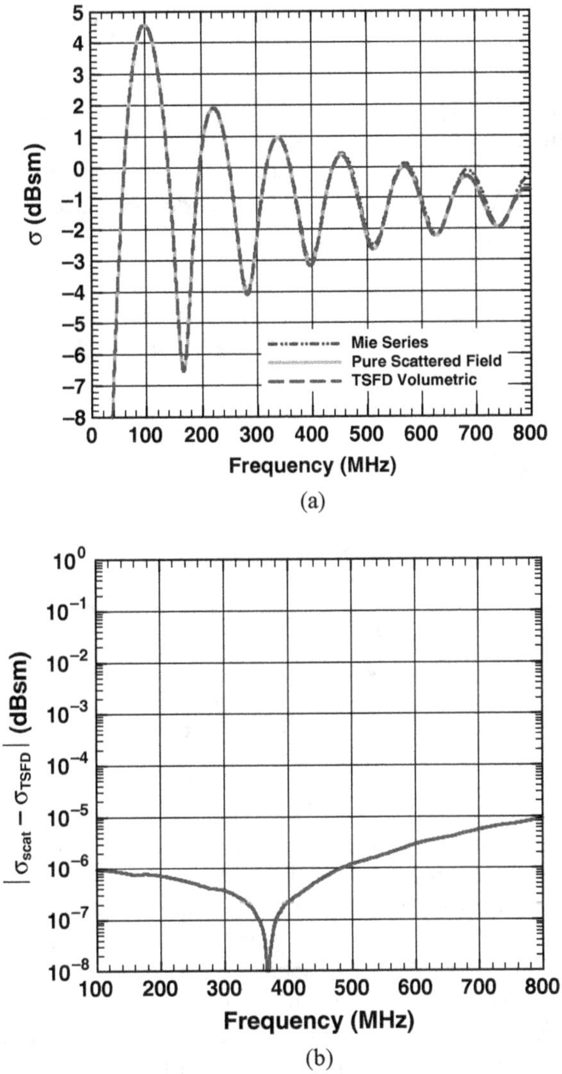

Figure 5.17 (a) Monostatic RCS for a PEC sphere using the TSFD volumetric excitation method compared to a pure scattered-field solution. (b) Absolute difference between the volumetric TSFD and scattered-field solutions. (After Riley et al. [26], Copyright © IEEE 2006.)

where $\mathbf{J} = \hat{n} \times \mathbf{H}$ and $\mathbf{M} = -\hat{n} \times \mathbf{E}$ are equivalent surface electric and magnetic currents, respectively. The near-to-far-field transformation surface S_{NTF} must be a closed surface containing all the sources inside. If an infinite ground plane is present (e.g., in the case of monopole antennas), the near-to-far-field transformation surface S_{NTF} and its image S'_{NTF} together comprise a closed surface. The equivalent surface currents on S'_{NTF} can easily be obtained by invoking the image theory.

In time-domain simulations, the far field can be calculated in either the frequency or the time domain. In the frequency-domain approach, the time-domain surface currents **J** and **M** are first transformed to the frequency domain via FFT, or alternatively, their frequency-domain representation at specific frequencies can be constructed concurrently with the time-stepping process via a DFT. Next, the frequency-domain far-field pattern is calculated using (5.59)–(5.62). In the time-domain approach, the far-field pattern is calculated directly from the time-domain surface currents **J** and **M** according to

$$E_\theta(\mathbf{r}, t) = -\frac{1}{4\pi r c_0}\left[\frac{\partial}{\partial t}L_\phi\left(\hat{\mathbf{r}}, t - \frac{\mathbf{r}\cdot\hat{\mathbf{k}}}{c_0}\right) + Z_0\frac{\partial}{\partial t}N_\theta\left(\hat{\mathbf{r}}, t - \frac{\mathbf{r}\cdot\hat{\mathbf{k}}}{c_0}\right)\right] \tag{5.63}$$

$$E_\phi(\mathbf{r}, t) = \frac{1}{4\pi r c_0}\left[\frac{\partial}{\partial t}L_\theta\left(\hat{\mathbf{r}}, t - \frac{\mathbf{r}\cdot\hat{\mathbf{k}}}{c_0}\right) - Z_0\frac{\partial}{\partial t}N_\phi\left(\hat{\mathbf{r}}, t - \frac{\mathbf{r}\cdot\hat{\mathbf{k}}}{c_0}\right)\right] \tag{5.64}$$

where

$$\mathbf{N}(\hat{\mathbf{r}}, \tau) = \iint_{S_{\mathrm{NTF}}} \mathbf{J}\left(\mathbf{r}', \tau + \frac{\mathbf{r}'\cdot\hat{\mathbf{r}}}{c_0}\right) dS' \tag{5.65}$$

$$\mathbf{L}(\hat{\mathbf{r}}, \tau) = \iint_{S_{\mathrm{NTF}}} \mathbf{M}\left(\mathbf{r}', \tau + \frac{\mathbf{r}'\cdot\hat{\mathbf{r}}}{c_0}\right) dS'. \tag{5.66}$$

Similar to the FDTD implementation in Section 4.4, the frequency-domain finite element near-to-far-field transformation is most suitable for computing far-field patterns with dense angular sampling at only a few frequencies, while the time-domain near-to-far-field transformation is most suitable for computing far-field patterns over a wide frequency range at only a few observation angles.

It is evident that to calculate the far field, we have to have both the tangential electric and magnetic fields on the near-to-far-field transformation surface S_{NTF}. When a boundary integral equation is used for mesh truncation, the truncation surface can be used as S_{NTF} since both the tangential electric and magnetic fields are computed in this approach. When approximate absorbing boundary conditions or perfectly matched layers are employed for mesh truncation, a surface tightly enclosing the antenna structure should be chosen as S_{NTF}. In this case, since the finite element solution usually computes one field, say the electric field, the other field has to be calculated using one of Maxwell's equations, which would result in a slight loss of accuracy because of differentiation in the curl operator. Special formulations can be designed for the finite element solution of both electric and magnetic fields on a surface; however, they are not yet used widely in practice. When a spherical surface is used as S_{NTF}, a formulation can be derived to calculate the far field based on only one field on S_{NTF}. However, this would require a large computational domain, making the finite element solution less efficient.

5.4 NEAR-FIELD VISUALIZATION

Antenna measurements typically provide knowledge of impedance, radiation pattern, and gain. Because measurements often reveal that improvements are necessary to a particular antenna's performance, it becomes important to understand the physical cause of the undesired performance characteristics. Visualization of near fields provides insight that is not typically available through measurement. In this way, analysis augments measurement, allowing antenna designers to perform analytical experiments.

Visualization can be used in numerous useful ways, each providing insight into the physics driving antenna performance. Each antenna application will drive the specific use of field visualizations. Although it is not possible to discuss all possible uses here, a few examples are given where field visualizations have proven their usefulness to practicing antenna engineers. For instance, strong magnetic field lines might unintentionally couple the horizontal and vertical ports of an antenna, causing degraded cross polarization. Alternatively, when designing antennas having anisotropic substrates, visualization of the fields provides physical insight into how the various field components are affected by the anisotropy. Visualization of the Poynting vector can be extremely useful in providing printed-antenna designers insight into the physical mechanisms related to substrate losses, including surface wave phenomena. In the following we present two specific examples to show how field visualization can help us to gain a better understanding of electromagnetic field radiation and propagation.

One example concerns the design of the Luneburg lens as a wideband antenna enabling wide-angle scan of beams at microwave frequencies. This lens was designed initially using geometrical optics [28], which concluded that when the relative permittivity of a dielectric sphere varies in the radial direction according to $\varepsilon_r\ (r) = 2 - (r/a)^2$, where a is the radius of the sphere, the rays produced by a point source placed on the surface of the sphere and entering into the sphere would be collimated into parallel rays when exiting at the other side of the sphere, as illustrated in Figure 5.18. Because of spherical symmetry, this unique feature made the Luneburg lens highly attractive as an antenna capable of wide-angle scanning and multibeam transmission. However, due to the difficulty of manufacturing the lens with a continuous variation of the permittivity, Luneburg lenses were made in a stepped-index model which consisted of several homogeneous dielectric spherical shells whose permittivities were different [29,30]. However, the experimental outcomes were disappointing; more specifically, the gain measured was reported to be lower than the value expected and in some cases worse than that of a simple homogeneous dielectric sphere. The poor performance was often attributed to the stepped-index model, which was necessary due to the limited capability of manufacturing materials with an arbitrary permittivity. With the development of numerical techniques for solving Maxwell's equations, we can now simulate the propagation of electromagnetic fields in a perfect Luneburg lens without a need for the stepped-index model [31]. If the fields are calculated in the frequency domain, the complex fields can be converted into the fields in the time domain by $\mathbf{E}(\mathbf{r}, t) = \mathrm{Re}[\mathbf{E}(\mathbf{r}, \omega)e^{j\omega t}]$, from which the field variation can be visualized. One snapshot of the electric field in the E- and H-planes calculated using the finite element method is shown in Figure 5.19, where a radiating

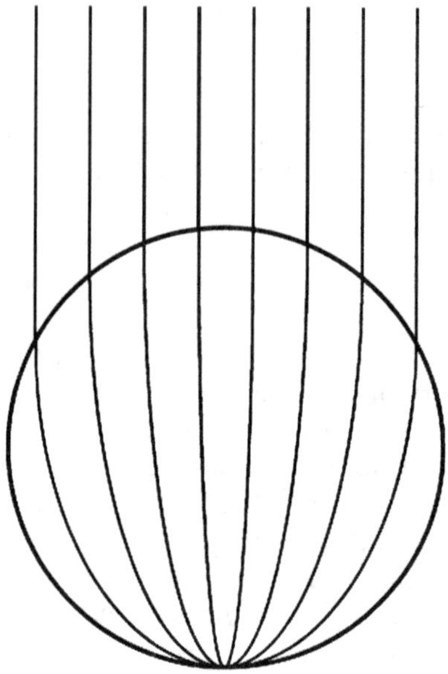

Figure 5.18 Ideal spherical Luneburg lens.

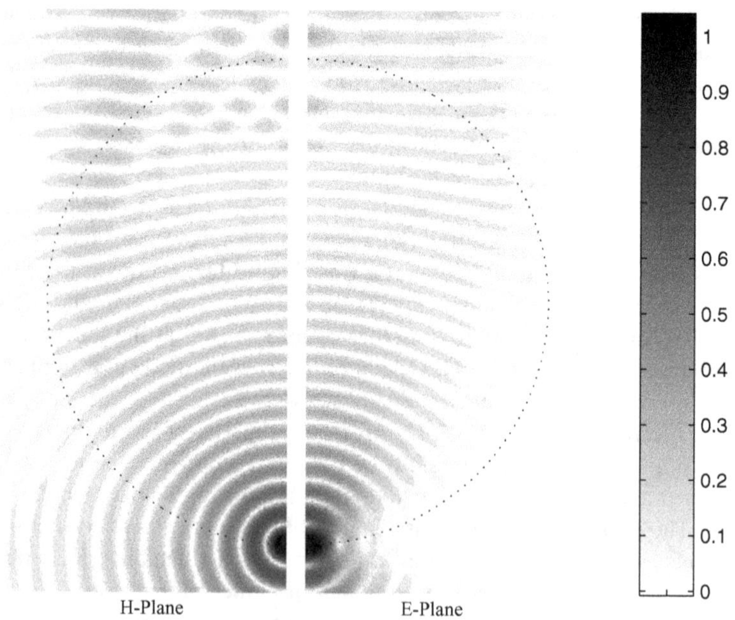

Figure 5.19 Snapshot of the electric field inside and nearby a Luneburg lens excited by a Hertzian dipole. (After Greenwood and Jin [31], Copyright © IEEE 1999.) (*See insert for color representation of figure.*)

Hertzian dipole is placed on the bottom surface of a Luneburg lens that has a diameter of $10\lambda_0$. From the figure it can be seen that the spherical wave radiated by the dipole is gradually transformed into a plane wave exiting at the top side of the lens. However, a more careful examination reveals that the exiting plane wave is not perfect; the amplitude is not perfectly uniform and the phase front is not perfectly planar. Numerical tests indicated that this undesired field variation was not due to numerical errors. To investigate this further, a reciprocal problem was simulated, where a plane wave was incident on the Luneburg lens from the bottom side [31]. According to the reciprocity and geometrical optics, this plane wave should be focused on a point at the top surface of the lens. A snapshot of the electric field from this simulation is shown in Figure 5.20, which shows that, indeed, the field is gradually focused to a small spot at the top surface of the lens. However, the focal point does not reside stationary on the surface of the lens. Instead, it is initially formed inside the lens, then moves across the surface of the lens, and eventually disappears when it moves farther away from the lens. This phenomenon can be observed more clearly from the animation of the field using a series of snapshots. Therefore, it is the movement or the oscillation of the focal point, which cannot be predicted by geometrical optics, rather than the stepped-index model, that is responsible for the poor performance of the Luneburg lens in the radiation case. Even if one can manufacture a perfect Luneburg lens, one would not be able to achieve the performance predicted by geometrical optics.

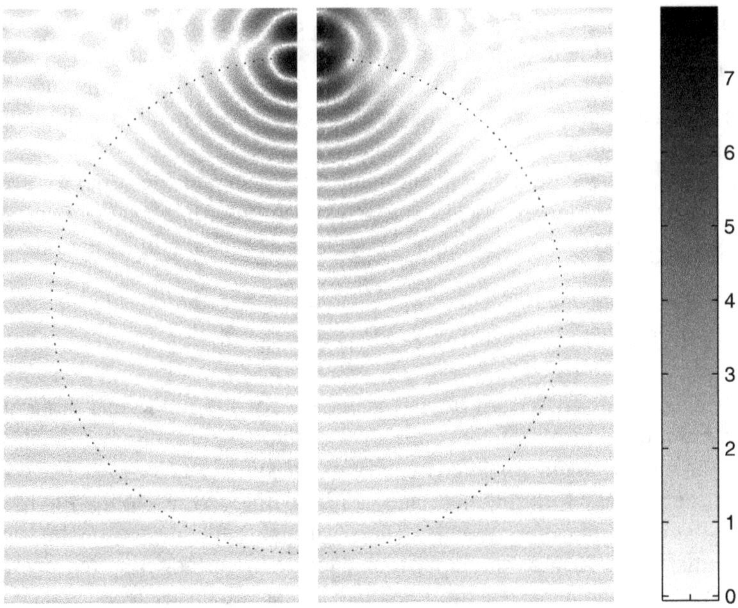

H-Plane E-Plane

Figure 5.20 Snapshot of the electric field inside and nearby a Luneburg lens excited by a plane wave incident from the bottom. (After Greenwood and Jin [31], Copyright © IEEE 1999.) (*See insert for color representation of figure.*)

Numerical simulations also revealed that the region that confines the oscillation of the focal point becomes smaller as the Luneburg lens becomes larger in terms of wavelength. Therefore, the performance of the Luneburg lens as an antenna would improve as the frequency goes higher. At the optical frequencies, the Luneburg lens would perform as expected.

When an antenna is deployed and installed on a platform such as an aircraft, it is critical to understand the interactions between the antenna and the platform. This subject is discussed in more detail in Chapter 11. A good understanding of the interactions can often be obtained by the visualization of the fields or the currents induced on the surface of the platform. Such visualization can easily reveal the spots that would seriously affect the radiation pattern of the antenna and the spot that would couple the antenna radiation into the interior of the host object. The antenna designer can then treat these specific spots, instead of the entire surface of the platform, to reduce the undesired impact on the radiation pattern and the undesired coupling into the interior electronic systems. To give an example simply for the purpose of illustration, Figure 5.21 shows the surface electric current induced on an aircraft (Beechcraft King Air 200) due to a Vivaldi antenna mounted underneath. The aircraft is approximately 13.3 m long and 16.6 m wide (including the wings), and the antenna excitation is a modulated Gaussian pulse with a central frequency of 200 MHz and a bandwidth of 100 MHz. The simulation was performed using the time-domain finite element method combined with the time-domain boundary integral equations discussed in Section 3.3.2. Evaluation of the boundary integrals was accelerated using the time-domain adaptive integral method [32]. In Figure 5.21, the surface electric current is shown on the decibel scale at several time instants sampled to cover the entire radiation period. Similar to the previous example, an animation movie made of the current distributions at each time step can easily be generated and would show the variation of the surface currents more clearly.

In summary, the near-field visualization allows the antenna designer to "see" how the field radiates, propagates, and interacts with the environment. As such, it often provides valuable physical insight that is not easily obtained with experiments.

5.5 SUMMARY

In this chapter we discussed an important aspect for the numerical simulation of antennas: the modeling of antenna sources. For the antenna radiation analysis, we discussed three types of antenna feed modeling. The simple current probe and voltage gap generator are easy to implement and can be employed to predict the antenna's far-field radiation patterns accurately. They can also generate reasonably accurate near-field quantities, such as the input impedance and mutual coupling coefficients for certain types of antennas. But their accuracy cannot be guaranteed for the analysis of general and more complex antennas. The third model, based on direct modeling of the antenna feed structure, is most accurate. In this model, the antenna waveguide feed is truncated and an accurate boundary condition is formulated and applied to the truncating waveguide port. Specific formulations were given for coaxial, circular, and

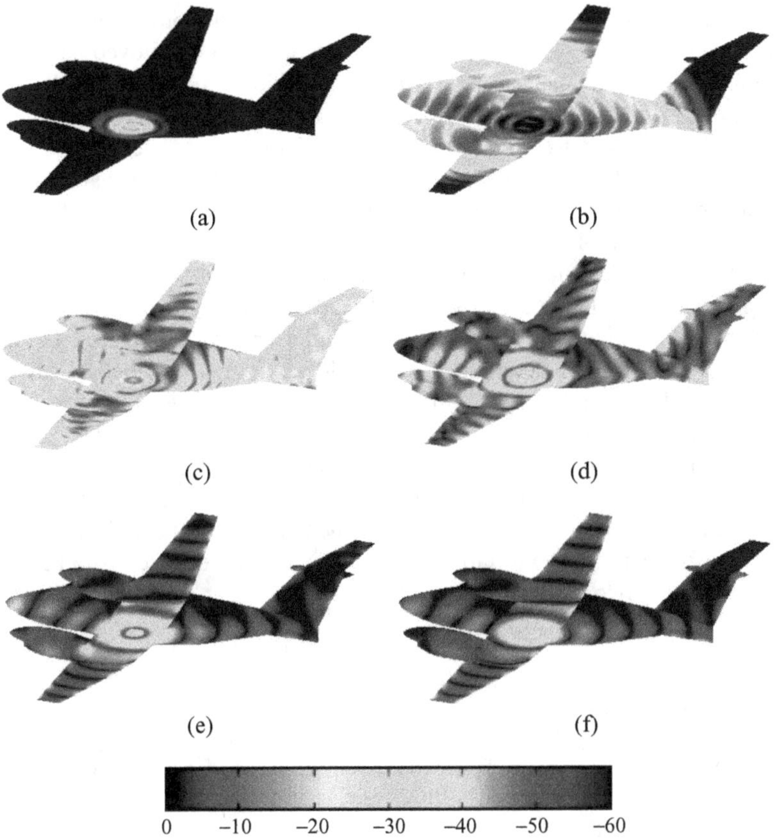

Figure 5.21 Snapshots of the surface electric current (on the decibel scale) on an aircraft at various time steps. (a) At the 120th time step. (b) At the 220th time step. (c) At the 320th time step. (d) At the 420th time step. (e) At the 520th time step. (f) At the 620th time step. (After Yilmaz et al. [32], Copyright © IEEE 2007.) (*See insert for color representation of figure.*)

rectangular waveguide ports in both the frequency and time domains. The formulation for more complex waveguide ports was also outlined in the frequency domain. This feed model has been found to yield excellent accuracy and is especially useful in the computation of near-field characteristics. The calculation of the input impedance was described for the three feed models.

For the antenna scattering analysis, we described three approaches to formulating the finite element solution and discussed their utilities. The total-field formulation is simple to implement and is best suited for analysis using a boundary integral equation for mesh truncation. When used in conjunction with an approximate absorbing boundary condition, it has, however, a slight loss of accuracy due to the grid dispersion error because it impresses the incident field at the absorbing boundary, which is usually placed some distance away from the scatterer. Moreover, the total-field formulation is not directly applicable to the case when perfectly matched layers are used for mesh truncation. The scattered-field formulation, on the other hand, impresses the incident field directly on the scatterer; hence, it can be employed easily in the simulations using either an absorbing boundary condition or perfectly matched layers for mesh truncation. It also has slightly better accuracy because it avoids the propagation of the incident wave from the truncation boundary to the scatterer. However, the direct impression of the incident field on the scatterer may be somewhat challenging for applications involving complex materials or specialized local boundary conditions. To retain the simplicity of the total-field formulation and the advantages of the scattered-field formulation, we then described the third approach, which is based on the decomposition of the total and scattered fields. This approach introduces a surface to divide the computational domain into an interior total-field region that tightly encloses the scatterer, and an exterior scattered-field region bounded by either the absorbing boundary or perfectly matched layers. Highly accurate numerical schemes were developed to impress the incident field.

After the discussion of the antenna source modeling, we addressed briefly the issue of far-field computation in both the frequency and time domains based on the near-to-far-field transformation, and finally, discussed the benefits of near-field visualization that can be used to gain insight into the underlying physics of antenna radiation and scattering by using two numerical examples.

REFERENCES

1. J. M. Jin and J. L. Volakis, "A hybrid finite element method for scattering and radiation by microstrip patch antennas and arrays residing in a cavity," *IEEE Trans. Antennas Propagat.*, vol. 39, pp. 1598–1604, Nov. 1991.
2. M. Feliziani and F. Maradei, "Modeling of electromagnetic fields and electrical circuits with lumped and distributed elements by the WETD method," *IEEE Trans. Magn.*, vol. 35, no. 3, pt. 1, pp. 1666–1669, May 1999.
3. D. Jiao and J. M. Jin, "Time-domain finite element simulation of cavity-backed microstrip patch antennas," *Microwave Opt. Tech. Lett.*, vol. 32, no. 4, pp. 251–254, Feb. 2002.
4. Z. Lou and J. M. Jin, "Modeling and simulation of broadband antennas using the time-domain finite element method," *IEEE Trans. Antennas Propagat.*, vol. 53, no. 12, pp. 4099–4110, Dec. 2005.
5. T. Hertel and G. Smith, "On the convergence of common FDTD feed models for antennas," *IEEE Trans. Antennas Propagat.*, vol. 51, no. 8, pp. 1771–1779, Aug. 2003.
6. W. L. Stutzman and G. A. Thiele, *Antenna Theory and Design*. New York: Wiley, 1981.
7. J.-M. Jin, *The Finite Element Method in Electromagnetics*, 2nd ed. Hoboken, NJ: Wiley, 2002.
8. J. Liu, J.-M. Jin, E. K. N. Yung, and R. S. Chen, "A fast three-dimensional higher-order finite element analysis of microwave waveguide devices," *Microwave Opt. Tech. Lett.*, vol. 32, pp. 344–352, Mar. 2002.
9. Z. Lou and J. M. Jin, "An accurate waveguide port boundary condition for the time-domain finite element method," *IEEE Trans. Microwave Theory Tech.*, vol. 53, no. 9, pp. 3014–3023, Sept. 2005.

10. K. Ise, K. Inoue, and M. Koshiba, "Three-dimensional finite-element method with edge elements for electromagnetic waveguide discontinuities," *IEEE Trans. Microwave Theory Tech.*, vol. 39, pp. 1289–1295, Aug. 1991.
11. S. H. Lee and J. M. Jin, "Adaptive solution space projection for fast and robust wideband finite-element simulation of microwave components," *IEEE Microwave Wireless Components Lett.*, vol. 17, pp. 474–476, July 2007.
12. J. F. Lee, D. K. Sun, and Z. J. Cendes, "Full-wave analysis of dielectric waveguides using tangential vector finite elements," *IEEE Trans. Microwave Theory Tech.*, vol. 39, pp. 1262–1271, Aug. 1991.
13. J. Tan and G. Pan, "A new edge element analysis of dispersive waveguiding structures," *IEEE Trans. Microwave Theory Tech.*, vol. 43, pp. 2600–2607, Nov. 1995.
14. J. Maloney, G. Smith, and W. Scott, "Accurate computation of the radiation from simple antennas using the finite difference time-domain method," *IEEE Trans. Antennas Propagat.*, vol. 38, pp. 1059–1068, July 1990.
15. K. Mao, J. K. Byun, and J. M. Jin, "Enhancing the modeling capability of the FE-BI method for simulation of cavity-backed antennas and arrays," *Electromagnetics,* vol. 26, no. 7, pp. 503–515, Oct. 2006.
16. T. Rylander and J. M. Jin, "Stable coaxial waveguide port algorithm for the time domain finite element method," *Microwave Opt. Tech. Lett.*, vol. 42, no. 2, pp. 115–119, July 2004.
17. C. Balanis, *Antenna Theory: Analysis and Design*, 2nd ed. Hoboken, NJ: Wiley, 2003.
18. K. Liu, C. Balanis, C. Birtcher, and G. Barber, "Analysis of pyramidal horn antennas using moment methods," *IEEE Trans. Antennas Propagat.*, vol. 41, no. 10, pp. 1379–1388, Oct. 1993.
19. A. Kirsch and P. Monk, "A finite element method for approximating electromagnetic scattering from a conducting object," *Numer. Math.*, vol. 92, no. 3, pp. 501–534, Sept. 2002.
20. D. Merewether, R. Fisher, and F. W. Smith, "On implementing a numerical Huygens' source scheme in a finite-difference program to illuminate scattering bodies," *IEEE Trans. Nucl. Sci.*, vol. 27, no. 6, pp. 1829–1833, Dec. 1980.
21. A. Taflove, *Computational Electrodynamics: The Finite-Difference Time-Domain Method.* Norwood, MA: Artech House, 1995.
22. T. Martin and L. Pettersson, "Dispersion compensation for Huygens' sources and far-zone transformations in FDTD," *IEEE Trans. Antennas Propagat.*, vol. 48, no. 4, pp. 494–501, Apr. 2000.
23. M. E. Watts and R. E. Diaz, "Perfect plane-wave injection into a finite FDTD domain through teleportation of fields," *Electromagnetics,* vol. 23, no. 2, pp. 187–201, 2003.
24. G. Ledfelt, "Hybrid time-domain methods and wire models for computational electromagnetics," Ph.D. thesis, TRITA-NA-0106, Department of Numerical Analysis and Computer Sciences, Royal Institute of Technology, Stockholm, Sweden, 2001.
25. W. Yao and Y. Wang, "An equivalence principle–based plane wave excitation in time/envelope-domain finite-element analysis," *IEEE Antennas Wireless Propagat. Lett.,* vol. 2, pp. 337–340, 2003.
26. D. Riley, J. M. Jin, Z. Lou, and L. E. R. Petersson, "Total- and scattered-field decomposition technique for the finite-element time-domain method," *IEEE Trans. Antennas Propagat.,* vol. 54, no. 1, pp. 35–41, Jan. 2006.

27. Z. Lou, L. E. R. Petersson, J. M. Jin, and D. J. Riley, "Total- and scattered-field decomposition technique for the finite-element time-domain modeling of buried scatterers," *IEEE Antennas Wireless Propagat. Lett.*, vol. 4, pp. 133–137, 2005.

28. K. S. Kelleher, "Designing dielectric microwave lenses," *Electronics*, vol. 29, no. 6, pp. 138–142, June 1956.

29. G. D. M. Peeler and H. P. Coleman, "Microwave stepped-index Luneburg lenses," *IRE Trans. Antennas Propagat.*, vol. 6, no. 2, pp. 202–207, Apr. 1958.

30. E. F. Buckley, "Stepped index Luneburg lenses," *Electron. Des.*, vol. 8, no. 8, pp. 86–89, Apr. 13, 1960.

31. A. D. Greenwood and J. M. Jin, "A field picture of wave propagation in inhomogeneous dielectric lenses," *IEEE Antennas Propagat. Mag.*, vol. 41, no. 5, pp. 9–18, Oct. 1999.

32. A. E. Yilmaz, Z. Lou, E. Michielssen, and J. M. Jin, "A single-boundary, implicit, and FFT-accelerated time-domain finite element–boundary integral solver," *IEEE Trans. Antennas Propagat.*, vol. 55, no. 5, pp. 1382–1397, May 2007.

6 Modeling of Complex Structures

The past five to 10 years have witnessed a remarkable improvement in our ability to provide accurate electromagnetic predictions for complex, practical geometry by using full-wave solution techniques that solve Maxwell's equations directly in both the frequency and time domains. This includes not only the resolution of diverse geometrical details that span a wide range of length scales, but also the ability to predict accurately high-frequency behavior that significantly increases the electrical size of the geometry. To accommodate these modeling challenges, sophisticated domain-decomposition and hybrid techniques have been developed. For example, an important aspect of electrical applications is the resolution of wires and wire cables. Although, in principle, individual wires can be resolved by finite element methods along with advanced solid modeling and meshing software, the implementation of this approach can lead to a tremendous increase in the number of discrete elements that fill the computational volume. Consequently, an alternative approach that models the physics of the wires locally can be beneficial. Similarly, the resolution of narrow apertures, thin-material layers, and even underlying electrical circuitry are all well suited to formulations that are specialized to incorporating the physics of these various system details instead of attempting their direct discretization. However, even though it is often straightforward to construct a hybridized solution procedure that interfaces specialized physical models into a broader volumetric finite element description, making these formulations robust and numerically stable often poses a significant challenge. In addition, interfacing various solution techniques is often done in a "weak-coupling" sense, where the feedback across the solution interface is not accounted for rigorously, and the resulting accuracy of this approach becomes highly application dependent.

In this chapter we discuss techniques to enhance the domain of applications that can be addressed by traditional finite element methods. The enhancements include the characterization of thin-material features, the resolution of narrow wires and narrow slot apertures, techniques to model complex lumped-element circuitry, and an approach to including distributed feeding networks in phased-array applications. These enhancements were selected because of their importance in antenna and array applications, which include the platform installation and supporting electronics. For example, antennas are often etched on thin-material sheets, and thin-material coatings are commonly used for electrical insulation, scattering mitigation, and environmental protection. Electrically thin wires and cables typically provide the means for

---

Finite Element Analysis of Antennas and Arrays, By Jian-Ming Jin and Douglas J. Riley

component interconnection, and various types of narrow slot apertures naturally occur as part of the system fabrication process.

A final topic discussed in this chapter is a validated example of predicting electromagnetic coupling into an electronic subsystem. Predicting and understanding the phenomenology of coupling down to the circuit-trace level is important to backdoor coupling assessments, where the term *backdoor* denotes electromagnetic penetration through nonintentional entry points, as well as co-site mitigation. Co-site mitigation is increasingly important, due to higher levels of integration of antennas and phased arrays into platforms such as airframes, ground vehicles, ships, and satellites. Although many factors affect a complete co-site analysis, the magnitude of the mutual coupling between the antennas and the direct radio-frequency (RF) coupling into the electronics compartments represent two of the more significant concerns. Predicting electromagnetic coupling into electronic subsystems poses many practical challenges, due to the often imprecise knowledge of the necessary geometrical and electrical parameters. Nevertheless, advanced finite element techniques represent a powerful predictive approach to this challenging application area.

6.1 THIN-MATERIAL LAYERS AND SHEETS

To include the influence of thin-material sheets and coatings in the finite element setting, we initially discuss the scalar impedance boundary condition (IBC) [1–4]. It will be seen that although the IBC approach has a straightforward formulation, the technique affects only the tangential component of the electric field. The jump in the normal component of the electric field can be accommodated by adopting degenerate three-dimensional finite elements so that they take the form of two-dimensional surface elements, and this is discussed in the second part of this section.

6.1.1 Impedance Boundary Conditions

Impedance boundary conditions have been used widely in integral-equation formulations [5,6], the finite-difference time-domain (FDTD) technique [7–10], the frequency-domain finite element method [11,12], and high-frequency asymptotic formulations [13]. In this section we examine the application of the IBC to the finite element method for several practical cases: (1) thin dielectric sheets, (2) resistive sheets and coatings, and (3) capacitive and inductive surface coatings. The formulations for incorporating these boundary conditions into the finite element method will initially be done in the frequency domain and then will be inverse Fourier transformed into the finite element time-domain (FETD) setting.

The approach adopted is based on the first-order IBC, which is attributed to both Leontovich [1–4] and Shchukin [3,4]. A finite element implementation [14] requires the adaptation of these impedance boundary conditions to the mixed boundary condition that was incorporated into the weak-form solution discussed in Section 2.1. For edge-based formulations, the IBC directly affects the electric field component tangential to the thin sheet. For many practical applications this is the primary component

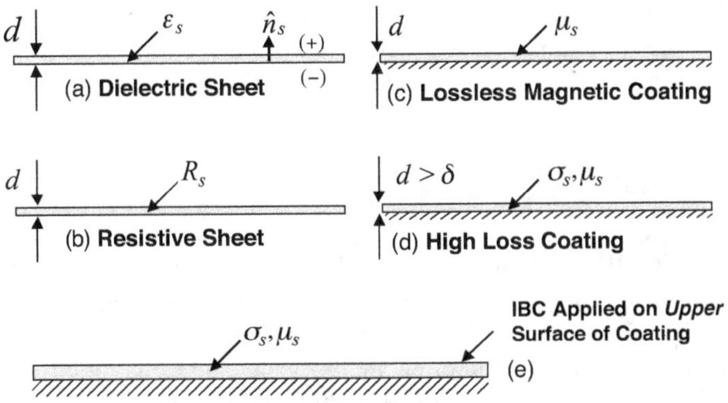

Figure 6.1 Configurations for thin-material sheets and coatings. (a) Dielectric sheet. (b) Resistive sheet. (c) Lossless magnetic coating on a PEC. (d) Lossy magnetic coating on a PEC. (e) Surface location for application of IBC.

affected by the presence of an electrical thin-material layer; however, the jump in the normal component of the electric field across the sheet is ignored in this approach. For example, for edge-on incidence on a planar thin dielectric sheet with the electric field perpendicular to a sheet's surface, the scattered field will be predicted to be zero based upon this formulation, whereas the true solution will typically be nonzero, albeit small.

Figure 6.1 shows four typical geometrical configurations involving thin sheets and coatings on a perfect electric conductor (PEC). The sheet resistivity will be denoted by R_s with units of $\Omega/\square$ (ohms per square). The permittivity of a dielectric sheet will be denoted by ε_s, with σ_s and μ_s used to denote conductivity and permeability, respectively. The material layer thickness is denoted by d. To incorporate an IBC, d is generally assumed to be less than 1/10 to 1/20 of the wavelength in the material layer, and in the case of the lossy magnetic coating shown in Figure 6.1(d), the thickness of the coating is assumed to be at least one or two skin depths over the bandwidth of interest, where the skin depth δ is given by $\delta = \sqrt{1/\pi f \mu_s \sigma_s}$.

The surface impedance boundary condition is defined by [4,10]

$$\mathbf{E}_{\text{tan}} = \overleftrightarrow{Z}_s \cdot (\hat{n}_s \times \mathbf{H}) \qquad \mathbf{r} \in S_s \tag{6.1}$$

where $\hat{n}_s$ defines an outward normal to the interface surface S_s. By applying the equivalence principle to a planar interface between two material regions, where the upper region is free space and the lower region is a lossy dielectric, it is possible to replace the dielectric region by the exact surface impedance boundary condition [10]

$$\overleftrightarrow{Z}_s = Z_s(\omega) \frac{k_s\left(\overleftrightarrow{I}_{\text{tan}} + \nabla_{\text{tan}}\nabla_{\text{tan}}/k_s^2\right)}{\sqrt{k_s^2 + \nabla_{\text{tan}}^2}} \tag{6.2}$$

where $\overleftrightarrow{I}_{\tan}$ is a two-dimensional dyadic, $\nabla_{\tan}$ is a two-dimensional gradient operator, and k_s denotes the wavenumber in the dielectric region. The Leontovich IBC is based on the scalar coefficient $Z_s(\omega)$ and leads to the following approximate representation for (6.1) [1–4]:

$$\mathbf{E}_{\tan} = Z_s(\omega)\hat{n}_s \times \mathbf{H} \qquad \mathbf{r} \in S_s. \tag{6.3}$$

The Leontovich IBC formally assumes planar interfaces, and its application requires that the fields internal to a thin sheet or coating propagate normal to the material interface for all frequencies and also remain constant along the interface. However, the Leontovich IBC can often yield acceptable results for curved surfaces provided that the electrical thickness of the sheet or coating is small relative to the local curvature. In addition, as the refractive index of the material region increases, the requirement for normal propagation is approximately satisfied even for the case of oblique incidence. Only the scalar Leontovich IBC is discussed in this section; however, it is noted that higher-order IBCs can be constructed from (6.2) and have been used to provide increased generality and accuracy in the FDTD method [9,10] and other techniques [15].

In the finite element formulation, the mixed boundary condition corresponding to (6.3) on the interface surface S_s is given by

$$\hat{n}_s \times \left(\frac{1}{\mu_0}\nabla \times \mathbf{E}\right) + j\omega Y_s(\omega)\hat{n}_s \times (\hat{n}_s \times \mathbf{E}) = 0 \qquad \mathbf{r} \in S_s \tag{6.4}$$

where Y_s denotes a surface admittance such that $Y_s = 1/Z_s$. The surface impedances for the specific geometries shown in Figure 6.1 are discussed next along with their adaptation to (6.4).

To first order, the surface impedance for the two-sided dielectric sheet shown in Figure 6.1(a) can be written in terms of the complex relative permittivity as [2]

$$Z_s(\omega) = \frac{1}{j\omega\varepsilon_0(\varepsilon_{rs} - 1)d} \tag{6.5}$$

where

$$\varepsilon_{rs} = \varepsilon'_{rs} - j\frac{\sigma_s}{\omega\varepsilon_0} \tag{6.6}$$

and $\varepsilon_s = \varepsilon_0\varepsilon_{rs}$. Because the sheet has two sides, the frequency-domain impedance boundary condition (6.4) is [1]

$$\hat{n}_s \times \left(\frac{1}{\mu_0}\nabla \times \mathbf{E}^+ - \frac{1}{\mu_0}\nabla \times \mathbf{E}^-\right) + j\omega\frac{1}{Z_s}\hat{n}_s \times (\hat{n}_s \times \mathbf{E}^\pm) = 0 \qquad \mathbf{r} \in S_s^\pm \tag{6.7}$$

where $\mathbf{E}^{\pm}$ denotes the electric field above and below the sheet and $S_s^{\pm}$ denotes surfaces above (+) and below (−) a midsurface represented by S_s. The tangential component of the electric field is continuous across the sheet and $\hat{n}_s$ is directed toward the "+" side. It is noted that a two-sided magnetic sheet with complex permeability μ_s is described by a dual representation of boundary condition (6.7) [1].

For the special case of a lossless dielectric sheet, the permittivity is real and $\varepsilon_s = \varepsilon_0 \varepsilon'_{rs}$. With this assumption, the surface impedance is a pure capacitive reactance and is given by

$$Z_s(\omega) = \frac{1}{j\omega\varepsilon_0(\varepsilon'_{rs} - 1)\,d}. \tag{6.8}$$

Substituting (6.8) into (6.7) and adopting the inverse Fourier transform leads to the following transient form for the mixed boundary condition for a lossless dielectric sheet:

$$\hat{n}_s \times \left(\frac{1}{\mu_0}\nabla \times \mathbf{E}^+ - \frac{1}{\mu_0}\nabla \times \mathbf{E}^-\right) + \varepsilon_0(\varepsilon'_{rs} - 1)\,d\hat{n}_s \times \left(\hat{n}_s \times \frac{\partial^2 \mathbf{E}^{\pm}}{\partial t^2}\right) = 0 \qquad \mathbf{r} \in S_s^{\pm}. \tag{6.9}$$

Note, that the contribution from the lossless dielectric sheet will enter into a transient finite element formulation through a second-order time derivative of the electric field.

Next, we consider the similar case of a resistive sheet as shown in Figure 6.1(b). In (6.5), we assume that $\varepsilon'_{rs} = 1$, $\sigma_s \gg \omega\,\varepsilon_0$, and $d < 2\delta$. The surface impedance Z_s is now purely resistive and given by

$$R_s = \frac{1}{\sigma_s d}. \tag{6.10}$$

The two-sided mixed boundary condition similar to (6.7) becomes

$$\hat{n}_s \times \left(\frac{1}{\mu_0}\nabla \times \mathbf{E}^+ - \frac{1}{\mu_0}\nabla \times \mathbf{E}^-\right) + j\omega\frac{1}{R_s}\hat{n}_s \times (\hat{n}_s \times \mathbf{E}^{\pm}) = 0 \qquad \mathbf{r} \in S_s^{\pm} \tag{6.11}$$

and the corresponding transient form is simply

$$\hat{n}_s \times \left(\frac{1}{\mu_0}\nabla \times \mathbf{E}^+ - \frac{1}{\mu_0}\nabla \times \mathbf{E}^-\right) + \frac{1}{R_s}\hat{n}_s \times \left(\hat{n}_s \times \frac{\partial \mathbf{E}^{\pm}}{\partial t}\right) = 0 \qquad \mathbf{r} \in S_s^{\pm}. \tag{6.12}$$

As expected, a resistive sheet adds dissipative loss to the wave equation and hence contributes through the first-order time derivative of the electric field. It is noted that the first-order ABC described by (3.1) can be recovered by defining R_s to be the characteristic impedance of the medium, $R_s = \sqrt{\mu_0/\varepsilon_0}$, when applying condition (6.4).

We now consider the lossless magnetic coating on a PEC shown in Figure 6.1(c). By using the transmission-line theory, the frequency-domain form for Z_s is well known and given by [1]

$$Z_s(\omega) = j\omega\mu_s d \frac{\tan k_s d}{k_s d} \tag{6.13}$$

where k_s denotes the wavenumber in the coating. For the case of an electrically thin coating such that $k_s d \ll 1$, then $Z_s \approx j\omega\mu_s d$. With this approximation, and further noting that μ_s is purely real, the transient form of (6.4) becomes

$$\hat{n}_s \times \left(\frac{1}{\mu_0}\nabla \times \mathbf{E}\right) + \frac{1}{\mu_s d}\hat{n}_s \times (\hat{n}_s \times \mathbf{E}) = 0 \qquad \mathbf{r} \in S_s. \tag{6.14}$$

Extensions to (6.13) to accommodate surface curvature as well as coatings that consist of multiple absorptive layers are available [1].

The single lossy magnetic coating shown in Figure 6.1(d) is considered next. Under the good-conductor assumption, $\sigma_s \gg \omega\varepsilon_0$, and also assuming that the thickness of the coating is greater than one to two skin depths, the surface impedance is given by [16]

$$Z_s(\omega) = \sqrt{\frac{j\omega\mu_s}{\sigma_s}} = (1+j)\sqrt{\frac{\omega\mu_s}{2\sigma_s}} = \frac{1+j}{\sigma_s\delta}. \tag{6.15}$$

Note that the wavelength λ_s in a good conducting medium is

$$\lambda_s \approx 2\sqrt{\frac{\pi}{\sigma_s\mu_s f}} = 2\pi\delta. \tag{6.16}$$

In terms of resistive and inductive components, (6.15) can be written as

$$Z_s(\omega) = R_s(\omega) + j\omega L_s(\omega) \tag{6.17}$$

with

$$R_s(\omega) = \sqrt{\frac{\omega\mu_s}{2\sigma_s}}, \qquad L_s(\omega) = \sqrt{\frac{\mu_s}{2\sigma_s\omega}} \tag{6.18}$$

where L_s denotes the surface inductance. In a frequency-domain finite element formulation, (6.17) and (6.18) are easily incorporated into (6.4). However, the dispersive nature of (6.15) will introduce a convolution into the time-domain representation for (6.4), where the inverse transform of $Y_s = \sqrt{\sigma_s/j\omega\mu_s}$ is $\sqrt{\sigma_s/\pi\mu_s t},\ t \geq 0$. For simplicity, we consider a narrow bandlimited excitation with center frequency $f = f_0$.

With this assumption, the resulting surface admittance has the approximate representation

$$Y_s(\omega)|_{\omega=\omega_0} \approx \frac{1}{2}\sqrt{\frac{2\sigma_s}{\omega_0\mu_s}}\left(1 - j\frac{\omega}{\omega_0}\right) \tag{6.19}$$

where $\omega_0 = 2\pi f_0$, and the transient form of (6.4) becomes

$$\hat{n}_s \times \left(\frac{1}{\mu_0}\nabla \times \mathbf{E}\right) + \frac{1}{2}\sqrt{\frac{2\sigma_s}{\omega_0\mu_s}}\,\hat{n}_s \times \left(\hat{n}_s \times \frac{\partial \mathbf{E}}{\partial t}\right) + \frac{\omega_0}{2}\sqrt{\frac{2\sigma_s}{\omega_0\mu_s}}\,\hat{n}_s \times (\hat{n}_s \times \mathbf{E}) = 0 \qquad \mathbf{r} \in S_s \tag{6.20}$$

which is valid for bandlimited waveforms with the center frequency $f = f_0$.

An additional case of interest for (6.17) is where both R_s and L_s are assumed to be frequency independent. To obtain a time-domain representation for (6.4) with this assumption, we introduce the Fourier transform pair

$$\frac{1}{R_s + j\omega L_s} \leftrightarrow \frac{u(t)}{L_s}e^{-R_s t/L_s} \tag{6.21}$$

where $u(t) \equiv 1$ for $t \geq 0$ and 0 otherwise. Even though R_s and L_s are assumed to be invariant with frequency, the time-domain form of (6.4) will involve a convolution given by

$$\hat{n}_s \times \left(\frac{1}{\mu_0}\nabla \times \mathbf{E}\right) + \frac{1}{L_s}\int_0^t \hat{n}_s \times \left(\hat{n}_s \times \frac{\partial \mathbf{E}}{\partial \tau}\right) e^{-R_s(t-\tau)/L_s}\,d\tau = 0 \qquad \mathbf{r} \in S_s. \tag{6.22}$$

In the limiting case $R_s = 0$ with $L_s \neq 0$, the second term in (6.22) simplifies to $(1/L_s)\,\hat{n}_s \times (\hat{n}_s \times \mathbf{E})$.

Finally, we consider a capacitive coating defined by $Y_s \equiv G_s + j\omega C_s$, where G_s denotes a surface conductance and C_s denotes the surface capacitance. In a frequency-domain finite element formulation, arbitrary frequency dependency for G_s and C_s is easily accommodated in (6.4). However, to obtain a time-domain representation that avoids convolutions, we assume that these terms are frequency independent over the bandwidth of interest. With this assumption, the transient form for (6.4) is simply

$$\hat{n}_s \times \left(\frac{1}{\mu_0}\nabla \times \mathbf{E}\right) + G_s\,\hat{n}_s \times \left(\hat{n}_s \times \frac{\partial \mathbf{E}}{\partial t}\right) + C_s\,\hat{n}_s \times \left(\hat{n}_s \times \frac{\partial^2 \mathbf{E}}{\partial t^2}\right) = 0 \qquad \mathbf{r} \in S_s. \tag{6.23}$$

We now incorporate these boundary conditions in the finite element formulation for the time-domain wave equation. By generalizing the matrices associated with the second-order ordinary differential equation (2.27), we can easily accommodate the derived time-domain thin-material boundary conditions, with the frequency-domain representations being similar. For this purpose, we rewrite the semidiscrete equation (2.27) as

$$([T]+[G^{(a)}])\frac{d^2\{E\}}{dt^2}+([R]+[G^{(b,d1)}])\frac{d\{E\}}{dt}+([S]+[G^{(c,d2)}])\{E\}+\{K\}=\{f\}. \tag{6.24}$$

The standard matrices $[T]$, $[R]$, and $[S]$ are defined by (2.28), (2.29), and (2.30), respectively, while the additional matrices $[G^{(a)}]$, $[G^{(b)}]$, $[G^{(c)}]$, $[G^{(d1)}]$, and $[G^{(d2)}]$, as well as the vector $\{K\}$, will incorporate the IBCs derived previously.

The lossless dielectric sheet boundary condition (6.9) has the matrix elements

$$G_{ij}^{(a)} = \iint_{S_s} \varepsilon_0(\varepsilon'_{rs}-1)\,d\,(\hat{n}_s\times\mathbf{N}_i)\cdot(\hat{n}_s\times\mathbf{N}_j)\,dS \tag{6.25}$$

and the matrix elements for the resistive sheet boundary condition (6.12) are given by

$$G_{ij}^{(b)} = \iint_{S_s} \frac{1}{R_s}(\hat{n}_s\times\mathbf{N}_i)\cdot(\hat{n}_s\times\mathbf{N}_j)\,dS. \tag{6.26}$$

The lossless magnetic coating boundary condition (6.14) gives rise to the matrix elements

$$G_{ij}^{(c)} = \iint_{S_s} \frac{1}{\mu_s d}(\hat{n}_s\times\mathbf{N}_i)\cdot(\hat{n}_s\times\mathbf{N}_j)\,dS \tag{6.27}$$

while the lossy magnetic coating (6.20) has two matrix contributions with elements

$$G_{ij}^{(d1)} = \frac{1}{2}\iint_{S_s}\sqrt{\frac{2\sigma_s}{\omega_0\mu_s}}(\hat{n}_s\times\mathbf{N}_i)\cdot(\hat{n}_s\times\mathbf{N}_j)\,dS \tag{6.28}$$

$$G_{ij}^{(d2)} = \frac{\omega_0}{2}\iint_{S_s}\sqrt{\frac{2\sigma_s}{\omega_0\mu_s}}(\hat{n}_s\times\mathbf{N}_i)\cdot(\hat{n}_s\times\mathbf{N}_j)\,dS. \tag{6.29}$$

The capacitive boundary condition (6.23) with frequency-independent conductance and capacitance has the matrix elements

$$G_{ij}^{(b)} = \iint_{S_s} G_s\,(\hat{n}_s\times\mathbf{N}_i)\cdot(\hat{n}_s\times\mathbf{N}_j)\,dS \tag{6.30}$$

$$G_{ij}^{(a)} = \iint_{S_s} C_s\,(\hat{n}_s\times\mathbf{N}_i)\cdot(\hat{n}_s\times\mathbf{N}_j)\,dS. \tag{6.31}$$

Finally, the vector $\{K\}$ is used to accommodate the inductive boundary condition (6.22) and is given in semidiscrete form by

$$\{K\} = [G^{(e)}] \int_0^t e^{-R_s(t-\tau)/L_s} \frac{d\{E\}}{d\tau} d\tau \tag{6.32}$$

where

$$G_{ij}^{(e)} = \iint_{S_s} \frac{1}{L_s} (\hat{n}_s \times \mathbf{N}_i) \cdot (\hat{n}_s \times \mathbf{N}_j) \, dS. \tag{6.33}$$

For the limiting case $R_s = 0$ with $L_s \neq 0$, $\{K\} \rightarrow [G^{(e)}]\{E\}$.

We now consider several examples based on the first-order impedance boundary conditions discussed in this section. The first example is the microstrip bandpass filter shown in Figure 6.2. The particular design utilizes a thin dielectric sheet, or pad, under the central region to shift the passband associated with the filter. The application is modeled by using the FETD technique for three configurations. First, the relative permittivity of the pad will be defined equal to that of the dielectric

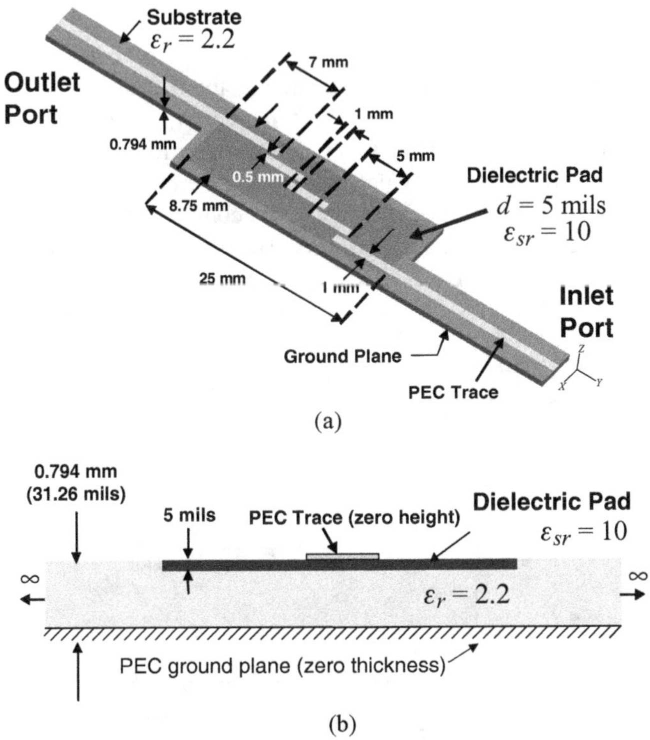

Figure 6.2 Microstrip bandpass filter with a high-contrast thin dielectric layer. (a) Three-dimensional geometry. (b) Cross section. [*See insert for color representation of* (a).]

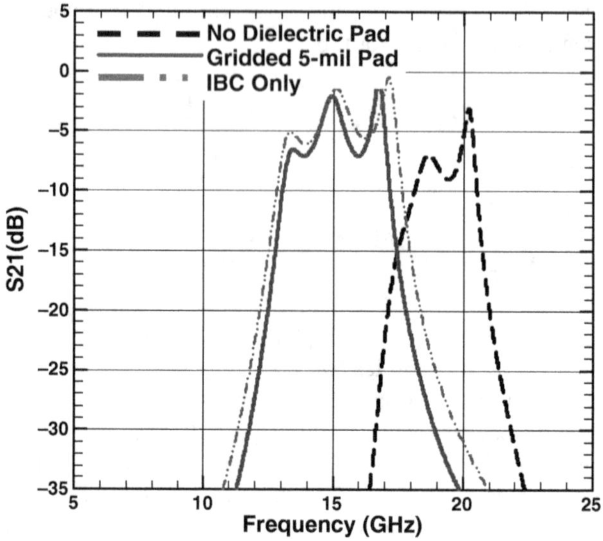

Figure 6.3 Insertion loss (S21) for a bandpass filter showing the influence of a thin-material layer.

substrate; second, the 5-mil pad will be fully gridded using linear tetrahedral finite elements; and third, the 5-mil pad will be treated by the IBC described by (6.9). The results for the insertion loss (S21) for the three cases are shown in Figure 6.3. The presence of the thin dielectric sheet is seen to affect the center frequency for the passband significantly, as evidenced by both the gridded and IBC solutions. It is interesting to note that because the tangential component of the electric field is dominate within the pad region, the first-order IBC is effective for this application.

The second example is a thin resistive sheet placed above a PEC ground plane as shown in Figure 6.4. For the case of normal incidence along with a sheet

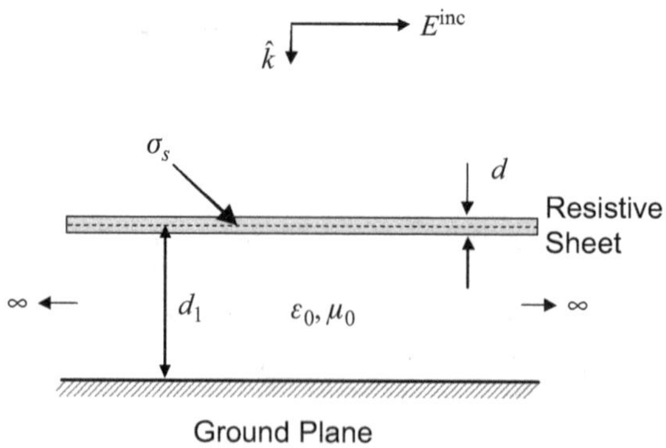

Figure 6.4 Cross section of a resistive sheet (R-card) placed above a PEC ground plane for the case of normal incidence.

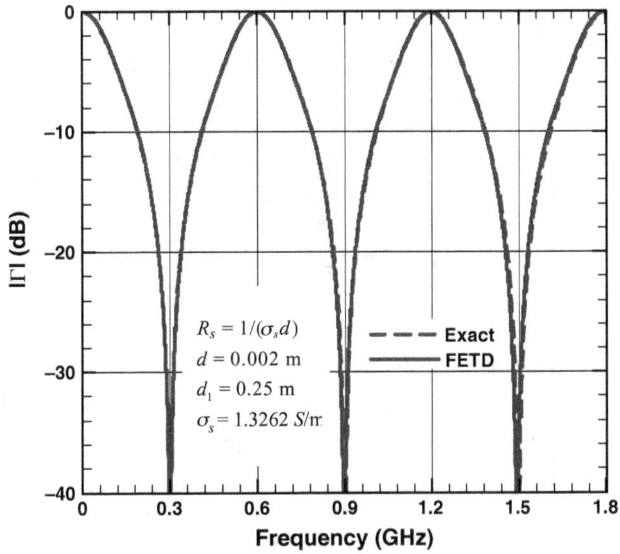

Figure 6.5 Magnitude of the reflection coefficient for a resistive sheet above a ground plane. The sheet surface resistivity is 377 Ω/□.

impedance that is matched to free space, the exact reflection coefficient is given by $\Gamma = -e^{j2k_0 d_1}/(1 + j2 \tan k_0 d_1)$, where d_1 denotes the height of the sheet above the ground plane. For application of (6.12), either $R_s = 377\ \Omega/\square$ can be defined directly, or alternatively, a thin conductive sheet with a finite thickness d can be defined according to (6.10). If we assume a thickness of 2 mm, a conductivity of 1.33 S/m will provide the desired sheet resistance. A comparison of an FETD solution with the exact equation for $|\Gamma|$ is provided in Figure 6.5.

The next two examples are based on applying inductive and capacitive coatings to a perfectly conducting sphere. For the FETD solution method, the appropriate boundary conditions are given by (6.22) and (6.23), respectively. The Mie series provides an exact reference solution. The frequency-dependent monostatic radar cross section (RCS) for the inductive sphere is shown in Figure 6.6, whereas the capacitive case is shown in Figure 6.7. For comparison, the result for a noncoated PEC sphere is also shown in the figures.

6.1.2 Shell Element Formulation

Thin-material sheets and coatings can be resolved by finite volumetric methods through local refinements to the mesh. In this way, the thin-material feature is fully characterized in three-dimensional space. However, this approach is typically impractical except for isolated instances, and therefore alternative approaches, such as the first-order Leontovich IBC technique described in the preceding section, are usually adopted. As noted previously, the simple Leontovich IBC has limitations,

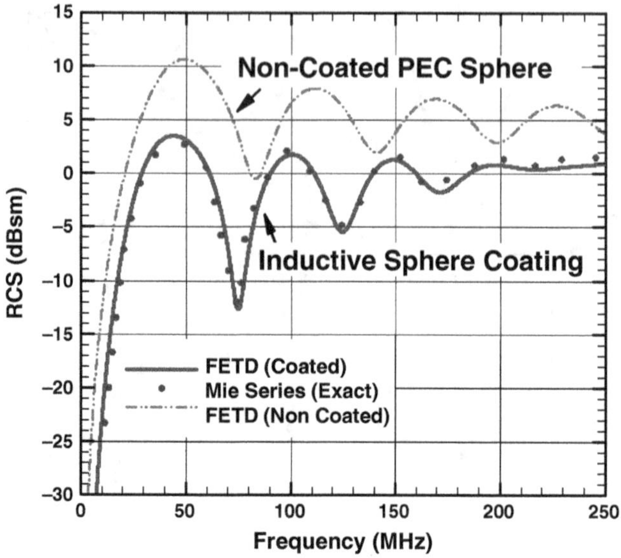

Figure 6.6 Inductive surface coating on a PEC sphere. The surface impedance is given by $Z_s = R_s + j\omega L_s = 377/2 + j\omega\, 377/(4\pi \times 10^8)$. (After Riley and Riley [12], Copyright © IEEE 2004.)

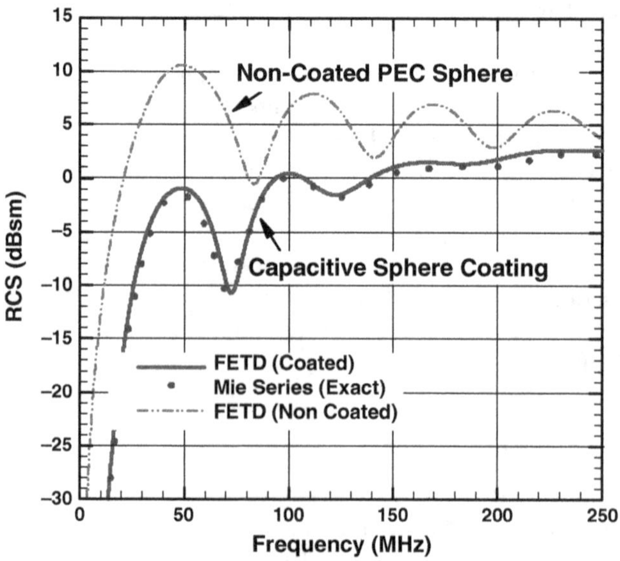

Figure 6.7 Capacitive surface coating on a PEC sphere. The surface impedance is given by $Z_s = 1/(G_s + j\omega C_s)$ with $G_s = 0.00531$ S/□ and $C_s = 8.44$ pF/□. (After Riley and Riley [12], Copyright © IEEE 2004.)

however, because it only affects the tangential field components, and in the case of lossless dielectric coatings on a conducting surface, a nonunity relative permittivity is ignored.

A more general approach that does not increase the total number of volumetric cells within the finite element domain is to adopt *degenerate finite elements*, constructed by collapsing the height of a three-dimensional element such that a two-dimensional surface element is obtained. The degrees of freedom associated with the original volumetric element are preserved in the process, and thus the technique is three-dimensional with regard to the finite element expansion coefficients. More specifically, possible discontinuities in the normal component of the electric field across the material interface can be resolved. However, because the physical thickness of the layer is now treated as a mathematical parameter, the technique inherently requires that the material layer be electrically thin, although this can be relaxed to some extent by using higher-order basis functions. Obviously, only certain types of volumetric finite elements can be used for this purpose, and these include hexahedral and prismatic elements. Because unstructured finite element meshes are typically based on tetrahedral elements, the class of prismatic elements is then well suited to be used to model the thin-material regions. Degenerate prismatic elements have previously been used for eddy current applications [17] as well as high-frequency scattering based on the FETD method [18]. Higher-order interpolatory vector basis functions are readily available for prism elements [19].

A prism element placed between two tetrahedra is shown in Figure 6.8, where the actual thickness d of the material layer is shown in Figure 6.8(a) and the degenerate form is shown in Figure 6.8(b). In the degenerate form, note that the degrees of freedom associated with the edges on the interior triangular faces are now doubled

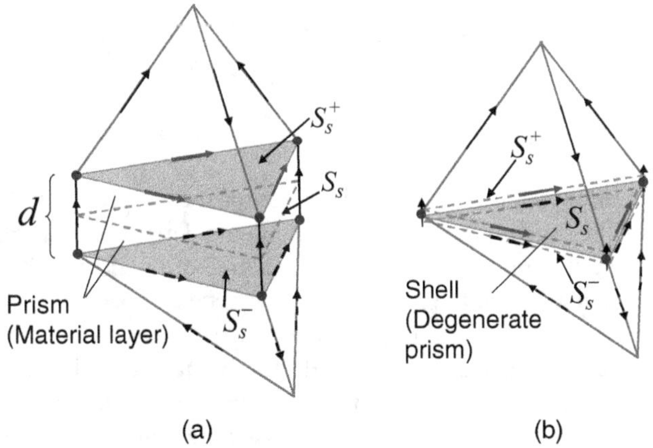

Figure 6.8 Use of degenerate prism elements to model thin-material sheets and coatings. (a) Physical three-dimensional model of a prism interfaced between two tetrahedral elements. (b) Simulation model that has collapsed the three-dimensional prism into a two-dimensional shell element such that the original degrees of freedom are retained.

such that there are edge unknowns associated with the tetrahedra both above and below the material layer. In addition, even though the vertical edges of the prism have been collapsed, the degrees of freedom associated with these edges have been accounted for properly in the construction of the finite element system matrices. For example, when using a right prism in the material layer, the matrices $[T]$, $[R]$, and $[S]$ in (2.27) have additional elemental contributions from the prism element given by [18]

$$T_{ij} = \begin{cases} \dfrac{1}{3}\displaystyle\iint_{S_s} \varepsilon_s d\mathbf{N}_i^p \cdot \mathbf{N}_j^p \, dS, & i, j \text{ on same face} \\ \dfrac{1}{6}\displaystyle\iint_{S_s} \varepsilon_s d\mathbf{N}_i^p \cdot \mathbf{N}_j^p \, dS, & i, j \text{ on opposite faces} \\ \displaystyle\iint_{S_s} \varepsilon_s d\lambda_i^p \lambda_j^p \, dS, & i, j \text{ both vertical} \end{cases} \tag{6.34}$$

$$R_{ij} = \begin{cases} \dfrac{1}{3}\displaystyle\iint_{S_s} \sigma_s d\mathbf{N}_i^p \cdot \mathbf{N}_j^p \, dS, & i, j \text{ on same face} \\ \dfrac{1}{6}\displaystyle\iint_{S_s} \sigma_s d\mathbf{N}_i^p \cdot \mathbf{N}_j^p \, dS, & i, j \text{ on opposite faces} \\ \displaystyle\iint_{S_s} \sigma_s d\lambda_i^p \lambda_j^p \, dS, & i, j \text{ both vertical} \end{cases} \tag{6.35}$$

$$S_{ij} = \begin{cases} \dfrac{1}{3}\displaystyle\iint_{S_s} \frac{d}{\mu_s}(\nabla \times \mathbf{N}_i^p) \cdot (\nabla \times \mathbf{N}_j^p) \, dS, & i, j \text{ on same face} \\ \dfrac{1}{6}\displaystyle\iint_{S_s} \frac{d}{\mu_s}(\nabla \times \mathbf{N}_i^p) \cdot (\nabla \times \mathbf{N}_j^p) \, dS, & i, j \text{ on opposite faces} \\ \displaystyle\iint_{S_s} \frac{d}{\mu_s}(\hat{n}_s \times \nabla\lambda_i^p) \cdot (\hat{n}_s \times \nabla\lambda_j^p) \, dS, & i, j \text{ both vertical} \end{cases} \tag{6.36}$$

where the $\mathbf{N}^p$'s denote the prism vector edge basis functions and the λ^p's denote the prism nodal basis functions [19].

Around the perimeter of the thin-material layer some care is required with regard to the interface with the surrounding tetrahedral elements. This is because termination of the layer requires transitioning back to a single degree of freedom for the tetrahedral elements that are connected to the perimeter edges. One approach is to collapse these degrees of freedom into a single edge. For the special case of a thin coating on a PEC surface, these collapsed edges would then be treated with a PEC boundary condition; that is, the electric field would be set to zero. For other configurations, the collapsed perimeter edges are free unknowns associated with the connecting edges of the surrounding tetrahedra.

This shell-element approach to modeling thin-material sheets and coatings is, of course, an approximate solution technique. The primary reasons for this are the following: (1) When using linear basis functions there is only one degree of freedom through the layer thickness; (2) the physical thickness is described mathematically instead of being resolved directly by the mesh; and (3) the perimeter of the layer region may not be defined adequately because it is necessary to combine the perimeter unknowns so that surrounding tetrahedral elements can be interfaced properly. Nevertheless, this technique has been found to provide similar results for those applications that are well characterized by the first-order IBC, yet provides increased generality and accuracy for applications where the first-order IBC is inadequate [18].

6.2 THIN WIRES AND SLOTS

Practical applications often possess many wires, cables, and apertures. Wires are typically very thin in both their physical and electrical radii, and narrow apertures often occur through the interconnection of metal faceplates. As noted previously, although an unstructured finite element mesh can resolve these features by transitioning the element size, this approach is typically impractical for all but a few isolated instances of these features, due to the resulting large number of finite elements. An alternative approach that exploits and incorporates the fundamental physics of the wires and slots into the finite element method is therefore desirable, and this approach has been developed and applied for many years within the FDTD method [20–29]. In this section we present techniques to resolve thin wires and narrow slot apertures within the FETD method. Because of the increased geometrical flexibility of FETD relative to FDTD, the paths followed by the wires and slots can be quite general. In addition, it is possible to construct hybrid algorithms that preserve the unconditional stability of the FETD method.

6.2.1 Thin Wires

Electrically thin wires and cables can be treated accurately by integral equation techniques as well as the telegrapher's (transmission-line) equations. The approach adopted in this section will be based on the telegrapher's equations interfaced with the FETD method. To account properly for full three-dimensional coupling between the telegrapher's equations and possible surrounding geometry will require the development of a specialized hybridization technique that couples the three-dimensional finite element method with the one-dimensional wire model.

We consider a conducting cylindrical wire with radius a and local path direction characterized by the position-dependent unit vector $\hat{s}$, as illustrated in Figure 6.9. Although randomly oriented in three-dimensional space, the current distribution is assumed to be void of azimuthal variation and hence will be described by one-dimensional equations such that the resulting description is often referred to as a *filament model*. For convenience, we assume that the wire(s) are edge-aligned with

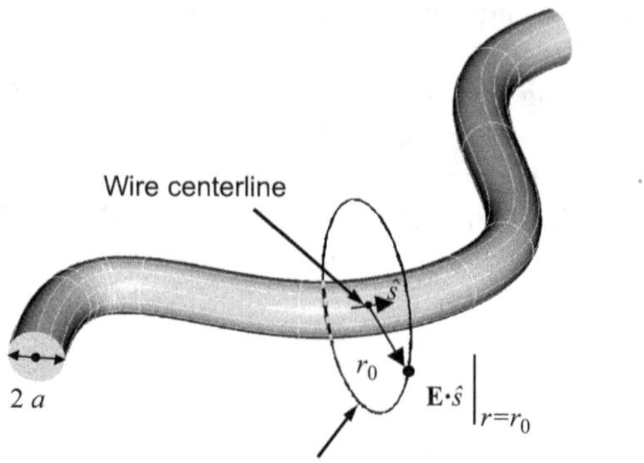

Figure 6.9 Physical wire oriented in three-dimensional space. Simulation model replaces the wire with a filament placed along the centerline. A hybrid field-wire solution technique is used.

the surrounding three-dimensional finite element cells [30], although this restriction can be relaxed [31]. The governing telegrapher's equations are given by [20]

$$C_w \frac{\partial V}{\partial t} = -C_w \frac{\sigma}{\varepsilon} V - \frac{\partial I}{\partial s} \tag{6.37}$$

$$L_w \frac{\partial I}{\partial t} = -\frac{\partial V}{\partial s} + \left(\mathbf{E} \cdot \hat{s}|_{r=r_0} + V_{\text{imp}} - IR_w\right) \tag{6.38}$$

where I denotes the wire current, V denotes the voltage (or charge) along the wire, V_{imp} denotes a possible impressed voltage source with unit of volts per unit length, R_w denotes possible resistive loading with unit of ohms per unit length, and the material parameters σ and ε are those associated with the medium supporting the wire.

As illustrated in Figure 6.9, the radial distance $r = r_0$ in (6.38) is measured from the centerline of the wire and defines a local cylindrical tube where the surrounding three-dimensional electric fields are evaluated. These electric fields are represented by $\mathbf{E} \cdot \hat{s}|_{r=r_0}$ in (6.38) and provide the field coupling to the wire. For example, for the simple case of a straight wire that resides along the edges of a uniform grid of cubical hexahedral cells with edge length Δ, $r_0 = \Delta$ would be used. In this case, an edge of the wire is surrounded by four parallel edges each a distance Δ away from the wire edge. These parallel edges define the electric field locations used to drive the corresponding wire edge, where a single average value is then constructed to provide the source term to be used in (6.38) for the specific wire edge. In the more complicated case of unstructured tetrahedral elements surrounding the wire, we define $r_0 = \Delta_{\text{av}}$, where Δ_{av} denotes the average edge length of the three-dimensional cells locally surrounding a given wire edge. The corresponding vector electric field values within the surrounding cells can then be evaluated through interpolation of the vector basis functions, and a single average value of these interpolated fields is used in (6.38) for the specific wire edge with direction $\hat{s}$. Each edge of the wire is evaluated similarly.

The parameters C_w and L_w are referred to as the *in-cell* capacitance and inductance, respectively, and are given by [20]

$$L_w = \frac{\mu}{2\pi} \ln \frac{r_0}{a}, \qquad C_w = \frac{\varepsilon\mu}{L_w}. \tag{6.39}$$

These expressions for L_w and C_w are based on an assumed $1/r$ variation for both the radially directed electric field and the circumferential magnetic field local to the wire. In particular, L_w has units of inductance per unit length and C_w has units of capacitance per unit length.

It is convenient to rewrite the first-order equations (6.37) and (6.38) in a wave-equation form given by [30]

$$\begin{aligned} L_w C_w \frac{\partial^2 I}{\partial t^2} &+ \left(L_w C_w \frac{\sigma}{\varepsilon} + R_w C_w\right) \frac{\partial I}{\partial t} - \frac{\partial^2 I}{\partial s^2} + C_w R_w \frac{\sigma}{\varepsilon} I \\ &= C_w \frac{\partial}{\partial t} (\mathbf{E} \cdot \hat{s}|_{r=r_0} + V_{\text{imp}}) + C_w \frac{\sigma}{\varepsilon} \left(\mathbf{E} \cdot \hat{s}|_{r=r_0} + V_{\text{imp}}\right). \end{aligned} \tag{6.40}$$

For simplicity, we assume that the medium around the wire is nonconducting such that $\sigma = 0$. In this case, (6.40) simplifies to

$$L_w \frac{\partial^2 I}{\partial t^2} + R_w \frac{\partial I}{\partial t} - \frac{1}{C_w} \frac{\partial^2 I}{\partial s^2} = \frac{\partial}{\partial t} \left(\mathbf{E} \cdot \hat{s}|_{r=r_0} + V_{\text{imp}}\right). \tag{6.41}$$

For a one-dimensional finite element discretization of (6.41) defined along a path denoted by s, the weak-form representation is given by

$$\int_s \left(\frac{L_w}{\mu\,\varepsilon} \frac{\partial \Phi}{\partial s} \frac{\partial I}{\partial s} + L_w \Phi \frac{\partial^2 I}{\partial t^2} + R_w\, \Phi \frac{\partial I}{\partial t} \right) ds = \int_s \Phi \frac{\partial}{\partial t} \left(\mathbf{E} \cdot \hat{s}|_{r=r_0} + V_{\text{imp}}\right) ds \tag{6.42}$$

where we have used $C_w = \varepsilon\mu/L_w$ and the basis functions Φ for I will be the traditional linear nodal functions [14] such that

$$I(s,t) = \sum_{i=1}^{N_{\text{node}}} \Phi_i(s) I_i(t). \tag{6.43}$$

The corresponding semidiscrete finite element representation of (6.42) is then

$$[T_w] \frac{d^2\{I\}}{dt^2} + [R_w] \frac{d\{I\}}{dt} + [S_w]\{I\} = \{f_w\} \tag{6.44}$$

and the matrix entries are given by

$$T_{w,ij} = \int_s L_w \, \Phi_i \Phi_j \, ds \tag{6.45}$$

$$R_{w,ij} = \int_s R_w \, \Phi_i \Phi_j \, ds \tag{6.46}$$

$$S_{w,ij} = \int_s \frac{L_w}{\mu\varepsilon} \frac{\partial \Phi_i}{\partial s} \frac{\partial \Phi_j}{\partial s} \, ds \,. \tag{6.47}$$

In addition, $\{I\} = [I_1, I_2, \ldots, I_{N_{\text{node}}}]^{\mathrm{T}}$ and the elements of the excitation vector $\{f_w\}$ are

$$f_{w,i}(t) = \int_s \Phi_i \frac{\partial}{\partial t} (\mathbf{E} \cdot \hat{s}|_{r=r_0} + V_{\text{imp}}) \, ds. \tag{6.48}$$

Equation (6.44) will be referred to as the *finite element thin-wire equation.*

The basic strategy for the hybridization of (6.44) with a three-dimensional finite element method is initially to use the local three-dimensional electric fields $\mathbf{E} \cdot \hat{s}|_{r=r_0}$ surrounding the wire as distributed source terms for the thin-wire problem. The current on the wire segments is then obtained through the solution of (6.44), and these currents are used subsequently as source terms for the three-dimensional finite element method.

More specifically, for the FETD method the weak-form representation of the electric field wave equation is

$$\iiint_V \left[(\nabla \times \mathbf{T}) \cdot \left(\frac{1}{\mu} \nabla \times \mathbf{E} \right) + \varepsilon \, \mathbf{T} \cdot \frac{\partial^2 \mathbf{E}}{\partial t^2} \right] dV + Y_0 \oiint_{S_o} (\hat{n} \times \mathbf{T}) \cdot \left(\hat{n} \times \frac{\partial \mathbf{E}}{\partial t} \right) dS$$
$$= - \iiint_V \mathbf{T} \cdot \left[\frac{\partial \mathbf{J}_{\text{imp}}}{\partial t} + \nabla \times \left(\frac{1}{\mu} \mathbf{M}_{\text{imp}} \right) \right] dV \tag{6.49}$$

and the semidiscrete form for (6.49) is given by (2.27). With standard first-order vector edge basis functions, the elements of the excitation vector (2.31) simply become $-\partial \tilde{I}_i / \partial t$, where $\tilde{I}_i$ denotes an arithmetic average of the nodal wire current values to obtain an average edge current for the ith edge, and $\mathbf{M}_{\text{imp}} = 0$. A self-consistent time-stepping algorithm that advances both (6.44) and (6.49) can be formulated in a straightforward manner [30]. However, we recall from Chapter 4 that symmetry is generally a requirement in the construction of stable hybrid formulations in the time domain, and this simple coupling technique will generally not be symmetric with regard to the source terms $\mathbf{E} \cdot \hat{s}|_{r=r_0}$ for (6.44) and the wire current I for (6.49). The lack of symmetry is apparent because the electric fields driving the wire equation form a distribution around the wire (Figure 6.9), yet the wire currents driving the volumetric region are constrained to the wire edges. Consequently, this hybrid technique may be susceptible to late-time weak instabilities on a general unstructured mesh. It is interesting to note that if the computational domain is discretized with brick-shaped elements and we adopt the two-point trapezoidal integration formula discussed in Chapter 4 for both (6.44) and (6.49), this simple wire-field coupling approach is, in fact, equivalent to the traditional FDTD thin-wire algorithm [20].

A more general and robust hybridization scheme that preserves symmetry between (6.44) and (6.49) has been formulated in Ref. 31. The basic strategy with this approach is to introduce a radial weighting function $g(r)$ around the wire such that $g(r)$ has the property

$$\int_{r\geq a} 2\pi r g(r) dr = 1. \tag{6.50}$$

A suitable representation for $g(r)$ is

$$g(r) = \begin{cases} 0, & r < a \\ \dfrac{1 + \cos(\pi r / r_0)}{\pi\left(r_0^2 - a^2\right) - 2r_0^2/\pi\,(1 + \cos(\pi a/r_0) + \pi a/r_0 \sin(\pi a/r_0))}, & a \leq r \leq r_0 \\ 0, & r > r_0. \end{cases} \tag{6.51}$$

The distance r_0 is shown in Figure 6.9 and was defined previously by $r_0 = \Delta_{\text{av}}$, where Δ_{av} denotes the average edge length of the three-dimensional cells locally surrounding a given wire edge. By using this weighting function and integrating over a volume V, an alternative weak-form representation for (6.41) is obtained:

$$\int_s \left(\frac{L_w}{\mu\varepsilon} \frac{\partial \Phi}{\partial s} \frac{\partial I}{\partial s} + L_w \Phi \frac{\partial^2 I}{\partial t^2} + R_w \Phi \frac{\partial I}{\partial t} \right) ds = \int_s \Phi \frac{\partial V_{\text{imp}}}{\partial t} ds + \iiint_V g(r) \Phi\, \hat{s} \cdot \frac{\partial \mathbf{E}}{\partial t} dV \tag{6.52}$$

where the property (6.50) was used to obtain the two simplified line integral representations in (6.52). The electric field $\mathbf{E}$ is now expanded in vector edge basis functions according to (2.26):

$$\mathbf{E}(\mathbf{r}, t) = \sum_{i=1}^{N_{\text{edge}}} \mathbf{N}_i(\mathbf{r})\, E_i(t). \tag{6.53}$$

With this expansion, the second term on the right-hand side of (6.52) becomes

$$\sum_{i=1}^{N_{\text{edge}}} \frac{\partial E_i(t)}{\partial t} \iiint_V \mathbf{N}_i(\mathbf{r}) \cdot \hat{s} g(r) \Phi\, dV. \tag{6.54}$$

The only change in the semidiscrete finite element representation (6.44) is with regard to the excitation vector $\{f_w\}$, and with (6.54) the elements now become

$$f_{w,i}(t) = \int_s \Phi_i \frac{\partial V_{\text{imp}}}{\partial t} ds + \sum_{j=1}^{N_{\text{edge}}} \frac{\partial E_j(t)}{\partial t} \iiint_V \Phi_i(s) g(r) \mathbf{N}_j(\mathbf{r}) \cdot \hat{s}\, dV. \tag{6.55}$$

To enable a symmetric excitation between (6.44) and (6.49), the current density $\mathbf{J}_{\text{imp}}$ is written as $\mathbf{J}_{\text{imp}}(\mathbf{r}, t) = I(s, t)\, g(r)\, \hat{s}$, where $g(r)$ is recalled to have the property (6.50). By using (6.43), $\mathbf{J}_{\text{imp}}$ becomes

$$\mathbf{J}_{\text{imp}}(\mathbf{r}, t) = \sum_{i=1}^{N_{\text{node}}} \Phi_i(s) g(r) I_i(t) \hat{s}. \tag{6.56}$$

The elements of the excitation vector for the semidiscrete representation of (6.49) are given by (2.31), and with (6.56) these become (assuming that $\mathbf{M}_{\text{imp}} = 0$)

$$f_i(t) = -\sum_{j=1}^{N_{\text{node}}} \frac{\partial I_j(t)}{\partial t} \iiint_V \Phi_j(s) g(r) \mathbf{N}_i(\mathbf{r}) \cdot \hat{s}\, dV. \tag{6.57}$$

The elements of the excitation vector (6.55) for the finite element thin-wire equation (6.44) and the excitation vector (6.57) for the volumetric region (2.27) lead to the following coupled system of equations:

$$\begin{bmatrix} [T] & 0 \\ 0 & [T_w] \end{bmatrix} \frac{d^2}{dt^2} \begin{Bmatrix} \{E\} \\ \{I\} \end{Bmatrix} + \begin{bmatrix} [R] & [P_w]^{\mathrm{T}} \\ -[P_w] & [R_w] \end{bmatrix} \frac{d}{dt} \begin{Bmatrix} \{E\} \\ \{I\} \end{Bmatrix} + \begin{bmatrix} [S] & 0 \\ 0 & [S_w] \end{bmatrix} \begin{Bmatrix} \{E\} \\ \{I\} \end{Bmatrix} = \begin{Bmatrix} 0 \\ \{\tilde{f}_w\} \end{Bmatrix} \tag{6.58}$$

where we have introduced the matrix $[P_w]$ and the excitation vector $\{\tilde{f}_w\}$, with elements given, respectively, by

$$P_{w,ij} = \iiint_V \Phi_i(s) g(r) \mathbf{N}_j(\mathbf{r}) \cdot \hat{s}\, dV \tag{6.59}$$

$$\tilde{f}_{w,i} = \int_s \Phi_i \frac{\partial V_{\text{imp}}}{\partial t} ds. \tag{6.60}$$

A formal proof is available of the unconditional stability for the coupled wire and field system (6.58) discretized using the Newmark-beta method [31].

An example of a thin-wire Archimedean spiral antenna is shown in Figure 6.10. The overall diameter of the antenna is 2.4 inches. Archimedean spirals can be

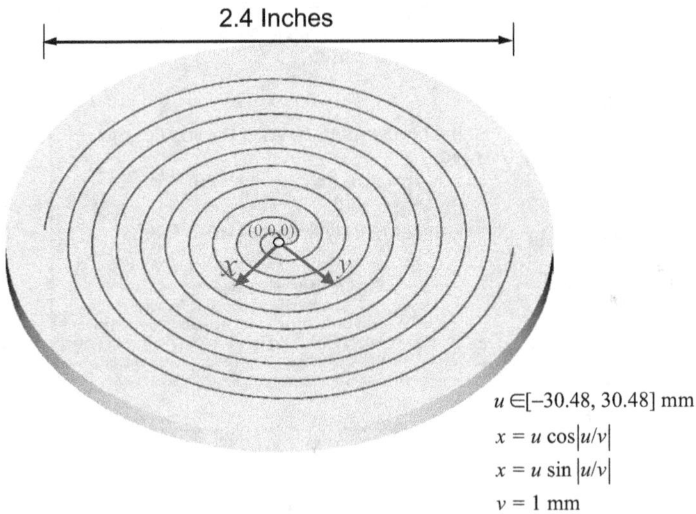

Figure 6.10 Thin-wire Archimedean spiral antenna. An ideal voltage source is placed at the origin. (*See insert for color representation of figure.*)

constructed easily within an xy-plane according to the equations $x = u \cos u$ and $y = u \sin u$ where u is specified over an interval. Because the specific radius a of the wire is a mathematical parameter when using the FETD field-wire hybrid formulation, a variety of cases for the performance of the Archimedean spiral can be studied rapidly. An interesting case is when the wire and air have approximately equal surface areas within the xy-plane of the spiral. The antenna is then approximately self-complementary, and as described in Section 7.2.2, we would expect an input impedance of 188.5 Ω between the two arms of the antenna. For the geometry in Figure 6.10, this situation occurs when the wire radius is $a = \pi/8$ mm. The input impedance calculated is shown in Figure 6.11, and we see that the FETD predictions for the spiral agree well with the self-complementary theory within the broadband operating range of the antenna.

An example of a rectangular resonator fed by a coaxial waveguide is shown in Figure 6.12. Although the geometry for this example was originally developed to validate thin-wire algorithms for the Cartesian grid FDTD method [29], application of the FETD method using unstructured tetrahedral elements is used here. A thin wire of radius $a = 0.8$ mm was extended from the lower surface to the upper surface of the resonator and terminated with 50- and 47-Ω lumped resistors on the respective surfaces. A voltage source with the 50-Ω impedance was used to represent the excitation from the coaxial waveguide. The real power delivered to the input port is given by $P = \frac{1}{2}\mathrm{Re}[\tilde{V}(\omega)\tilde{I}^*(\omega)]$, where $\tilde{V}$ and $\tilde{I}$ denote the deconvolved port voltage and current, respectively, and the asterisk denotes complex conjugation. A comparison of the FETD prediction and the data measured for the power delivered is shown in Figure 6.13.

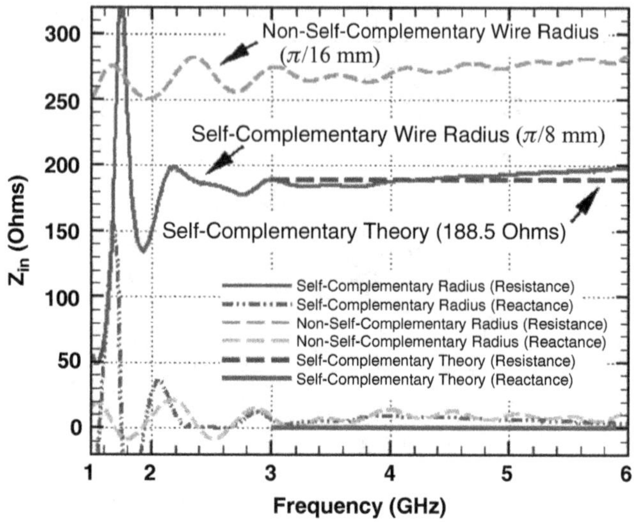

Figure 6.11 Input impedance of a thin-wire Archimedean spiral antenna. When the wire radius a is $\pi/8$ mm, the spiral is approximately self-complementary with an input impedance of 188.5 Ω.

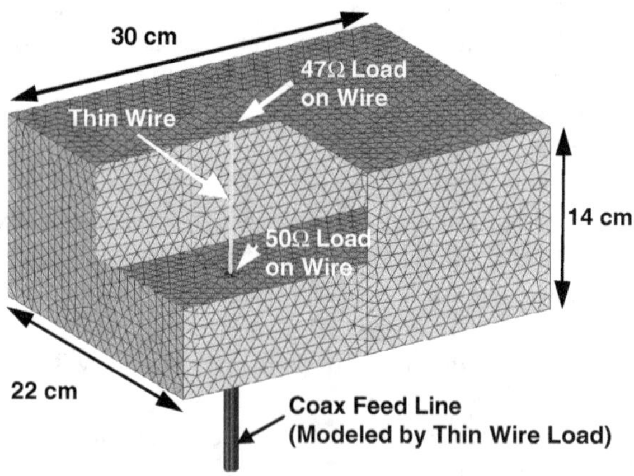

Figure 6.12 Rectangular cavity with a thin wire and coaxial feed. The FETD solution method is based on unstructured tetrahedral elements. The geometry is based on the original FDTD test object by Li et al. [29].

6.2.2 Thin Slots

Slot apertures that are electrically narrow in both width and depth occur frequently in practical applications. Similar to the case of electrically thin wires, unstructured finite element techniques can be used to resolve narrow slots. However, because the resolution of even a single narrow aperture could possibly lead to the creation of a large

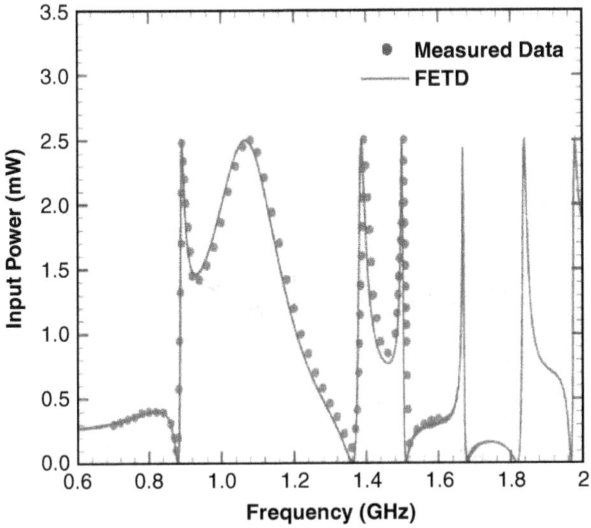

Figure 6.13 Input power for the geometry in Figure 6.12. Measured data are extracted from Li et al. [29].

number of finite elements, alternative approaches to accounting accurately for the physics of the aperture without its complete discretization are again desirable. Within the FDTD setting, many such techniques to characterize these types of apertures have been reported, including local modifications of the FDTD path integrals [26] and the integration of an integral equation for a slot in an infinite plane [28,29]. More recently, an alternative approach is to use the local telegrapher's equations cast in a form dual to those described previously for thin-wire applications [30,32]. By approaching the narrow slot application from a perspective dual to the wire case, a unified treatment of the two configurations can be formulated, and this is the approach taken in this section.

The geometry of a slot in a conducting wall is shown in Figure 6.14. From the perspective of the three-dimensional finite element method that supports this conducting wall, the slot is assumed to be short-circuited and hence identified only by a virtual line segment that defines the path of the slot, which is described by the position-dependent unit vector $\hat{s}$. The primary attribute of interest will be the voltage across the slot aperture, which can also be interpreted mathematically as a magnetic current. For the case of an electrically narrow slot with width w and depth d, where $w \ll \lambda$ and $d \ll \lambda$, the slot voltage is described by the following time-harmonic transmission-line equations [33]:

$$\frac{\partial I_s(\omega)}{\partial s} = -Y_{\text{slot}}(\omega)\, V_s(\omega) + \mathbf{H}_{\text{diff}}(\omega) \cdot \hat{s} \tag{6.61}$$

$$\frac{\partial V_s(\omega)}{\partial s} = -Z_{\text{slot}}(\omega)\, I_s(\omega) \tag{6.62}$$

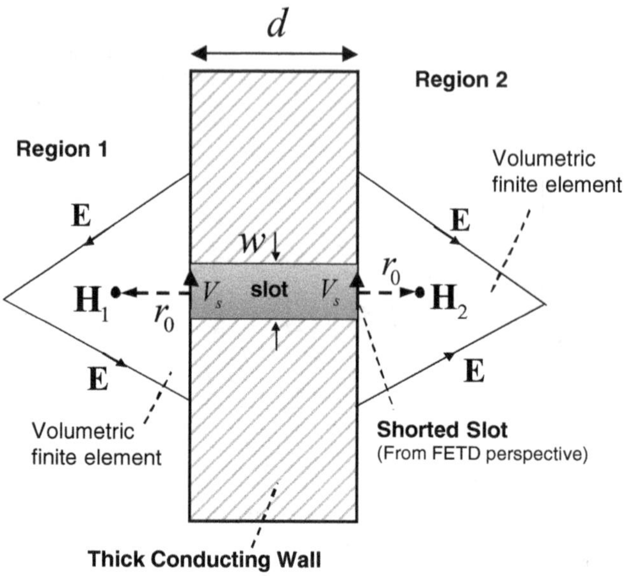

(a)

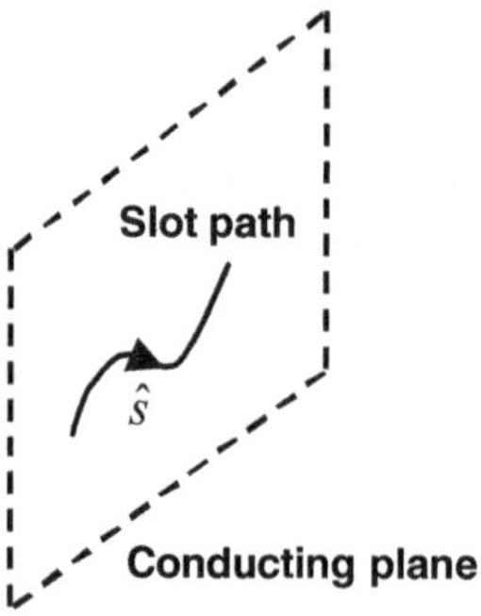

(b)

Figure 6.14 Slot aperture in a conducting wall. (a) Physical geometry of a slot with an electrically narrow width and depth, showing driving local magnetic fields offset from the wall by a distance r_0. (b) Slot path in the conducting plane with a local direction $\hat{s}$. The slot depth is accounted for mathematically in the local transmission-line formulation.

where V_s denotes the voltage across the slot width (or the axial magnetic current), I_s can be interpreted as magnetic charge, and $\mathbf{H}_{\text{diff}}$ denotes the difference in the external local magnetic field across the (short-circuited) surface(s) containing the slot; specifically, $\mathbf{H}_{\text{diff}} \equiv \mathbf{H}_2 - \mathbf{H}_1$, where regions 1 and 2 are as defined in Figure 6.14. The impedance per unit length, $Z_{\text{slot}}(\omega)$, and the admittance per unit length, $Y_{\text{slot}}(\omega)$, both depend on the physical properties associated with the slot, such as internal gaskets and/or internal wall losses [33]. For the simple lossless case with air in the interior of the slot, $Z_{\text{slot}}(\omega)$ and $Y_{\text{slot}}(\omega)$ are given by

$$\frac{1}{Z_{\text{slot}}(\omega)} = \frac{1}{j\omega L_s^{\text{extr}}} + \frac{1}{j\omega L_s^{\text{intr}}} \tag{6.63}$$

$$Y_{\text{slot}}(\omega) = j\omega\,(C_s^{\text{extr}} + C_s^{\text{intr}}) \tag{6.64}$$

where the capacitance and inductance internal to the slot, C_s^{intr} and L_s^{intr}, respectively, have the representations

$$C_s^{\text{intr}} = \varepsilon_0 \frac{d}{w}, \qquad L_s^{\text{intr}} = \mu_0 \frac{w}{d} \tag{6.65}$$

while the capacitance and inductance external to the slot, C_s^{extr} and L_s^{extr}, respectively, are

$$C_s^{\text{extr}} = \frac{2}{\pi}\varepsilon_0 \ln \frac{4r_0}{w}, \qquad L_s^{\text{extr}} = \frac{\varepsilon_0 \mu_0}{C_s^{\text{extr}}}. \tag{6.66}$$

Here, r_0 denotes an offset distance in the direction normal to the slot, which will typically be the distance to the location where the local magnetic fields will be evaluated (Figure 6.14). The definition of r_0 is similar to the previous case of thin wires, and more specific discussion on its application to thin slots will be provided later.

Because of the simple form of (6.65) and (6.66), we can define a single slot capacitance C_s and inductance L_s such that

$$C_s = \frac{2}{\pi}\varepsilon_0 \ln \frac{r_0}{a_s}, \qquad L_s = \frac{\varepsilon_0 \mu_0}{C_s} \tag{6.67}$$

where

$$C_s \equiv C_s^{\text{extr}} + C_s^{\text{intr}}, \qquad \frac{1}{L_s} \equiv \frac{1}{L_s^{\text{extr}}} + \frac{1}{L_s^{\text{intr}}} \tag{6.68}$$

and the equivalent radius a_s is defined by $a_s = (w/4)\exp(-\pi d/2w)$ [33].

The time-domain form for (6.61) and (6.62) is then

$$C_s \frac{\partial V_s}{\partial t} = -\frac{\partial I_s}{\partial s} + \mathbf{H}_{\text{diff}} \cdot \hat{s} \tag{6.69}$$

$$L_s \frac{\partial I_s}{\partial t} = -\frac{\partial V_s}{\partial s} \tag{6.70}$$

and the wave equation for the slot voltage V_s is given by

$$C_s \frac{\partial^2 V_s(t)}{\partial t^2} - \frac{1}{L_s} \frac{\partial^2 V_s(t)}{\partial s^2} = \frac{\partial}{\partial t} \mathbf{H}_{\text{diff}}(t) \cdot \hat{s} \bigg|_{r=r_0} . \tag{6.71}$$

As illustrated in Figure 6.14, r_0 is the distance perpendicular from the wall containing the slot to the location where the local magnetic fields that excite the slot are evaluated. For example, for the simple case of a straight slot that resides within a uniform FDTD grid of cubical hexahedral cells with edge length Δ, $r_0 = \Delta/2$ would be used on each side of the wall containing the slot. Note that this is the distance to the closest magnetic field edges that are parallel to a particular slot edge in both regions 1 and 2, as shown in Figure 6.14. The magnetic field difference required in (6.71) is then constructed by forming $\mathbf{H}_{\text{diff}} = \mathbf{H}_2 - \mathbf{H}_1$. In the more complicated case of unstructured tetrahedral elements surrounding the slot, we define $r_0 = \Delta_{\text{av}}/2$, where Δ_{av} denotes the average edge length of the three-dimensional cells locally surrounding a given slot edge. The corresponding vector magnetic field values within the surrounding cells can then be evaluated through interpolation of the vector basis functions, and a single average value of these interpolated fields is used in (6.71) for the specific $\hat{s}$-directed slot edge. Each edge of the slot is evaluated similarly. Note that in the case with first-order vector edge basis functions, only the magnetic field at the barycenter of the local cells will be available since taking the curl of these basis functions will lead to a constant value for the magnetic field.

Equation (6.71) corresponds to a dual representation to the thin-wire equation (6.41). A simple field-slot finite element formulation can be realized by the following: (1) V_s is obtained from a one-dimensional finite element discretization of (6.71); (2) because V_s corresponds to a magnetic current, it is used in (2.31) to update the electric fields using (2.36); and (3) Maxwell's equations are used to construct $\mathbf{H}_{\text{diff}}$ based on the magnetic fields local to the slot in regions 1 and 2, as shown in Figure 6.14. Although an implementation based on this concept can provide accurate results [30], it will generally not be symmetric with regard to the field-based forcing function for the slot equation and the slot-based forcing function for the field equation. Consequently, weak instabilities may occur.

Similar to the field-wire formulation, a symmetric field-slot implementation can be constructed that is provably unconditionally stable [32]. However, the field-slot case is more complicated because the excitation is based on the local magnetic fields instead of the local electric fields, and consequently, a symmetric formulation is currently based on the use of the vector facet basis functions introduced in Section 4.4 as the expansion functions for $\mathbf{H}_{\text{diff}}$ in (6.71). The following provides a brief derivation of the symmetric field-slot coupled finite element system, which is patterned after Ref. 32.

By using the weighting function $g(r)$ defined in (6.51), the weak-form representation for (6.71) becomes

$$\int_s \left(\frac{1}{L_s} \frac{\partial \Phi}{\partial s} \frac{\partial V_s}{\partial s} + C_s \Phi \frac{\partial^2 V_s}{\partial t^2} \right) ds = \iiint_V g(r)\, \Phi \, \frac{\partial}{\partial t} \mathbf{H}_{\text{diff}}(t) \cdot \hat{s} \, dV \tag{6.72}$$

where Φ denotes the same linear nodal basis functions used for the thin-wire application. Because of the two-sided nature of the slot configuration, and in particular the difference in the local magnetic fields associated with the two sides, we redefine the weighting function $g(r)$ from (6.51) such that

$$\tilde{g}(r) \equiv \begin{cases} 2g(r), & r \in \text{region } 2 \\ -2g(r), & r \in \text{region } 1 \end{cases} \tag{6.73}$$

where the two regions are shown in Figure 6.14. With this definition, we can conveniently write the right-hand side of (6.72) as

$$\iiint_V \tilde{g}(r)\,\Phi\,\frac{\partial}{\partial t}\mathbf{H}(t)\cdot\hat{s}\,dV. \tag{6.74}$$

From (4.93), we now expand the magnetic field $\mathbf{B}(\mathbf{r}, t)$ in terms of the vector facet basis functions $\mathbf{N}_i^f(\mathbf{r})$, and for the case of simple materials such that $\mathbf{H} = (1/\mu_0)\mathbf{B}$, this expansion becomes

$$\mathbf{H}(\mathbf{r}, t) = \frac{1}{\mu_0}\sum_i B_i(t)\,\mathbf{N}_i^f(\mathbf{r}). \tag{6.75}$$

In terms of the nodal basis functions, the slot voltage V_s is expanded as

$$V_s(s, t) = \sum_{i=1}^{N_{\text{node}}} \Phi_i(s)\,V_{s,i}(t). \tag{6.76}$$

With (6.75) and (6.76), the semidiscrete form for (6.72) becomes

$$[T_s]\frac{d^2\{V_s\}}{dt^2} + [S_s]\{V_s\} = \{f_s\} \tag{6.77}$$

where the matrix entries are given by

$$T_{s,ij} = \int_s C_s\,\Phi_i\,\Phi_j\,ds \tag{6.78}$$

$$S_{s,ij} = \int_s \frac{1}{L_s}\,\frac{\partial\Phi_i}{\partial s}\,\frac{\partial\Phi_j}{\partial s}\,ds. \tag{6.79}$$

In addition, $\{V_s\} = [V_{s,1}, V_{s,2}, \ldots, V_{s,N_{\text{node}}}]^{\text{T}}$ and the elements of the excitation vector $\{f_s\}$ are

$$f_{s,i}(t) = \sum_{j=1}^{N_{\text{face}}} \frac{\partial B_j(t)}{\partial t}\iiint_V \frac{1}{\mu_0}\mathbf{N}_j^f(\mathbf{r})\cdot\hat{s}\,\tilde{g}(r)\,\Phi_i\,dV. \tag{6.80}$$

Alternatively, we can write (6.80) in matrix form and obtain

$$[T_s]\frac{d^2\{V_s\}}{dt^2} + [S_s]\{V_s\} = [P_s]\frac{d\{B\}}{dt} \tag{6.81}$$

where the elements of $[P_s]$ are

$$P_{s,ij} = \iiint_V \frac{1}{\mu_0}\mathbf{N}_j^f(\mathbf{r}) \cdot \hat{s}\,\tilde{g}(r)\,\Phi_i\,dV. \tag{6.82}$$

From (4.103), we eliminate $d\{B\}/dt$ from (6.81) to obtain

$$[T_s]\frac{d^2\{V_s\}}{dt^2} + [S_s]\{V_s\} = -[P_s][C]\{E\} - [P_s][M^f]^{-1}\{g\} \tag{6.83}$$

where $[C]$ is the incidence matrix with nonzero entries consisting simply of ± 1 [34] and the elements of $\{g\}$ are given by (4.101).

We now expand $\mathbf{M}_{\text{imp}}$ in terms of vector facet basis functions such that

$$\mathbf{M}_{\text{imp}}(\mathbf{r}, t) = \sum_i J_{m,i}(t)\,\mathbf{N}_i^f(\mathbf{r}). \tag{6.84}$$

With (6.84), $[M^f]^{-1}\{g\}$ in (6.83) simply becomes $\{J_m\}$, and (6.83) simplifies to

$$[T_s]\frac{d^2\{V_s\}}{dt^2} + [S_s]\{V_s\} = -[P_s][C]\{E\} - [P_s]\{J_m\}. \tag{6.85}$$

Because the slot voltage $V_s\,\hat{s}$ corresponds to $\mathbf{M}_{\text{imp}}$, (6.84) can also be represented by

$$\mathbf{M}_{\text{imp}}(\mathbf{r}, t) = \sum_{i=1}^{N_{\text{node}}} V_{s,i}(t)\,\tilde{g}(r)\,\Phi_i(s)\,\hat{s}\,. \tag{6.86}$$

By setting (6.84) equal to (6.86), multiplying by $\mathbf{N}_j^f$, and integrating over all space, we obtain

$$\iiint_V \sum_{i=1}^{N_{\text{node}}} V_{s,i}(t)\,\tilde{g}(r)\,\Phi_i(s)\,\hat{s} \cdot \mathbf{N}_j^f\,dV = \iiint_V \sum_i J_{m,i}(t)\,\mathbf{N}_i^f \cdot \mathbf{N}_j^f\,dV. \tag{6.87}$$

In matrix form, (6.87) can be written as

$$[P_s]^{\mathrm{T}}\{V_s\} = [T^f]\{J_m\} \tag{6.88}$$

where $[T^f]$ is defined by (4.98).

We now focus on the three-dimensional field equations described by (6.49). By using vector edge basis functions for the electric field and using the vector facet expansion (6.84), the semidiscrete representation for (6.49) becomes

$$[T]\frac{d^2\{E\}}{dt^2} + [R]\frac{d\{E\}}{dt} + [S]\{E\} = \{f\} \tag{6.89}$$

where the elements of the excitation vector $\{f\}$ are given by

$$f_i(t) = -\iiint_V \mathbf{N}_i \cdot \frac{\partial \mathbf{J}_{\text{imp}}}{\partial t}\, dV - \sum_{j=1}^{N_{\text{face}}} J_{m,j}(t) \iiint_V \mathbf{N}_j^f \cdot \left[\frac{1}{\mu_0}\nabla \times \mathbf{N}_i\right] dV. \tag{6.90}$$

We now use (6.88) and (6.90) and rewrite the right-hand side of (6.89) in matrix form such that

$$[T]\frac{d^2\{E\}}{dt^2} + [R]\frac{d\{E\}}{dt} + [S]\{E\} = -[C]^{\mathrm{T}}[P_s]^{\mathrm{T}}\{V_s\} + \{\tilde{f}\} \tag{6.91}$$

where we have used $[A]^{\mathrm{T}}[T^f]^{-1}=[C]^{\mathrm{T}}$, with $[A]$ defined by (4.99) and $[C]$ again denoting the incidence matrix, and the elements of $\{\tilde{f}\}$ are

$$\tilde{f}_i(t) = -\iiint_V \mathbf{N}_i \cdot \frac{\partial \mathbf{J}_{\text{imp}}}{\partial t}\, dV. \tag{6.92}$$

The final field-slot coupled system of equations is given by (6.85) and (6.91). In matrix form, this system can be written as

$$\begin{bmatrix} [T] & 0 \\ 0 & [T_s] \end{bmatrix} \begin{Bmatrix} \{\ddot{E}\} \\ \{\ddot{V}_s\} \end{Bmatrix} + \begin{bmatrix} [R] & 0 \\ 0 & 0 \end{bmatrix} \begin{Bmatrix} \{\dot{E}\} \\ \{\dot{V}_s\} \end{Bmatrix} + \begin{bmatrix} [S] & [C]^{\mathrm{T}}[P_s]^{\mathrm{T}} \\ [P_s][C] & [S_s] + [P_s][T^f]^{-1}[P_s]^{\mathrm{T}} \end{bmatrix}$$
$$\times \begin{Bmatrix} \{E\} \\ \{V_s\} \end{Bmatrix} = \begin{Bmatrix} \{\tilde{f}\} \\ 0 \end{Bmatrix} \tag{6.93}$$

where we have used the dot notation to indicate time differentiation. A formal proof of the unconditional stability of system (6.93) discretized using the Newmark-beta method is available [32].

An example of a rectangular resonator with a thin wire and a corner-mounted thin slot is illustrated in Figure 6.15. Similar to the thin-wire example shown in Figure 6.12, this test object was originally developed to validate thin-wire and thin-slot algorithms for the Cartesian grid FDTD method [29]. Application of the FETD method using unstructured tetrahedral elements is alternatively applied here. The slot length was 12 cm, while the slot width was 1.58 mm and the depth was

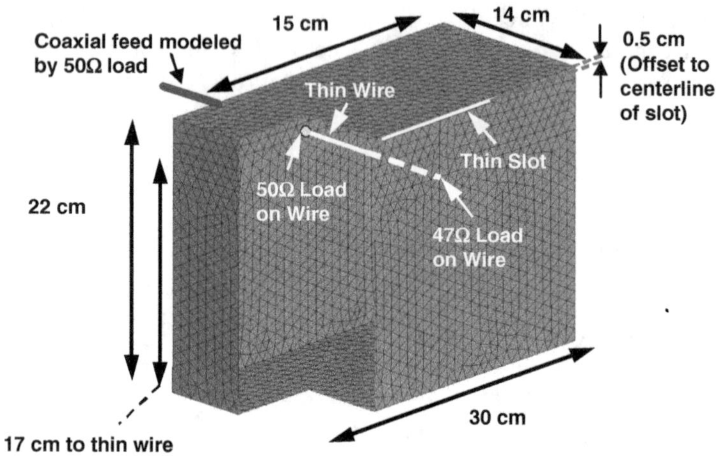

Figure 6.15 Rectangular cavity with a thin wire, thin slot, and coaxial feed. The FETD solution method is based on unstructured tetrahedral elements. The geometry is based on the original FDTD test object by Li et al. [29].

0.5 mm The radius of the thin wire was 0.8 mm and was terminated with 50- and 47-Ω resistors. A voltage source with a 50-Ω impedance was used similarly to represent excitation from the coaxial waveguide. A comparison of the FETD prediction and the data measured for the power delivered for the geometry in Figure 6.15 is shown in Figure 6.16.

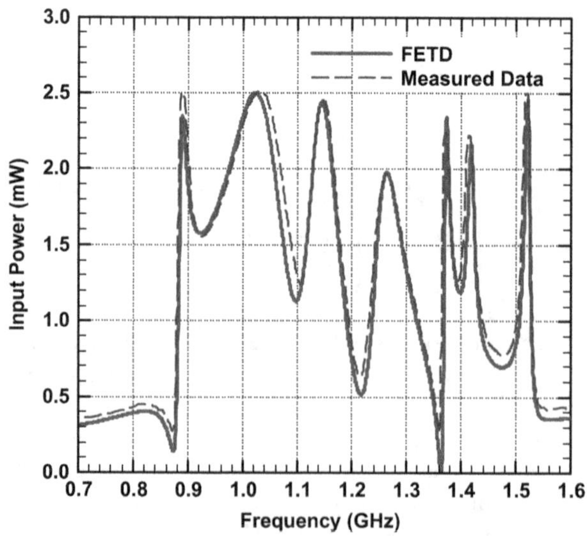

Figure 6.16 Input power for the geometry in Figure 6.15. Measured data are extracted from Li et al. [29].

Figures 6.13 and 6.16 both show remarkable correlations between measured and predicted data for these simple internal applications. Unfortunately, most practical interior applications have far greater complexity and often require accurate domain knowledge, such as geometrical, electrical, and material parameters, that may be difficult to obtain precisely.

6.3 LUMPED-CIRCUIT ELEMENTS

Within the FDTD method, techniques to include isolated lumped elements, as well as the integration of external circuit simulators, are well known [35]. In the finite element setting, it is straightforward to include isolated lumped-circuit elements such as resistors, capacitors, and inductors directly into the finite element system of equations [14]. More specifically, the contribution of a lumped resistor with static resistance R on the kth edge is obtained by making the substitution $R_s \rightarrow R/\delta(\mathbf{r} - \mathbf{r}_k)$ in (6.26), where $\delta(\mathbf{r})$ denotes the Dirac delta function, while a lumped capacitor C is included by the substitution $C_s \rightarrow C\delta(\mathbf{r} - \mathbf{r}_k)$ in (6.31), and a lumped inductor L is included by the substitution $L_s \rightarrow L/\delta(\mathbf{r} - \mathbf{r}_k)$ in (6.33). Note that these contributions will affect the diagonal entries of the primary finite element matrices. An example of an isolated resistor placed on a finite element edge is shown in Figure 6.17(a).

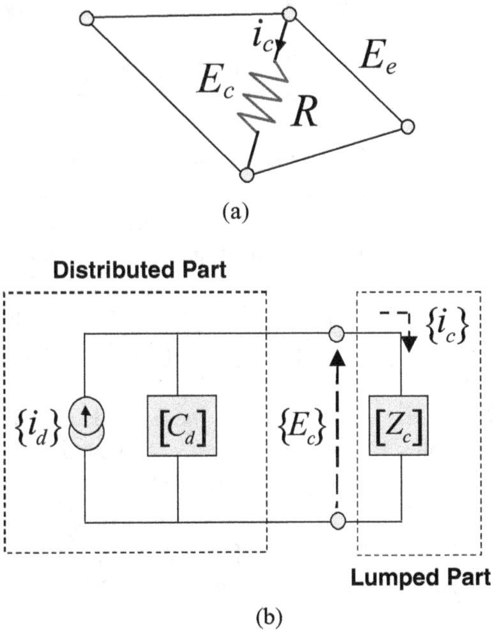

Figure 6.17 Lumped-circuit modeling in the finite element method. (a) Isolated resistor on a finite element edge. (b) Circuit diagram to accommodate complex circuit networks in the FETD method. Field data are accommodated within the distributed region, and the lumped-circuit network is accommodated within the lumped region.

To include more extensive circuit networks into the finite element method, or to include the more complex behavior of isolated lumped elements, such as nonlinearities, it is convenient to isolate the edges of the finite element domain that are associated with the circuitry. In this way, the circuit aspects can be treated as a separate application, yet remain fully coupled to the surrounding noncircuit edges within the finite element domain. The basic strategy is to create a separate schematic representation for the circuit edges that can be solved by an external circuit simulator. In this section we discuss formulations for both the conditionally stable FETD representation based on the first-order Maxwell's equations described by (4.104) and (4.105), as well as the unconditionally stable FETD representation based on the wave equation (2.27). The development based on the first-order Maxwell's equations follows Refs. 36 and 37, and the development based on the wave equation is patterned after Ref. 38.

6.3.1 Coupled First-Order Equations

A derivation of an FETD algorithm that is based on the first-order Maxwell's equations was provided in Section 4.4. Recall that this formulation is conditionally stable, although it is implicit with regard to the electric field advancement. The principal semidiscrete equations are (4.95) and (4.96), which are restated here for convenience:

$$[T^f]\frac{d\{B\}}{dt} + [A]\{E\} = -\{g\} \tag{6.94}$$

$$[T]\frac{d\{E\}}{dt} - [A]^{\mathrm{T}}\{B\} = \{f\} \tag{6.95}$$

where $\{H\}$ and $\{B\}$ are related through $\{H\}$=$[T^f]\{B\}$ and the matrices $[T]$, $[T^f]$, $[A]$, and the incidence matrix $[C]$= $[T^f]^{-1}[A]$ are as defined in Section 4.4.

It is convenient to separate the finite element edges that are associated with the circuitry from the noncircuit finite element edges and write (6.95) compactly as

$$\begin{Bmatrix} \{b_e\} \\ \{b_c\} \end{Bmatrix} = \begin{bmatrix} [T_{ee}] & [T_{ec}] \\ [T_{ce}] & [T_{cc}] \end{bmatrix} \frac{d}{dt} \begin{Bmatrix} \{E_e\} \\ \{E_c\} \end{Bmatrix} + \begin{Bmatrix} \{i_e\} \\ \{i_c\} \end{Bmatrix} \tag{6.96}$$

where the variables with the subscript e are those associated with the noncircuit edges, whereas the variables with the subscript c correspond to the edges connected to circuitry, and we have introduced the definition

$$\begin{Bmatrix} \{b_e\} \\ \{b_c\} \end{Bmatrix} \equiv \begin{bmatrix} [A_{ee}] & [A_{ec}] \\ [A_{ce}] & [A_{cc}] \end{bmatrix}^{\mathrm{T}} \begin{Bmatrix} \{B_e\} \\ \{B_c\} \end{Bmatrix}. \tag{6.97}$$

The two rows of (6.96) can be written as

$$\frac{d\{E_e\}}{dt} = [T_{ee}]^{-1}\left(-[T_{ec}]\frac{d\{E_c\}}{dt} - \{i_e\} + \{b_e\}\right) \tag{6.98}$$

and

$$\{i_d\} = [C_d]\frac{d\{E_c\}}{dt} + \{i_c\} \tag{6.99}$$

where

$$\{i_d\} \equiv \{b_c\} - [T_{ce}][T_{ee}]^{-1}(\{b_e\} - \{i_e\}) \tag{6.100}$$

$$[C_d] \equiv -[T_{ce}][T_{ee}]^{-1}[T_{ec}] + [T_{cc}]. \tag{6.101}$$

In (6.99), the current $\{i_c\}$ can often be written conveniently as $\{i_c\} = [Y_c]\{E_c\}$, where $[Y_c]$ denotes the admittance matrix for the lumped-circuit elements within the finite element mesh. However, to increase generality, (6.99) can be combined with the state equations for complex networks that may be connected to these finite element edges. Note that the electric field coefficients will be scaled by their corresponding edge lengths to obtain voltage coefficients.

The solution procedure is to obtain $\{E_c\}$ from (6.99) and then construct $\{E_e\}$ from (6.98). A circuit representation for (6.98) and (6.99) is shown in Figure 6.17(b). When the lumped-circuit network contains nonlinear elements, general nonlinear equation solvers such as Newton–Raphson methods [40] can be used to obtain $\{E_c\}$. Alternatively, an external network simulator such as SPICE [39] can be used for this purpose. However, as with all hybridizations of dissimilar solution methods, it is not generally possible to make definitive statements with regard to the stability of the resulting algorithm unless the interface region is defined precisely.

6.3.2 Wave Equation

A decomposition that isolates the finite element edges that are connected to circuitry can similarly be developed for the second-order wave equation. The semidiscrete wave equation (2.27) is restated as

$$[T]\frac{d^2\{E\}}{dt^2} + [S]\{E\} = \{f\} \tag{6.102}$$

where for simplicity we have set $[R] = 0$. We now separate the finite element edges that are associated with the circuitry from the noncircuit finite element edges; consequently, (6.102) can be written equivalently as

$$\begin{bmatrix} [T_{ee}] & [T_{ec}] \\ [T_{ce}] & [T_{cc}] \end{bmatrix} \frac{d^2}{dt^2} \begin{Bmatrix} \{E_e\} \\ \{E_c\} \end{Bmatrix} + \begin{bmatrix} [S_{ee}] & [S_{ec}] \\ [S_{ce}] & [S_{cc}] \end{bmatrix} \begin{Bmatrix} \{E_e\} \\ \{E_c\} \end{Bmatrix} = \begin{Bmatrix} \{f_e\} \\ \{f_c\} \end{Bmatrix} \tag{6.103}$$

where $\{E_e\}$ denotes the scaled electric-field expansion coefficients associated with noncircuit edges, while $\{E_c\}$ denotes the coefficients associated with the circuit edges. The excitation vector $\{f\}$ is decomposed similarly and the interpretation of the matrix decompositions is apparent.

By applying the Newmark-beta scheme with $\beta = 1/4$ to the $\{E_e\}$ equation, we can rewrite (6.103) as the following two equations:

$$[M_{ee}]\{E_e\}^{n+1} = 2[M_{ee}]\{E_e\}^n - [M_{ee}]\{E_e\}^{n-1} - [S_{ee}]\{E_e\}^n - \left([T_{ec}]\frac{d^2\{E_c\}}{dt^2} + [S_{ec}]\{E_c\}\right)\Bigg|_{t=n\Delta t} + \{f_e\}|_{t=n\Delta t} \tag{6.104}$$

$$\left([T_{cc}]\frac{d^2\{E_c\}}{dt^2} + [S_{cc}]\{E_c\}\right)\Bigg|_{t=n\Delta t} = -[M_{ce}]\{E_e\}^{n+1} + 2[M_{ce}]\{E_e\}^n - [S_{ce}]\{E_e\}^n - [M_{ce}]\{E_e\}^{n-1} + \{f_c\}|_{t=n\Delta t} \tag{6.105}$$

where

$$[M_{ee}] \equiv \frac{1}{\Delta t^2}[T_{ee}] + \frac{1}{4}[S_{ee}]; \qquad [M_{ce}] \equiv \frac{1}{\Delta t^2}[T_{ce}] + \frac{1}{4}[S_{ce}]. \tag{6.106}$$

We now substitute (6.104) into (6.105) to obtain the following coupled system of equations:

$$\left([C_d]\frac{d^2\{E_c\}}{dt^2} + [L_d]^{-1}\{E_c\} - \{f_c\}\right)\Bigg|_{t=n\Delta t} = -[M_{ce}]\{\tilde{E}_e\}^{n+1} - [M_{ce}]\{E_e\}^{n-1} + (2[M_{ce}] - [S_{ce}])\{E_e\}^n \tag{6.107}$$

$$\{E_e\}^{n+1} = \{\tilde{E}_e\}^{n+1} - [M_{ee}]^{-1}\left([T_{ec}]\frac{d^2\{E_c\}}{dt^2} + [S_{ec}]\{E_c\}\right)\Bigg|_{t=n\Delta t} \tag{6.108}$$

where $\{\tilde{E}_e\}^{n+1}$ is obtained from the traditional (noncircuit) FETD formulation

$$\{\tilde{E}_e\}^{n+1} = 2\{E_e\}^n - \{E_e\}^{n-1} - [M_{ee}]^{-1}([S_{ee}]\{E_e\}^n - \{f_e\}|_{t=n\Delta t}) \tag{6.109}$$

and

$$[C_d] \equiv [T_{cc}] - [M_{ce}]([M_{ee}]^{-1}[T_{ec}]) \tag{6.110}$$

$$[L_d]^{-1} \equiv [S_{cc}] - [M_{ce}]([M_{ee}]^{-1}[S_{ec}]). \tag{6.111}$$

By defining $\{f_c\} = -d\{i_c\}/dt$, (6.107) can be rewritten equivalently as

$$\{i_d\} = [C_d]\frac{d\{E_c\}}{dt} + \{i_c\} \tag{6.112}$$

$$\frac{d\{i_d\}}{dt} = -[L_d]^{-1}\{E_c\} - [M_{ce}](\{\tilde{E}_e\}^{n+1} + \{E_e\}^{n-1}) + (2[M_{ce}] - [S_{ce}])\{E_e\}^n. \tag{6.113}$$

Note that in a fully discrete form, (6.112) and (6.113) become coupled when a traditional Newmark-beta average for $\{E_c\}$ is adopted. Equation (6.112) can be combined with the state equations for complex connecting networks, if necessary.

If we assume that $\{i_c\} = [Y_c]\{E_c\}$, where $[Y_c]$ denotes a (possibly time-dependent) admittance matrix for the lumped elements, an unconditionally stable time-stepping algorithm for (6.112) and (6.113) is given by

$$\left[\frac{2}{\Delta t}[C_d] + [Y_c] + \frac{\Delta t}{2}[L_d]^{-1}\right]\{E_c\}^{n+1} = \left[\frac{2}{\Delta t}[C_d] - [Y_c] - \frac{\Delta t}{2}[L_d]^{-1}\right]\{E_c\}^{n-1} - (2[Y_c] + \Delta t[L_d]^{-1})\{E_c\}^n + \{\tilde{i}_d\}^n \tag{6.114}$$

$$\{i_d\}^{n+1} = \{i_d\}^{n-1} - \frac{\Delta t}{2}[L_d]^{-1}(\{E_c\}^{n+1} + \{E_c\}^{n-1} + 2\{E_c\}^n) + 2\Delta t\left[(2[M_{ce}] - [S_{ce}])\{E_e\}^n - [M_{ce}](\{\tilde{E}_e\}^{n+1} + \{E_e\}^{n-1})\right] \tag{6.115}$$

where

$$\{\tilde{i}_d\}^n = 2\{i_d\}^n + 2\{i_d\}^{n-1} + 2\Delta t[(2[M_{ce}] - [S_{ce}])\{E_e\}^n - [M_{ce}](\{\tilde{E}_e\}^{n+1} + \{E_e\}^{n-1})] \tag{6.116}$$

and $\{\tilde{E}_e\}^{n+1}$ is obtained from (6.109). The solution procedure is to solve (6.109) initially to obtain $\{\tilde{E}_e\}^{n+1}$, which is then used to construct $\{\tilde{i}_d\}^n$. Equation (6.114) can then be solved for $\{E_c\}^{n+1}$, and finally, $\{i_d\}^{n+1}$ is advanced from (6.115). The circuit diagram is similar to Figure 6.17(b).

For simple lumped-circuit components, the solution of (6.114)–(6.116) will yield results that are numerically identical to a direct solution of the finite element formulation (2.27). The advantage of separating the circuit edges from the noncircuit edges is that the circuit problem becomes a subdomain application with the global finite element solution. For example, a simple time-dependent resistor will affect only the small matrices in (6.114) associated with the local circuit edges instead of introducing a time dependency into the large global finite element matrices. The more complicated situation of a nonlinear Gunn diode is considered in the next section.

6.3.3 Example

An active patch antenna that contains two Gunn diodes is shown in Figure 6.18. This example was originally used for the validation of nonlinear lumped-circuit models for the FDTD method [41], and was subsequently used for validation of the FETD lumped-circuit modeling technique described in Section 6.3.1 [36]. A discussion of the oscillation modes associated with this geometry can be found in Ref. 42. When the two antenna structures are connected by a metal strip, a difference mode occurs at 11.84 GHz, whereas when the two structures are connected by a chip resistor, a sum mode occurs at 11.04 GHz. The difference mode is considered in this example.

The circuit model for the Gunn diodes shown in Figure 6.18(b) leads to the following nonlinear differential equation:

$$C_1 \frac{dV_d}{dt} + \frac{V_d}{R_1} + F(V_d) = -\frac{V_c}{R_1} \tag{6.117}$$

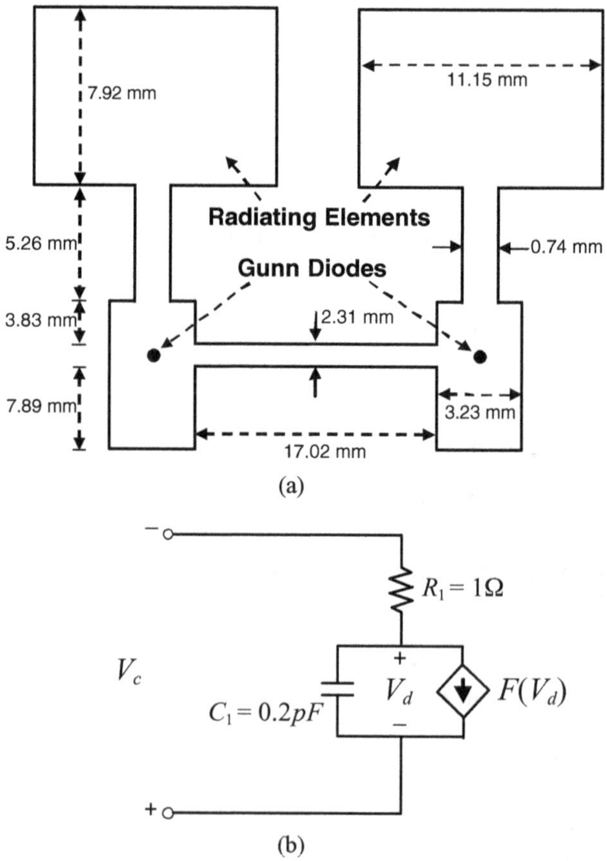

Figure 6.18 Active patch antenna on a dielectric substrate with a thickness of 0.789 mm and a relative permittivity of 2.33. (a) Geometry. (b) Circuit diagram for Gunn diodes.

where the active current source is given by [41]

$$F(V_d) = -0.0252V_d + 0.0265(V_d)^3. \tag{6.118}$$

When each of the two Gunn diodes is connected to a single finite element edge of length Δ, the electric field along this edge is given by $E_c = -V_c/\Delta$, where E_c is as shown in Figure 6.17(b). By observing in Figure 6.18(b) that $V_d = -i_c R_1 - V_c$, (6.117) can be written equivalently in terms of the current i_c as

$$C_1 \frac{dV_c}{dt} + C_1 R_1 \frac{di_c}{dt} + i_c = F(V_d). \tag{6.119}$$

When combining (6.119) with either (6.99) or (6.112), it is apparent that a time-marching algorithm will require the solution of a nonlinear system of equations to obtain the port parameter $\{E_c\}^{n+1}$. As noted previously, this can be accomplished by using either the Newton–Raphson method or a network simulation tool such as SPICE. By adopting an FETD–SPICE solution approach [36], the steady-state voltage appearing across each diode of the active antenna is shown in Figure 6.19, and the H-plane pattern at 11.84 GHz is shown in Figure 6.20. Although good results have been obtained with an FETD–SPICE interface, no known proof of the stability of this interface has been established.

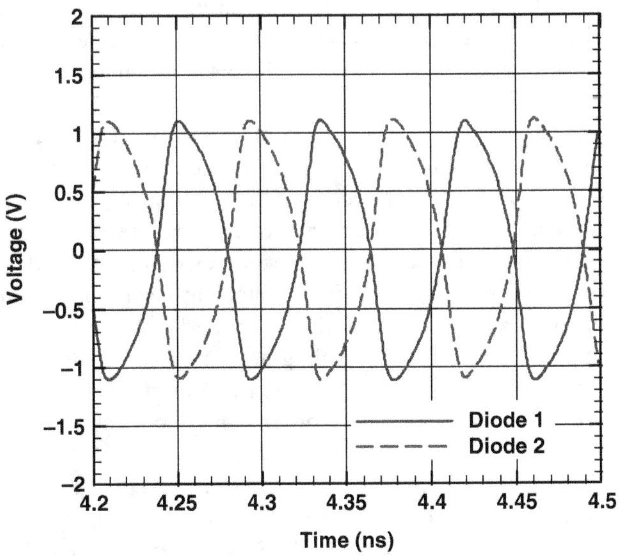

Figure 6.19 Steady-state voltage across each diode based on the FETD-SPICE method. (After Guillouard et al. [36], Copyright © IEEE 1999.)

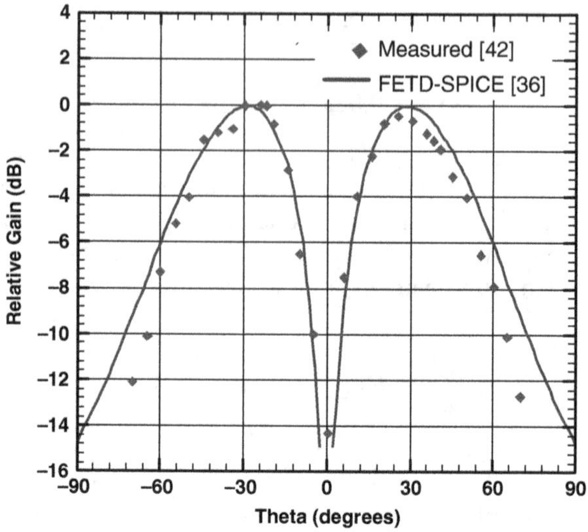

Figure 6.20 *H*-plane difference-mode pattern at 11.84 GHz. (After Guillouard et al. [36], Copyright © IEEE 1999.)

6.4 DISTRIBUTED FEED NETWORK

In the preceding section we described techniques to decompose a finite element application into a field portion and a circuit portion that are solved concurrently in time. The approach was based fundamentally on lumped-circuit concepts, where complex networks are connected to individual edges of the finite element mesh and solved concurrently with the fields by using linear or nonlinear equation solution techniques.

In this section we consider a related decomposition for the case of distributed ports [43]. Because phased-array antennas are often fed by waveguides, such as coaxial transmission lines, a port-based decomposition can provide increased accuracy and generality compared to feed models that are based on lumped-circuit techniques. The basic strategy is based on a bidirectional decomposition of the fields as well as the port voltages and port currents, and this formulation has been found to provide an accurate and numerically stable hybrid algorithm for interfacing the FETD method with a secondary simulation solver for the feeding network. We note that a bidirectional decomposition approach has also been reported for interfacing the FDTD method with embedded circuit networks based on the multiconductor transmission-line theory [44].

Figure 6.21 illustrates the relationship between the antennas, the feeding network, and the port interfaces. The port surfaces between the antennas and the feeding network are represented by S_p, where $p = 1, 2 \ldots, N_p$, and we assume that the feeding network is connected to a source through an additional port surface denoted by S_0. The transverse electric fields on the port surfaces are denoted by $\mathbf{E}_0, \mathbf{E}_1, \ldots, \mathbf{E}_{N_p}$,

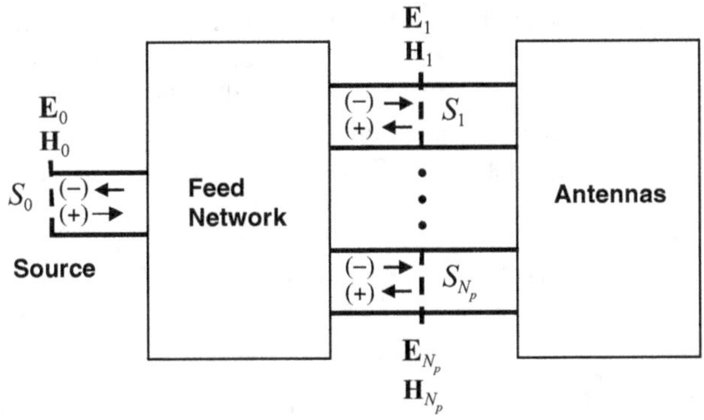

Figure 6.21 Port interfaces between source, feeding network, and antennas with bi-directional propagation.

and the magnetic fields are defined similarly. The $(-)$ convention is used to represent fields propagating from the feeding network toward the antennas, whereas the $(+)$ convention is used to represent fields propagating from the antennas toward the feeding network. Similarly, across the source port S_0, the $(-)$ convention is used to represent fields propagating from the feeding network toward the source, whereas the $(+)$ convention corresponds to propagation from the source toward the feeding network. In terms of these bidirectional decompositions, the total transverse electric and magnetic fields on the pth port are represented in time-harmonic form by

$$\mathbf{E}_p(\omega) = \mathbf{E}_p^{(+)}(\omega) + \mathbf{E}_p^{(-)}(\omega) \tag{6.120}$$

$$\mathbf{H}_p(\omega) = \mathbf{H}_p^{(+)}(\omega) + \mathbf{H}_p^{(-)}(\omega). \tag{6.121}$$

The port modal voltages and currents can be decomposed similarly, and in the case of an assumed TEM-dominant mode, we can represent the TEM modal coefficients for the voltage and current at the pth port by

$$V_p(\omega) = V_p^{(+)}(\omega) + V_p^{(-)}(\omega) \tag{6.122}$$

$$I_p(\omega) = I_p^{(+)}(\omega) - I_p^{(-)}(\omega). \tag{6.123}$$

The corresponding time-domain representations of (6.122) and (6.123) are

$$V_p(t) = V_p^{(+)}(t) + V_p^{(-)}(t) \tag{6.124}$$

$$I_p(t) = I_p^{(+)}(t) - I_p^{(-)}(t). \tag{6.125}$$

To construct the field-network hybrid algorithm, the implementation strategy will be to rigorously impress the $V_p^{(-)}(t)$ port voltages associated with the feeding network into the FETD method that is used to describe the antenna domain, and impress the

$V_p^{(+)}(t)$ port voltages associated with the antenna domain into the network circuit solver associated with the feeding network. To realize this strategy, we focus initially on the relationship between $V_p^{(-)}(\omega)$ and $V_p^{(+)}(\omega)$ within the feeding network, which is easily established through the S-parameter matrix.

The bidirectional modal voltage coefficients $V_p^{(\pm)}(\omega)$, $p = 0,1,\ldots, N_p$, are described by [43]

$$\{\tilde{V}^{(-)}(\omega)\} = [S(\omega)]\{\tilde{V}^{(+)}(\omega)\} \tag{6.126}$$

where $\{\tilde{V}^{(\pm)}(\omega)\} = [V_0^{(\pm)}(\omega)/\sqrt{Z_0}, V_1^{(\pm)}(\omega)/\sqrt{Z_1}, \ldots, V_{Np}^{(\pm)}(\omega)/\sqrt{Z_{Np}}]^{\mathrm{T}}$, with Z_p denoting the TEM port impedance of the pth port, and $[S(\omega)]$ denotes the traditional S-matrix description of the feeding network.

The time-domain representation of (6.126) is

$$\{\tilde{V}^{(-)}(t)\} = [S(t)] * \{\tilde{V}^{(+)}(t)\} \tag{6.127}$$

where $*$ denotes the time convolution. An efficient recursive evaluation for the convolution operator can be implemented by adopting a rational function representation for the elements of the S-matrix $[S(\omega)]$; specifically,

$$S_{ij}(\omega) \approx \sum_{k=1}^{N_k} \frac{c_{ij,k}}{j\omega - a_k} + d_{ij} \qquad i, j = 0, 1, 2, \ldots, N_p \tag{6.128}$$

where a_k and $c_{ij,k}$ denote the poles and residues, respectively, and d_{ij} denotes an optional real function. To construct (6.128), the VECTFIT algorithm [45] can be used. The time-domain representation for (6.128) is obtained by inverse Fourier transformation,

$$S_{ij}(t) \approx \sum_{k=1}^{N_k} \mathrm{Re}(c_{ij,k}\, e^{a_k t})u(t) + d_{ij}\delta(t) \tag{6.129}$$

where $u(t) \equiv 1$ for $t \geq 0$ and 0 otherwise, and $\delta(t)$ denotes the Dirac delta function. With (6.129) the convolution operation in (6.127) can be evaluated efficiently by a recursive convolution. It is noted that for the case that the feeding network can be described by the multiconductor transmission-line theory, modal analysis techniques can alternatively be used to construct $V_p^{(\pm)}$ directly in the time domain [46].

We now focus our attention on adapting the modal port voltages $V_p^{(-)}(t)$ to the FETD algorithm, which is facilitated by using the waveguide port boundary condition (WPBC) described by (5.30)–(5.34) in Section 5.1.3. By considering the case that $\mathbf{E}_p^{(-)}(t)$ is incident on the pth port, we can rewrite (5.32)–(5.34) such that

$$\hat{n}_p \times (\nabla \times \mathbf{E}) + P_p(\mathbf{E}) = \mathbf{U}_p^{\mathrm{inc}}(\mathbf{E}^{(-)}) \tag{6.130}$$

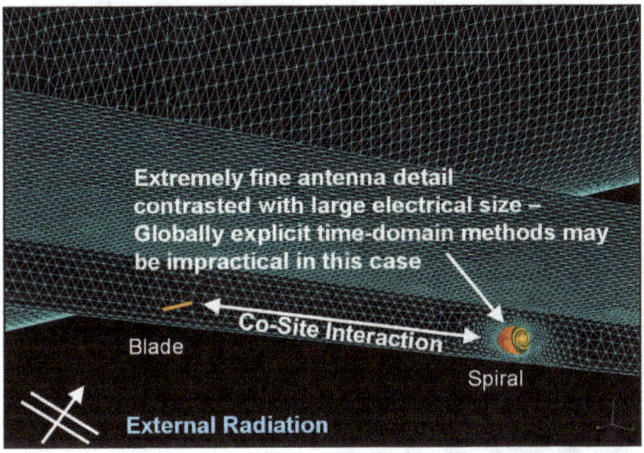

Figure 1.1 Example showing very small finite elements to model fine structures on a large object. Such a problem is challenging for explicit, conditionally stable time-domain methods and can be better handled by either implicit, unconditionally stable time- or frequency-domain techniques.

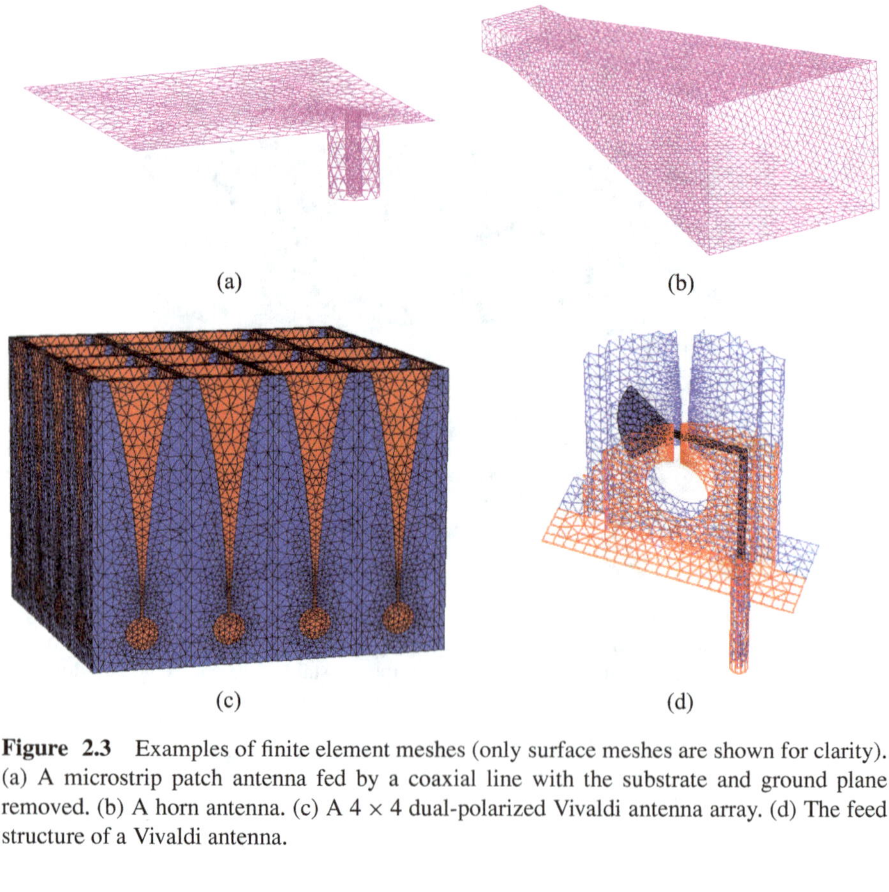

Figure 2.3 Examples of finite element meshes (only surface meshes are shown for clarity). (a) A microstrip patch antenna fed by a coaxial line with the substrate and ground plane removed. (b) A horn antenna. (c) A 4×4 dual-polarized Vivaldi antenna array. (d) The feed structure of a Vivaldi antenna.

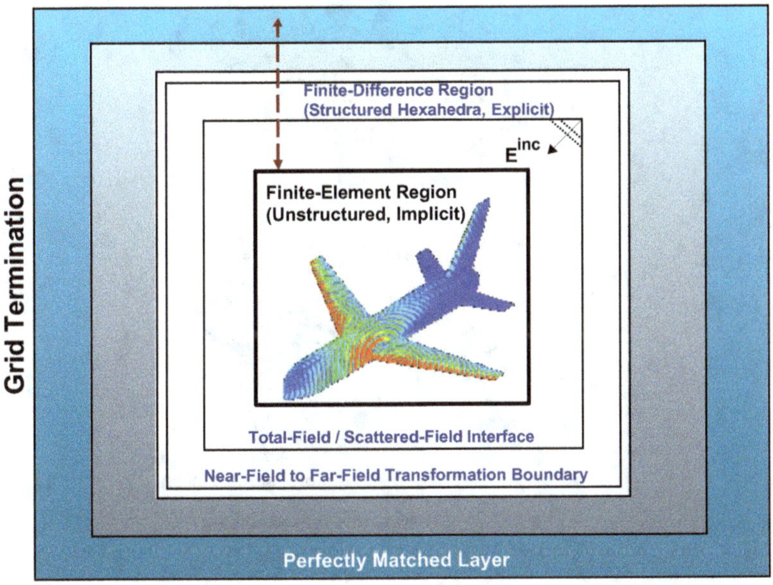

Figure 4.8 Locations of the FETD, FDTD, and PML regions, as well as the near-to-far-field transformation surface used for a complete FETD–FDTD implementation.

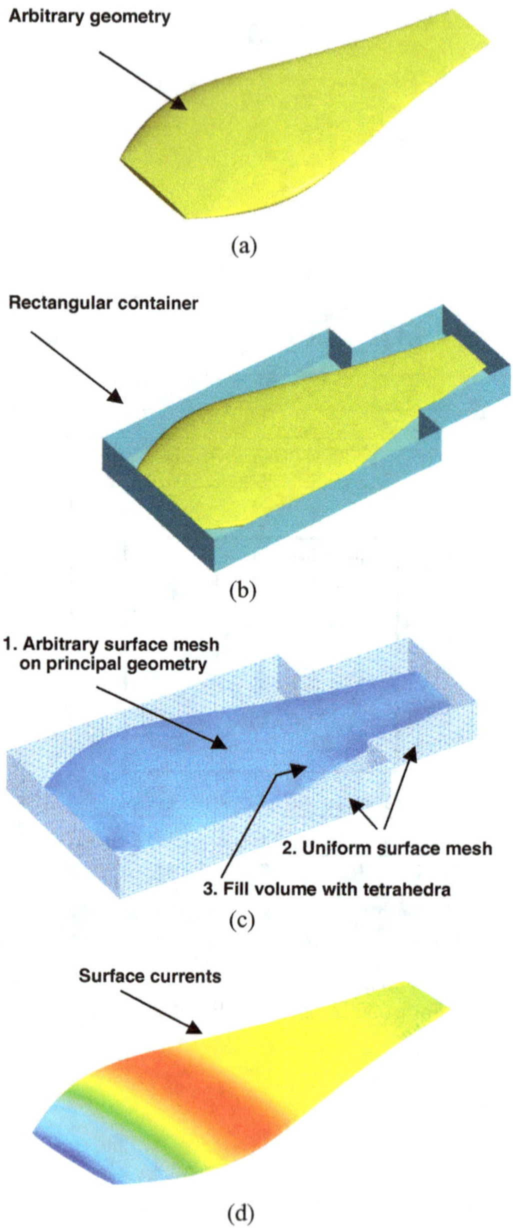

Figure 4.9 Procedure to construct a simple unstructured tetrahedral grid interfaced to a structured brick-element grid. (a) Solid geometry desired. (b) Suitably sized rectangular box to hold the geometry. (c) Uniform triangular surface mesh on the box and arbitrary triangular mesh on the geometry (filled with a tetrahedral mesh in between). (d) Unstructured mesh embedded in a background FDTD mesh (not shown) and the solution.

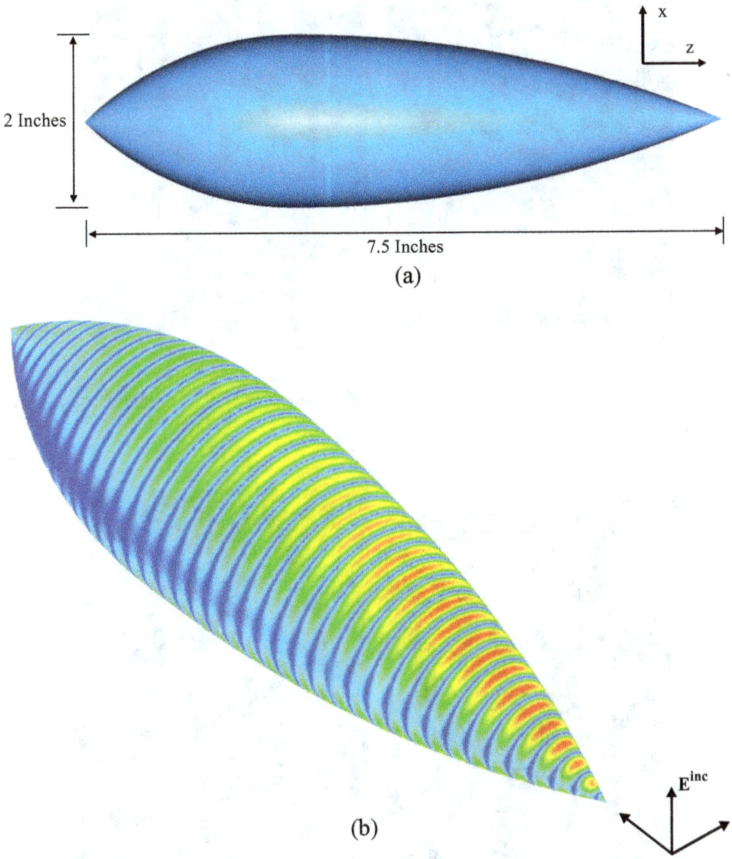

Figure 4.14 Metallic double ogive. (a) Geometry. (b) Surface current density for a sinusoidal plane-wave excitation at 30 GHz.

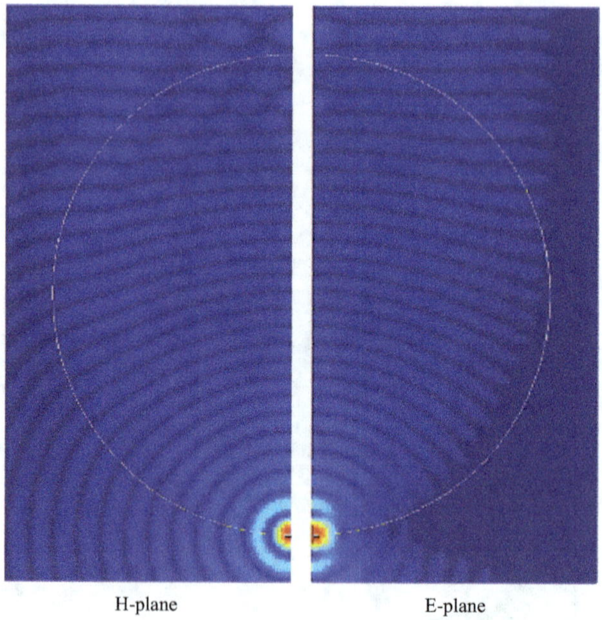

Figure 5.19 Snapshot of the electric field inside and nearby a Luneburg lens excited by a Hertzian dipole.

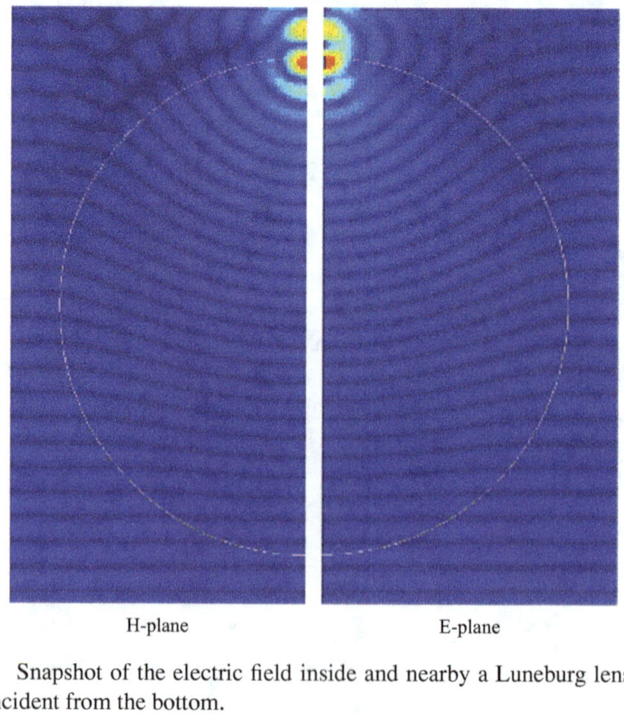

Figure 5.20 Snapshot of the electric field inside and nearby a Luneburg lens excited by a plane wave incident from the bottom.

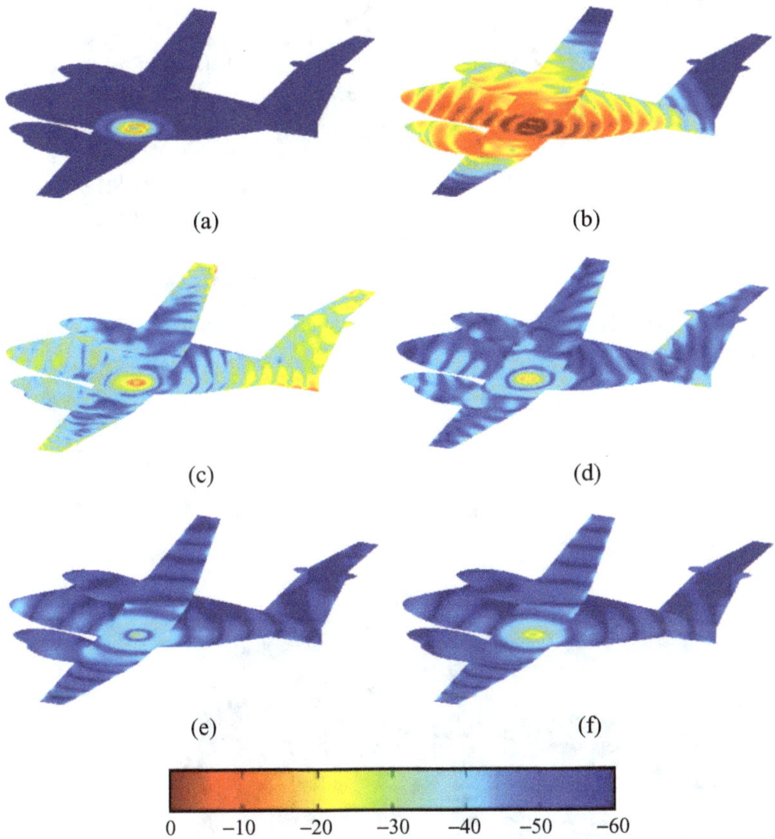

Figure 5.21 Snapshots of the surface electric current (on the decibel scale) on an aircraft at various time steps. (a) At the 120th time step. (b) At the 220th time step. (c) At the 320th time step. (d) At the 420th time step. (e) At the 520th time step. (f) At the 620th time step.

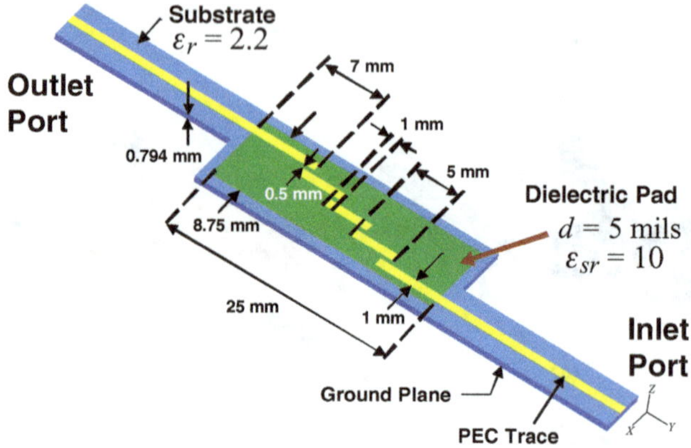

Figure 6.2(a) Three-dimensional geometry of a microstrip bandpass filter with a high-contrast thin dielectric layer.

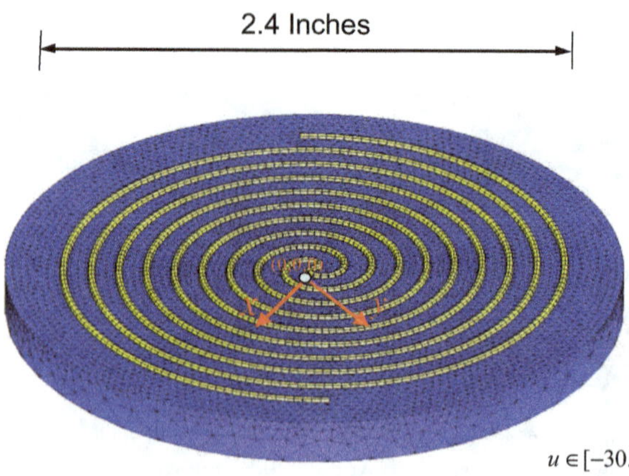

$u \in [-30.48, 30.48]$ mm

$x = u \cos|u/v|$

$x = u \sin|u/v|$

$v = 1$ mm

Figure 6.10 Thin-wire Archimedean spiral antenna. An ideal voltage source is placed at the origin.

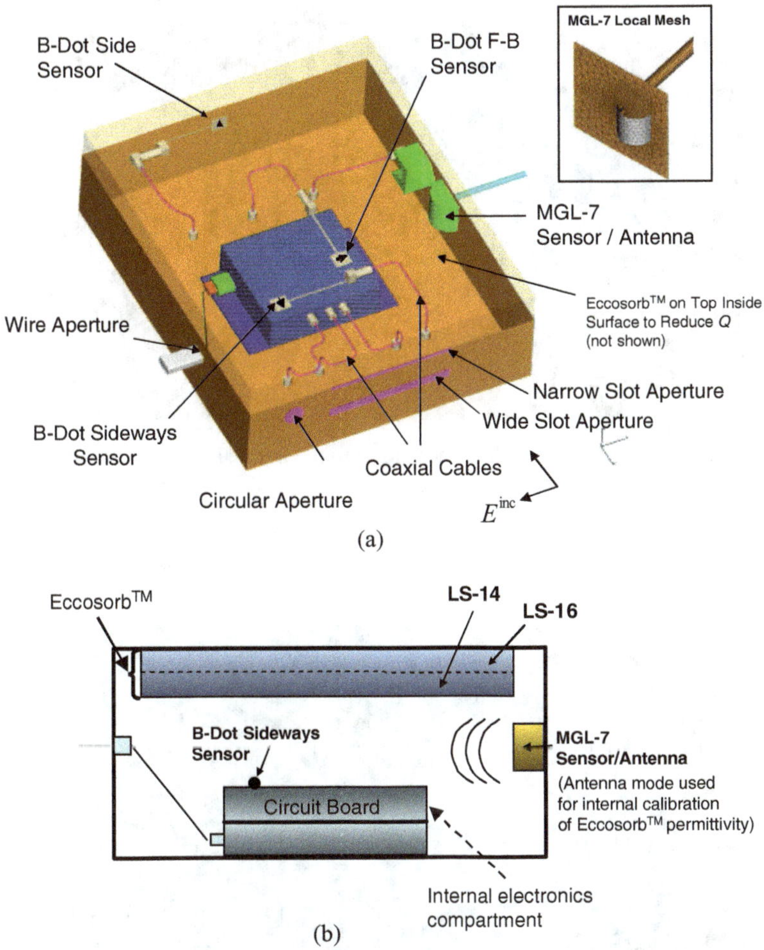

Figure 6.24 (a) Device-under-test (DUT) geometry showing various sensors and aperture configurations. (b) Cross section showing two Eccosorb absorptive layers (LS-14 and LS-16) mounted on the top surface of DUT.

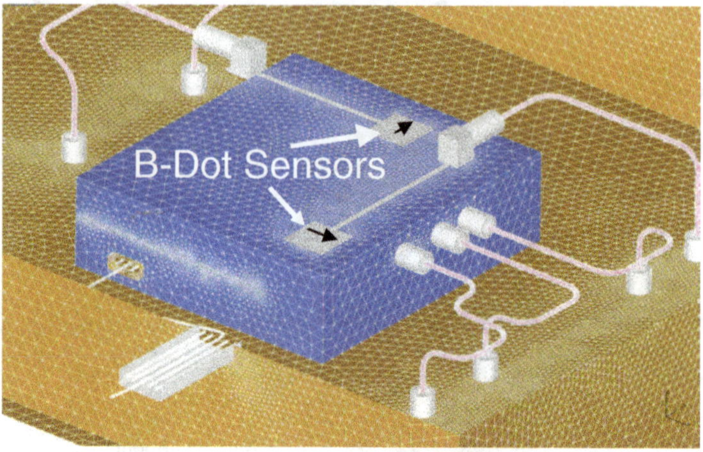

Figure 6.25 Finite element surface mesh local to the internal electronics compartment.

Figure 6.26 Finite element surface mesh on the internal circuit board.

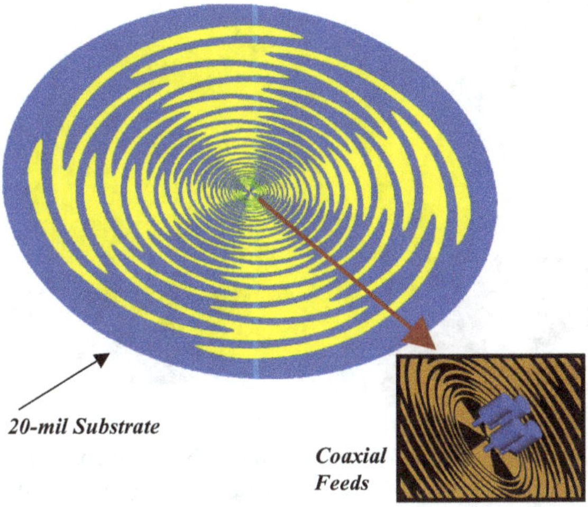

Figure 7.11(a) Geometry of a four-arm sinuous antenna etched on a 20-mil dielectric substrate.

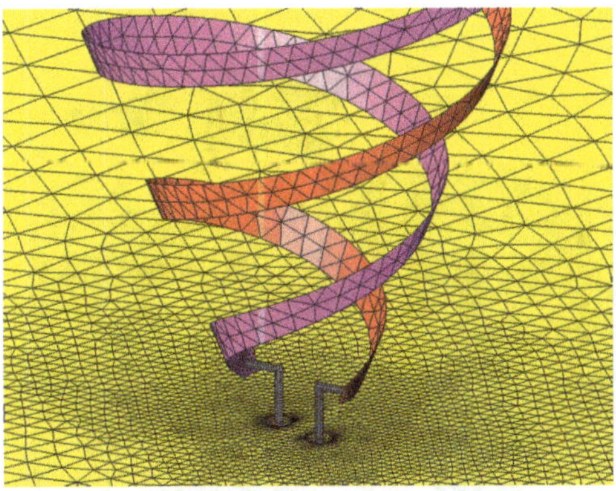

Figure 7.14(b) Finite element surface mesh local to the inverted conical spiral feed region for an inverted two-arm conical spiral over a ground plane.

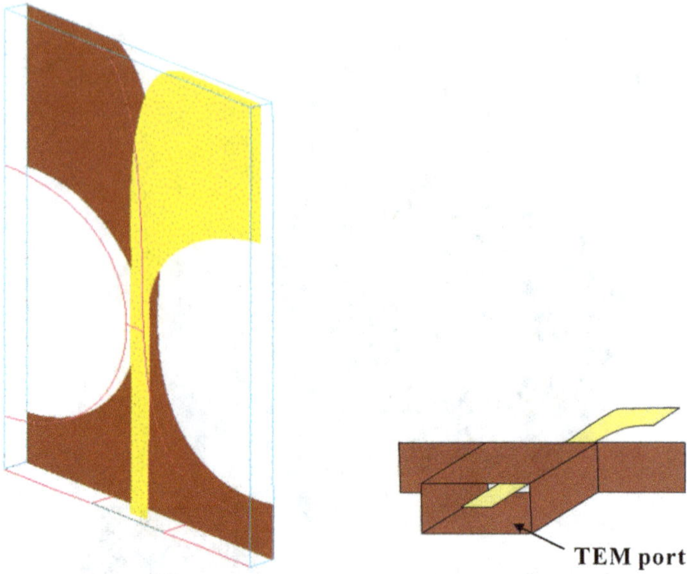

Figure 7.16(a) Three-dimensional representation of a balanced antipodal Vivaldi antenna. Yellow is stripline track/flare (sandwiched by two 1.575-mm-thick Duroid layers), brown is back ground plane transition/flare, and purple (unshaded) is front ground plane transition/flare, which is the same as the back ground plane. Cyan is the substrate outline. The figure on the right shows the input TEM port.

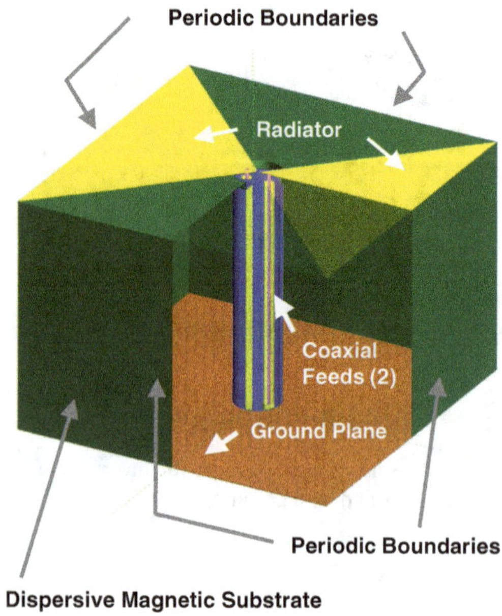

Figure 9.16(a) Unit cell geometry of an ultrawideband phased array. The surface of the right-front corner is made semitransparent to show the internal structure of the unit cell.

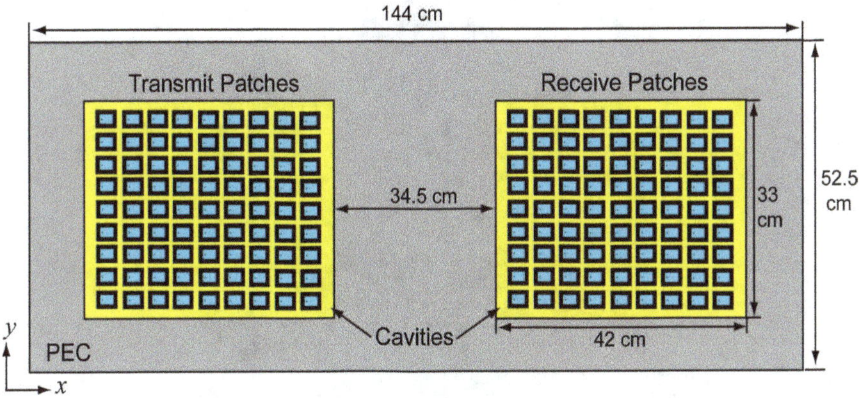

Figure 10.9 Geometry of two 9 × 9 microstrip patch antenna arrays placed on a ground plane. Each array is housed in a shallow cavity recessed in the ground plane so that it is flush-mounted.

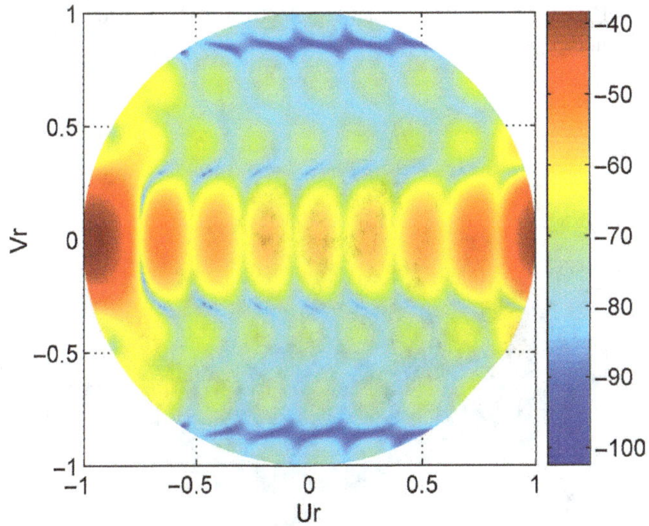

Figure 10.10 Normalized power (in decibels) coupled from the transmitting array with broadside scan to the receiving array as a function of the scan angle of the receiving array.

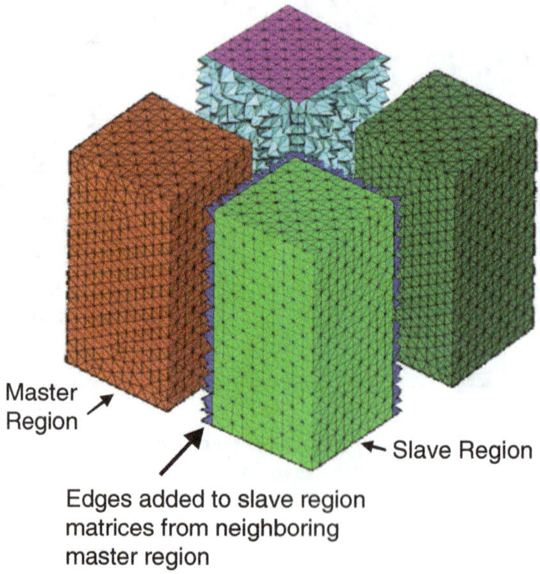

Figure 10.13 Two-dimensional decomposition of a three-dimensional tetrahedral mesh for application of the IDD technique. The master region edges that are used to augment the slave region are highlighted.

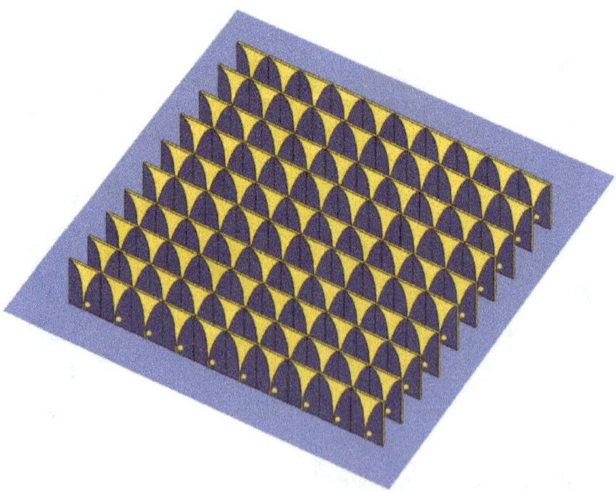

Figure 10.14(a) Geometry of a 10×10 Vivaldi array on a ground plane.

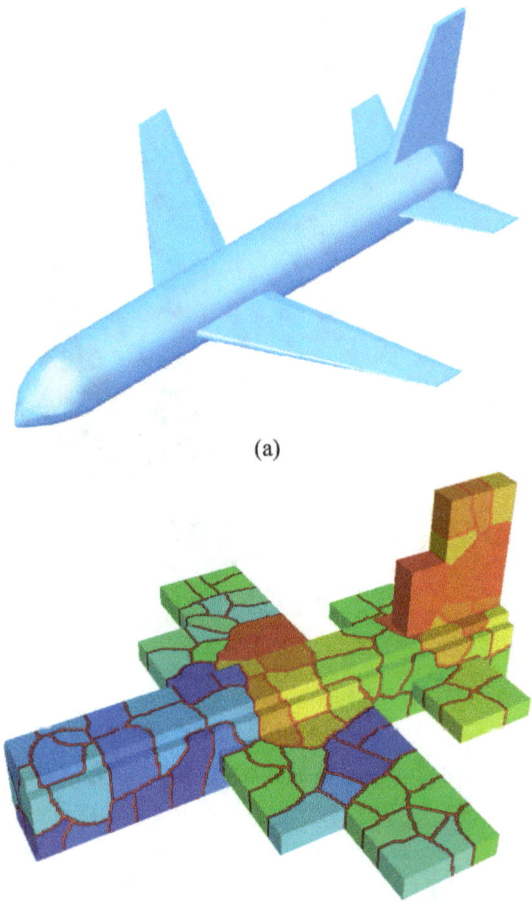

(a)

(b)

Figure 11.5 (a) Solid model of a 15-m-long airframe. (b) Enclosure of the finite element discretization region decomposed into subdomains.

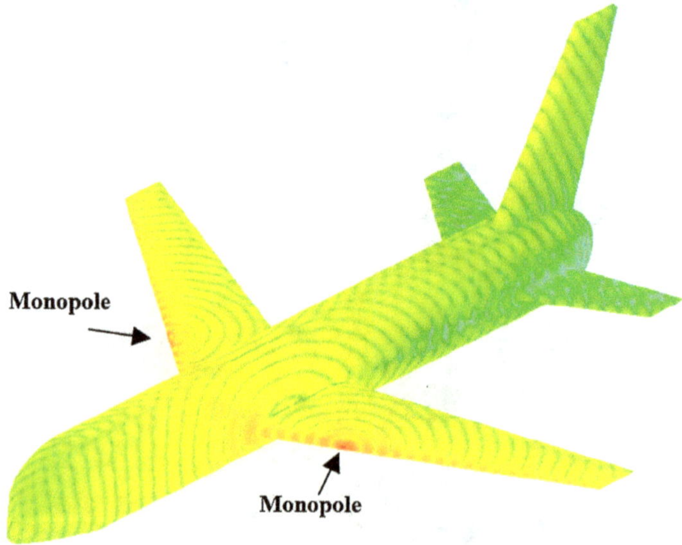

Figure 11.6 Surface electric current distribution due to radiation by two monopole antennas mounted on a simple airplane model.

where

$$P_p(\mathbf{E}) = -\mathbf{e}_{0,p}^{\text{TEM}} \iint_{S_p} \mathbf{e}_{0,p}^{\text{TEM}} \cdot \frac{1}{c}\frac{\partial}{\partial t}\mathbf{E}\, dS - \sum_{m=1}^{\infty} \mathbf{e}_{m,p}^{\text{TE}} \iint_{S_p} \mathbf{e}_{m,p}^{\text{TE}} \cdot \left(\frac{1}{c}\frac{\partial}{\partial t}\mathbf{E} + h_m * \mathbf{E}\right) dS$$

$$- \sum_{m=1}^{\infty} \mathbf{e}_{tm,p}^{\text{TM}} \iint_{S_p} \mathbf{e}_{tm,p}^{\text{TM}} \cdot \left(\frac{1}{c}\frac{\partial}{\partial t}\mathbf{E} + g_m * \mathbf{E}\right) dS \tag{6.131}$$

$$\mathbf{U}_p^{\text{inc}}(\mathbf{E}^{(-)}) = \hat{n}_p \times (\nabla \times \mathbf{E}^{(-)}) - \mathbf{e}_{0,p}^{\text{TEM}} \iint_{S_p} \mathbf{e}_{0,p}^{\text{TEM}} \cdot \frac{1}{c}\frac{\partial}{\partial t}\mathbf{E}^{(-)}\, dS$$

$$- \sum_{m=1}^{\infty} \mathbf{e}_{m,p}^{\text{TE}} \iint_{S_p} \mathbf{e}_{m,p}^{\text{TE}} \cdot \left(\frac{1}{c}\frac{\partial}{\partial t}\mathbf{E}^{(-)} + h_m * \mathbf{E}^{(-)}\right) dS$$

$$- \sum_{m=1}^{\infty} \mathbf{e}_{tm,p}^{\text{TM}} \iint_{S_p} \mathbf{e}_{tm,p}^{\text{TM}} \cdot \left(\frac{1}{c}\frac{\partial}{\partial t}\mathbf{E}^{(-)} + g_m * \mathbf{E}^{(-)}\right) dS. \tag{6.132}$$

Because (6.131) and (6.132) accommodate higher-order modes, the WPBC can be applied arbitrarily close to the antenna termination at each port. However, if we assume only TEM propagation and place the WPBC sufficiently far from the antenna aperture such that higher-order evanescent modes have decayed sufficiently, both (6.131) and (6.132) simplify significantly. With this assumption, $\mathbf{E}_p^{(-)}(t) = V_p^{(-)}(t)\,\mathbf{e}_{0,p}^{\text{TEM}}$, which results in

$$P_p(\mathbf{E}) \approx -\mathbf{e}_{0,p}^{\text{TEM}} \iint_{S_p} \mathbf{e}_{0,p}^{\text{TEM}} \cdot \frac{1}{c}\frac{\partial}{\partial t}\mathbf{E}\, dS \tag{6.133}$$

$$\mathbf{U}_p^{\text{inc}}(\mathbf{E}^{(-)}) = -2\,\mathbf{e}_{0,p}^{\text{TEM}} \frac{1}{c}\frac{dV_p^{(-)}(t)}{dt} \tag{6.134}$$

where $V_p^{(-)}$ is obtained from a discrete representation of (6.127) and (6.129). Equations (6.130), (6.131), and (6.132) can easily be incorporated into the time-domain weak-form wave equation (2.25), after which a time-marching system for the total electric field $\{E\}$ can be constructed [43,47]. From the electric field we can obtain the total TEM modal voltage V_p on the pth port by forming

$$V_p = \sum_i E_i \Phi_{i0,p}^{\text{TEM}} \tag{6.135}$$

where

$$\Phi_{i0,p}^{\text{TEM}} = \iint_{S_p} \mathbf{N}_i \cdot \mathbf{e}_{0,p}^{\text{TEM}}\, dS \tag{6.136}$$

and E_i are the usual expansion coefficients for the electric field. The incident TEM modal voltage on the feeding network, $V_p^{(+)}$, is subsequently obtained by using (6.124) such that $V_p^{(+)}(t) = V_p(t) - V_p^{(-)}(t)$. Consequently, a time-stepping algorithm that couples $V_p^{(+)}$ from the antenna region with $V_p^{(-)}$ from the network region can be developed in a straightforward manner [43].

An example that applies the technique described in this section to a single-polarization 8×8 printed Vivaldi antenna array with a multistage feeding network is shown in Figure 6.22. Feeding networks typically consist of power dividers and phase shifters. The power dividers split the source power, where port 0 is assumed to be the

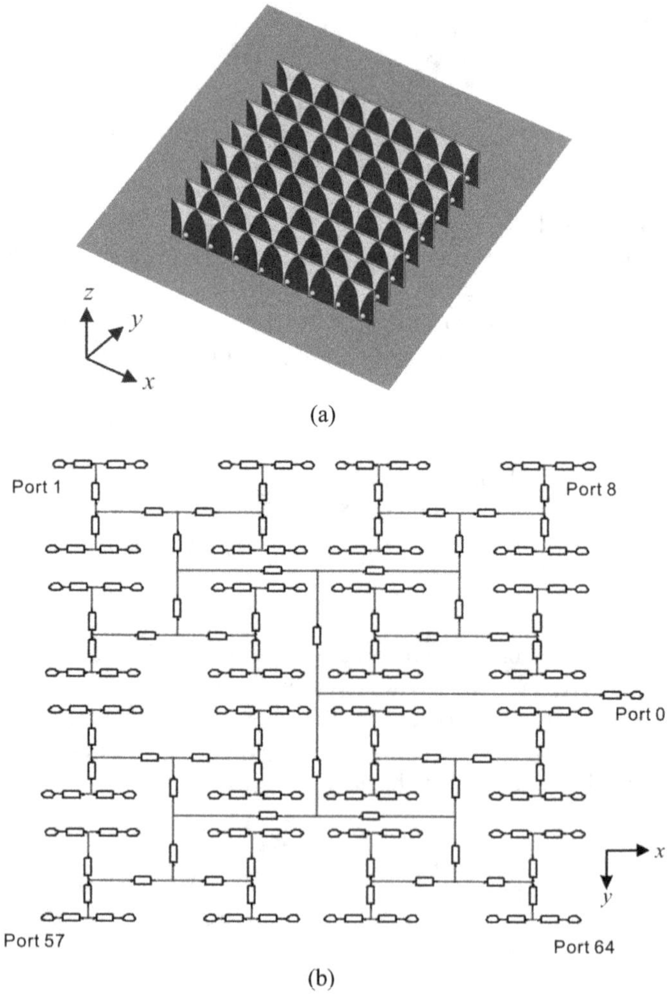

Figure 6.22 (a) Single-polarization 8 × 8 printed Vivaldi antenna array. (b) Multistage feeding network. Transmission lines are represented schematically by the rectangular blocks in the distribution diagram. (After Wang et al. [43], Copyright © IEEE 2008.)

source port, into the several paths required to feed the various array elements. Multistage dividers often consist of several lossless transmission lines and T-junctions that are arranged in multiple stages. The phase shifters control the relative elemental phase, which enables the array to scan to various directions. True time delay between the elements can be accomplished by using transmission lines with appropriate electrical length. In Figure 6.22(b), transmission lines are represented schematically by the rectangular blocks shown in the distribution diagram. Radiation patterns for the scan angle $\theta_s = 45°$ and $\phi_s = 0°$ are shown in Figure 6.23 for both the multistage

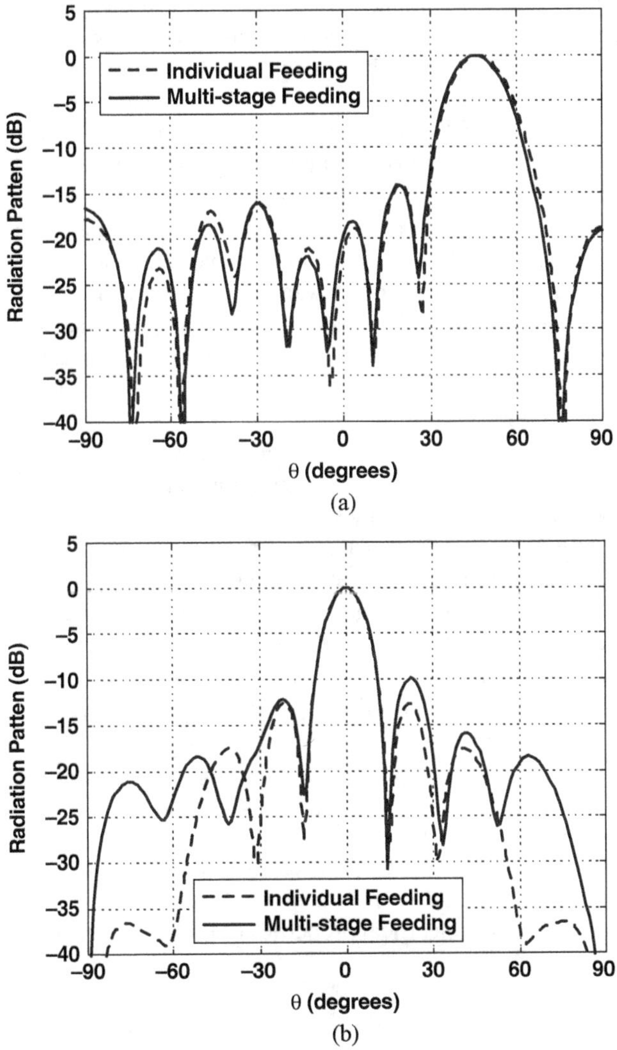

Figure 6.23 Scanning radiated patterns for the 8×8 Vivaldi array with individual and multistage feeding for the case $\theta_s = 45°$ and $\phi_s = 0°$. (a) E-plane. (b) H-plane. (After Wang et al. [43], Copyright © IEEE 2008.)

feeding network described above as well as feeding each array element individually. The multistage feeding approach is seen to affect primarily the amplitude of the sidelobes. More examples involving different feeding networks can be found in Ref. 43.

6.5 SYSTEM-LEVEL COUPLING EXAMPLE

Practical antenna systems require electronic components behind the antenna aperture structure. Depending on the shielding and circuit isolation design, the performance of these components can be affected through inadvertent coupling mechanisms. When very high power is transmitted by the antenna apertures, the coupling into nearby electronics can be of interest. For example, modern systems often place numerous antennas in close proximity to one another, and therefore mutual coupling may adversely affect the performance of the underlying circuit components, due to RF penetration through apertures and/or coupling onto wires and cables.

In this section we examine the use of finite element techniques to predict L-band (1–2 GHz) and S-band (2–4 GHz) coupling into complex electronics enclosures where the coupling mechanisms can be through various types of apertures. A solid model of a chassis with an internally shielded electronic compartment that contains a circuit board is shown in Figure 6.24. This particular electronics chassis used commercially available Eccosorb to reduce internal reflections, and a cross section of the chassis depicting the Eccosorb material layers is shown in Figure 6.24(b). Note that the RF circuits are placed within the inner metal box, and that a second metal box surrounds the first. Finite element surface mesh representations of the internal regions of the chassis and also the internal electronics box are shown in Figures 6.25 and 6.26, respectively. A current observer was placed on one of the traces of the internal circuit board. To couple energy onto these traces first requires coupling onto the cable wire within the chassis, which then propagates through the multipin connector into the small internal electronics box, where the central pin connects to the circuit trace containing the current observer.

The electronics chassis has the following dimensions: 0.432, 0.358, and 0.13 m along the x-, y-, and z-directions. For an empty rectangular cavity of these dimensions the first few resonant modal frequencies are given by $TE_{011} = 0.544$ GHz, $TE_{101} = 1.2$ GHz, $TE_{110} = 1.23$ GHz, $TE_{111} = 1.275$ GHz, and $TM_{111} = 1.275$ GHz. However, and as noted previously, to suppress internal resonances the electronics boxes used Eccosorb absorptive material layers, and therefore the true internal cavity resonances are not predicable by a simple analysis. Consequently, the first step of this study was to characterize the electrical properties of these absorptive material layers.

6.5.1 Internal Dispersive Material Calibration

Practical systems often contain important features where the electrical properties are not readily known across the entire bandwidth of interest. This is particularly true

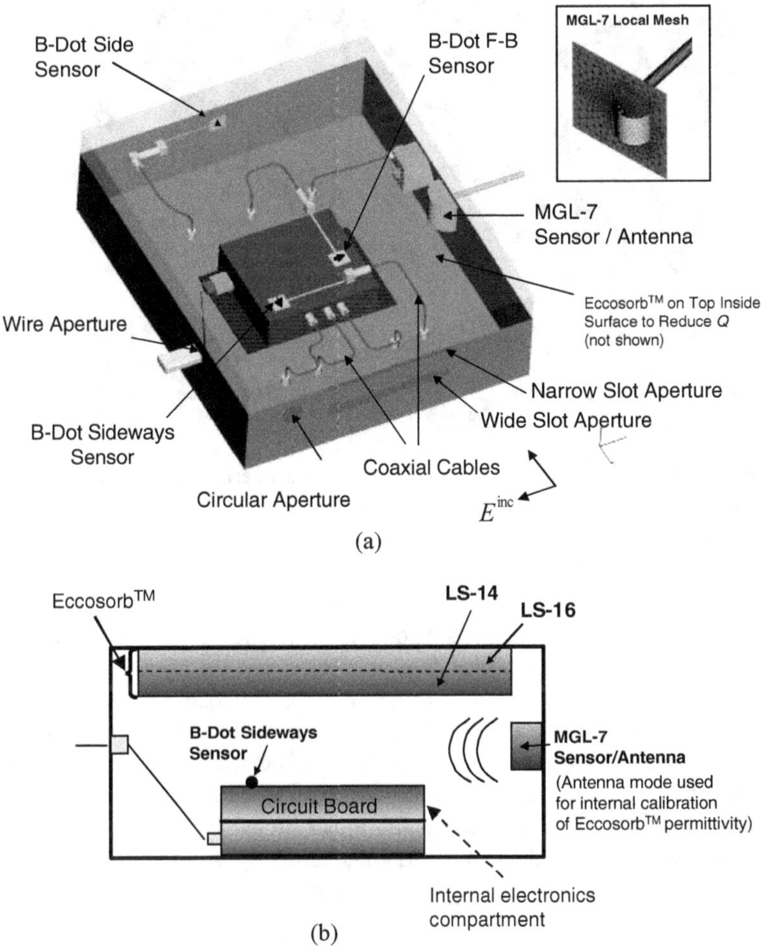

Figure 6.24 (a) Device under test (DUT) geometry showing various sensors and aperture configurations. (b) Cross section showing two Eccosorb absorptive layers (LS-14 and LS-16) mounted under the top surface of DUT. (*See insert for color representation of figure.*)

in the characterization of materials. The fidelity of a detailed finite element analysis is reduced when material properties cannot be specified precisely in the simulation. Because detailed material properties may not be readily available from manufacturers, empirical characterization over the bandwidth of interest may be required.

To examine the interior response of the chassis shown in Figure 6.24, the dispersive electrical properties of the two Eccosorb absorptive material layers (LS-14 and LS-16) were assumed to follow the complex permittivity profiles shown in Figure 6.27. All apertures in the chassis were sealed for this configuration, and the interior response at the B-Dot Sideways sensor due to the MGL-7 sensor driven in an antenna mode was predicted by the FETD method and also measured. The cavity response is shown

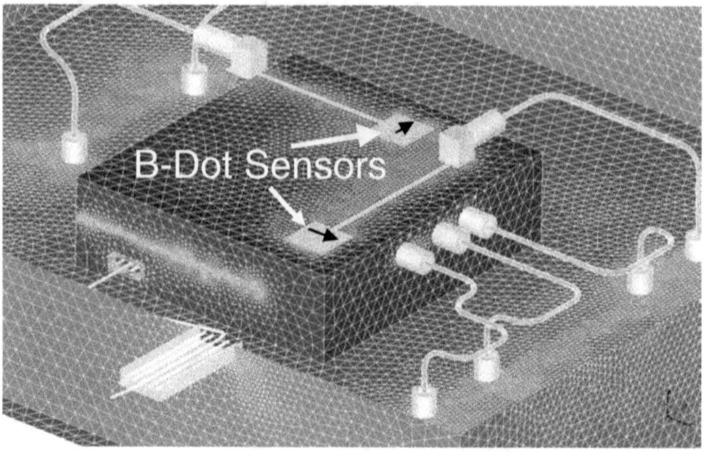

Figure 6.25 Finite element surface mesh local to the internal electronics compartment. (After Riley et al. [48], Copyright © IEEE 2006.) (*See insert for color representation of figure.*)

in Figure 6.28, where the data correspond to a transfer function that relates the sensor power received to the power transmitted. The noise floor for the measured data for this configuration was approximately −100 dB. Although the locations of the peaks and nulls do not align in frequency precisely between the predictions and the measurements, the overall correlation in the frequency trend, amplitude, and bandwidth of the internal resonances observed is considered to be good. For purposes of estimating shielding needs, an upper bound is important, whereas the exact frequency location of the peaks and nulls generally is not.

Figure 6.26 Finite element surface mesh on the internal circuit board. (After Riley et al. [48], Copyright © IEEE 2006.) (*See insert for color representation of figure.*)

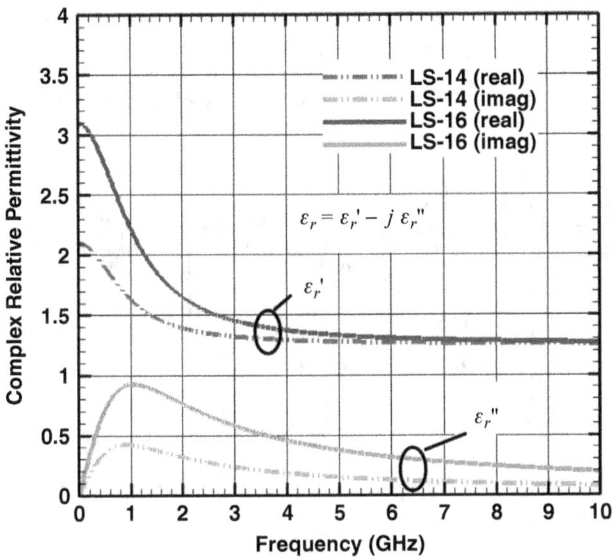

Figure 6.27 Estimated complex permittivity for Eccosorb material layers placed on the interior top surface of the test object.

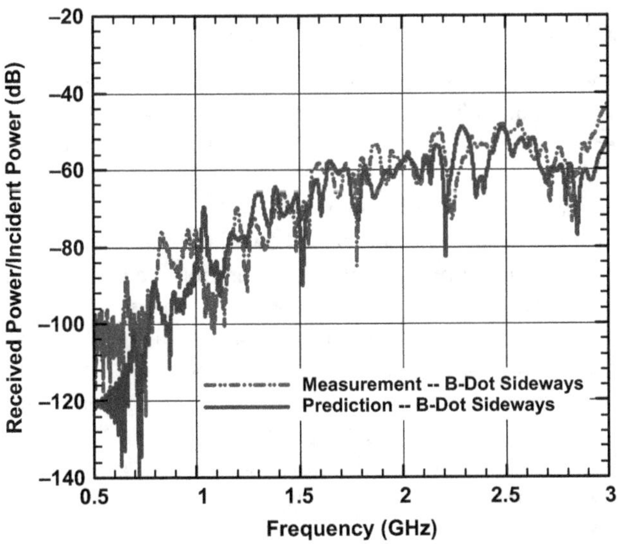

Figure 6.28 Comparison of the FETD predicted and measured data for coupling between the MGL-7 and the B-Dot Sideways sensors with all apertures closed.

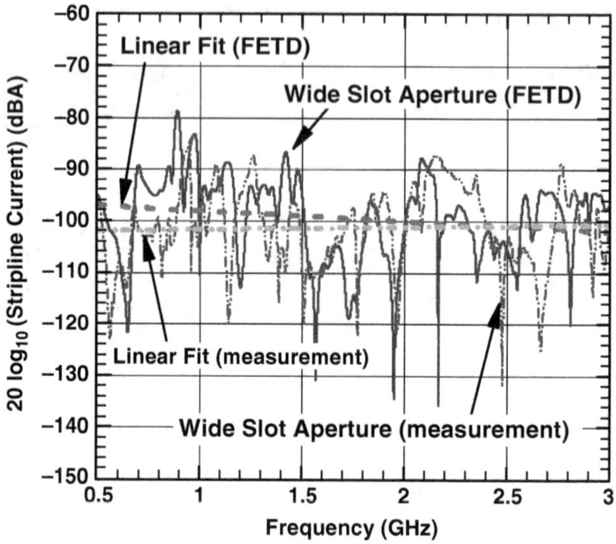

Figure 6.29 Comparison of the FETD and measured data for the current trace observer shown in Figure 6.26. Coupling is through the wide slot aperture for the case of an external plane wave with normal incidence and horizontal polarization. Linear data fits are also provided. (After Riley et al. [48], Copyright © IEEE 2006.)

6.5.2 External Illumination and Aperture Coupling

A horn antenna was used to illuminate the electronics chassis externally with the wide slot aperture open (Figure 6.24). Energy could then couple onto the internal wires and multipin connectors and into the internal electronics box containing the circuit-board traces. Figure 6.29 shows FETD predictions and experimental data for the trace current observer on the circuit board within the secondary internal compartment, where the location of the trace-current observer is shown in Figure 6.26. Additional results for coupling based on other open aperture configurations can be found in Ref. 48.

Unlike the analysis of isolated antennas, practical internal RF coupling assessments involve numerous uncertainties in both geometrical and electrical parameters. Because internal electronics compartments can exhibit extensive resonance characteristics, small variations in the physical placement of components and the material properties can affect the overall internal system response significantly. Nevertheless, this example has demonstrated that high-fidelity finite element techniques provide useful data to system integrators with regard to the coupling levels that may be present in a complex system assessment.

6.6 SUMMARY

In this chapter we have discussed several techniques to enhance finite element formulations so that their application scope can be extended. The techniques addressed included the modeling of thin-material sheets and coatings, thin wires and thin slots, lumped-circuit networks, and distributed port feeding systems for antennas and arrays. The thin-material models were based on both the formulation of appropriate impedance boundary conditions and the adaptation of degenerate prism finite elements. The first-order impedance boundary condition is easily incorporated into both the frequency- and time-domain finite element formulations, although it affects only the tangential component of the electric field and requires certain other conditions for its application, such as local uniformity of the surface fields and restrictions on the angle of incidence. Alternatively, thin-material layers can often be rigorously modeled by using extruded finite elements such as prisms and hexahedra. The three-dimensional degrees of freedom associated with these elements can be preserved by collapsing their original three-dimensional forms into two-dimensional surface elements. Consequently, this approach can provide increased generality and accuracy because the impact of the material layer on both the tangential and normal field components is accounted for, in an approximate sense, within the finite element solution.

The modeling of thin wires and thin slots within the finite element method adopted a different approach that combined two independent mathematical models. More specifically, in the case of wires, the physics of the wire was treated by a local transmission-line model, while its supporting environment was described by the traditional three-dimensional finite element formulation of the wave equation. A field-wire coupling technique was described where the local electric fields from the FETD method provided distributed source terms for the transmission-line model, which in turn provided the current distribution along the wire that supplied source terms for the volumetric FETD model. Thin slots were modeled similarly using a dual form for the thin-wire transmission-line equations. As with all hybridization schemes that interface different solution techniques, care is required to preserve symmetry within the global solution method; otherwise, numerical instabilities are likely to occur within time-domain formulations. Techniques were described that provide unconditionally stable interfaces between both the FETD method and the thin-wire and thin-slot transmission-line equations.

A technique to incorporate complex lumped-circuit networks into the FETD method was also discussed. The approach was based on isolating those edges within the finite element domain that are interfaced to circuitry. By decomposing the global formulation into a distributed field portion and a lumped-element portion, alternative solution methods for the circuit network can be solved concurrently with the volumetric FETD technique. Formulations to include lumped-circuit elements in the conditionally stable FETD method based on the two first-order Maxwell's equations, as well as the unconditionally stable FETD method based on the wave equation, were discussed. Subsequently, a generalization of this field-circuit concept to the case of rigorously interfacing antenna feeding networks to the antenna ports was discussed. The approach was based on a bidirectional decomposition of the fields at the port interfaces between the antenna and network feeding portions of the antenna system. Because the output of the feeding region is impressed into the finite element model for the antennas through waveguide modes at the ports, high solution accuracy and generality can be obtained. An example of a scanning 64-element Vivaldi array with a multistage feeding network was provided.

The final topic considered in this chapter was electromagnetic coupling into an electronic subsystem. In cases where modern platforms support high-power antenna systems, the potential interaction of these antennas may represent an important concern in the overall system design. In particular, the possible RF coupling into electronics boxes may require additional attention to shielding. We presented a validated example of using the FETD method to predict free-field coupling into an electronics chassis that contained a shielded electronics compartment with simple internal circuitry. One of the challenges associated with rigorously modeling certain system features is to obtain accurate domain knowledge of the various system components, such as the specification of complex material properties over a wide frequency band. For complex internal coupling applications, this limited domain knowledge often restricts our ability to obtain pointwise correlation between measurements and predictions over wide bandwidths. Nevertheless, advanced finite element techniques provide a useful tool for establishing upper bounds on the coupling levels that are important to the integration of high-power antenna systems into platforms.

REFERENCES

1. K. M. Mitzner, "Effective boundary conditions for reflection and transmission by an absorbing shell of arbitrary shape," *IEEE Trans. Antennas Propagat.*, vol. 16, pp. 706–712, Nov. 1968.
2. T. B. A. Senior, "Combined resistive and conductive sheets," *IEEE Trans. Antennas Propagat.*, vol. 33, pp. 577–579, May 1985.
3. G. Pelosi and P. Ya. Ufimtsev, "The impedance boundary condition," *IEEE Antennas Propagat. Mag.*, vol. 38, no. 1, pp. 31–35, Feb. 1996.
4. I. V. Lindell and A. H. Sihvola, "Realization of impedance boundary," *IEEE Trans. Antennas Propagat.*, vol. 54, pp. 3669–3676, Dec. 2006.
5. L. N. Medgyesi-Mitschang and D.-S. Y. Wang, "Hybrid solutions for large-impedance coated bodies of revolution," *IEEE Trans. Antennas Propagat.*, vol. 34, pp. 1319–1329, Nov. 1986.
6. A. Bendali, M'B. Fares, and J. Gay, "A boundary-element solution of the Leontovich problem," *IEEE Trans. Antennas Propagat.*, vol. 47, pp. 1597–1605, Oct. 1999.
7. J. H. Beggs, R. J. Luebbers, K. S. Yee, and K. S. Kunz, "Finite-difference time-domain implementation of surface impedance boundary conditions," *IEEE Trans. Antennas Propagat.*, vol. 40, pp. 49–56, Jan. 1992.
8. J. G. Maloney and G. S. Smith, "A comparison of methods for modeling electrically thin dielectric and conducting sheets in the finite-difference time-domain (FDTD) method," *IEEE Trans. Antennas Propagat.*, vol. 41, pp. 690–694, May 1993.
9. N. Farahat, S. Yuferev, and N. Ida, "High order surface impedance boundary conditions for the FDTD method," *IEEE Trans. Magn.*, vol. 37, pp. 3242–3245, May 2001.
10. M. K. Kärkkäinen and S. A. Tetryakov, "Finite-difference time-domain model of interfaces with metals and semiconductors based on a higher order surface impedance boundary condition," *IEEE Trans. Antennas Propagat.*, vol. 51, pp. 2448–2455, Sept. 2003.

11. J.-M. Jin, J. L. Volakis, C. L. Yu, and A. C. Woo, "Modeling of resistive sheets in finite element solutions," *IEEE Trans. Antennas Propagat.*, vol. 40, pp. 727–731, June 1992.
12. D. J. Riley and N. W. Riley, "First order models for thin-material sheets and coatings in the finite element time-domain method," *IEEE AP-S Int. Symp. Dig.*, vol. 4, pp. 3489–3492, June 2004.
13. Ç. Tokgöz and R. J. Marhefka, "A UTD based asymptotic solution for the surface magnetic field on a source excited circular cylinder with an impedance boundary condition," *IEEE Trans. Antennas Propagat.*, vol. 54, pp. 1750–1757, June 2006.
14. J.-M. Jin, *The Finite Element Method in Electromagnetics*, 2nd ed. Hoboken, NJ: Wiley, 2002.
15. D. J. Hoppe and Y. Rahmat-Samii, "Scattering by superquadric dielectric-coated cylinders using higher order impedance boundary conditions," *IEEE Trans. Antennas Propagat.*, vol. 40, pp. 1513–1523, Dec. 1992.
16. C. A. Balanis, *Advanced Engineering Electromagnetics*. New York: Wiley, 1989.
17. Z. Ren, "Degenerated prism element-general nodal and edge shell elements for field computation in thin structures," *IEEE Trans. Magn.*, vol. 34, pp. 2547–2550, May 1998.
18. E. Abenius and F. Edelvik, "Thin sheet modeling using shell elements in the finite element time-domain method," *IEEE Trans. Antennas Propagat.*, vol. 54, pp. 28–34, Jan. 2006.
19. R. D. Graglia, D. R. Wilton, A. F. Peterson, and I.-L. Gheorma, "Higher order interpolatory vector bases on prism elements," *IEEE Trans. Antennas Propagat.*, vol. 46, pp. 442–450, Mar. 1998.
20. R. Holland and L. Simpson, "Finite-difference analysis of EMP coupling to thin struts and wires," *IEEE Trans. Electromagn. Compat.*, vol. 23, pp. 88–97, May 1981.
21. K. R. Umashankar, A. Taflove, and B. Beker, "Calculation and experimental validation of induced currents on coupled wires in an arbitrary shaped cavity," *IEEE Trans. Antennas Propagat.*, vol. 35, pp. 1248–1257, Nov. 1987.
22. J.-P. Berenger, "A multiwire formalism for the FDTD method," *IEEE Trans. Electromagn. Compat.*, vol. 42, pp. 257–264, Aug. 2000.
23. F. Edelvik, "A new technique for accurate and stable modeling of arbitrarily oriented thin wires in the FDTD method," *IEEE Trans. Electromagn. Compat.*, vol. 45, pp. 416–423, May 2003.
24. J. Gilbert and R. Holland, "Implementation of the thin-slot formalism in the finite-difference EMP code THREDII," *IEEE Trans. Nucl. Sci.*, vol. 28, pp. 4269–4274, Dec. 1981.
25. C. D. Turner and L. D. Bacon, "Evaluation of a thin-slot formalism for finite-difference time-domain electromagnetics codes," *IEEE Trans. Electromagn. Compat.*, vol. 30, pp. 523–528, Nov. 1988.
26. A. Taflove, K. R. Umashankar, B. Beker, F. Harfoush, and K. S. Yee, "Detailed FD-TD analysis of electromagnetic fields penetrating narrow slots and lapped joints in thick conducting screens," *IEEE Trans. Antennas Propagat.*, vol. 36, pp. 247–257, Feb. 1988.
27. D. J. Riley and C. D. Turner, "Hybrid thin-slot algorithm for the analysis of narrow apertures in finite-difference time-domain calculations," *IEEE Trans. Antennas Propagat.*, vol. 38, pp. 1943–1950, Dec. 1990.
28. D. J. Riley and C. D. Turner, "The inclusion of wall loss in finite-difference time-domain algorithms," *IEEE Trans. Electromagn. Compat.*, vol. 33, pp. 304–311, Nov. 1991.

29. M. Li, K.-P. Ma, D. M. Hockanson, J. L. Drewniak, T. H. Hubing, and T. P. Van Doren, "Numerical and experimental corroboration of an FDTD thin-slot model for slots near corners of shielding enclosures," *IEEE Trans. Electromagn. Compat.*, vol. 39, pp. 225–232, Aug. 1997.
30. D. Riley, "Transient finite elements for computational electromagnetics: hybridization with finite differences, modeling thin wire and thin slots, and parallel processing," *Proc. 17th Annu. Revi. Prog. Appl. Comput. Electromagn. (ACES)*, Monterey, CA, pp. 128–138, Mar. 2001.
31. F. Edelvik, G. Ledfelt, P. Lötstedt, and D. J. Riley, "An unconditionally stable subcell model for arbitrarily oriented thin wires in the FETD method," *IEEE Trans. Antennas Propagat.*, vol. 51, pp. 1797–1805, Aug. 2003.
32. F. Edelvik and T. Weiland, "Stable modeling of arbitrary thin slots in the finite element time-domain method," *Int. J. Numer. Model.*, vol. 17, pp. 365–383, June 2004.
33. L. K. Warne and K. C. Chen, "A simple transmission line model for narrow slot apertures having depth and losses," *IEEE Trans. Electromagn. Compat.*, vol. 34, pp. 173–182, Aug. 1992.
34. A. Bossavit, *Computational Electromagnetism: Variational Formulations, Complementarity, Edge Elements*. San Diego, CA: Academic Press, 1998.
35. Y.-S. Tsuei, A. C. Cangellaris, and J. L. Prince, "Rigorous electromagnetic modeling of chip-to-package (first level) interconnections," *IEEE Trans. Components Hybrids Manuf. Technol.*, vol. 16, pp. 876–883, Aug. 1993.
36. K. Guillouard, M. F. Wong, V. Fouad Hanna, and J. Citerne, "A new global time-domain electromagnetic simulator of microwave circuits including lumped elements based on finite element method," *IEEE Trans. Microwave Theory Tech.*, vol. 47, pp. 2045–2048, Oct. 1999.
37. S.-H. Chang, R. Coccioli, Y. Qian, and T. Itoh, "A global finite element time-domain analysis of active nonlinear microwave circuits," *IEEE Trans. Microwave Theory Tech.*, vol. 47, pp. 2410–2416, Dec. 1999.
38. H.-P. Tsai, Y. Wang, and T. Itoh, "An unconditionally stable extended (USE) finite element time-domain solution of active nonlinear microwave circuits using perfectly matched layers," *IEEE Trans. Microwave Theory Tech.*, vol. 50, pp. 2226–2232, Oct. 2002.
39. SPICE, Department of Electrical Engineering and Computer Sciences, University of California, Berkeley, CA, http://bwrc.eecs.berkeley.edu/Classes/IcBook/SPICE/.
40. W. H. Press, B. P. Flannery, S. A. Teukolsky, and W. T. Vetterling, *Numerical Recipes: The Art of Scientific Computing*. Cambridge, UK: Cambridge University Press, 2007.
41. B. Toland, J. Lin, B. Houshmand, and T. Itoh, "FDTD analysis of an active antenna," *IEEE Microwave Guided Wave Lett.*, vol. 3, pp. 423–425, Nov. 1993.
42. J. Lin, T. Itoh, and S. Nogi, "Mode switch in a two-element active array," *IEEE AP-S Int. Symp. Dig.*, pp. 664–667, July 1993.
43. R. Wang, H. Wu, A. C. Cangellaris, and J.-M. Jin, "Incorporation of feed-network into the time-domain finite element modeling of antenna arrays," *IEEE Trans. Antennas Propagat.*, vol. 56, no. 8, pp. 2599–2612, Aug. 2008.
44. D. J. Riley, C. D. Turner, L. D. Bacon, and J. D. Kotulski, "Numerical modeling tools for transient electromagnetic problems," SAND90-0519, Sandia National Laboratories, Albuquerque, NM, Apr. 1990.

45. B. Gustavsen and A. Semlyen, "Rational approximation of frequency domain responses by vector fitting," *IEEE Trans. Power Deliv.*, vol. 14, pp. 1052–1061, July 1999.
46. A. R. Djordjevic, T. K. Sarkar, and R. F. Harrington, "Time-domain response of multiconductor transmission lines," *IEEE Proc.*, vol. 75, pp. 743–764, June 1987.
47. Z. Lou and J. M. Jin, "An accurate waveguide port boundary condition for the time-domain finite element method," *IEEE Trans. Microwave Theory Tech.*, vol. 53, no. 9, pp. 3014–3023, Sept. 2005.
48. D. Riley, G. Sower, D. Schafer, and G. Baca, "Validated electromagnetic coupling predictions into an electronic subsystem based on the finite element time-domain method," *IEEE AP-S Int. Symp. Dig.*, pp. 2789–2792, July 2006.

7 Antenna Simulation Examples

With the development of the basic finite element formulation in Chapters 2, the mesh truncation techniques in Chapters 3 and 4, the antenna feed modeling and parameter calculation in Chapter 5, and the treatment of fine features in Chapter 6, computer codes can be developed to analyze single antennas and small arrays, where the resulting matrix equation can be solved without resorting to special solution techniques (discussed in Chapters 8 to 11). In this chapter we present a variety of antenna simulation examples to demonstrate the capability and versatility of the finite element method. All the simulations were performed using our in-house-developed computer codes based on the techniques discussed in the preceding chapters.

7.1 NARROWBAND ANTENNAS

Narrowband antennas are often based on resonant configurations. Because of resonance, the electromagnetic field in a narrowband antenna decays very slowly in time and varies considerably as a function of frequency. For a rigorous solution, a small frequency step has to be employed in a frequency-domain simulation to capture this frequency variation, and a long time duration has to be simulated in the time domain for the fields to either vanish or to reach steady state. In this section we examine two types of narrowband antennas that are widely used in practice: the coaxial-fed monopole antenna and the coaxial-fed patch antenna.

7.1.1 Coaxial-Fed Monopole Antenna

Monopole antennas have a typical bandwidth of approximately 10% and are used widely in communication systems. A coaxial-fed monopole antenna mounted on a small cylinder residing on an infinite ground plane is shown in Figure 7.1(a). The height h of the antenna is 20.68 cm and the radius a is 1.62 mm, which leads to the antenna thickness factor, $2\ \ln(2h/a) = 11$, corresponding to a moderately thin monopole. Both the radius and height of the small mounting cylinder are 24.31 mm. All metal surfaces were assumed to be perfect electrical conductors, and the coaxial feed model was based on the techniques described in Section 5.1.3. Measurements [1] and predictions for the input admittance are shown in Figure 7.1(b),

Finite Element Analysis of Antennas and Arrays, By Jian-Ming Jin and Douglas J. Riley

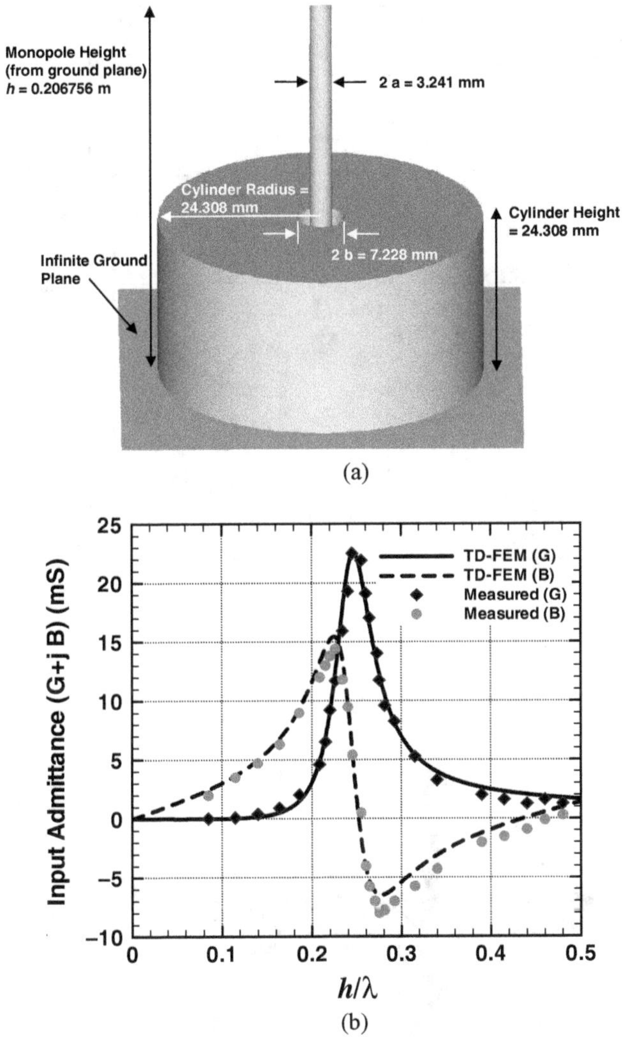

Figure 7.1 Coaxially fed monopole antenna on a ground plane. (a) Geometry. (b) Input admittance. Measured data are adapted from Liu and Grimes [1].

where the predictions were based on the hybrid FETD–FDTD algorithm discussed in Chapter 4.

7.1.2 Monopole Antennas on a Plate

This example consists of five monopole antennas mounted on a 30.48-cm × 30.48-cm finite ground plane of thickness 3.175 mm, with the location of each antenna shown in Figure 7.2(a) [2]. Each monopole has a radius of 0.635 mm and a length of

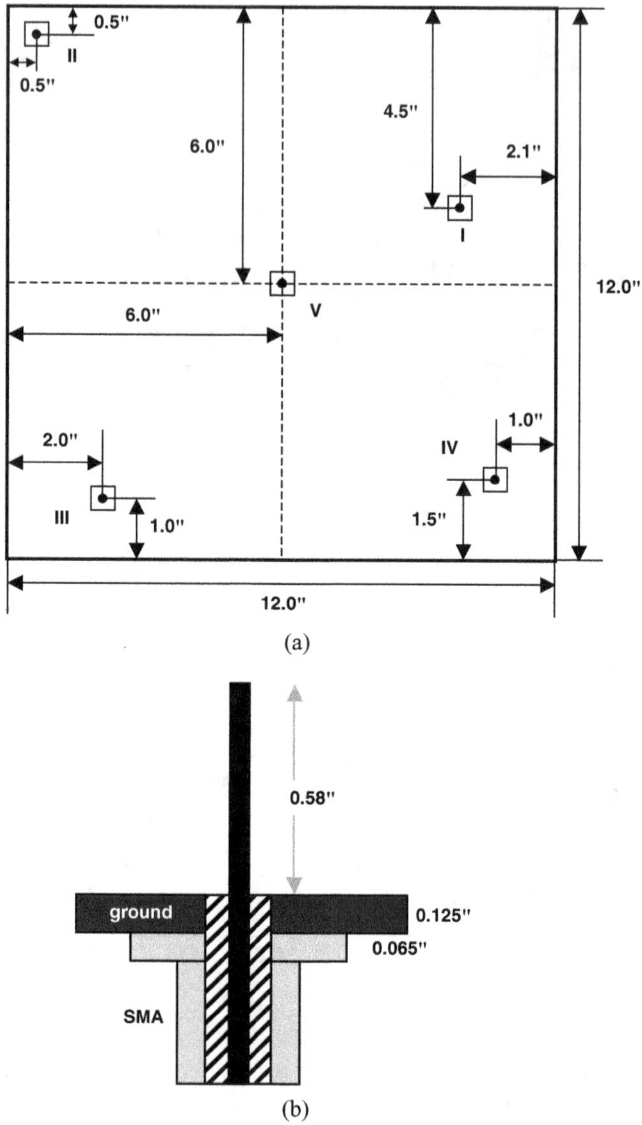

Figure 7.2 Five monopole antennas mounted on a finite ground plane. (a) Top view showing the size of the plate and locations of the monopoles (unit in inches). (b) Cross-sectional view of one monopole with an SMA connector.

14.68 mm, which is the extension of the inner conductor of the coaxial feed, whose outer radius is 2.057 mm, and is filled with Teflon that has a relative permittivity of 2.0. A cross-sectional view of one monopole and its SMA connector is displayed in Figure 7.2(b). Figure 7.3 gives the impedance matrix from 3 to 6 GHz. To be more specific, the graph on the *i*th row and *j*th column plots the real part (solid line)

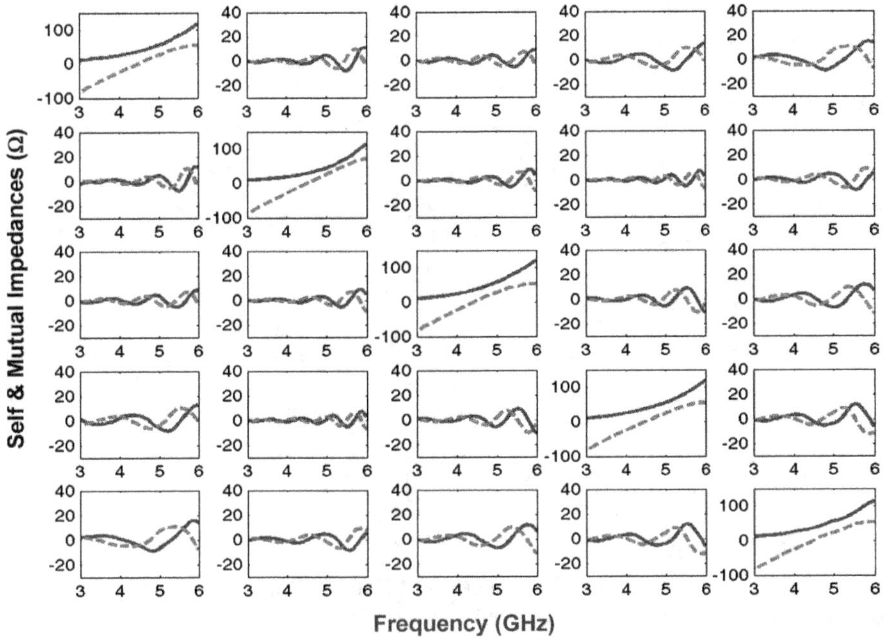

Figure 7.3 Impedance matrix (solid lines for real part and dashed lines for imaginery part) for the five monopole antennas on a finite plate. The graph on the ith row and jth column shows Z_{ij}.

and imaginary part (dashed line) of the impedance Z_{ij} observed at the aperture of the coaxial feed. The results were computed using the time-domain finite element method with the first-order absorbing boundary condition for mesh truncation and the coaxial feed model for excitation. To compute Z_{ij}, we first excite the jth antenna and match-load all other antennas and then calculate the S-parameters. The symmetry of the impedance matrix is clear in the numerical solution. Figure 7.4 displays the radiations patterns at 4.7 GHz in the two principal planes and two diagonal planes when the center antenna is excited and all other antennas are terminated with a match load.

7.1.3 Patch Antennas on a Plate

Microstrip patch antennas are typically narrowband antennas, which are compact, easy and inexpensive to fabricate, and can be placed conformally on host platforms. A small 2 × 2 patch antenna array is shown in Figure 7.5 [2]. The patches are printed on a 1.52-mm-thick substrate having a relative permittivity of 3.38. Both the substrate and its ground plane have a finite size of 30.48 cm × 30.48 cm. The patches are fed with coaxial lines whose dimensions and dielectric filling are the same as the one shown in Figure 7.2(b). The edge-to-edge distance between the adjacent patches is 6.86 mm and the size of the patches is 2.54 cm × 2.54 cm. Figure 7.6

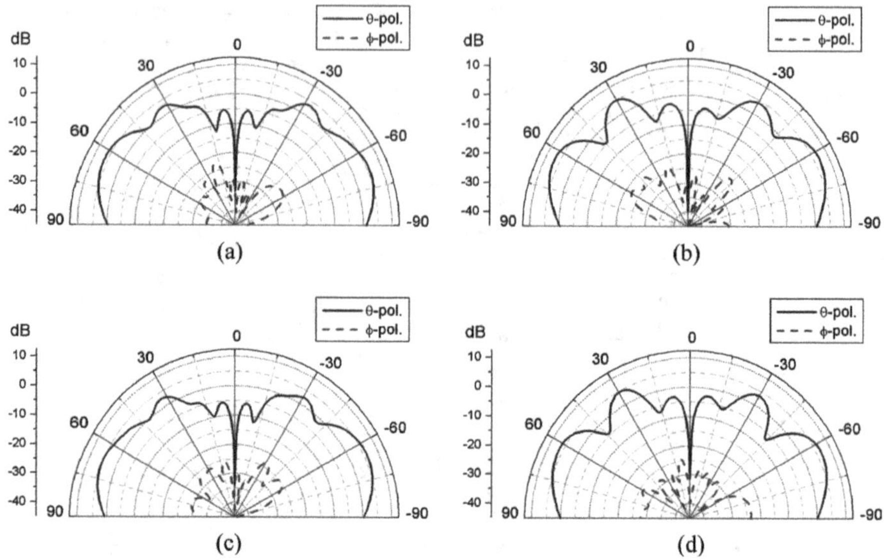

Figure 7.4 Radiation patterns in four planes at 4.7 GHz when only the center antenna is excited. (a) $\phi = 0°$. (b) $\phi = 45°$. (c) $\phi = 90°$. (d) $\phi = 135°$.

gives the impedance matrix from 2.8 to 3.2 GHz, with the graph on the ith row and jth column plotting the real part (solid line) and the imaginary part (dashed line) of the impedance Z_{ij} observed at the aperture of the coaxial feed. The results were computed using the time-domain finite element method with the first-order absorbing boundary condition for mesh truncation and the coaxial feed model for excitation. In this case, because of the perfect symmetry of the array configuration, many of the graphs in Figure 7.6 are identical. For example, all the self-impedances are the same, and also, $Z_{14} = Z_{12}, Z_{23} = Z_{12}$, and $Z_{24} = Z_{13}$. This example was also simulated using the frequency-domain finite element method combined with the boundary integral equation for mesh truncation, and the results are very similar to those in Figure 7.6.

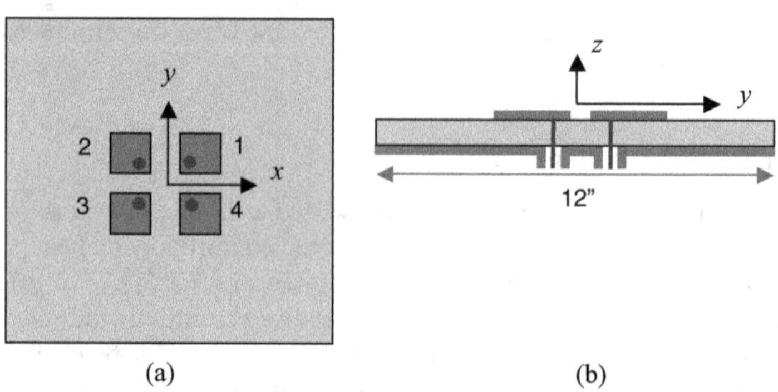

Figure 7.5 Four patch antennas mounted on a finite ground plane. (a) Top view. (b) Side view.

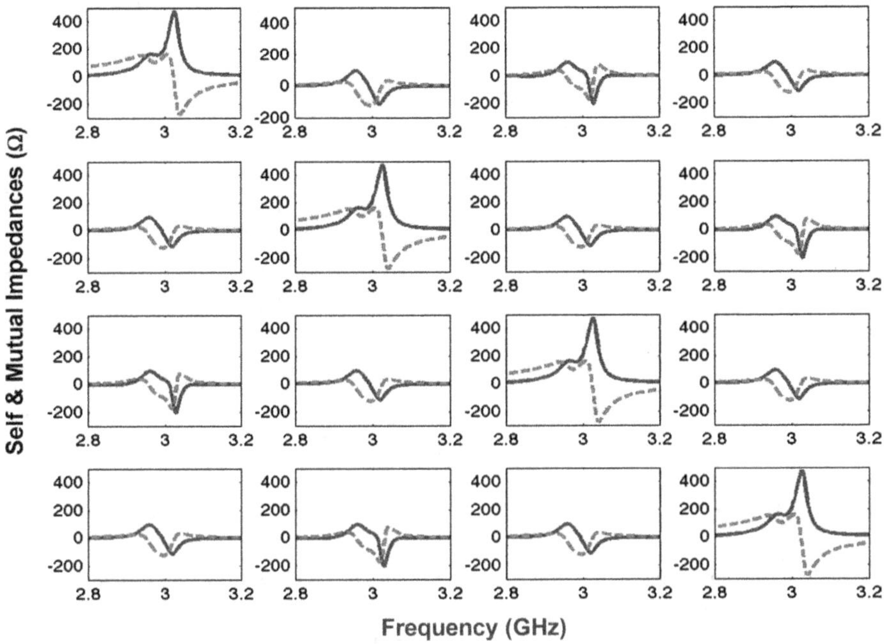

Figure 7.6 Impedance matrix (solid lines for real part and dashed lines for imaginery part) for the four patch antennas on a finite plate. The graph on the ith row and jth column shows Z_{ij}.

Figure 7.7 displays the radiation patterns at 3.0 GHz in the two principal planes when the two diagonal antennas are excited with the same phase and the other two diagonal antennas are fed with the same phase, which is 180° out-of-phase from the other two. The asymmetry in the radiation patterns is due to the feeding mode.

7.1.4 Conformal Patch Antenna Array

This example consists of an 11 × 11 microstrip patch antenna array with the specific geometry described in Figure 7.8. The patches have the dimensions 3.0 cm × 3.56 cm, and the edge-to-edge distance between adjacent patches is 1.4 cm in both the x- and y-directions. The coaxial line feed has inner and outer radii $r_i = 0.48$ mm and $r_o = 1.5$ mm and is filled with dielectric material of relative permittivity 1.86. The feed is placed symmetrically in the x-direction but is offset 0.86 cm from the center of a given patch in the y-direction. The entire array is printed on a 52-cm × 58-cm substrate having a thickness of 7.0 mm and dielectric constants of $\varepsilon_r = 2.67$ and $\mu_r = 1.0$ and is flush-mounted on an infinite ground plane. The frequency-domain finite element method combined with the boundary integral equation for mesh truncation is employed for the simulation, and the time-consuming boundary integral equation is evaluated using acceleration based on the adaptive integral method [3]. Discretization of the entire problem resulted in 26,513 unknowns

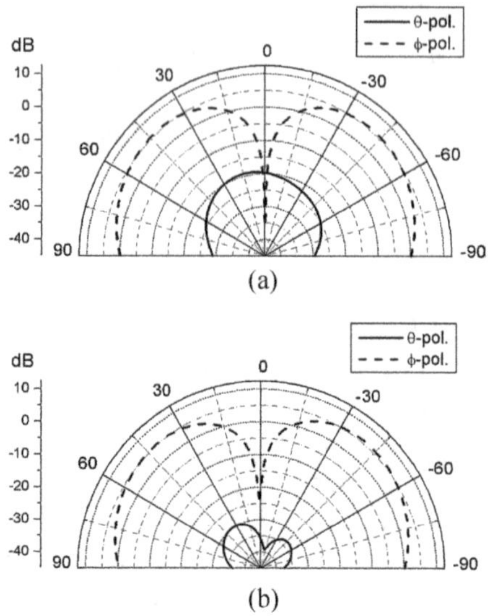

Figure 7.7 Radiation patterns in the *xz*- and *yz*-planes at 3.0 GHz. (a) In the *xz*-plane. (b) In the *yz*-plane.

Figure 7.8 Geometry of an 11 × 11 microstrip patch antenna array.

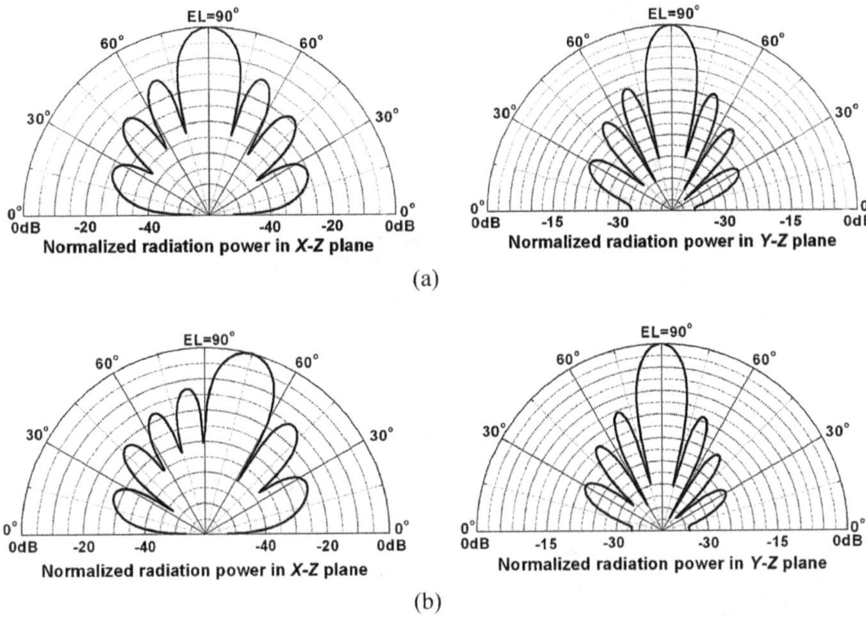

Figure 7.9 Radiation patterns of an 11 × 11 microstrip patch antenna array. (b) The array is fed with the same phase. (b) The array is fed with $-30°$ progressive phase shift resulting in a scan toward the x-direction. (After Mao et al. [3], Copyright © Taylor & Francis 2006.)

on the top surface of the antenna array and 133,025 unknowns inside the substrate. Figure 7.9 shows the radiation patterns of the microstrip patch antenna array at 2.25 GHz when the antenna elements are fed, respectively, with the same phase, which produces the main radiation lobe in the z-direction, and with a $-30°$ progressive phase shift, which results in a scan toward the x-direction.

7.2 BROADBAND ANTENNAS

Broadband antennas are often more geometrically complex and their meshing more challenging than that for narrowband antennas. Electromagnetic fields in a broadband antenna are characterized by a relatively short duration in time and a rapid spatial variation, although the far fields vary slowly as a function of frequency. In this section we examine five broadband antennas solved by the finite element method.

7.2.1 Ridged Horn Antenna

Ridged horn antennas are broadband radiators that are used widely as standard gain horns for calibrated gain measurements as well as for feed elements in satellite tracking or communication systems. Ridges are introduced to increase the bandwidth of regular horn antennas, which are known to have a large gain but a limited

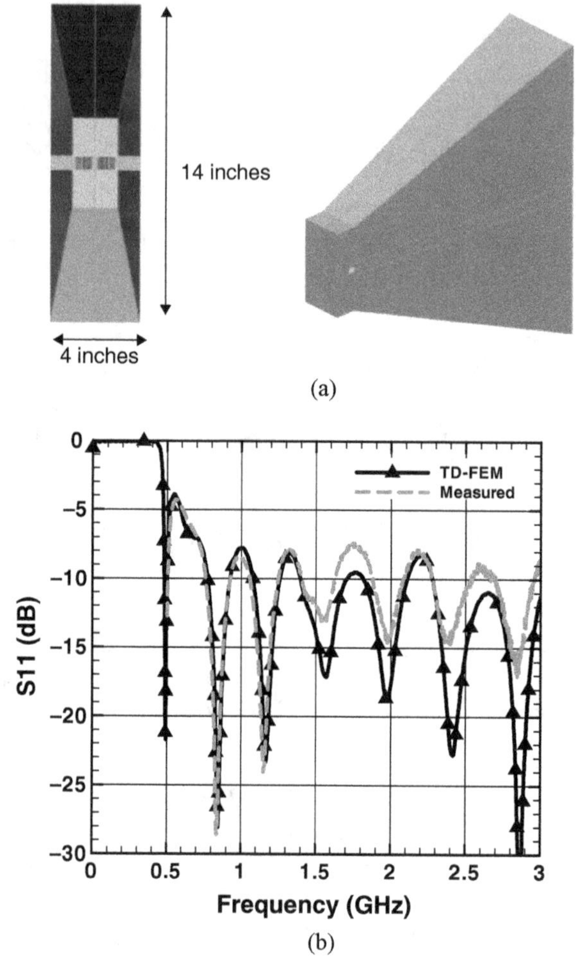

Figure 7.10 Coaxial-fed ridged horn antenna. (a) Geometry. (b) S_{11} versus frequency.

bandwidth [4]. A 10.16-cm × 35.56-cm single-polarization design is shown in Figure 7.10(a). The antenna is fed by a 50-Ω coaxial transmission line. Predictions and measurements for S_{11} versus frequency are shown in Figure 7.10(b). The hybrid FETD–FDTD algorithm described in Chapter 4 was used to predict performance and the measurements were performed by the ESL Division of Northrop Grumman Corporation. The antenna and coaxial feed were modeled with the FETD algorithm, while the region external to the antenna was modeled using the FDTD algorithm with a mesh termination by perfectly matched layers. The coaxial feed model was based on the technique described in Section 5.1.3. The global time step was based on the cell size in the FDTD region, and even though the maximum-to-minimum element size in the finite element region was approximately 100 : 1, no further reduction in the time step was required in the finite element region, due to its local unconditional stability.

7.2.2 Sinuous Antenna

A planar sinuous antenna is a wideband, low-profile antenna that was invented by DuHamel [5]. It can provide dual linear or dual circular polarizations and the planar sinuous radiates equivalently in both the forward and backward directions unless a reflector or an absorber is placed on one side. Without loss, planar reflectors will short out the antenna at frequencies determined by transmission-line theory, thereby reducing the useful bandwidth. Consequently, practical applications often place the antenna in a cavity with absorptive material, although this reduces the gain by 3 dB, due to the suppression of the backlobe. Alternative implementations project the planar antenna onto a cone that can largely preserve both the operational bandwidth and the efficiency of the planar structure, although this may not be a feasible option in surface conformal applications.

The sinuous antenna is typically based on two or four arms rotationally centered about the feed point. Although less common, six- and eight-arm sinuous antennas have also been developed. The arms are generated in a zigzag fashion [5], and a representative four-arm structure is shown in Figure 7.11(a). When the generating zigzag curve for each arm is rotated $\pm 22.5^\circ$ about its centerline, an infinite four-arm sinuous antenna in a homogeneous environment will be a self-complementary structure and falls formally into the class of frequency-independent antennas with a resulting theoretically infinite bandwidth. However, because the current vanishes precisely at the tip location where the sinuous arms change direction, small oscillations with frequency are normally observed in the active input impedance, as well as frequency variations observed in the radiated beamwidth. Consequently, the sinuous tends to scale with frequency in a more discrete sense, similar to the log-periodic antenna. The sinuous antenna is generally believed to radiate in active regions that are one wavelength in circumference.

Depending on the number of arms, both the sinuous and related spiral antennas are fed by baluns or coaxial transmission lines, along with beamforming circuitry to extract different modes of operation [6]. The arm-to-ground input impedance for an N-arm ideal self-complementary antenna in free space is given by the expression $30\pi/\sin(\pi M/N)$, where $M = 1$ and $M = N - 1$ correspond to the *sum mode* and $M = 2$ and $M = N - 2$ correspond to the *difference mode*. The radiation pattern for the sum mode has a peak on the boresight, whereas the difference mode has a null. For an ideal four-arm self-complementary antenna in free space, the sum-mode, arm-to-ground active input impedance is 133.3 Ω. For the two-arm structure, the arm-to-arm input impedance is 188.5 Ω.

Because the sinuous antenna can have a very wide range in length scales from the outer boundary of the antenna into the feed region, it presents significant challenges to any predictive method. In addition, these structures are often etched on thin dielectric substrates that can affect the input impedance significantly, and therefore these thin-material layers must also be modeled accurately. For very wide bandwidth sinuous antenna designs that are on the order of 9 : 1, for example, from 2 to 18 GHz, the feed region is electrically very small, and the variation of the edge lengths required to mesh this type of antenna can be greater than 100 : 1. Lower-bandwidth

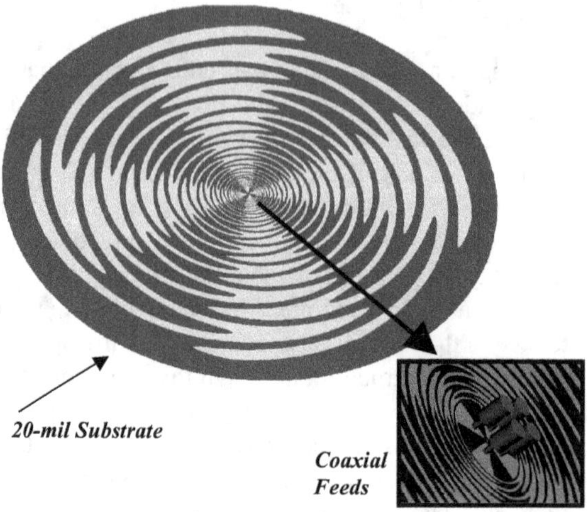

(a)

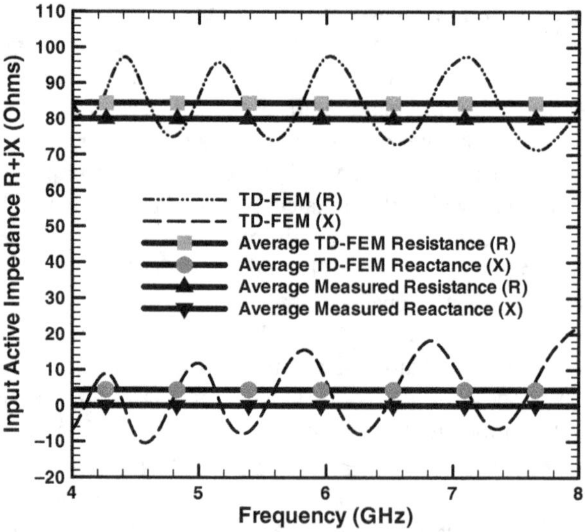

(b)

Figure 7.11 Four-arm sinuous antenna etched on a 20-mil dielectric substrate. (a) Geometry. (b) Predicted arm-to-common active input impedance and average measured impedance. [*See insert for color representation of* (a).]

designs normally expand the physical size of the feed region, which reduces the high-frequency performance, and these designs are therefore simpler to analyze by numerical methods.

As noted above, sinuous antennas can be driven in either dual linear or dual circular polarizations. To obtain circular polarization for the sum mode of the four-arm antenna, the feeds are usually phased progressively 90° relative to each other.

For the four-arm design that is etched on a 20-mil dielectric substrate and illustrated in Figure 7.11(a), the computed sum mode, arm-to-ground input impedance over a bandwidth of 4 to 8 GHz is shown in Figure 7.11(b). As shown in this figure, the presence of the thin dielectric substrate significantly lowers the nominal input impedance compared to the ideal free-space configuration of 133.3 Ω, and this effect has been confirmed by measurements. As noted above and observed in Figure 7.11(b), oscillations are seen in the input impedance of practical sinuous antennas, which are again believed to be due to small reflections from the corner regions associated with the alternating directions of the arms.

The hybrid FETD–FDTD technique described in Chapter 4 was used to analyze this geometry. Time-domain methods are well suited to the analysis of broadband antennas [7] because their entire bandwidth can be recovered through a single simulation. The use of a local, unconditionally stable solution technique is particularly beneficial for this class of antenna because the time step does not need to be reduced for numerical stability as the element size becomes progressively smaller into the feed region. In addition, an interesting observation with regard to the time-domain modeling of self-complementary antennas is that due to the infinite bandwidth and constant real active impedance of ideal structures, the ratio of the nonzero transient voltage and current in the feed must be constant for all time. Consequently, in the very early time of the simulation, this impedance will be established.

7.2.3 Logarithmic Spiral Antenna

The logarithmic spiral antenna is a broadband antenna that is commonly used in communication and direction-finding applications [8–10]. Due to the smooth variation of the arms along their entire length, the logarithmic spiral has continuous scaling with frequency in a nontruncated configuration. Because the infinite logarithmic spiral in a homogeneous environment is self-complementary, it has the same active input impedance characteristics as a function of the number of arms that were described in Section 7.2.2.

Figure 7.12(a) shows the top view of a logarithmic spiral antenna consisting of two free-standing conducting arms. The centerline of the arms is prescribed by $r = r_0\tau^{\phi/2\pi}$ in the polar coordinate system, where r_0 is a constant and the scaling factor τ is 1.588. The inner and outer radii of the spiral are 0.22 and 3.5 cm, respectively. The inner radius determines the upper limit of the bandwidth, and the outer radius determines the lower limit. This antenna radiates in two broad lobes whose directions are perpendicular to the plane of the antenna. In the simulation, the antenna is fed by the probe feed model described in Section 5.1.1, and the solution is obtained using the time-domain finite element method with the first-order absorbing boundary condition for mesh truncation [11]. A detailed view of the feed region is shown in Figure 7.12(b). A 0.56-mm-long current probe is placed at the center to excite the antenna, and the voltage across the feed is measured along the two observation probes 0.28 mm away from the current probe. The input resistance (solid line) and reactance (dashed line) versus frequency calculated are shown in Figure 7.13. The curves exhibit a transition from the resonant region to the broadband region as the

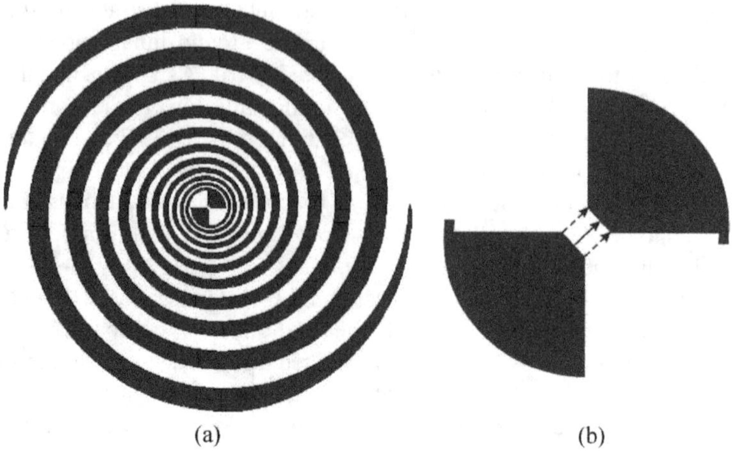

Figure 7.12 Two-arm logarithmic spiral antenna. (a) Geometry of the arms. (b) Enlarged feed region.

frequency increases and good broadband behavior for frequencies beyond 8 GHz. We also observe that the calculated input impedance converges to the theoretical value (188.5 Ω arm to arm for the two-arm self-complementary antenna) [10] at the higher-frequency end. In practice, many logarithmic spiral antennas are constructed on a dielectric substrate and some might be backed with a cavity filled with an absorbing material to eliminate one of the two broad lobes in the radiation pattern [6]. All these configurations can be simulated effectively using the finite element method.

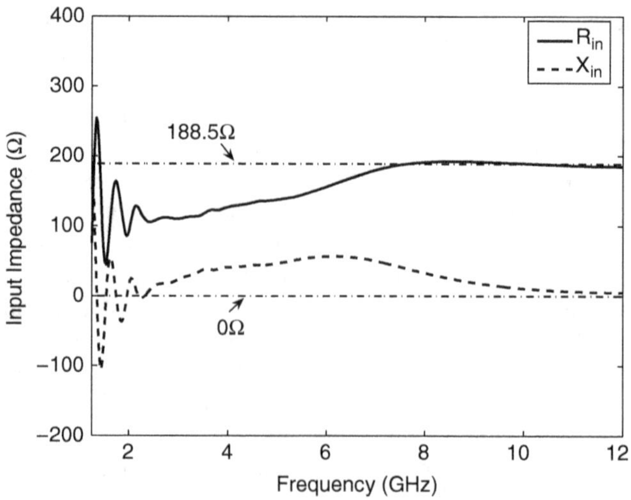

Figure 7.13 Input impedance of a two-arm logarithmic spiral antenna. (After Lou and Jin [11], Copyright © IEEE 2005.)

7.2.4 Inverted Conical Spiral Antenna

Unlike planar spiral and sinuous antennas, conical spiral antennas and their inverted versions above a ground plane radiate unidirectionally along their axes [12]. Figure 7.14(a) shows a two-arm conical spiral that has been inverted and fed against a ground plane to support wide-bandwidth antenna performance. An inset of the finite element mesh of the coaxial feed region is shown in Figure 7.14(b). As the frequency

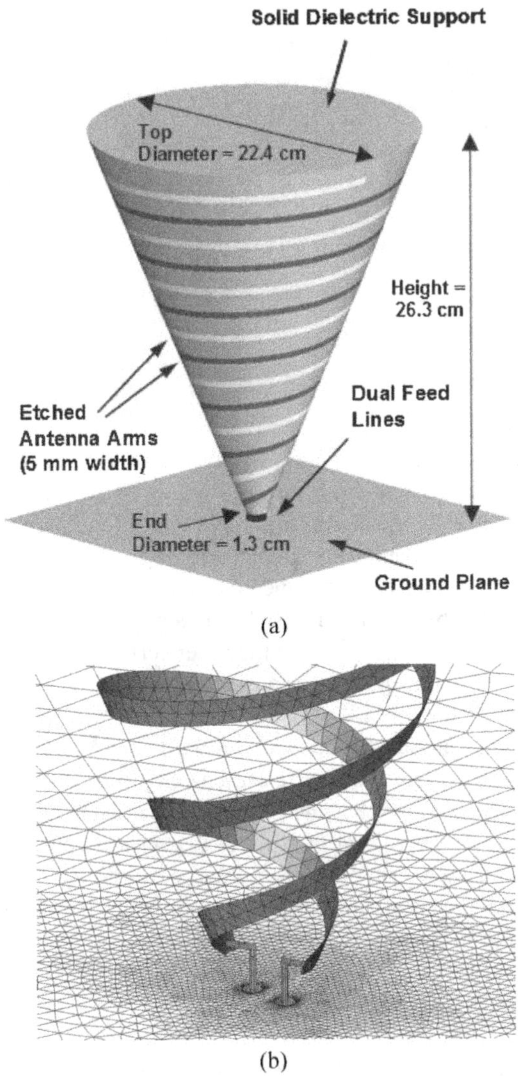

Figure 7.14 Inverted two-arm conical spiral over a ground plane. (a) Geometry. (b) Finite element surface mesh local to the inverted conical spiral feed region. [*See insert for color representation of* (b).]

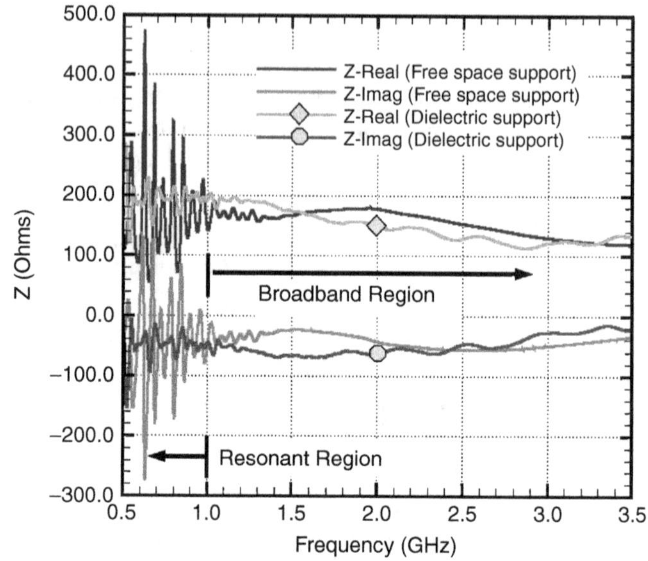

Figure 7.15 Input impedance of an inverted two-arm conical spiral over a ground plane for a free-space and a solid dielectric ($\varepsilon_r = 4$) cone.

decreases, significant current moves out along the arms, thereby increasing the physical distance of the peak current from the feed region and ground plane. Preserving the electrical distance of the peak current above the ground plane as the frequency is varied is important in obtaining the broadband performance of conical multiarm antennas in the presence of a ground plane, and this consideration affects the choice of the cone angle for the best wideband performance.

Although this early model is etched on the outer surface of a solid dielectric cone, it is also possible to use a thin-material approximation inside the finite element region to analyze the spiral etched on a thin dielectric shell. In practice, multiarm printed antennas such as spiral, sinuous, helical, and multiarmed log-periodic antennas can be etched in a similar way on a thin dielectric shell having three-dimensional shapes that include not only cones, but also truncated cones and hemispheres. Figure 7.15 shows the arm-to-arm input impedance of this inverted two-arm conical spiral for both a free-space and a solid dielectric cone of relative permittivity 4. With the presence of the dielectric cone, the oscillation in the resonant region is reduced significantly compared to the spiral in free space.

7.2.5 Antipodal Vivaldi Antenna

Vivaldi antennas belong to the class of traveling-wave tapered slot antennas that have been noted for their wideband performance [13,14]. These endfire antennas are often printed on a dielectric substrate, although machined three-dimensional implementations are also widely used. The term *flared-notch antenna* is also commonly used to describe this class of endfire antenna. The free-space electrical width of the open end of these antennas is typically designed to be about one-half wavelength at the lower end of the operating bandwidth. Both the printed Vivaldi and the machined flared-notch antennas are commonly used as either isolated radiators or in an array configuration. Because of their compact size for a given low frequency, flared-notch antennas are excellent choices for wideband arrays. In the case of arrays, flared notches must be designed in the array to obtain acceptable active impedance over their wide bandwidth.

In this example we compute the impedance and radiation characteristics of a single balanced antipodal Vivaldi antenna fed by a stripline [15]. As shown in Figure 7.16(a), the antenna consists of a 3-mm-wide conducting strip sandwiched between two dielectric boards with a combined thickness of 3.15 mm. The conducting strip, which is connected to the center conductor of a stripline, is gradually flared to form the radiating arm shown by the dashed lines in Figure 7.16(a). Ground planes are printed on the outer sides of the dielectric boards and are connected to the outer conductor of the stripline, and both are tapered to form the other two radiating arms. The shape of all the flares is described by elliptical arcs. In our numerical simulation, the stripline input is modeled as a 12-mm $\times$ 3.15-mm TEM port. The results were computed using the time-domain finite element method with perfectly matched layers for mesh truncation and the waveguide port boundary condition to model the stripline feed [11]. The return loss (S_{11}) calculated at the TEM port is plotted in Figure 7.16(b) over the frequency range of 1 to 10 GHz. The simulation result by the Ansoft HFSS commercial software package is plotted in the same figure for comparison [15]. The time-domain signal reveals that the strongest reflection occurs at the interface between the flare and the external free space, which points out possible ways to improve the design. Figure 7.17 displays the E- and H-plane co-polarized radiation patterns at 10 GHz. The E-plane cut exhibits a dip at boresight, while the H-plane cut shows a broadside main lobe.

7.2.6 Vlasov Antenna

The Vlasov antenna [16,17], a coaxial-fed circular waveguide antenna, is shown in cross-sectional view in Figure 7.18(a). The antenna is fed by a 50-Ω coaxial cable, which is connected to a segment of a coaxial waveguide where its radius is gradually increased to connect to a segment of coaxial waveguide of constant radius. This segment is then followed by a coaxial/circular waveguide transition, which launches the TM_{01} mode into the circular waveguide. The radiating aperture is formed by cutting the hollow circular waveguide at an oblique angle of 26.18°. The geometry of this antenna is challenging for numerical simulations because the radius of the circular waveguide is approximately 50 times larger than the inner radius of the coaxial feed. The spatial discretization results in a very nonuniform mesh, which in turn yields an ill-conditioned linear system to solve.

The Vlasov antenna of Figure 7.18(a) was simulated using the time-domain finite element method (TDFEM, which is an alternative acronym to FETD) in conjunction with the first-order absorbing boundary condition [11,18]. It was also analyzed using

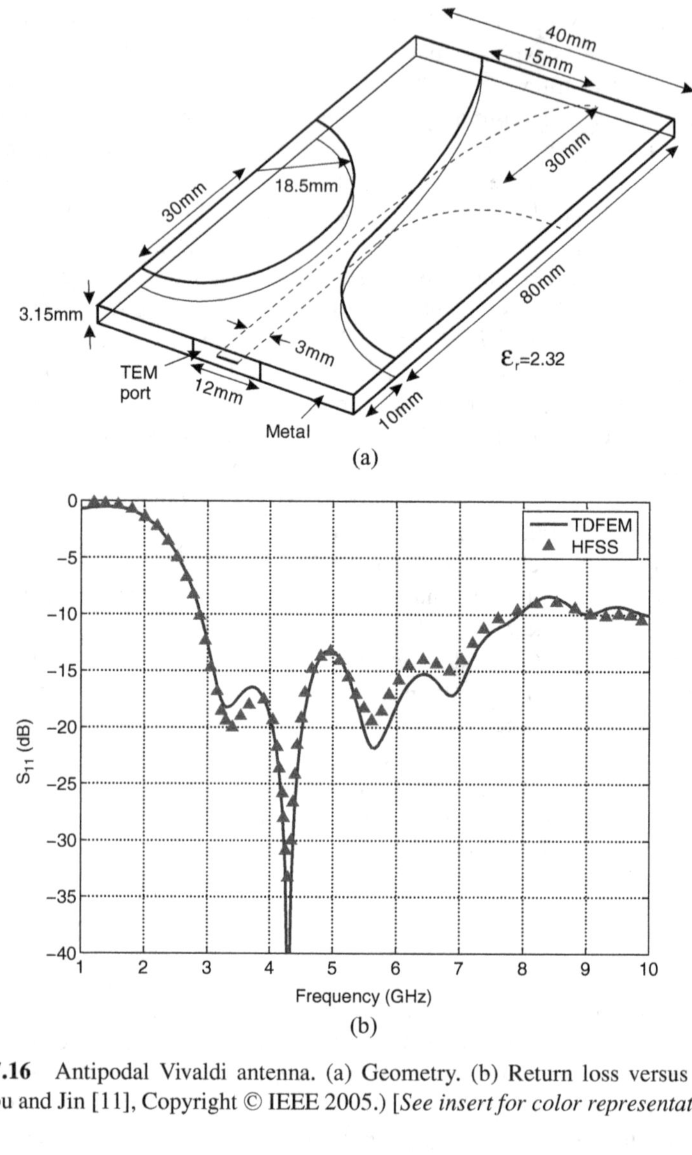

Figure 7.16 Antipodal Vivaldi antenna. (a) Geometry. (b) Return loss versus frequency. (After Lou and Jin [11], Copyright © IEEE 2005.) [*See insert for color representation of* (a).]

the hybrid FETD–FDTD technique described in Chapter 4 to further verify the solution. The total number of unknowns in the TDFEM simulation is approximately 1 million. The return loss computed at the coaxial port is plotted in Figure 7.18(b). The agreement between the two numerical simulations is truly remarkable except at the higher end of the frequency range. It is noted that at a return loss of −20 dB, the power reflection is only about 1%. The antenna gains versus frequency at various far-field angles are displayed in Figure 7.19. Good agreement is observed between the TDFEM predictions and the measured data [19].

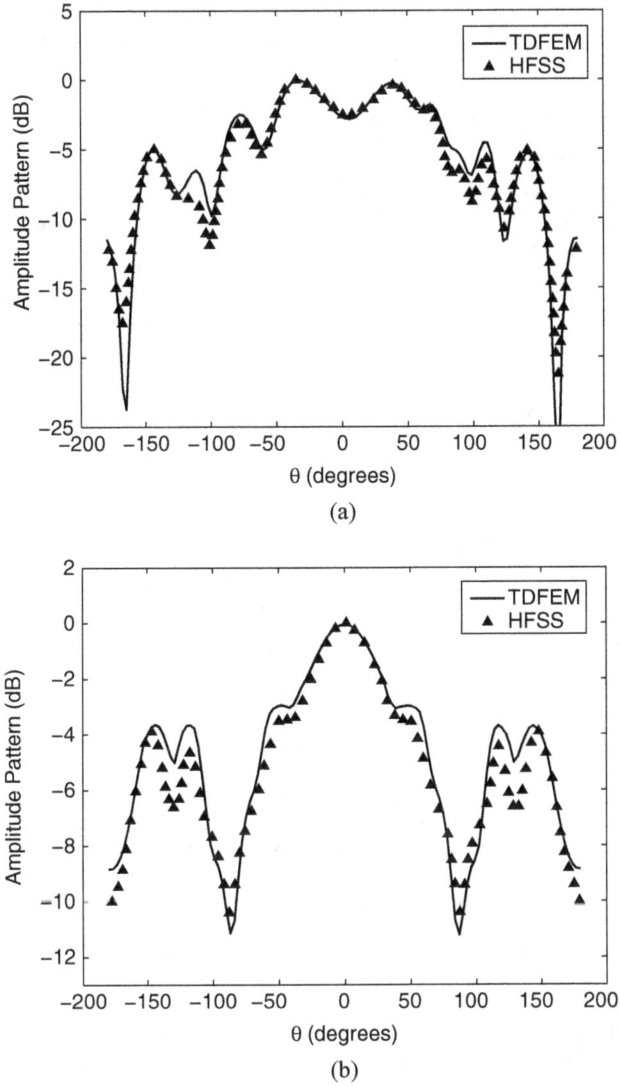

Figure 7.17 Radiation patterns of a Vivaldi antenna at 10 GHz. (a) E-plane cut. (b) H-plane cut. (After Lou and Jin [11], Copyright © IEEE 2005.)

7.3 ANTENNA RCS SIMULATIONS

In addition to the prediction of antenna radiation characteristics, such as the input impedance and radiation patterns, simulation of the antenna scattering characteristics is also important for the design of antennas used in low-observable applications. The finite element method can easily be adapted to perform this important simulation by

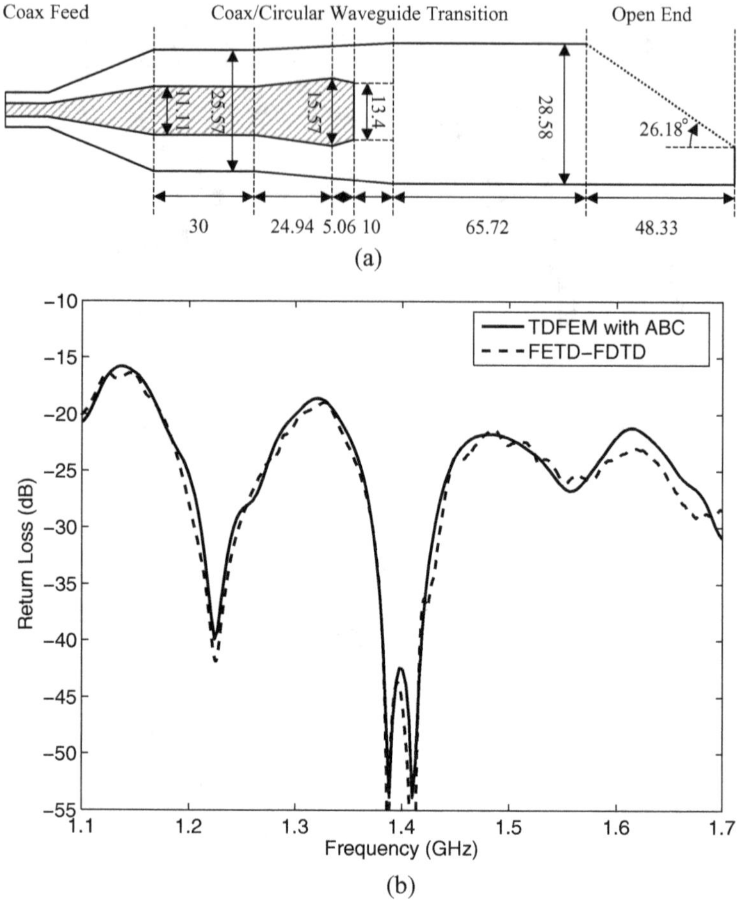

Figure 7.18 (a) Cross-sectional view of a Vlasov antenna. (b) Return loss at the coaxial port for the Vlasov antenna. (After Lou and Jin [18], Copyright © IEEE 2006.)

using the techniques in Section 5.2. Here we present two examples to demonstrate the scattering simulation of two relatively simple antennas.

7.3.1 Microstrip Patch Antenna

The first example is scattering by a microstrip patch antenna consisting of a 3.66-cm × 2.6-cm rectangular conducting patch printed on a finite dielectric substrate of size 7.32 cm × 5.2 cm × 0.158 cm, relative permittivity 2.17, and loss tangent 0.001, as sketched in Figure 7.20(a). The entire antenna is recessed in an infinitely large ground plane such that the patch lies in the same surface of the ground plane. Figure 7.20(b) gives the monostatic radar cross section (RCS) for the antenna as a function of frequency for a plane wave incident at $\phi^{\text{inc}} = 45°$ and $\theta^{\text{inc}} = 60°$. The results are computed by the frequency-domain finite element method using a boundary

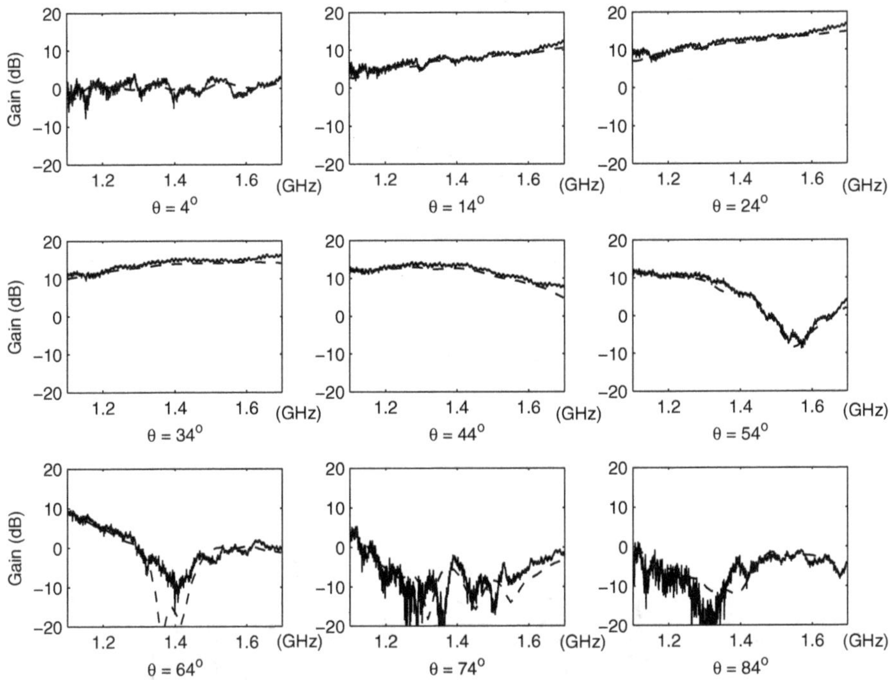

Figure 7.19 *E*-plane antenna gain for a Vlasov antenna versus frequency at various angles. Dashed line: TDFEM prediction. Solid line: measurement [19]. (After Lou and Jin [11], Copyright © IEEE 2005.)

integral equation for mesh truncation [20] and the time-domain finite element method using perfectly matched layers for mesh truncation [21]. The calculations agree with each other very well. These results also agree well with experimental data and the moment-method solutions [22–24] for an infinitely large substrate, because the RCS for this structure is dominated by the contribution of the conducting patch. Without the patch, the RCS of the structure is below −44 dBsm for the entire frequency band of 2 to 8 GHz. It is interesting to observe that the monostatic RCS is characterized by a series of peaks, each corresponding to a resonant frequency of the patch. Numerical simulation can not only verify our intuitive prediction of the sources of scattering, but also quantify the level of the scattering by each source. This, in turn, can be very useful to the antenna designer in developing various schemes to reduce the RCS.

7.3.2 Standard Gain Horn Antenna

The second example is scattering by a standard gain horn with the feed removed and the end of the waveguide terminated by a conductor. The front and side views of the antenna are shown in Figure 7.21, together with the detailed dimensions. The measurement was conducted in the anechoic chamber of Lockheed Martin

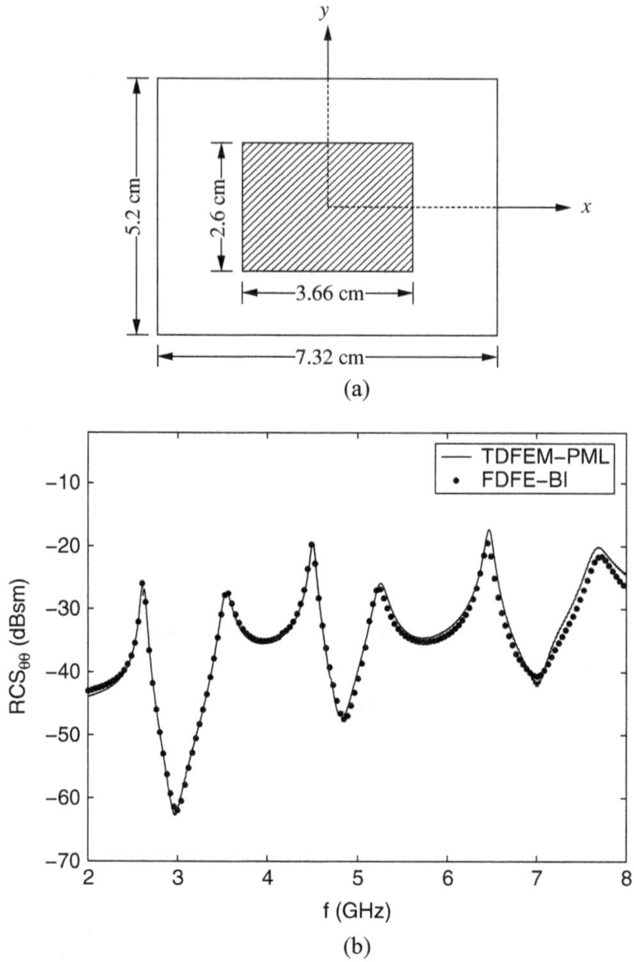

Figure 7.20 Scattering by a rectangular patch printed on the top of a substrate housed in a 0.158-cm-deep cavity. (a) Geometry of the patch and the cavity filled with a substrate of relative permittivity 2.17 and loss tangent 0.001. (b) Monostatic RCS versus frequency for the plane-wave incidence at $\phi^{\text{inc}} = 45°$ and $\theta^{\text{inc}} = 60°$. (After Jiao and Jin [21], Copyright © Wiley 2002.)

Aeronautics Company, Fort Worth, Texas. The monostatic RCS was measured from 2 to 18 GHz and from $-90°$ to $90°$ in the azimuthal xy-plane for both HH and VV polarizations. Simulation results were performed at several frequencies using the frequency-domain finite element method in conjunction with a boundary integral equation for mesh truncation [25]. It is noted that the RCS of this geometry exhibits a variety of physical phenomena, such as that associated with the cutoff of the feeding waveguide [25]. The data calculated and measured at 10 GHz are provided in Figure 7.22, which show excellent agreement. Accurate numerical predictions can

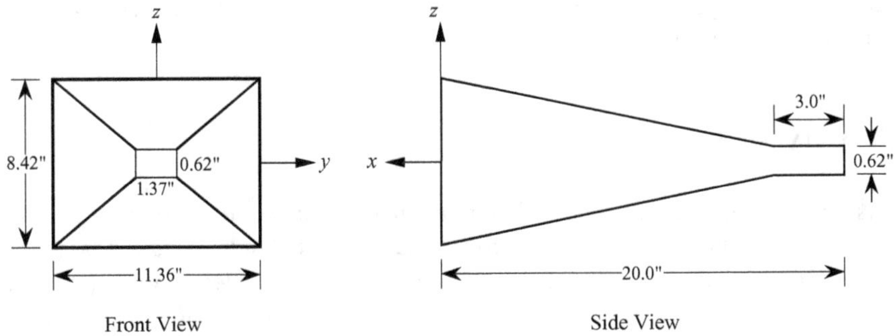

Figure 7.21 Geometry of a standard gain horn antenna.

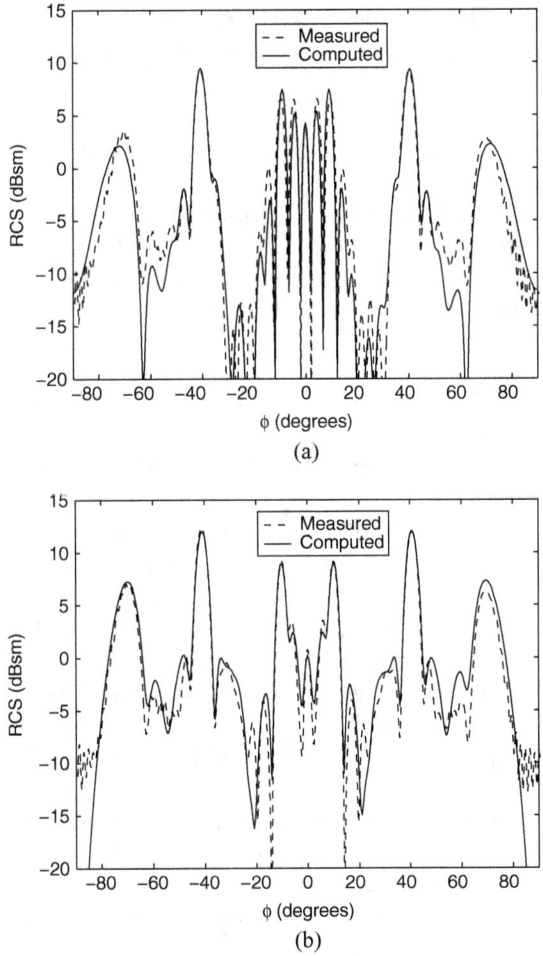

Figure 7.22 Monostatic RCS of a horn antenna at 10 GHz. (a) HH polarization. (b) VV polarization. (After Jin et al. [25], Copyright © IEEE 2003.)

reveal complex physical scattering phenomena that can be used to reduce the cost and timeline of measurements.

7.4 SUMMARY

In this chapter we demonstrated the finite element analysis of a variety of narrow- and broadband antennas. The narrowband antennas included a single coaxial-fed monopole, multiple monopoles mounted on a plate, multiple patch antennas mounted on a plate, and a conformal patch antenna array. The broadband antennas included a ridged horn antenna, a sinuous antenna, the logarithmic spiral antenna, an inverted conical spiral antenna, an antipodal Vivaldi antenna, and the Vlasov antenna. In addition to the radiation case, we also examined scattering by a microstrip patch antenna as well as by a horn antenna. All the examples showed clearly that the finite element method can model and simulate very complex antennas with high accuracy.

REFERENCES

1. G. Liu and C. A. Grimes, "A quasistatic FDTD source model for coaxially driven monopole antennas," *Microwave Opt. Tech. Lett.*, vol. 26, pp. 30–34, July 2000.
2. K. Hill, C. Macon, T. Van, B. Welsh, P. Beyerle, A. MacFarland, and E. Thiele, "EMCC antenna benchmarks: part I," Air Force Research Laboratory and ATK-Mission Research, Ohio, Apr. 2005.
3. K. Mao, J. K. Byun, and J. M. Jin, "Enhancing the modeling capability of the FE-BI method for simulation of cavity-backed antennas and arrays," *Electromagnetics*, vol. 26, no. 7, pp. 503–515, Oct. 2006.
4. K. L. Walton and V. C. Sundberg, "Broadband ridged horn design," *Microwave J.*, pp. 96–101, Mar. 1964.
5. R. H. Duhamel, "Dual polarized sinuous antennas," U.S. patent 4,658,262, Apr. 14, 1987.
6. R. G. Corzine and J. A. Mosko, *Four-Arm Spiral Antennas*. Norwood, MA: Artech House, 1990.
7. N. Montgomery, R. Hutchins, and D. J. Riley, "Thin-wire hybrid FETD/FDTD broadband antenna predictions," *USNC/URSI Natl. Radio Sci. Mtg. Dig.*, p. 194, July 2001.
8. P. E. Mayes, "Frequency-independent antennas and broadband derivatives thereof," *Proc. IEEE*, vol. 80, pp. 103–112, Jan. 1992.
9. J. D. Dyson, "The equiangular spiral antenna," *IRE Trans. Antennas Propagat.*, vol. AP-7, pp. 181–187, Apr. 1959.
10. G. Deschamps, "Impedance properties of complementary multiterminal planar structures," *IRE Trans. Antennas Propagat.*, vol. AP-7, pp. S371–S379, Dec. 1959.
11. Z. Lou and J. M. Jin, "Modeling and simulation of broadband antennas using the time-domain finite element method," *IEEE Trans. Antennas Propagat.*, vol. 53, no. 12, pp. 4099–4110, Dec. 2005.
12. J. D. Dyson, "The unidirectional equiangular spiral antenna," *IRE Trans. Antennas Propagat.*, vol. AP-7, pp. 329–334, Oct. 1959.

13. P. Gibson, "The Vivaldi aerial," *Proc. 9th European Microwave Conf.*, pp. 101–105, Brighton, UK, 1979.
14. N. Fourikis, N. Lioutas, and N. Shuley, "Parametric study of the co- and cross-polarization characteristics of tapered planar and antipodal slotline antennas," *IEE Proc. Microwave Antennas Propagat.*, vol. 140, no. 1, pp. 17–22, Feb. 1993.
15. "The 2000 CAD benchmark unveiled," *Microwave Engineering Online*, July 2001; see http://img.cmpnet.com/edtn/europe/mwee/pdf/CAD.pdf.
16. S. N. Vlasov and I. M. Orlova, "Quasioptical transformer which transforms waveguide having a circular cross section into a highly directional wave beam," *Radiofizika*, vol. 17, pp. 148–154, Jan. 1974.
17. B. G. Ruth, R. K. Dalstrom, C. D. Schlesiger, and L. F. Libelo, "Design and low-power testing of a microwave Vlasov mode converter," *IEEE MTT-S Int. Microwave Symp. Dig.*, vol. 3, pp. 1277–1280, 1989.
18. Z. Lou and J. M. Jin, "A novel dual-field time-domain finite-element domain-decomposition method for computational electromagnetics," *IEEE Trans. Antennas Propagat.*, vol. 54, no. 6, pp. 1850–1862, June 2006.
19. A. Greenwood and K. Hendricks, "Vlasov antenna data for electromagnetic validation data," Report AFRL-DE-TR-2003-1092, Air Force Research Laboratory, June 2003.
20. J. M. Jin and J. L. Volakis, "A hybrid finite element method for scattering and radiation by microstrip patch antennas and arrays residing in a cavity," *IEEE Trans. Antennas Propagat.*, vol. 39, pp. 1598–1604, Nov. 1991.
21. D. Jiao and J. M. Jin, "Time-domain finite element simulation of cavity-backed microstrip patch antennas," *Microwave Opt. Tech. Lett.*, vol. 32, no. 4, pp. 251–254, Feb. 2002.
22. E. H. Newman and D. P. Forrai, "Scattering from a microstrip patch," *IEEE Trans. Antennas Propagat.*, vol. 35, pp. 245–251, Mar. 1987.
23. D. P. Forrai and E. H. Newman, "Radiation and scattering from loaded microstrip antennas over a wide bandwidth," Tech. Rep. 719493-1, Ohio State University, Columbus, OH, Sept. 1988.
24. D. M. Pozar, "Radiation and scattering from a microstrip patch on a uniaxial substrate," *IEEE Trans. Antennas Propagat.*, vol. 35, pp. 613–621, June 1987.
25. J. M. Jin, J. Liu, Z. Lou, and C. S. Liang, "A fully high-order finite element simulation of scattering by deep cavities," *IEEE Trans. Antennas Propagat.*, vol. 51, no. 9, pp. 2420–2429, Sept. 2003.

8 Axisymmetric Antenna Modeling

The term *axisymmetric antenna* refers to a certain class of antennas whose geometry and material property (but not necessarily the feed excitation) exhibit rotational symmetry. Such an object is also often called a *body of revolution* (BOR). Although these antennas can be analyzed using a three-dimensional finite element method, the rotational symmetry present in them allows a much more efficient solution with two-dimensional finite element analysis. Since axisymmetric antennas have wide applications in practice, we devote this chapter to the finite element analysis of this type of antenna.

8.1 METHOD OF ANALYSIS

Because of its rotational symmetry, an axisymmetric antenna can be simulated in a two-dimensional domain corresponding to its angular cross section with the aid of the Fourier expansion. The first finite element analysis of axisymmetric antennas was developed by Mei [1], which employed the finite element method to formulate the field inside a spherical surface that encloses the entire antenna and then the wavefunction expansion to represent the radiated field outside the spherical surface. Another approach to truncating the finite element mesh, proposed more recently [2], made use of cylindrical perfectly matched layers. The most accurate and general BOR method is one that combines the finite element and boundary integral (FE-BI) methods [3,4]. This method has recently seen a significant improvement, with correct modeling of the boundary conditions along the rotational axis in both the finite element and boundary integral formulations [5,6].

8.1.1 Finite Element Formulation

Consider an axisymmetric object whose rotational axis coincides with the z-axis. The relative permittivity and permeability of its material composition are position-dependent diagonal tensors

$$\overleftrightarrow{\varepsilon}_r = \varepsilon_{\rho\rho}(\rho, z)\hat{\rho}\hat{\rho} + \varepsilon_{zz}(\rho, z)\hat{z}\hat{z} + \varepsilon_{\phi\phi}(\rho, z)\hat{\phi}\hat{\phi} \tag{8.1}$$

$$\overleftrightarrow{\mu}_r = \mu_{\rho\rho}(\rho, z)\hat{\rho}\hat{\rho} + \mu_{zz}(\rho, z)\hat{z}\hat{z} + \mu_{\phi\phi}(\rho, z)\hat{\phi}\hat{\phi}. \tag{8.2}$$

Finite Element Analysis of Antennas and Arrays, By Jian-Ming Jin and Douglas J. Riley

We assume further that the object may contain anisotropic impedance and/or resistive surfaces with a normalized surface impedance/resistivity

$$\overleftrightarrow{\eta}_s = \eta_{tt}(\rho, z)\hat{t}\hat{t} + \eta_{\phi\phi}(\rho, z)\hat{\phi}\hat{\phi} \tag{8.3}$$

where $\hat{t}$ denotes the tangential unit vector in the transverse plane. At such a surface, the fields satisfy the standard impedance/resistive boundary condition. The weak-form representation of the vector wave equation for the electric field is given by

$$\begin{aligned}
&\iiint_V \left[(\nabla \times \mathbf{T}) \cdot \overleftrightarrow{\mu}_r^{-1} \cdot (\nabla \times \mathbf{E}) - k_0^2 \mathbf{T} \cdot \overleftrightarrow{\varepsilon}_r \cdot \mathbf{E}\right] dV \\
&\quad - jk_0 \iint_{S_R} (\hat{n} \times \mathbf{T}) \cdot \overleftrightarrow{\eta}_s^{-1} \cdot (\hat{n} \times \mathbf{E})\, dS + jk_0 \iint_{S_o} (\hat{n} \times \mathbf{T}) \cdot (\hat{n} \times \mathbf{E})\, dS \\
&= -\iiint_V \mathbf{T} \cdot \left[jk_0 Z_0 \mathbf{J}_{\text{imp}} + \nabla \times (\overleftrightarrow{\mu}_r^{-1} \cdot \mathbf{M}_{\text{imp}})\right] dV
\end{aligned} \tag{8.4}$$

where S_R denotes the impedance/resistive surfaces and V denotes the computational domain, which is bounded by surface S_o. Furthermore, the first-order absorbing boundary condition (2.7) is applied to S_o as illustrated in Figure 8.1. Because of the

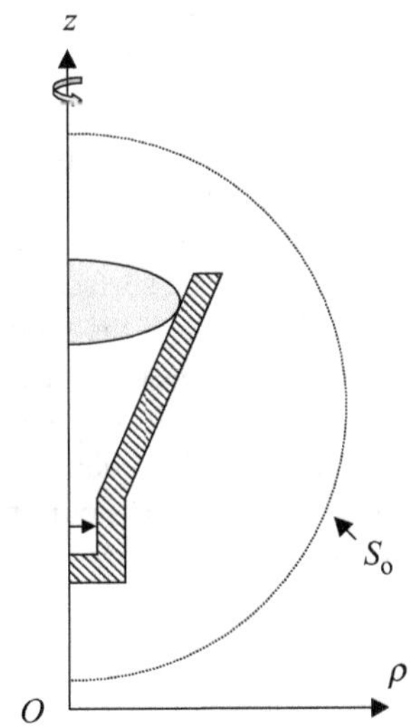

Figure 8.1 Cross-sectional view of an axisymmetric antenna with computational domain truncated by an absorbing boundary condition.

axisymmetry, both the field and source can be represented by the Fourier series

$$\mathbf{E}(\rho, \phi, z) = \sum_{m=-\infty}^{\infty} \mathbf{E}^{(m)}(\rho, z)e^{jm\phi} \tag{8.5}$$

$$\mathbf{J}_{\text{imp}}(\rho, \phi, z) = \sum_{m=-\infty}^{\infty} \mathbf{J}_{\text{imp}}^{(m)}(\rho, z)e^{jm\phi} \tag{8.6}$$

$$\mathbf{M}_{\text{imp}}(\rho, \phi, z) = \sum_{m=-\infty}^{\infty} \mathbf{M}_{\text{imp}}^{(m)}(\rho, z)e^{jm\phi}. \tag{8.7}$$

Substituting (8.5)–(8.7) into (8.4) and choosing the testing function as

$$\mathbf{T}(\rho, \phi, z) = \mathbf{T}^{(m)}(\rho, z)e^{-jm\phi} \tag{8.8}$$

we obtain

$$\begin{aligned}
&\iint_S \left\{\left[\left(\nabla_t - \hat{\phi}\frac{jm}{\rho}\right) \times \mathbf{T}^{(m)}\right] \cdot \overleftrightarrow{\mu}_r^{-1} \cdot \left[\left(\nabla_t + \hat{\phi}\frac{jm}{\rho}\right) \times \mathbf{E}^{(m)}\right] - k_0^2\mathbf{T}^{(m)} \cdot \overleftrightarrow{\varepsilon}_r \cdot \mathbf{E}^{(m)}\right\} dS \\
&\quad - jk_0 \int_{C_R} [\hat{n} \times \mathbf{T}^{(m)}] \cdot \overleftrightarrow{\eta}_s^{-1} \cdot [\hat{n} \times \mathbf{E}^{(m)}]\, dC + jk_0 \int_{C_o} [\hat{n} \times \mathbf{T}^{(m)}] \cdot [\hat{n} \times \mathbf{E}^{(m)}]\, dC \\
&= -\iint_S \mathbf{T}^{(m)} \cdot \left[jk_0Z_0\mathbf{J}_{\text{imp}}^{(m)} + \left(\nabla_t + \hat{\phi}\frac{jm}{\rho}\right) \times \left[\overleftrightarrow{\mu}_r^{-1} \cdot \mathbf{M}_{\text{imp}}^{(m)}\right]\right] dS
\end{aligned} \tag{8.9}$$

where S denotes the cross section of V in the xz-plane, C_R is the line of intersection between S_R and the xz-plane, and C_o is the line of intersection between S_o and the xz-plane. As a result, the volume integrals in (8.4) are reduced to surface integrals over the angular cross section, and accordingly, the surface integrals in (8.4) are reduced to line integrals. Because of the orthogonality of the Fourier modes, the analysis can then be carried out independently for each Fourier mode.

Next, we need to choose proper basis functions to expand $\mathbf{E}^{(m)}$ and proper testing functions for $\mathbf{T}^{(m)}$ to convert (8.9) into a matrix equation. The basis functions have to be able to represent $\mathbf{E}^{(m)}$ accurately. Note that the Fourier components of the electric and magnetic fields at the z-axis satisfy the following conditions:

$$E_\rho^{(0)} = E_\phi^{(0)} = 0 \tag{8.10}$$

$$E_\rho^{(\pm 1)} = \mp jE_\phi^{(\pm 1)}, \qquad E_z^{(\pm 1)} = 0 \tag{8.11}$$

$$E_\rho^{(m)} = E_\phi^{(m)} = E_z^{(m)} = 0 \qquad |m| > 1. \tag{8.12}$$

To satisfy these conditions, the Fourier component of the electric field can be expanded as [7]

$$\mathbf{E}^{(m)} = \begin{cases} \sum_{i=1}^{N_{\text{edge}}} \mathbf{N}_i(\rho, z) e_{t,i}^{(m)} + \hat{\phi} \sum_{i=1}^{N_{\text{node}}} N_i(\rho, z) e_{\phi,i}^{(m)} & m = 0 \\ \sum_{i=1}^{N_{\text{edge}}} \rho \mathbf{N}_i(\rho, z) e_{t,i}^{(m)} + (\hat{\phi} \mp j\hat{\rho}) \sum_{i=1}^{N_{\text{node}}} N_i(\rho, z) e_{\phi,i}^{(m)} & m = \pm 1 \\ \sum_{i=1}^{N_{\text{edge}}} \rho \mathbf{N}_i(\rho, z) e_{t,i}^{(m)} + \hat{\phi} \sum_{i=1}^{N_{\text{node}}} N_i(\rho, z) e_{\phi,i}^{(m)} & |m| > 1 \end{cases} \tag{8.13}$$

where $\mathbf{N}_i(\rho, z)$ denotes the edge-based vector basis function whose unknown expansion coefficient is $e_{t,i}^{(m)}$, and $N_i(\rho, z)$ denotes the nodal basis function whose unknown expansion coefficient is $e_{\phi,i}^{(m)}$. The conditions (8.10) and (8.12) on $E_\phi^{(0)}$ and $E_\phi^{(m)}$ can be imposed either by setting $e_{\phi,i}^{(m)} = 0$ for the nodes on the z-axis or by excluding those nodes in the count of N_{node} for $m = 0$ and $|m| > 1$. By using the same basis functions in (8.13) as the testing function $\mathbf{T}^{(m)}$, (8.9) can be converted into a matrix equation for each Fourier mode. Once $\mathbf{E}^{(m)}$ is solved for, the electric field can be obtained from (8.5).

8.1.2 Mesh Truncation Using Perfectly Matched Layers

The main limitation on the accuracy of the finite element solution described above is the mesh truncation by the first-order absorbing boundary condition. The accuracy of this mesh truncation can be improved significantly by using a layer of perfectly matched absorbing material just inside the truncation boundary where the absorbing boundary condition is applied. In this case we can use either a spherical or a cylindrical perfectly matched layer since both have been developed in the past [8–10]. For most problems, a finite cylindrical perfectly matched layer is preferred because compared with a spherical layer, a cylindrical mesh truncation usually yields a smaller computational domain. For a cylindrical perfectly matched layer (Figure 8.2), the anisotropic permittivity and permeability are given by

$$\overleftrightarrow{\varepsilon} = \varepsilon \overleftrightarrow{\mathbf{\Lambda}}, \qquad \overleftrightarrow{\mu} = \mu \overleftrightarrow{\mathbf{\Lambda}} \tag{8.14}$$

where

$$\overleftrightarrow{\mathbf{\Lambda}} = \begin{bmatrix} \frac{\tilde{\rho}}{\rho} \frac{s_z}{s_\rho} & 0 & 0 \\ 0 & \frac{\rho}{\tilde{\rho}} s_\rho s_z & 0 \\ 0 & 0 & \frac{\tilde{\rho}}{\rho} \frac{s_\rho}{s_z} \end{bmatrix}. \tag{8.15}$$

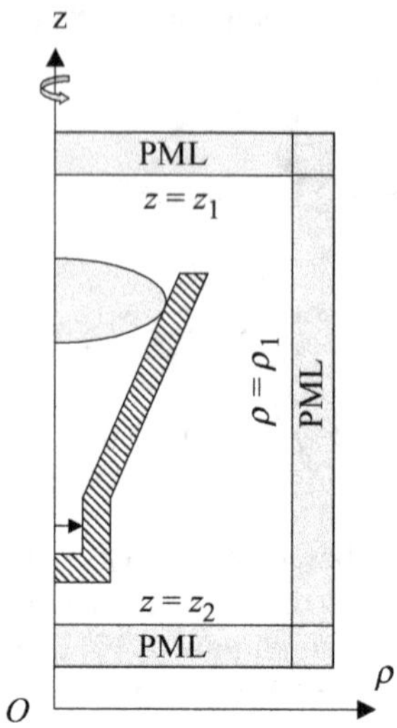

Figure 8.2 Cross-sectional view of an axisymmetric antenna with computational domain truncated by perfectly matched layers.

In (8.15), $\tilde{\rho}$ is defined as

$$\tilde{\rho} = \int_0^{\rho} s_\rho(\rho')\, d\rho' \tag{8.16}$$

where $s_\rho(\rho)$ is a parameter characterizing the property of the perfectly matched layer. For a pth-order profile, $s_\rho(\rho)$ is expressed as

$$s_\rho = \begin{cases} 1 & 0 \le \rho \le \rho_1 \\ 1 + \dfrac{\sigma_{\max}}{j\omega\varepsilon_0}\left(\dfrac{\rho - \rho_1}{L}\right)^p & \rho > \rho_1 \end{cases} \tag{8.17}$$

where ρ_1 denotes the interface between the air and matched layer in the ρ-direction and L denotes the thickness of the matched layer. Accordingly, $\tilde{\rho}$ is given by

$$\tilde{\rho} = \begin{cases} \rho & 0 \le \rho \le \rho_1 \\ \rho + \dfrac{\sigma_{\max}}{j\omega\varepsilon_0}\dfrac{L}{p+1}\left(\dfrac{\rho - \rho_1}{L}\right)^{p+1} & \rho > \rho_1. \end{cases} \tag{8.18}$$

For the top and bottom perfectly matched layers that truncate the computational domain in the z-direction, the corresponding $s_z(z)$ is given by

$$s_z = \begin{cases} 1 + \dfrac{\sigma_{\max}}{j\omega\varepsilon_0}\left(\dfrac{z - z_1}{L}\right)^p & z > z_1 \\ 1 & z_1 \geq z \geq z_2 \\ 1 + \dfrac{\sigma_{\max}}{j\omega\varepsilon_0}\left(\dfrac{z_2 - z}{L}\right)^p & z < z_2 \end{cases} \tag{8.19}$$

where z_1 denotes the interface between the air and the upper matched layer and z_2 denotes the interface between the air and the lower matched layer. As for any other perfectly matched layers, for a fixed layer thickness, an increasing value of $\sigma_{\max}/\omega\varepsilon_0$ increases the absorption, thus reducing reflection of the matched layer, but it also increases the spurious reflection due to the spatial discretization. A smaller value of $\sigma_{\max}/\omega\varepsilon_0$ reduces the spurious reflection, but it also reduces the absorption of the layer. Therefore, there is a trade-off in the selection of the value for $\sigma_{\max}/\omega\varepsilon_0$. Numerical experiments found that for a mesh density of 20 first-order elements per wavelength and a thickness $L = 0.25\lambda$, the optimal value for $\sigma_{\max}/\omega\varepsilon_0$ is between 5.5 to 6.0 for $p = 2$ [11].

8.1.3 Mesh Truncation Using Boundary Integral Equations

As discussed in Chapter 3, mesh truncation using perfectly matched layers is still approximate because of the fact that perfectly matched layers have a finite thickness and have to be terminated by an impenetrable surface. To obtain highly accurate finite element solutions of open-region scattering and radiation, we have to use more accurate mesh truncation techniques such as the one based on boundary integral equations (Figure 8.3). In this case, the weak-form representation of the vector wave equation for the electric field is given by

$$\begin{aligned} &\iiint_V \left[(\nabla \times \mathbf{T}) \cdot \overleftrightarrow{\mu}_r^{-1} \cdot (\nabla \times \mathbf{E}) - k_0^2 \mathbf{T} \cdot \overleftrightarrow{\varepsilon}_r \cdot \mathbf{E}\right] dV \\ &\quad - jk_0 \iint_{S_R} (\hat{n} \times \mathbf{T}) \cdot \overleftrightarrow{\eta}_s^{-1} \cdot (\hat{n} \times \mathbf{E})\, dS - jk_0 Z_0 \iint_{S_o} \mathbf{T} \cdot (\hat{n} \times \mathbf{H})\, dS \\ &= -\iiint_V \mathbf{T} \cdot \left[jk_0 Z_0 \mathbf{J}_{\text{imp}} + \nabla \times \left(\overleftrightarrow{\mu}_r^{-1} \cdot \mathbf{M}_{\text{imp}}\right)\right] dV \end{aligned} \tag{8.20}$$

which contains the unknown tangential magnetic field on the truncation surface. Similar to the electric field, the magnetic field can be expanded into a Fourier series as

$$\mathbf{H}(\rho, \phi, z) = \sum_{m=-\infty}^{\infty} \mathbf{H}^{(m)}(\rho, z) e^{jm\phi} \tag{8.21}$$

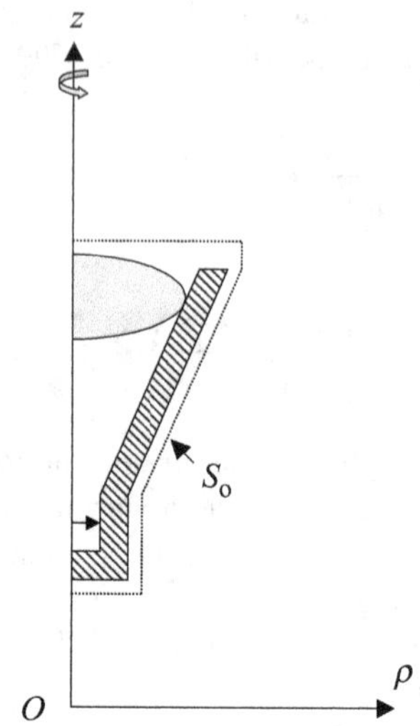

Figure 8.3 Cross-sectional view of an axisymmetric antenna with computational domain truncated by a boundary integral equation.

which, together with (8.5)–(8.8), reduces (8.20) to the two-dimensional expression

$$\iint_S \left\{ \left[\left(\nabla_t - \hat{\phi}\frac{jm}{\rho} \right) \times \mathbf{T}^{(m)} \right] \cdot \overleftrightarrow{\mu}_r^{-1} \cdot \left[\left(\nabla_t + \hat{\phi}\frac{jm}{\rho} \right) \times \mathbf{E}^{(m)} \right] - k_0^2 \mathbf{T}^{(m)} \cdot \overleftrightarrow{\varepsilon}_r \cdot \mathbf{E}^{(m)} \right\} dS$$

$$- jk_0 \int_{C_R} [\hat{n} \times \mathbf{T}^{(m)}] \cdot \overleftrightarrow{\eta}_s^{-1} \cdot [\hat{n} \times \mathbf{E}^{(m)}]\, dC - jk_0 Z_0 \int_{C_o} \mathbf{T}^{(m)} \cdot [\hat{n} \times \mathbf{H}^{(m)}]\, dC$$

$$= -\iint_S \mathbf{T}^{(m)} \cdot \left[jk_0 Z_0 \mathbf{J}_{\text{imp}}^{(m)} + \left(\nabla_t + \hat{\phi}\frac{jm}{\rho} \right) \times \left[\overleftrightarrow{\mu}_r^{-1} \cdot \mathbf{M}_{\text{imp}}^{(m)} \right] \right] dS. \qquad (8.22)$$

The Fourier components of the magnetic field have to satisfy the same conditions as (8.10)–(8.12), or more explicitly,

$$H_\rho^{(0)} = H_\phi^{(0)} = 0 \qquad (8.23)$$

$$H_\rho^{(\pm 1)} = \mp j H_\phi^{(\pm 1)}, \qquad H_z^{(\pm 1)} = 0 \qquad (8.24)$$

$$H_\rho^{(m)} = H_\phi^{(m)} = H_z^{(m)} = 0 \qquad |m| > 1. \qquad (8.25)$$

To satisfy these conditions and couple correctly to the boundary integral equations, we can expand the transverse tangential magnetic field along each segment of the exterior boundary using the zeroth-order basis function; that is, the transverse tangential magnetic field is approximated as a constant along each segment. Mathematically, this is expressed as

$$H_{t,s}^{(m)} = \frac{h_{t,s}^{(m)}}{\rho_s} \tag{8.26}$$

where the subscript s denotes the segment number, ρ_s is the cylindrical radius of the segment taken at its midpoint, and $h_{t,s}^{(m)}$ is the unknown coefficient on the segment. Furthermore, we expand the angular component of the magnetic field along the exterior boundary using the modified linear basis functions

$$H_{\phi,i}^{(m)} = \begin{cases} \dfrac{N_i(t)}{\rho} h_{\phi,i}^{(m)} & |m| = 1 \\ \dfrac{N_i(t)}{\rho_i} h_{\phi,i}^{(m)} & |m| \neq 1 \end{cases} \tag{8.27}$$

where i denotes a node on the exterior boundary (except for the two nodes on the axis), $N_i(t)$ is a linear (rooftop) basis function that has a value of 1 at node i and decreases linearly to zero at the adjacent nodes, ρ_i is the cylindrical radius of node i, and $h_{\phi,i}^{(m)}$ is the unknown expansion coefficient corresponding to node i. Substituting (8.13), (8.26), and (8.27) into (8.22), for each Fourier mode, we obtain the matrix equation

$$\left[A^{(m)}\right]\left\{e^{(m)}\right\} + \left[B^{(m)}\right]\left\{h_S^{(m)}\right\} = \left\{b_{\text{imp}}^{(m)}\right\} \tag{8.28}$$

where $[A^{(m)}]$ is an $N_T \times N_T$ sparse and symmetric matrix, $\{e^{(m)}\}$ is an $N_T \times 1$ vector that stores all the unknown coefficients $e_{t,i}^{(m)}$ and $e_{\phi,i}^{(m)}$, $[B^{(m)}]$ is an $N_T \times N_S$ sparse rectangular matrix, and $\{h_S^{(m)}\}$ is an $N_S \times 1$ vector that stores all the unknown coefficients $h_{t,s}^{(m)}$ and $h_{\phi,i}^{(m)}$. Here N_T is the sum of the total numbers of the edges and nodes in the entire finite element mesh and N_S is the sum of the numbers of the edges and nodes on the exterior boundary. Since $[A^{(m)}]$ is a highly sparse and symmetric matrix, it can be solved very efficiently using a robust sparse solver. As a result, (8.28) can be reduced to

$$\left\{e_S^{(m)}\right\} + \left[C^{(m)}\right]\left\{h_S^{(m)}\right\} = \left\{\tilde{b}_{\text{imp}}^{(m)}\right\} \tag{8.29}$$

where $[C^{(m)}]$ is now an $N_S \times N_S$ full matrix, which relates the surface electric and magnetic fields, and $\{e_S^{(m)}\}$ is an $N_S \times 1$ vector that stores all the unknown coefficients $e_{t,s}^{(m)}$ and $e_{\phi,i}^{(m)}$ associated with the electric field on the exterior boundary. The $e_{t,s}^{(m)}$ and

$e_{\phi,i}^{(m)}$ are defined similarly to (8.26) and (8.27), that is,

$$E_{t,s}^{(m)} = \frac{e_{t,s}^{(m)}}{\rho_s} \tag{8.30}$$

$$E_{\phi,i}^{(m)} = \begin{cases} \dfrac{N_i(t)}{\rho} e_{\phi,i}^{(m)} & |m| = 1 \\ \dfrac{N_i(t)}{\rho_i} e_{\phi,i}^{(m)} & |m| \neq 1. \end{cases} \tag{8.31}$$

Note that $[C^{(m)}]$ contains all the information about the electromagnetic properties of the object.

Next, we consider the formulation of the fields outside the object. It was shown in Chapter 3 that the field has to satisfy the combined field integral equation (CFIE)

$$\begin{aligned} &\tfrac{1}{2}\bar{\mathbf{J}}_S + \hat{n} \times \tilde{\mathcal{K}}(\bar{\mathbf{J}}_S) + \hat{n} \times \mathcal{L}(\mathbf{M}_S) + \hat{n} \times \left[\tfrac{1}{2}\mathbf{M}_S + \hat{n} \times \tilde{\mathcal{K}}(\mathbf{M}_S) - \hat{n} \times \mathcal{L}(\bar{\mathbf{J}}_S)\right] \\ &= Z_0 \hat{n} \times \mathbf{H}^{\text{inc}} - \hat{n} \times (\hat{n} \times \mathbf{E}^{\text{inc}}) \qquad \text{on } S_o \end{aligned} \tag{8.32}$$

where

$$\mathbf{M}_S = \mathbf{E} \times \hat{n}, \qquad \bar{\mathbf{J}}_S = Z_0 \hat{n} \times \mathbf{H} \tag{8.33}$$

and the operators $\mathcal{L}$ and $\tilde{\mathcal{K}}$ are the same as defined in (3.109) and (3.110) with (3.114). In (8.32), $\mathbf{E}^{\text{inc}}$ and $\mathbf{H}^{\text{inc}}$ denote the incident electric and magnetic fields, which are absent in a radiation case. By using the same expansion as (8.26) and (8.27) for the magnetic field and (8.30) and (8.31) for the electric field, we can employ Galerkin's method to discretize (8.32) into the matrix equation

$$\left[P^{(m)}\right]\left\{e_S^{(m)}\right\} + \left[Q^{(m)}\right]\left\{h_S^{(m)}\right\} = \left\{b_{\text{inc}}^{(m)}\right\} \tag{8.34}$$

where both $[P^{(m)}]$ and $[Q^{(m)}]$ are $N_S \times N_S$ full matrices and $\{b_{\text{inc}}^{(m)}\}$ is an $N_S \times 1$ known vector that can be calculated from the incident field. The computation of (8.34) is similar to the moment-method computation for a body of revolution [12, 13]. The angular integration can be performed using a Gaussian quadrature and the integration along each segment can be performed analytically for the singular part and numerically for the nonsingular part. Equation (8.34) can then be solved together with (8.29) for the surface electric and magnetic fields $\{e_S^{(m)}\}$ and $\{h_S^{(m)}\}$.

8.1.4 Far-Field Computation

Once the surface electric and magnetic near-fields are calculated, we can evaluate the far fields by integrating them on a closed surface, as discussed in Section 5.3. For the axisymmetric case, the angular integral can be evaluated analytically; consequently, the far field can be calculated by evaluating a line integral [11]. The θ- and ϕ-components of the far field are given by

$$E_\theta(r,\theta,\phi) = \frac{k_0 e^{-jk_0 r}}{2r} \sum_{m=-\infty}^{\infty} j^m e^{jm\phi} \int_{C'} \rho' e^{jk_0 z'\cos\theta} \Bigg\{ j\bar{\mathbf{J}}_S^{(m)} \cdot \Bigg[\hat{\rho}\cos\theta\, j J'_m(k_0\rho'\sin\theta) + \hat{\phi}\cos\theta \frac{m J_m(k_0\rho'\sin\theta)}{k_0\rho'\sin\theta} + \hat{z}\sin\theta J_m(k_0\rho'\sin\theta)\Bigg] - \mathbf{M}_S^{(m)} \cdot \Bigg[\hat{\rho}\frac{jm J_m(k_0\rho'\sin\theta)}{k_0\rho'\sin\theta} + \hat{\phi} J'_m(k_0\rho'\sin\theta)\Bigg]\Bigg\} dl' \quad (8.35)$$

$$E_\phi(r,\theta,\phi) = \frac{k_0 e^{-jk_0 r}}{2r} \sum_{m=-\infty}^{\infty} j^m e^{jm\phi} \int_{C'} \rho' e^{jk_0 z'\cos\theta} \Bigg\{ j\bar{\mathbf{J}}_S^{(m)} \cdot \Bigg[-\hat{\rho}\frac{m J_m(k_0\rho'\sin\theta)}{k_0\rho'\sin\theta} + \hat{\phi} j J'_m(k_0\rho'\sin\theta)\Bigg] + \mathbf{M}_S^{(m)} \cdot \Bigg[\hat{\rho}\cos\theta J'_m(k_0\rho'\sin\theta) - \hat{\phi}\cos\theta\frac{jm J_m(k_0\rho'\sin\theta)}{k_0\rho'\sin\theta} - \hat{z}\sin\theta\, j J_m(k_0\rho'\sin\theta)\Bigg]\Bigg\} dl' \quad (8.36)$$

where C' denotes the integration contour, $J_m(x)$ denotes the Bessel function of order m, and $J'_m(x)$ denotes the derivative of $J_m(x)$ with respect to its argument. In the following we give a few examples that have been analyzed using the numerical method described in this section.

8.2 APPLICATION EXAMPLES

In this section we consider three axisymmetric antenna configurations to demonstrate the application of the finite element analysis described in the preceding section. These three configurations are a Luneburg lens, a corrugated horn antenna, and a current loop sandwiched between a conducting sphere and a dielectric spherical shell.

8.2.1 Luneburg Lens

The Luneburg lens is a dielectric sphere with the relative permittivity given by

$$\varepsilon_r(r) = 2 - \left(\frac{r}{a}\right)^2 \qquad 0 \le r \le a \quad (8.37)$$

where a denotes the radius of the sphere. As mentioned in Section 5.4, the Luneburg lens offers aberration-free scanning at all aspects and over an extremely wide bandwidth at microwave frequencies. The lens has the property that energy propagated from a surface feed point through the lens is collimated into parallel rays emerging from the other side. Numerous studies and investigations were conducted in the

1950s with the goal of using the spherical Luneburg lens as a wideband, large-sector multiple-beam radar antenna. However, because of the advent of phased-array technology, only the geodesic analog of the Luneburg lens found applications in several limited-scan military radar systems. One area of recent interest relates to potential applications involving low-cost, wideband, multibeam airborne communications antennas over a ground plane.

The Luneburg lens is commonly analyzed using Fermat's principle from geometrical optics. However, this asymptotic analysis cannot accurately describe the wave propagation through the lens. In contrast, the finite element method can compute the field distribution inside and outside the lens and provide a much better understanding of the unique properties and performance characteristics of the lens [14]. As an example, the radiation patterns of a Hertzian dipole placed on the surface of both $7\lambda_0$- and $10\lambda_0$-diameter Luneburg lenses are shown in Figure 8.4. Clearly, the radiation pattern becomes more directive as the size of the lens increases, and the sidelobe level is at -15 dB. However, this directivity is lower than that predicted with geometrical

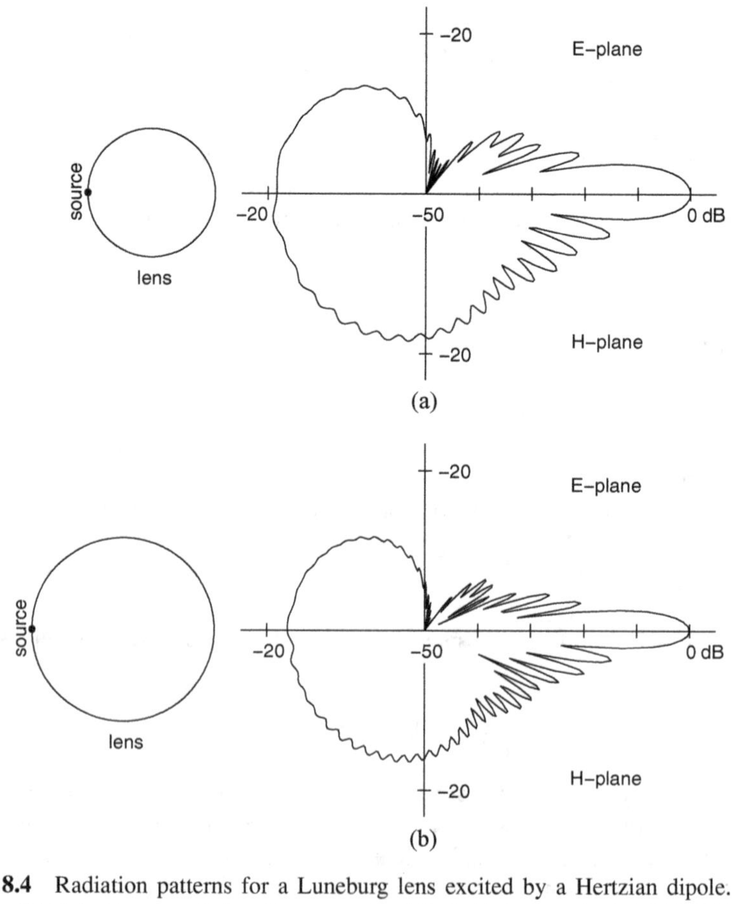

Figure 8.4 Radiation patterns for a Luneburg lens excited by a Hertzian dipole. (a) $7\lambda_0$ diameter. (b) $10\lambda_0$ diameter. (After Greenwood and Jin [14], Copyright © IEEE 1999.)

optics, and it was found recently that this is caused by the nonstationary oscillating focal point of the Luneburg lens [14], as illustrated in Figure 5.19.

Although the Luneburg lens has not found widespread applications as an antenna, Luneburg lens reflectors, which provide passive wideband and wide-angle radar signature augmentation, have found steady applications in small military target drones, decoys, and marine vessels. In this case the lens collects the energy that impinges on a hemispherical lens surface, refracts it through the lens, and focuses it onto the center of the opposite hemispherical surface. If the energy is reflected at the focal point, it will be reradiated back in the original direction. The Luneburg lens becomes an extremely efficient retroreflector with the simple addition of a spherical metallic cap and provides coverage over a solid angle roughly equal to that subtended by the cap. Figure 8.5 shows the geometrical dimensions and the respective indices of refraction of the Rozendal RA-2850 monostatic Luneburg lens reflector, with a three-layer dielectric lens, including a 60° half-cone angle spherical reflector cap. The HH- or VV-polarized monostatic RCS of this reflector at 10 GHz was computed by using the finite element method together with the boundary integral equation for mesh truncation. The result is displayed in Figure 8.6 and compared with the data measured [15]. It is seen that the peak amplitude achieved by the RA-2850 with a standard 120° spherical reflector cap spans over an extremely wide angular sector of nearly 120°, defined at the −3-dB points. In addition, the computed results are seen to be in very good agreement with the test data.

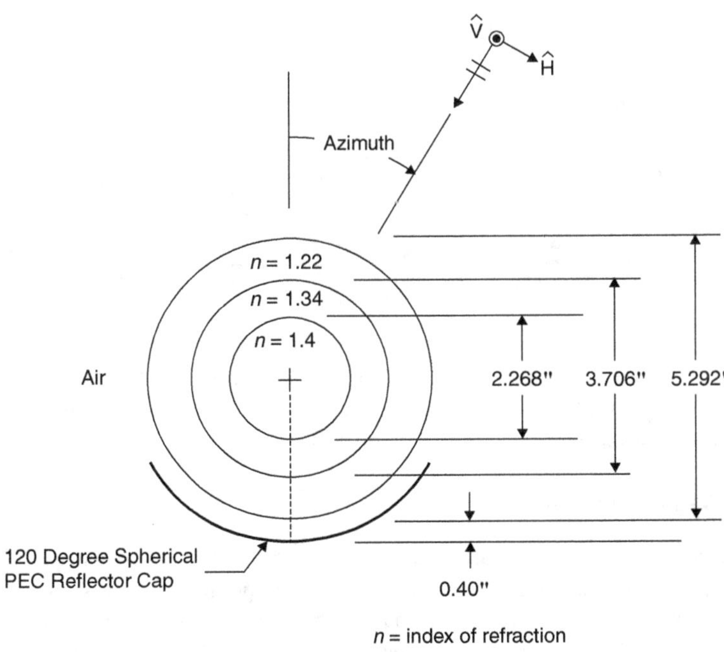

Figure 8.5 Three-layer stepped-index RA-2850 Luneburg lens reflector with a fourth virtual layer of air.

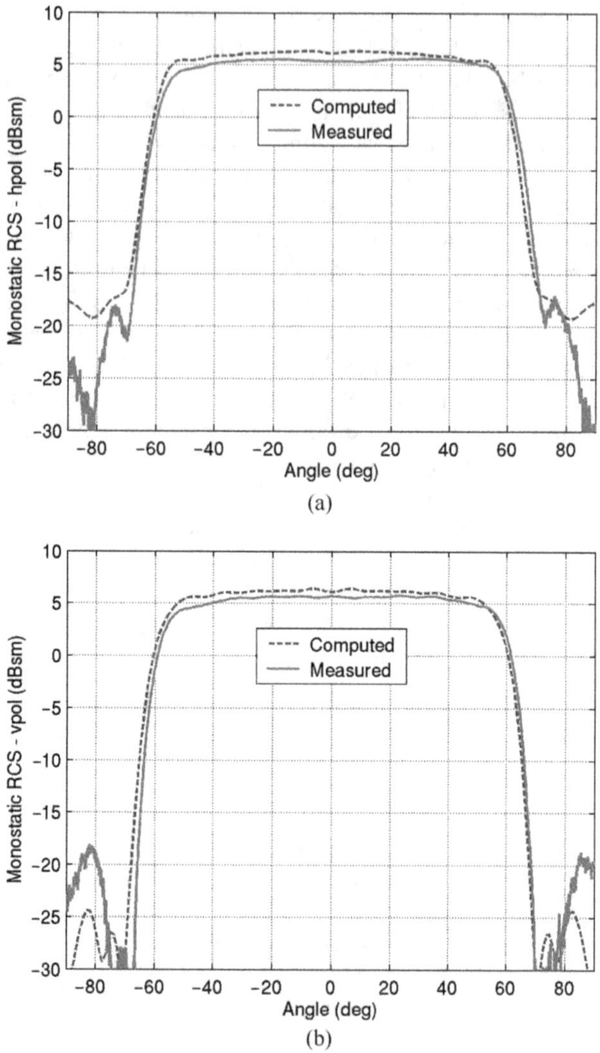

Figure 8.6 Comparison between the computed and measured RCS for a three-layer Luneburg lens with a 120° spherical PEC reflector cap. (a) HH polarization. (b) VV polarization. (After Liang et al. [15], Copyright © IEEE 2005.)

8.2.2 Corrugated Horn

Circular horn antennas with corrugated boundaries have been investigated extensively [16] as antennas that can radiate circularly polarized waves over a wide beamwidth and wide bandwidth. As part of the investigation, measurements on several corrugated horn antennas have been carried out. A diagram showing one such antenna is given in Figure 8.7(a). The antenna is constructed by bolting together metal washers with thicknesses alternating between 0.07938 and 0.3175 cm. In the

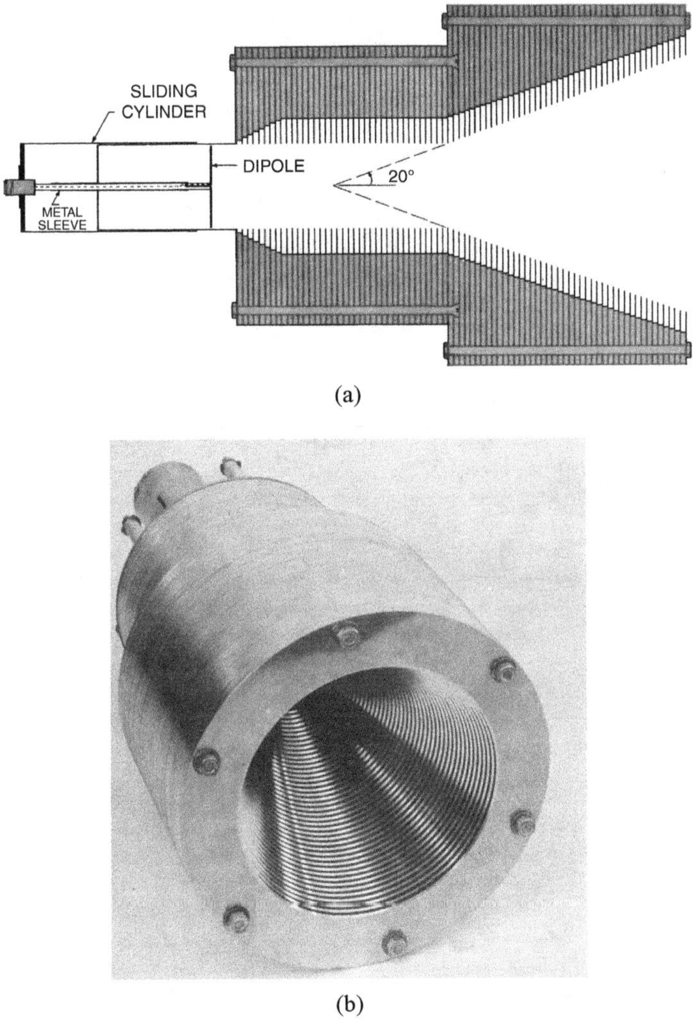

Figure 8.7 Corrugated horn antenna. (a) Cross-sectional diagram. (b) Photo.

section of the waveguide in front of the 20° flare, the washers have inner radii of 3 and 4.8 cm, respectively. A photograph of the antenna is shown in Figure 8.7(b). Measurements of the radiation pattern of this antenna are given in Ref. 16 at 4.5, 5.0, 5.2, 5.5, 6.0, and 6.5 GHz. The radiation pattern was computed by the finite element method using cylindrical perfectly matched layers at these same frequencies [2], and some of the simulation and measured results are shown in Figures 8.8 to 8.10. The agreement between the simulation results and the measured results is excellent. At power levels above −30 dB, note the agreement between the finite element simulation and measurement in predicting slight offsets in the E- and H-plane patterns

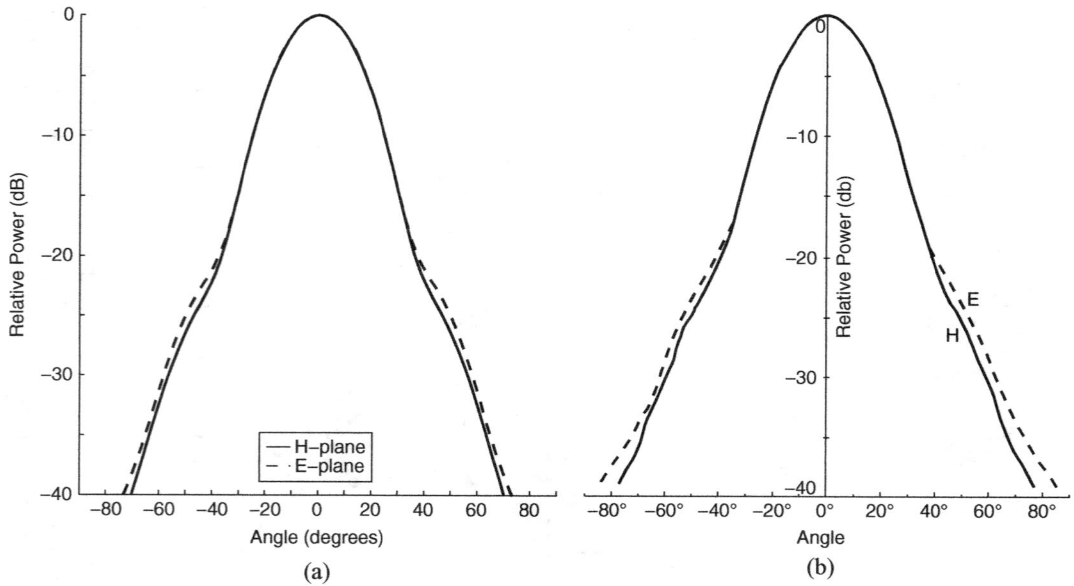

Figure 8.8 Radiation pattern of a corrugated horn antenna at 4.5 GHz: (a) Calculation. (b) Measured results. (After Greenwood and Jin [2], Copyright © IEEE 1999.)

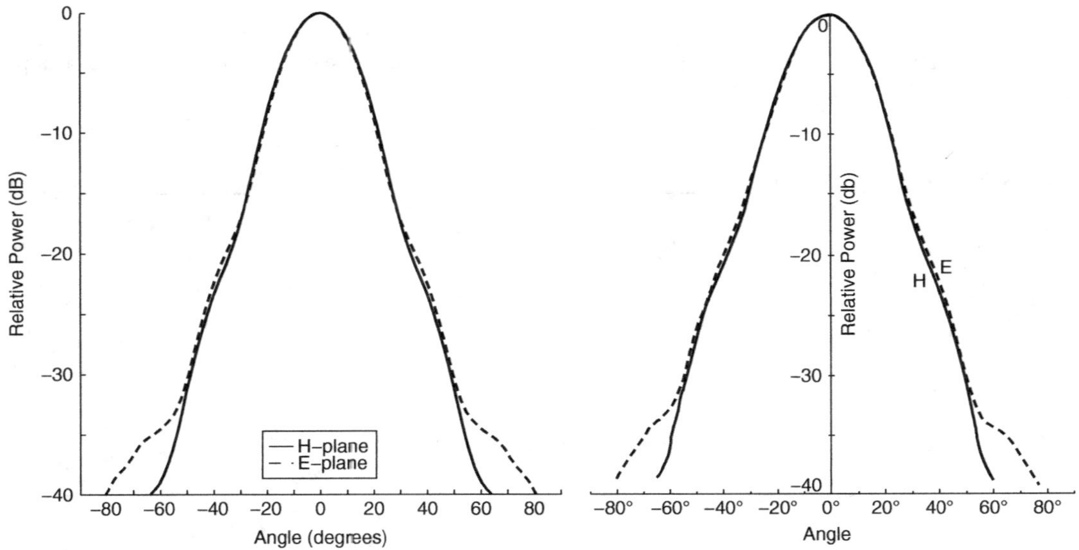

(a)

(b)

Figure 8.9 Radiation pattern of a corrugated horn antenna at 5.2 GHz: (a) Calculation. (b) Measured results. (After Greenwood and Jin [2], Copyright © IEEE 1999.)

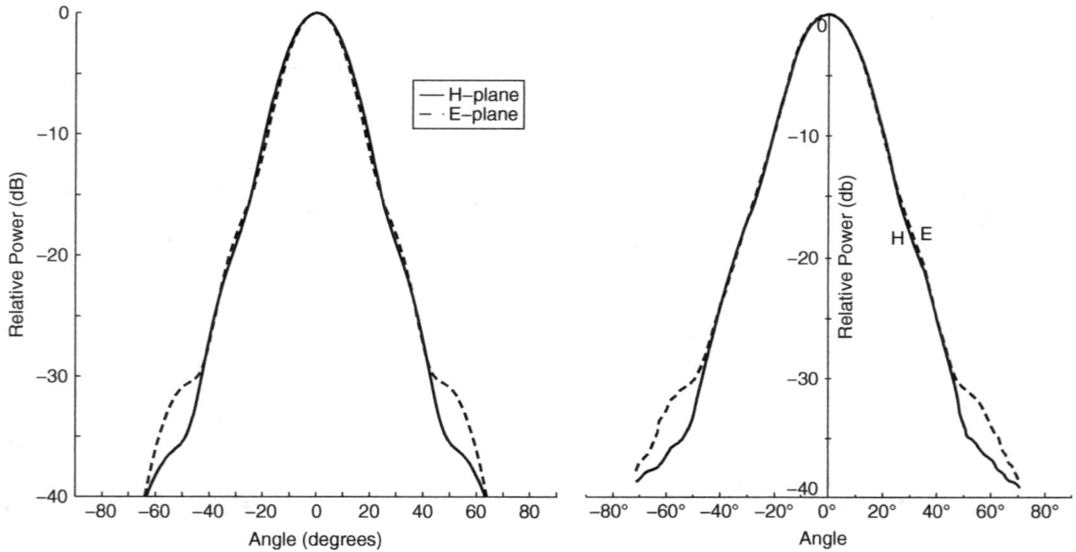

(a)

(b)

Figure 8.10 Radiation pattern of a corrugated horn antenna at 6.0 GHz: (a) Calculation. (b) Measured results. (After Greenwood and Jin [2], Copyright © IEEE 1999.)

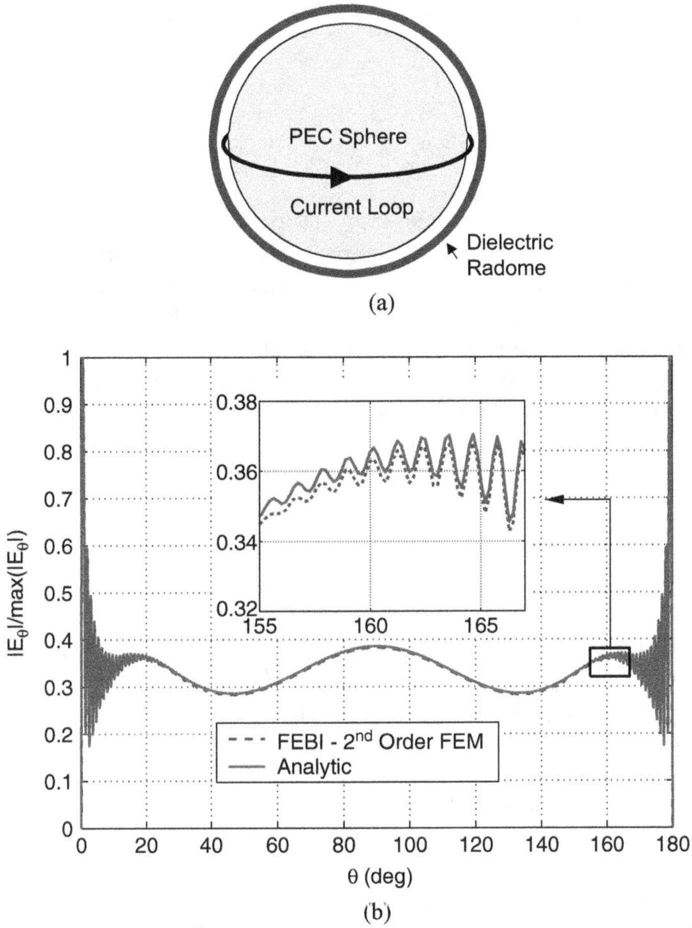

Figure 8.11 Radiation of a current loop in the presence of a conducting sphere and dielectric radome. (a) Geometry. (b) Far-field radiation pattern. (After Dunn et al. [6], Copyright © IEEE 2006.)

with respect to each other. The similar E- and H-plane patterns indicate that when properly fed, this antenna can be used to radiate circularly polarized waves.

8.2.3 Current Loop Inside a Radome

The final example consists of a perfectly conducting sphere of radius $25\lambda_0$ surrounded by a uniform current loop of radius $25.2\lambda_0$ and a spherical radome shell of inner radius $25.4\lambda_0$, outer radius $25.9\lambda_0$, and relative permittivity 2. This example was simulated using the finite element method with a boundary integral equation [6]; the result is shown in Figure 8.11 and is compared with the analytical solution presented by Partal et al. [17]. Because of the large electrical size of the problem, the radiated field oscillates rapidly especially near the z-axis; yet the finite element solution captures these oscillations very accurately.

8.3 SUMMARY

In this chapter we described an efficient approach to analyzing axisymmetric antennas. By adopting the Fourier expansion, the approach reduces a three-dimensional analysis problem to a two-dimensional problem. We started with a detailed finite element formulation using the first-order absorbing boundary condition and then discussed more accurate mesh truncation using perfectly matched layers and boundary integral equations. We presented three examples to illustrate use of the finite element analysis of axisymmetric antennas. Because axisymmetric antennas are very common in practical applications, numerical formulations that exploit this rotational symmetry lead to powerful and efficient simulation tools.

REFERENCES

1. K. K. Mei, "Unimoment method of solving antenna and scattering problems," *IEEE Trans. Antennas Propagat.*, vol. 22, pp. 760–766, Nov. 1974.
2. A. D. Greenwood and J. M. Jin, "Finite element analysis of complex axisymmetric radiating structures," *IEEE Trans. Antennas Propagat.*, vol. 47, no. 8, pp. 1260–1266, Aug. 1999.
3. G. C. Chinn, L. W. Epp, and D. J. Hoppe, "A hybrid finite-element method for axisymmetric waveguide-fed horns," *IEEE Trans. Antennas Propagat.*, vol. 44, no. 3, pp. 280–285, Mar. 1996.
4. C. Zuffada, T. Cwik, and V. Jamnejad, "Modeling radiation with an efficient hybrid finite-element integral-equation waveguide mode-matching technique," *IEEE Trans. Antennas Propagat.*, vol. 45, no. 1, pp. 34–39, Jan. 1997.
5. J. M. Jin, "A highly robust and versatile finite element–boundary integral hybrid code for scattering by BOR objects," *IEEE Trans. Antennas Propagat.*, vol. 53, no. 7, pp. 2274–2281, July 2005.
6. E. Dunn, J. K. Byun, E. Branch, and J. M. Jin, "Numerical simulation of BOR scattering and radiation using a higher-order FEM," *IEEE Trans. Antennas Propagat.*, vol. 54, no. 3, pp. 945–952, Mar. 2006.
7. M. F. Wong, M. Park, and V. Fouad-Hanna, "Axisymmetric edge-based finite element formulation for bodies of revolution: application to dielectric resonators," *IEEE MTT-S Microwave Symp. Dig.*, pp. 285–288, 1995.
8. W. C. Chew, J. M. Jin, and E. Michielssen, "Complex coordinate stretching as a generalized absorbing boundary condition," *Microwave Opt. Tech. Lett.*, vol. 15, no. 6, pp. 363–369, Aug. 1997.
9. F. L. Teixeira and W. C. Chew, "Systematic derivation of anisotropic PML absorbing media in cylindrical and spherical coordinates," *IEEE Microwave Guided Wave Lett.*, vol. 7, pp. 371–373, Nov. 1997.

10. J. Maloney, M. Kesler, and G. Smith, "Generalization of PML to cylindrical geometries," *13th Annu. Rev. Prog. Appl. Comput. Electromagn.* (*ACES*), Monterey, CA, vol. II, pp. 900–908, Mar. 1997.

11. A. D. Greenwood and J. M. Jin, "A novel efficient algorithm for scattering from a complex BOR using mixed finite elements and cylindrical PML," *IEEE Trans. Antennas Propagat.*, vol. 47, no. 4, pp. 620–629, Apr. 1999.

12. J. R. Mautz and R. F. Harrington, "An improved *E*-field solution for a conducting body of revolution," *Arch. Elektron. Uebertragungstech.*, vol. 36, no. 5, pp. 198–206, 1982.

13. A. Glisson and D. Wilton, "Simple and efficient numerical techniques for treating bodies of revolution," Engineering Experiment Station Tech. Rep. 105, University of Mississippi, Oxford, MS, Mar. 1979.

14. A. D. Greenwood and J. M. Jin, "A field picture of wave propagation in inhomogeneous dielectric lenses," *IEEE Antennas Propagat. Mag.*, vol. 41, no. 5, pp. 9–18, Oct. 1999.

15. C. S. Liang, D. A. Streater, J. M. Jin, E. Dunn, and T. Rozendal, "A quantitative study of Luneberg lens reflectors," *IEEE Antennas Propagat. Mag.*, vol. 47, no. 2, pp. 30–42, Apr. 2005.

16. M. J. Al-Hakkak and Y. T. Lo, "Circular waveguides and horns with anisotropic and corrugated boundaries," Tech. Rep. 73-3, Antenna Laboratory, Department of Electrical Engineering, University of Illinois, Urbana, IL, 1973.

17. H. P. Partal, J. R. Mautz, and E. Arvas, "Radiation from a circular loop in the presence of spherically symmetric conducting or dielectric objects," *IEEE Trans. Antennas Propagat.*, vol. 48, no. 10, pp. 1646–1652, Oct. 2000.

9 Infinite Phased-Array Modeling

Phased-array antennas are of great importance in modern radar and communication systems [1]. Accurate prediction of array performance using numerical methods not only reduces the development cost and design timeline but also provides invaluable physical insight to design engineers. Increasing demands on array performance may require nontraditional designs using anisotropic dispersive materials and/or complex radiating elements. Furthermore, a successful antenna array design may require a complicated feeding and matching network, which may impose certain numerical analysis challenges. Among the various numerical methods, the finite element method is well suited to perform such analyses, due to its versatility in geometry and material modeling. In fact, much work has been carried out during the past two decades on the development of the finite element method for modeling three-dimensional doubly periodic phased arrays for a variety of configurations [2–10]. In particular, Jin and Volakis [2] developed the first three-dimensional finite element analysis for infinitely periodic arrays of cavity-backed antennas. McGrath and Pyati [3] and Lucas and Fontana [4] extended this method to more general periodic arrays consisting of more complicated antennas. Eibert et al. [6] applied a special technique to accelerate the evaluation of the periodic Green's function employed in the boundary integral equation for the truncation of the finite element computational domain. Recently, Petersson and Jin [9,10] developed a numerical scheme to carry out finite element analysis of periodic arrays in the time domain.

In this chapter we consider the finite element analysis of arrays that are infinitely periodic in the two-dimensional xy-plane. Although not a physically realistic configuration, the infinite-array model provides a reasonably good approximation to the performance of the interior elements in a large finite array. The numerical analysis of an infinite array with a uniform progressively phased excitation is relatively easy because the computational domain can be confined to a single array element (unit cell) due to the periodicity in the electromagnetic fields. In the interior region of the unit cell, the standard finite element discretization is applied. On the boundaries of the unit cell, proper boundary conditions are imposed to model the field behavior correctly in both the periodic and nonperiodic directions. Such boundary conditions have been developed in both the frequency and time domains.

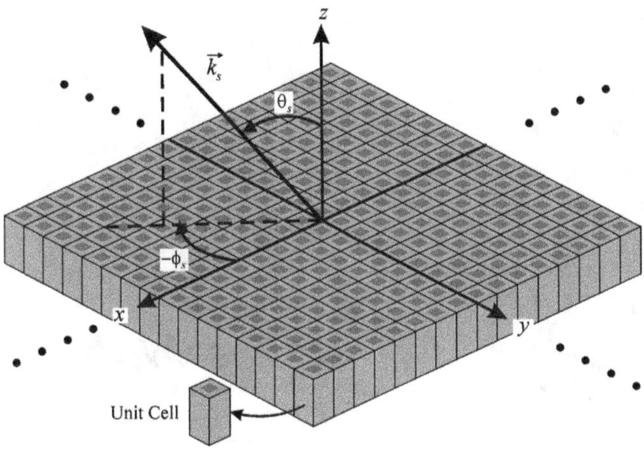

Figure 9.1 Infinite periodic array in the *xy*-plane that extends to infinity in four directions. (After Petersson and Jin [10], Copyright © IEEE 2006.)

9.1 FREQUENCY-DOMAIN MODELING

When a phased array is operating with a harmonic time dependence (frequency domain), each antenna element is excited by an individual source with a specific phase difference relative to the other sources, resulting in a main beam being radiated in a desired direction. To steer the beam, the relative phase shift between the elements is changed such that the beam radiates in a new direction. An infinite antenna array can be analyzed using a single unit cell, as shown in Figure 9.1, where the field at one periodic surface of a unit cell is related to the field at the opposite parallel surface of the unit cell through a simple phase shift. This fact makes it straightforward to utilize a unit cell to numerically analyze infinitely periodic phased-array antenna structures in the frequency domain, and two approaches to realize these periodic boundary conditions in the finite element method are discussed in Section 9.1.1. The more difficult task is to truncate the computational domain along the nonperiodic direction (the z-direction in this case), and this topic is discussed in Section 9.1.2.

9.1.1 Periodic Boundary Conditions

Consider a planar array extending infinitely in the *xy*-plane as shown in Figure 9.1. According to the Floquet theorem [11], the electric and magnetic fields satisfy the following periodic relations:

$$\mathbf{E}(x + mT_x, y + nT_y, z) = \mathbf{E}(x, y, z)e^{-j(mk_x^s T_x + nk_y^s T_y)} \tag{9.1}$$

$$\mathbf{H}(x + mT_x, y + nT_y, z) = \mathbf{H}(x, y, z)e^{-j(mk_x^s T_x + nk_y^s T_y)} \tag{9.2}$$

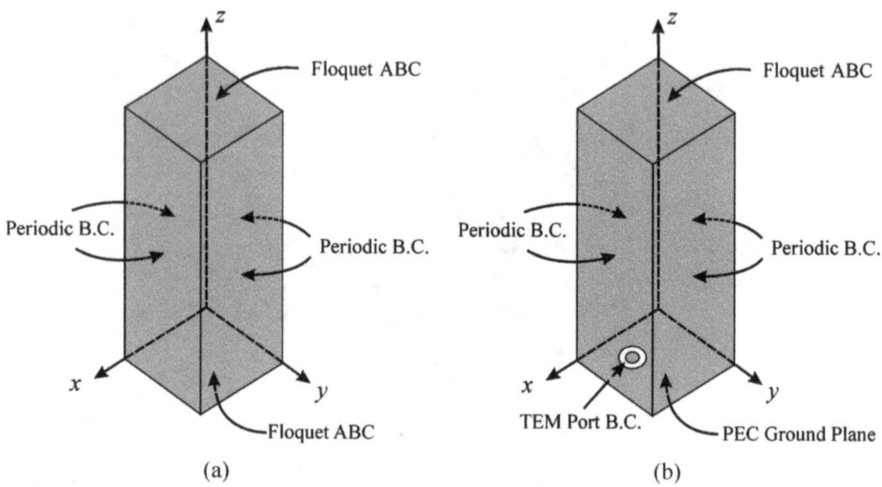

Figure 9.2 (a) Unit cell with an open bottom. (b) Unit cell assuming a ground plane and waveguide port excitation. (After Petersson and Jin [10], Copyright © IEEE 2006.)

where T_x and T_y are the unit cell spacings in the x- and y-directions, and $k_x^s = k_0 \sin\theta_s \cos\phi_s$ and $k_y^s = k_0 \sin\theta_s \sin\phi_s$, with (θ_s, ϕ_s) being the scan angle of the array. Applying (9.1) and (9.2) to the four side surfaces of the unit cell results in periodic boundary conditions that relate the electric and magnetic fields on the opposite side surfaces with a phase-shift term determined from the scan angle of the array (Figure 9.2).

To formulate the finite element equations and implement the periodic boundary conditions, we assume initially that perfectly matched layers are used to truncate the computational domain in the nonperiodic direction. In other words, if the original problem domain is infinite in the z-direction, we use a perfectly matched layer above the unit cell and another one below the unit cell (if necessary) to make the computational domain finite in the z-direction. In Section 9.1.2 we address alternative boundary conditions in the nonperiodic direction. For simplicity, the perfectly matched layers are backed by a perfect electric conductor. As we have shown in Chapter 3, the uniaxial perfectly matched layers can be implemented using a tensorial relative permittivity $\overleftrightarrow{\varepsilon}_r$ and permeability $\overleftrightarrow{\mu}_r$. According to the formulation presented in Chapter 2, the weak-form representation of the vector wave equation for the electric field is given by

$$\begin{aligned}&\iiint_V \left[(\nabla \times \mathbf{T}) \cdot \overleftrightarrow{\mu}_r^{-1} \cdot (\nabla \times \mathbf{E}) - k_0^2 \mathbf{T} \cdot \overleftrightarrow{\varepsilon}_r \cdot \mathbf{E}\right] dV \\ &\quad = j\omega\mu_0 \iint_S \mathbf{T} \cdot (\hat{n} \times \mathbf{H})\, dS - \iiint_V \mathbf{T} \cdot \left[jk_0 Z_0 \mathbf{J}_{\text{imp}} + \nabla \times \left(\overleftrightarrow{\mu}_r^{-1} \cdot \mathbf{M}_{\text{imp}}\right)\right] dV\end{aligned} \tag{9.3}$$

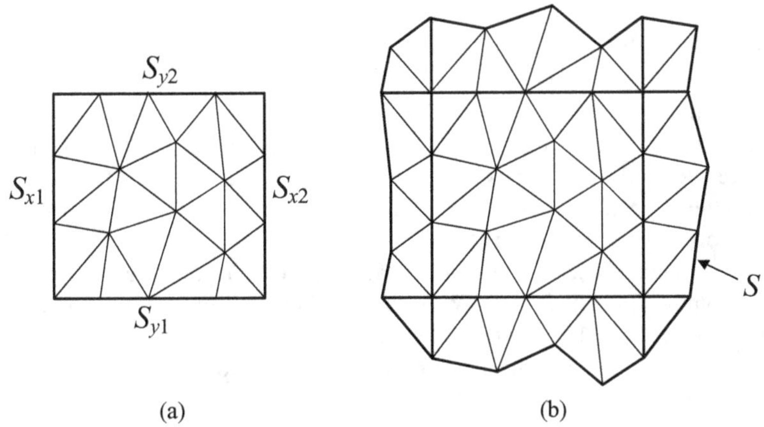

Figure 9.3 (a) Two-dimensional illustration of a finite element mesh for a unit cell. (b) Unit cell enclosed in a larger surface S.

where S denotes a surface that encloses V. Since we do not have a known boundary condition on S, we have to treat the surface integral in (9.3). There are two approaches to dealing with this surface integral. In the first approach we add a layer of finite elements around the perimeter of the unit cell so that S becomes offset from the periodic boundaries. By offsetting this surface, the surface integral does not contribute to the finite element matrix for the unit cell. In the second approach, no additional elements will be added and S will then coincide with the periodic boundaries of the unit cell. By enforcing the proper phase relationships for the fields on the periodic boundaries, it will be seen that the two approaches are equivalent.

As mentioned above, the first approach employs a surface S that encloses a volume larger than the unit cell such that when we assemble the finite element matrix over the unit cell, the surface integral has no contribution to this matrix. For the unit cell shown in Figure 9.3(a), the enlarged volume is illustrated in Figure 9.3(b). The corresponding finite element matrix equation can be written as

$$[K]\{E\} = \{b\} \tag{9.4}$$

where

$$K_{ij} = \iiint_{V_{\text{uc}}} \left[(\nabla \times \mathbf{N}_i) \cdot \overleftrightarrow{\mu}_r^{-1} \cdot (\nabla \times \mathbf{N}_j) - k_0^2 \mathbf{N}_i \cdot \overleftrightarrow{\varepsilon}_r \cdot \mathbf{N}_j\right] dV \tag{9.5}$$

$$b_i = -\iiint_{V_{\text{uc}}} \mathbf{N}_i \cdot \left[jk_0 Z_0 \mathbf{J}_{\text{imp}} + \nabla \times \left(\overleftrightarrow{\mu}_r^{-1} \cdot \mathbf{M}_{\text{imp}}\right)\right] dV. \tag{9.6}$$

In the equations above, V_{uc} denotes the volume of the unit cell, and $\{E\}$ represents the column vector whose entries are the expansion coefficients of the electric field inside the unit cell as well as on the side surfaces of the unit cell. Whereas the equations in (9.4) for any E_i's that reside inside the unit cell are complete, the equations for E_i's that reside on the side surfaces on the unit cell are incomplete because these E_i's are also associated with basis functions residing outside the specific unit cell. To make these equations complete, we have to consider the contributions from these exterior basis functions.

To account for the contributions from the outside basis functions associated with E_i's that reside on the side surfaces, we can mesh the unit cell such that the surface meshes on the opposite side surfaces are identical. Furthermore, we can also number the expansion coefficients on the opposite side surfaces in the same manner such that for an expansion coefficient on one side surface, we have a corresponding expansion coefficient on the opposite surface. With such a volumetric mesh for the unit cell, we can repeat the mesh in the x- and y-directions to generate the mesh for the infinitely periodic array. For the convenience of description, let us denote the side surface on the left side as S_{x1} and the one on the right side as S_{x2}. We can similarly define S_{y1} and S_{y2}, as illustrated in Figure 9.3(a). Based on the specific mesh described above, it is evident that for an E_i that resides on S_{x1}, the contribution from the basis function on the left side of S_{x1} is the same as the contribution of the left-side basis function with the corresponding expansion coefficient, denoted by E_j, on S_{x2}, except for the phase shift represented by $e^{-j\Psi_x}$, where $\Psi_x = k_x^s T_x$. Denoting the equations for E_i and E_j as

$$\sum_{l=1}^{N} K_{il} E_l = b_i \qquad \forall E_i \in S_{x1} \tag{9.7}$$

$$\sum_{l=1}^{N} K_{jl} E_l = b_j \qquad \forall E_j \in S_{x2} \tag{9.8}$$

where N denotes the total number of the unknown expansion coefficients in the unit cell, we can make the equation for E_i complete by adding (9.8) to (9.7) with a proper compensation of the phase shift. This gives us

$$\sum_{l=1}^{N} (K_{il} + K_{jl} e^{j\Psi_x}) E_l = b_i + b_j e^{j\Psi_x} \qquad \forall E_i \in S_{x1}. \tag{9.9}$$

We can also add (9.7) to (9.8) with a phase shift to make the equation for E_j complete; however, this is unnecessary because when we enforce the periodic boundary condition, (9.8) will be replaced by the equation

$$E_j = E_i e^{-j\Psi_x} \qquad \forall E_i \in S_{x1}, E_j \in S_{x2}. \tag{9.10}$$

Similarly, for an E_i that resides on S_{y1} we can make its equation complete by adding the equation for the corresponding E_j on S_{y2} with a proper phase shift. This results in

$$\sum_{l=1}^{N} (K_{il} + K_{jl} e^{j\Psi_y}) E_l = b_i + b_j e^{j\Psi_y} \qquad \forall E_i \in S_{y1} \tag{9.11}$$

and then we apply the periodic boundary condition

$$E_j = E_i e^{-j\Psi_y} \qquad \forall E_i \in S_{y1}, E_j \in S_{y2} \tag{9.12}$$

where $\Psi_y = k_y^s T_y$. Equation (9.11) accounts for the contribution of the basis function below S_{y1} to the equation for E_i. However, for an E_i that resides on the corner edge intersected by S_{x1} and S_{y1}, its equation is contributed by basis functions in all four quadrants. The contribution from the first quadrant is already included in (9.4), the contribution from the second quadrant is accounted for by (9.9), and the contribution from the fourth quadrant is included by (9.11). What is missing is the contribution from the associated basis function in the third quadrant, and this contribution is the same, except for a phase shift, as that for the corresponding E_j that resides on the corner edge intersected by S_{x2} and S_{y2}, which was calculated in (9.4). To add this contribution, we further modify the equation for E_i by

$$\sum_{l=1}^{N} [K'_{il} + K_{jl} e^{j(\Psi_x+\Psi_y)}] E_l = b'_i + b_j e^{j(\Psi_x+\Psi_y)} \qquad \forall E_i \in S_{x1} \cap S_{y1} \tag{9.13}$$

and then apply the periodic boundary condition

$$E_j = E_i e^{-j(\Psi_x+\Psi_y)} \qquad \forall E_i \in S_{x1} \cap S_{y1}, E_j \in S_{x2} \cap S_{y2}. \tag{9.14}$$

In (9.13), K'_{il} and b'_i represent the matrix elements after the modifications through (9.9) and (9.11). After the application of (9.9)–(9.14) to (9.4), we obtain the complete finite element matrix equation, which can be solved for the electric field expansion coefficients in the unit cell.

In the approach described above, the surface integral in (9.3) is avoided by using a larger enclosure surface that does not contribute to the finite element equations in the unit cell, and the equations on the side surfaces are made complete by including the contributions from basis functions outside the unit cell. The same result can also be obtained by setting the volume in (9.3) to be the volume of the unit cell, that is, $V = V_{\text{uc}}$. In this second approach, (9.3) becomes

$$\begin{aligned}
&\iiint_{V_{\text{uc}}} \left[(\nabla \times \mathbf{T}) \cdot \overleftrightarrow{\mu}_r^{-1} \cdot (\nabla \times \mathbf{E}) - k_0^2 \mathbf{T} \cdot \overleftrightarrow{\varepsilon}_r \cdot \mathbf{E}\right] dV \\
&\quad = j\omega\mu_0 \iint_{S_{\text{uc}}} \mathbf{T} \cdot (\hat{n} \times \mathbf{H})\, dS - \iiint_{V_{\text{uc}}} \mathbf{T} \cdot \left[jk_0 Z_0 \mathbf{J}_{\text{imp}} + \nabla \times \left(\overleftrightarrow{\mu}_r^{-1} \cdot \mathbf{M}_{\text{imp}}\right)\right] dV
\end{aligned} \tag{9.15}$$

where S_{uc} consists of all side surfaces of that unit cell; that is, $S_{\text{uc}} = S_{x1} \cup S_{x2} \cup S_{y1} \cup S_{y2}$. In this case we have to consider the surface integral. The corresponding finite element matrix equation for (9.15) can be written as

$$[K]\{E\} = \{b\} + \{h\} \tag{9.16}$$

where $[K]$ and $\{b\}$ are the same as those defined by (9.5) and (9.6), respectively, and the elements of $\{h\}$ are given by

$$h_i = j\omega\mu_0 \iint_{S_{\rm uc}} \mathbf{N}_i \cdot (\hat{n} \times \mathbf{H})\, dS. \tag{9.17}$$

If E_i resides inside the unit cell, its basis function $\mathbf{N}_i$ has no tangential component on any side surfaces; hence, $h_i = 0$ for all E_i's inside the unit cell. Now consider an E_i with a basis function that has a tangential component only on the surface S_{x1}. In this case its equation can be written as

$$\sum_{l=1}^{N} K_{il} E_l = b_i + h_i \qquad \forall E_i \in S_{x1} \tag{9.18}$$

where

$$h_i = -j\omega\mu_0 \iint_{S_{x1}} \mathbf{N}_i \cdot (\hat{x} \times \mathbf{H})\, dS \tag{9.19}$$

since $\hat{n} = -\hat{x}$ on S_{x1}. Using the special mesh described earlier (i.e., that opposing surfaces have identical surface meshes), we can find the corresponding coefficient E_j with basis function $\mathbf{N}_j$ that has a tangential component only on S_{x2} and is identical to $\mathbf{N}_i$. The equation for E_j is given by

$$\sum_{l=1}^{N} K_{jl} E_l = b_j + h_j \qquad \forall E_j \in S_{x2} \tag{9.20}$$

with

$$h_j = j\omega\mu_0 \iint_{S_{x2}} \mathbf{N}_j \cdot (\hat{x} \times \mathbf{H})\, dS. \tag{9.21}$$

Both h_i in (9.18) and h_j in (9.20) are unknown. However, because of the periodic condition for the magnetic field given in (9.2), we find their relation as

$$h_j = -h_i e^{-j\Psi_x}. \tag{9.22}$$

Hence, we can cancel h_i in (9.18) by multiplying (9.20) by $e^{j\Psi_x}$ and adding the resulting equation to (9.18). This yields

$$\sum_{l=1}^{N} (K_{il} + K_{jl} e^{j\Psi_x}) E_l = b_i + b_j e^{j\Psi_x} \qquad \forall E_i \in S_{x1} \tag{9.23}$$

which is identical to (9.9). Equation (9.20) can then be replaced by the periodic boundary condition in (9.10). The treatment for an E_i where its basis function has a tangential component only on S_{y1} is the same, yielding the same results as given by (9.11) and (9.12).

It is interesting to look at an E_i that resides on the corner edge intersected by S_{x1} and S_{y1}, whose basis function has a tangential component on both S_{x1} and S_{y1}. In this case the equation is given by

$$\sum_{l=1}^{N} K_{il} E_l = b_i + h_i \qquad \forall E_i \in S_{x1} \cap S_{y1} \tag{9.24}$$

where

$$h_i = h_i^{(x)} + h_i^{(y)} = -j\omega\mu_0 \iint_{S_{x1}} \mathbf{N}_i \cdot (\hat{x} \times \mathbf{H})\, dS - j\omega\mu_0 \iint_{S_{y1}} \mathbf{N}_i \cdot (\hat{y} \times \mathbf{H})\, dS. \tag{9.25}$$

This unknown h_i can be eliminated by the equations for the corresponding expansion coefficients residing at the three other corner edges of the unit cell. For the corresponding E_m that resides on the corner edge intersected by S_{y1} and S_{x2}, its equation is given by

$$\sum_{l=1}^{N} K_{ml} E_l = b_m + h_m \qquad \forall E_m \in S_{y1} \cap S_{x2} \tag{9.26}$$

with

$$h_m = h_m^{(x)} + h_m^{(y)} = -j\omega\mu_0 \iint_{S_{y1}} \mathbf{N}_m \cdot (\hat{y} \times \mathbf{H})\, dS + j\omega\mu_0 \iint_{S_{x2}} \mathbf{N}_m \cdot (\hat{x} \times \mathbf{H})\, dS. \tag{9.27}$$

For the corresponding E_n that resides on the corner edge intersected by S_{x1} and S_{y2}, its equation is given by

$$\sum_{l=1}^{N} K_{nl} E_l = b_n + h_n \qquad \forall E_n \in S_{x1} \cap S_{y2} \tag{9.28}$$

with

$$h_n = h_n^{(x)} + h_n^{(y)} = -j\omega\mu_0 \iint_{S_{x1}} \mathbf{N}_n \cdot (\hat{x} \times \mathbf{H})\, dS + j\omega\mu_0 \iint_{S_{y2}} \mathbf{N}_n \cdot (\hat{y} \times \mathbf{H})\, dS. \tag{9.29}$$

Finally, for the corresponding E_j that resides on the corner edge intersected by S_{x2} and S_{y2}, its equation is given by

$$\sum_{l=1}^{N} K_{jl} E_l = b_j + h_j \qquad \forall E_j \in S_{x2} \cap S_{y2} \tag{9.30}$$

with

$$h_j = h_j^{(x)} + h_j^{(y)} = j\omega\mu_0 \iint_{S_{x2}} \mathbf{N}_j \cdot (\hat{x} \times \mathbf{H})\, dS + j\omega\mu_0 \iint_{S_{y2}} \mathbf{N}_j \cdot (\hat{y} \times \mathbf{H})\, dS. \tag{9.31}$$

Using the periodic condition for the magnetic field and the relation between the basis functions $\mathbf{N}_i$, $\mathbf{N}_j$, $\mathbf{N}_m$, and $\mathbf{N}_n$, we find that

$$h_m^{(y)} = -h_i^{(y)} e^{-j\Psi_x}, \qquad h_n^{(x)} = -h_i^{(x)} e^{-j\Psi_y} \tag{9.32}$$

$$h_j^{(x)} = -h_m^{(x)} e^{-j\Psi_y}, \qquad h_j^{(y)} = -h_n^{(y)} e^{-j\Psi_x}. \tag{9.33}$$

Using these relations, we can multiply (9.26) by $e^{j\Psi_x}$, (9.28) by $e^{j\Psi_y}$, and (9.30) by $e^{j(\Psi_x+\Psi_y)}$, and add the resulting equations to (9.24) to find

$$\sum_{l=1}^{N} [K_{il} + K_{ml} e^{j\Psi_x} + K_{nl} e^{j\Psi_y} + K_{jl} e^{j(\Psi_x+\Psi_y)}] E_l$$
$$= b_i + b_m e^{j\Psi_x} + b_n e^{j\Psi_y} + b_j e^{j(\Psi_x+\Psi_y)} \qquad \forall E_i \in S_{x1} \cap S_{y1}. \tag{9.34}$$

This equation can be shown to be identical to (9.13). Hence, the two approaches yield the identical finite element matrix equation, as can be expected.

Although the formulations described above seem quite complicated, the numerical implementation is very straightforward. Let us first denote the periodic boundary condition as

$$E_j = E_i e^{-j\Psi_{ij}} \tag{9.35}$$

where the phase-shift term is given by

$$\Psi_{ij} = \begin{cases} k_x^s T_x & \forall E_i \in S_{x1},\, E_j \in S_{x2} \\ k_y^s T_y & \forall E_i \in S_{y1},\, E_j \in S_{y2} \\ k_x^s T_x + k_y^s T_y & \forall E_i \in S_{x1} \cap S_{y1},\, E_j \in S_{x2} \cap S_{y2}. \end{cases} \tag{9.36}$$

Given (9.4), we can then complete the matrix equation and enforce the periodic boundary conditions by the following operations:

$$K_{il} \leftarrow K_{il} + K_{jl}e^{j\Psi_{ij}}, \quad b_i \leftarrow b_i + b_j e^{j\Psi_{ij}} \tag{9.37}$$

$$K_{jl} \leftarrow 0, \quad b_j \leftarrow 0, \quad K_{jj} \leftarrow 1, \quad K_{ji} \leftarrow -e^{-j\Psi_{ij}}. \tag{9.38}$$

Here, the backward arrow simply means to replace the value for the left-hand side with that of the right-hand side. The matrix equation so obtained contains all the unknown expansion coefficients residing either inside the unit cell or on the surface of the unit cell. Actually, we can simply eliminate all the dependent unknown E_j's on S_{x2} and S_{y2} to obtain a smaller matrix equation. This can be achieved by carrying out the following operations after (9.37):

$$K_{li} \leftarrow K_{li} + K_{lj}e^{-j\Psi_{ij}} \tag{9.39}$$

and then eliminating all the rows and columns corresponding to E_j's. These operations can be written in matrix form as the transformation [3]

$$[K'] = [R][K][R]^{\mathrm{H}} \tag{9.40}$$

where the transformation matrix $[R]$ has a dimension of $M \times N$, with N denoting the total number of unknowns in the unit cell and M denoting the total number of unknowns in the unit cell excluding those residing on S_{x2} and S_{y2}, and the superscript H denotes the conjugate transpose or Hermitian transpose. Assuming that the unknowns residing on S_{x2} and S_{y2} are numbered last, the first M columns of $[R]$ are the $M \times M$ identity matrix. The other entries for $[R]$ are given by

$$R_{ij} = \begin{cases} e^{jk_x^s T_x} & \forall E_i \in S_{x1}, E_j \in S_{x2} \\ e^{jk_y^s T_y} & \forall E_i \in S_{y1}, E_j \in S_{y2} \\ e^{j(k_x^s T_x + k_y^s T_y)} & \forall E_i \in S_{x1} \cap S_{y1}, E_j \in S_{x2} \cap S_{y2} \end{cases} \tag{9.41}$$

and all the rest unspecified entries are zeros. The final matrix equation can be written as

$$[K']\{E'\} = \{b'\} \tag{9.42}$$

where $\{E'\}$ contains all the unknowns in the unit cell except for those residing on S_{x2} and S_{y2} and $\{b'\} = [R]\{b\}$. Finally, we note that it is not necessary to generate $[K]$ and $\{b\}$ initially and then compute $[K']$ and $\{b'\}$ either through operations in (9.37)–(9.39) or through the matrix multiplication in (9.40). Instead, the unknowns residing on S_{x2} and S_{y2} will not even be assigned, and $[K']$ and $\{b'\}$ are assembled directly from the finite elements in the unit cell.

9.1.2 Mesh Truncation Techniques

As discussed in Chapter 3, the three commonly used mesh truncation techniques are based on the use of absorbing boundary conditions, perfectly matched layers, and boundary integral equations. The implementation of perfectly matched layers can be effected by the modeling of a uniaxial anisotropic medium, which was used in the preceding section to illustrate the finite element formulation. The standard first-order absorbing boundary condition for the upper truncation surface is given by

$$\hat{z} \times (\nabla \times \mathbf{E}) + jk_0\hat{z} \times (\hat{z} \times \mathbf{E}) \approx 0 \qquad \mathbf{r} \in S_T \tag{9.43}$$

where we assume that the truncation surface S_T resides in air. The implementation of this boundary condition in the finite element formulation is straightforward. However, this condition absorbs only the wave propagating along the z-direction (i.e., $\theta_s = 0$), and it has a significant reflection for waves propagating in other directions. The implementation of (9.43) in the finite element formulation yields the same matrix equation as (9.4) except that K_{ij} is modified as

$$\begin{aligned} K_{ij} = & \iiint_{V_{\text{uc}}} \left[(\nabla \times \mathbf{N}_i) \cdot \overleftrightarrow{\mu}_r^{-1} \cdot (\nabla \times \mathbf{N}_j) - k_0^2 \mathbf{N}_i \cdot \overleftrightarrow{\varepsilon}_r \cdot \mathbf{N}_j \right] dV \\ & + jk_0 \iint_{S_T} (\hat{z} \times \mathbf{N}_i) \cdot (\hat{z} \times \mathbf{N}_j)\, dS. \end{aligned} \tag{9.44}$$

As we discussed in Chapter 3, absorbing boundary conditions and perfectly matched layers are only approximate because they cannot absorb all kinds of waves impinging on the truncation boundary. To derive a more accurate truncation technique and also shed light on better use of absorbing boundary conditions and perfectly matched layers, let us examine the fields above the periodic array more carefully. For example, consider the electric field, which satisfies the periodic condition in (9.1). If we define a new field function as

$$\mathbf{P}(x, y, z) = \mathbf{E}(x, y, z) e^{j(k_x^s x + k_y^s y)} \tag{9.45}$$

this function then becomes a two-dimensional periodic function in any plane ($z = z_c$) parallel to the xy-plane with the periodicities T_x and T_y. Consequently, it can be expanded into a Fourier series as

$$\mathbf{P}(x, y, z_c) = \sum_{p=-\infty}^{\infty} \sum_{q=-\infty}^{\infty} \tilde{\mathbf{P}}_{pq}(z_c)\, e^{j(2\pi px/T_x + 2\pi qy/T_y)} \tag{9.46}$$

where $\tilde{\mathbf{P}}_{pq}(z_c)$ is the Fourier transform of $\mathbf{P}(x, y, z_c)$ and is given by

$$\tilde{\mathbf{P}}_{pq}(z_c) = \frac{1}{T_x T_y} \iint_{S_T} \mathbf{P}(x, y, z_c)\, e^{-j(2\pi px/T_x + 2\pi qy/T_y)}\, dS. \tag{9.47}$$

Substituting (9.45) into (9.46) and (9.47), we obtain the Fourier expansion for the electric field as

$$\mathbf{E}(x, y, z_c) = \sum_{p=-\infty}^{\infty} \sum_{q=-\infty}^{\infty} \tilde{\mathbf{E}}_{pq}(z_c)\, e^{-j(k_{xp}x + k_{yq}y)} \tag{9.48}$$

and

$$\tilde{\mathbf{E}}_{pq}(z_c) = \frac{1}{T_x T_y} \iint_{S_T} \mathbf{E}(x, y, z_c)\, e^{j(k_{xp}x + k_{yq}y)}\, dS \tag{9.49}$$

where

$$k_{xp} = k_x^s - \frac{2\pi p}{T_x}, \qquad k_{yq} = k_y^s - \frac{2\pi q}{T_y}. \tag{9.50}$$

By using the dispersion relation, the field above the plane $z = z_c$ is given by

$$\mathbf{E}(x, y, z) = \sum_{p=-\infty}^{\infty} \sum_{q=-\infty}^{\infty} \tilde{\mathbf{E}}_{pq}(z_c)\, e^{-j[k_{xp}x + k_{yq}y + k_{zpq}(z - z_c)]} \qquad \forall z \geq z_c \tag{9.51}$$

where

$$k_{zpq} = \begin{cases} \sqrt{k_0^2 - k_{xp}^2 - k_{yq}^2}, & \text{when } k_{xp}^2 + k_{yq}^2 \leq k_0^2 \\ -j\sqrt{k_{xp}^2 + k_{yq}^2 - k_0^2}, & \text{when } k_{xp}^2 + k_{yq}^2 > k_0^2 . \end{cases} \tag{9.52}$$

The result in (9.51) reveals that the field excited by a periodic phased array can be decomposed into an infinite number of modes, which are referred to as *Floquet modes*. Depending on the frequency, some modes can propagate and some modes decay along the z-direction. The former are usually referred to as *propagating waves* and the latter are often called *evanescent waves*. The number of propagating modes depends on the frequency, but there usually exists one propagating mode corresponding to $p = q = 0$ (the fundamental mode), which represents a plane wave propagating in the (θ_s, ϕ_s) direction. For the special case of $\theta_s = \phi_s = 0°$, the fundamental Floquet mode can be absorbed perfectly with (9.43); however, since (9.43) does not absorb any evanescent waves, we should always place the truncation surface S_T sufficiently far away to allow the evanescent waves to decay to an insignificant amplitude. When $(\theta_s, \phi_s) \neq (0°, 0°)$ or when the frequency is high enough to support more propagating waves, the use of (9.43) can introduce a significant reflection error. When perfectly matched layers are employed for mesh truncation, one should either use the complex-shifted-frequency or second-order models to absorb the evanescent waves effectively since these waves can be strong in some periodic structures. These perfectly matched layers have been demonstrated in the finite-difference time-domain simulations of periodic structures [12,13].

To absorb all the propagating and evanescent waves effectively, we have to employ an exact boundary integral equation. To derive such an equation for an infinitely periodic phased array, we first take the curl of (9.51) and then cross it with $\hat{z}$ to obtain

$$\begin{aligned}\hat{z} \times (\nabla \times \mathbf{E}) &= -j \sum_{p=-\infty}^{\infty} \sum_{q=-\infty}^{\infty} \left[\hat{z} \times (\mathbf{k}_{tpq} \times \tilde{\mathbf{E}}_{pq}) + k_{zpq}\hat{z} \times (\hat{z} \times \tilde{\mathbf{E}}_{pq})\right] \\ &\quad \cdot e^{-j[k_{xp}x + k_{yq}y + k_{zpq}(z-z_c)]} \\ &= -j \sum_{p=-\infty}^{\infty} \sum_{q=-\infty}^{\infty} \left[\mathbf{k}_{tpq}(\hat{z} \cdot \tilde{\mathbf{E}}_{pq}) + k_{zpq}\hat{z} \times (\hat{z} \times \tilde{\mathbf{E}}_{pq})\right] \\ &\quad \cdot e^{-j[k_{xp}x + k_{yq}y + k_{zpq}(z-z_c)]}\end{aligned} \tag{9.53}$$

where $\mathbf{k}_{tpq} = k_{xp}\hat{x} + k_{yq}\hat{y}$. Applying Gauss's electric law $\nabla \cdot \mathbf{E} = 0$ to (9.51), we find that

$$\mathbf{k}_{tpq} \cdot \tilde{\mathbf{E}}_{pq} + k_{zpq}\hat{z} \cdot \tilde{\mathbf{E}}_{pq} = 0 \tag{9.54}$$

and its substitution into (9.53) yields

$$\begin{aligned}\hat{z} \times (\nabla \times \mathbf{E}) &= \sum_{p=-\infty}^{\infty} \sum_{q=-\infty}^{\infty} \left[\frac{j\mathbf{k}_{tpq}}{k_{zpq}}(\mathbf{k}_{tpq} \cdot \tilde{\mathbf{E}}_{pq}) - jk_{zpq}\hat{z} \times (\hat{z} \times \tilde{\mathbf{E}}_{pq})\right] \\ &\quad \cdot e^{-j(k_{xp}x + k_{yq}y)} \qquad \forall z = z_c.\end{aligned} \tag{9.55}$$

Equation (9.55) can also be written in a more familiar form as

$$\hat{z} \times (\nabla \times \mathbf{E}) = 2k_0^2\hat{z} \times \sum_{p=-\infty}^{\infty} \sum_{q=-\infty}^{\infty} \tilde{\bar{\mathbf{G}}}(k_{xp}, k_{yq}) \cdot (\hat{z} \times \tilde{\mathbf{E}}_{pq})\, e^{-j(k_{xp}x + k_{yq}y)} \qquad \forall z = z_c \tag{9.56}$$

where

$$\tilde{\bar{\mathbf{G}}}(k_{xp}, k_{yq}) = \frac{1}{2jk_{zpq}k_0^2}\begin{bmatrix} k_0^2 - k_{xp}^2 & -k_{xp}k_{yq} \\ -k_{xp}k_{yq} & k_0^2 - k_{yq}^2 \end{bmatrix} \tag{9.57}$$

which is called the *periodic dyadic Green's function* in the spectral domain. Equation (9.55) or (9.56) provides a relation between the tangential electric and magnetic fields on the truncation surface S_T placed at $z = z_c$. Since it is derived rigorously, this condition is exact and can absorb all the Floquet modes without reflection. It can be referred to as the *periodic boundary integral equation* or the *periodic radiation condition.*

The application of (9.56) at the top truncation surface S_T yields the weak-form solution for the electric field as

$$\begin{aligned}
&\iiint_{V_{\text{uc}}} \left[(\nabla \times \mathbf{T}) \cdot \overleftrightarrow{\mu}_r^{-1} \cdot (\nabla \times \mathbf{E}) - k_0^2 \mathbf{T} \cdot \overleftrightarrow{\varepsilon}_r \cdot \mathbf{E}\right] dV \\
&\quad = j\omega\mu_0 \iint_{S_{\text{uc}}} \mathbf{T} \cdot (\hat{n} \times \mathbf{H})\, dS + 2k_0^2 \iint_{S_T} (\hat{z} \times \mathbf{T}) \cdot \sum_{p=-\infty}^{\infty} \sum_{q=-\infty}^{\infty} \tilde{\overline{\overline{\mathbf{G}}}}(k_{xp}, k_{yq}) \\
&\qquad \cdot (\hat{z} \times \tilde{\mathbf{E}}_{pq})\, e^{-j(k_{xp}x + k_{yq}y)}\, dS \\
&\qquad - \iiint_{V_{\text{uc}}} \mathbf{T} \cdot \left[jk_0 Z_0 \mathbf{J}_{\text{imp}} + \nabla \times \left(\overleftrightarrow{\mu}_r^{-1} \cdot \mathbf{M}_{\text{imp}}\right)\right] dV.
\end{aligned} \tag{9.58}$$

The resulting finite element matrix equation before the application of the periodic boundary conditions is the same as (9.4) except that K_{ij} now becomes

$$\begin{aligned}
K_{ij} = &\iiint_{V_{\text{uc}}} \left[(\nabla \times \mathbf{N}_i) \cdot \overleftrightarrow{\mu}_r^{-1} \cdot (\nabla \times \mathbf{N}_j) - k_0^2 \mathbf{N}_i \cdot \overleftrightarrow{\varepsilon}_r \cdot \mathbf{N}_j\right] dV \\
&- 2k_0^2 T_x T_y \sum_{p=-\infty}^{\infty} \sum_{q=-\infty}^{\infty} (\hat{z} \times \tilde{\mathbf{N}}_{i,pq}^*) \cdot \tilde{\overline{\overline{\mathbf{G}}}}(k_{xp}, k_{yq}) \cdot (\hat{z} \times \tilde{\mathbf{N}}_{j,pq})
\end{aligned} \tag{9.59}$$

where

$$\tilde{\mathbf{N}}_{i,pq} = \frac{1}{T_x T_y} \iint_{S_T} \mathbf{N}_i\, e^{j(k_{xp}x + k_{yq}y)}\, dS. \tag{9.60}$$

For the numerical implementation, the infinite summations in (9.59) are always truncated. A good truncation point is to include all the propagating Floquet modes and a few evanescent modes. Larger array periods or a higher frequency permit more propagating Floquet modes; hence, more terms should be included in the summation. It is also important that the spatial discretization on S_T be dense enough to resolve the spatial variation of the highest-order Floquet mode included. A similar treatment can be applied to the bottom truncation surface, if necessary.

It is interesting to consider the use of an approximation of (9.56) where we keep only the dominant term corresponding to $p = q = 0$. Such an approximation yields

$$\hat{z} \times (\nabla \times \mathbf{E}) = 2k_0^2 \hat{z} \times \tilde{\overline{\overline{\mathbf{G}}}}\left(k_x^s, k_y^s\right) \cdot (\hat{z} \times \tilde{\mathbf{E}}_{00})\, e^{-j(k_x^s x + k_y^s y)} \qquad \forall z = z_c. \tag{9.61}$$

This boundary condition will absorb perfectly a plane wave propagating in the (θ_s, ϕ_s) direction. Since $\tilde{\overline{\overline{\mathbf{G}}}}(k_x^s, k_y^s)$ represents a nondiagonal 2×2 tensor, this indicates that to absorb perfectly a wave incident at an oblique angle, the absorbing boundary must be anisotropic. Equation (9.61) can also be written as

$$\hat{z} \times (\nabla \times \mathbf{E}) = 2k_0^2 \hat{z} \times \tilde{\bar{\mathbf{G}}}(k_x^s, k_y^s) \cdot (\hat{z} \times \mathbf{E}) \qquad \forall z = z_c \tag{9.62}$$

which is slightly easier to implement than (9.61). The alternative expression of (9.62) is

$$\hat{z} \times (\nabla \times \mathbf{E}) = \frac{jk_0}{\cos\theta_s} \hat{\mathbf{k}}_t^s (\hat{\mathbf{k}}_t^s \cdot \mathbf{E}) - jk_0 \cos\theta_s \hat{z} \times (\hat{z} \times \mathbf{E}) \qquad \forall z = z_c \tag{9.63}$$

where $\hat{\mathbf{k}}_t^s = \sin\theta_s \cos\phi_s \hat{x} + \sin\theta_s \sin\phi_s \hat{y}$.

9.1.3 Extension to Skew Arrays

For the sake of simplicity, the finite element analysis described above applies to arrays with an orthogonal lattice. The formulation can be extended easily to analyze arrays with a skew lattice (Figure 9.4). Let us denote the principal axes of the skew lattice as η_1 and η_2 and the angle between these axes as α. The periodic relation becomes

$$\mathbf{E}(\eta_1 + mT_1, \eta_2 + nT_2, z) = \mathbf{E}(\eta_1, \eta_2, z)e^{-j(mk_1^s T_1 + nk_2^s T_2)} \tag{9.64}$$

where T_1 and T_2 are the periodicities measured along the η_1- and η_2-axes, and k_1^s and k_2^s are the projected wavenumbers in the two directions. Without loss of generality,

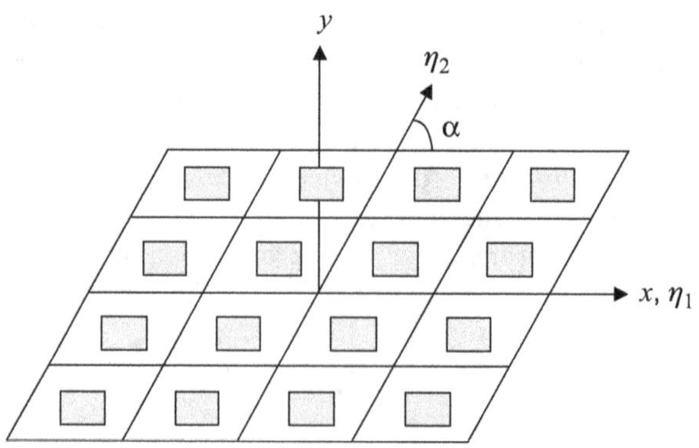

Figure 9.4 Array with a skew lattice. The principal axes of the skew lattice are labeled as η_1 and η_2 and the angle between these axes is labeled as α.

we can assume that the η_1-axis coincides with the x-axis so that $T_1 = T_x$ and $k_1^s = k_x^s$. It can be seen easily that

$$k_2^s = k_x \cos\alpha + k_y \sin\alpha, \qquad T_2 = \frac{T_y}{\sin\alpha}. \tag{9.65}$$

Substituting these into (9.64), we obtain the periodic relation in Cartesian coordinates as

$$\mathbf{E}(x + mT_x + nT_y \cot\alpha, y + nT_y, z) = \mathbf{E}(x, y, z)e^{-j[mk_x^s T_x + n(k_y^s + k_x^s \cot\alpha)T_y]}. \tag{9.66}$$

Therefore, the finite element formulation in Section 9.1.1 remains the same except that all k_y^s's should be replaced by $k_y^s + k_x^s \cot\alpha$. Consequently, (9.36) becomes

$$\Psi_{ij} = \begin{cases} k_x^s T_x & \forall E_i \in S_{x1}, E_j \in S_{x2} \\ \left(k_y^s + k_x^s \cot\alpha\right)T_y & \forall E_i \in S_{y1}, E_j \in S_{y2} \\ k_x^s T_x + \left(k_y^s + k_x^s \cot\alpha\right)T_y & \forall E_i \in S_{x1} \cap S_{y1}, E_j \in S_{x2} \cap S_{y2} \end{cases} \tag{9.67}$$

and (9.41) becomes [3]

$$R_{ij} = \begin{cases} e^{jk_x^s T_x} & \forall E_i \in S_{x1}, E_j \in S_{x2} \\ e^{j\left(k_y^s + k_x^s \cot\alpha\right)T_y} & \forall E_i \in S_{y1}, E_j \in S_{y2} \\ e^{j\left[k_x^s T_x + \left(k_y^s + k_x^s \cot\alpha\right)T_y\right]} & \forall E_i \in S_{x1} \cap S_{y1}, E_j \in S_{x2} \cap S_{y2}. \end{cases} \tag{9.68}$$

The implementation of the first-order absorbing boundary condition and perfectly matched layers also remain unchanged. The periodic boundary integral equation can be formulated using the Fourier expansion in η_1 and η_2, which is given by [11]

$$\mathbf{E}(\eta_1, \eta_2, z) = \sum_{p=-\infty}^{\infty} \sum_{q=-\infty}^{\infty} \tilde{\mathbf{E}}_{pq}(z_c)\, e^{-j[k_{1p}\eta_1 + k_{2q}\eta_2 + k_{zpq}(z - z_c)]} \qquad \forall z \geq z_c \tag{9.69}$$

where

$$k_{1p} = k_1^s - \frac{2\pi p}{T_1}, \qquad k_{2q} = k_2^s - \frac{2\pi q}{T_2}. \tag{9.70}$$

Since $\eta_1 = x - y \cot\alpha$ and $\eta_2 = y/\sin\alpha$, (9.69) can be written in the same form as (9.53) except that now

$$k_{yq} = k_y^s - \frac{2\pi q}{T_y} + \frac{2\pi p}{T_x}\cot\alpha. \tag{9.71}$$

Therefore, the formulation in Section 9.1.2 can be applied to a skew-lattice array by replacing all k_{yq}'s with the one given in (9.71).

9.1.4 Extension to Scattering Analysis

The formulation described above is pertinent to the radiation case where the sources are embedded in the computational domain. The formulation can be modified to deal with the case of scattering of an incident uniform plane wave. In fact, the formulations that we have been developing in Section 9.1 remain intact, except that $k_x^s = -k_0 \sin\theta^{\rm inc} \cos\phi^{\rm inc}$ and $k_y^s = -k_0 \sin\theta^{\rm inc} \sin\phi^{\rm inc}$, where $(\theta^{\rm inc}, \phi^{\rm inc})$ denotes the angle of incidence. For the mesh truncation, when the first-order absorbing boundary condition is employed, (9.43) becomes

$$\hat{z} \times (\nabla \times \mathbf{E}) + jk_0 \hat{z} \times (\hat{z} \times \mathbf{E}) \approx \mathbf{U}^{\rm inc} \qquad \mathbf{r} \in S_T \tag{9.72}$$

where $\mathbf{U}^{\rm inc} = \hat{z} \times (\nabla \times \mathbf{E}^{\rm inc}) + jk_0 \hat{z} \times (\hat{z} \times \mathbf{E}^{\rm inc})$. This yields the right-hand-side vector

$$b_i = j\omega\mu_0 \iint_{S_T} \mathbf{N}_i \cdot [\hat{z} \times \mathbf{H}^{\rm inc} - Y_0 \hat{z} \times (\hat{z} \times \mathbf{E}^{\rm inc})]\, dS \tag{9.73}$$

which replaces (9.6). When the periodic boundary integral equation is employed, (9.56) becomes

$$\hat{z} \times (\nabla \times \mathbf{E}) = -2j\omega\mu_0 \hat{z} \times \mathbf{H}^{\rm inc} + 2k_0^2 \hat{z} \times \sum_{p=-\infty}^{\infty} \sum_{q=-\infty}^{\infty} \tilde{\bar{\mathbf{G}}}(k_{xp}, k_{yq}) \cdot (\hat{z} \times \tilde{\mathbf{E}}_{pq}) e^{-j(k_{xp}x + k_{yq}y)} \tag{9.74}$$

which gives the right-hand-side vector

$$b_i = 2j\omega\mu_0 \iint_{S_T} \mathbf{N}_i \cdot (\hat{z} \times \mathbf{H}^{\rm inc})\, dS \tag{9.75}$$

to replace (9.6). When perfectly matched layers are employed for mesh truncation, we can either formulate the finite element analysis in terms of the scattered field or use the total- and scattered-field decomposition scheme to excite the incident field through a Huygens surface, as discussed in Section 5.2.3.

9.1.5 Application Examples

We first consider the finite element simulation of an infinite array of microstrip patch antennas. The microstrip patches, shown in Figure 9.5, are placed on a grounded dielectric substrate with height h and permittivity ε_r. The patches are fed with coaxial cables that enter the substrate from below the ground plane. In the simulation, the

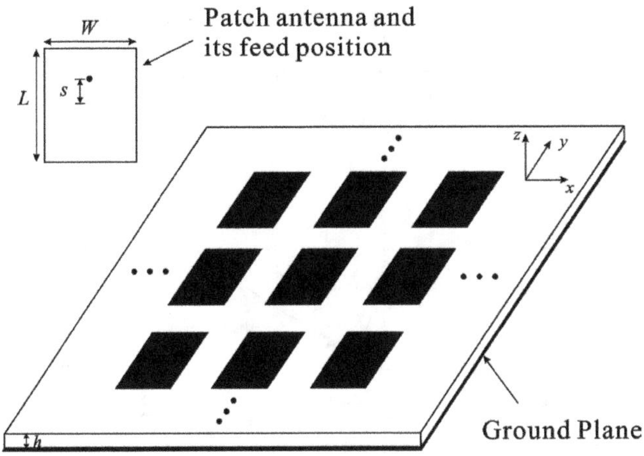

Figure 9.5 Infinite periodic microstrip patch array on a substrate. Each patch is fed by a coaxial line.

computational domain, or unit cell, contains only a single patch, and proper periodic boundary conditions are imposed on its boundaries. For a phased array, a critical performance parameter is the active reflection coefficient. One definition for this is given by

$$R(\theta_s,\phi_s) = \frac{Z_{\text{in}}(\theta_s,\phi_s) - Z_{\text{in}}(0,0)}{Z_{\text{in}}(\theta_s,\phi_s) + Z_{\text{in}}^*(0,0)} \tag{9.76}$$

where Z_{in} is the active input impedance of the antenna. The use of the conjugate operator in (9.76) discounts the imaginary part of the input impedance and more clearly reveals the scan properties of the array. The array under consideration has a period of $0.5\lambda_0$ in both the x- and y-directions. The patches are $0.3\lambda_0 \times 0.3\lambda_0$ in size and the relative permittivity of the substrate is 2.55. The E-plane active reflection coefficients are calculated for two different substrate heights: $h = 0.02\lambda_0$ and $h = 0.06\lambda_0$. The results calculated using the simplified probe and precise coaxial feed models are plotted in Figure 9.6. A moment-method analysis by Pozar and Schaubert [14] is used here for reference results. It is noted that all three results agree with one another very well. All results predict a scan blindness at $\theta = 68.8^\circ$ for $h = 0.06\lambda_0$ as a result of surface-wave excitation. Moreover, for this particular array, the agreement between the simplified and precise feed models suggests that scan performances such as active reflection coefficients are insensitive to the specific feed scheme, as indicated in Ref. 14.

Another independent verification is obtained for a microstrip patch array consisting of $0.25\lambda_0 \times 0.25\lambda_0$ rectangular patches with periodic spacings of $T_x = T_y = 0.5\lambda_0$ and a dielectric substrate with a relative permittivity of 2.5. Figure 9.7 shows the active input resistance versus the scan angle for two different dielectric substrate thicknesses calculated using the current probe feed (Section 5.1.1), and the results

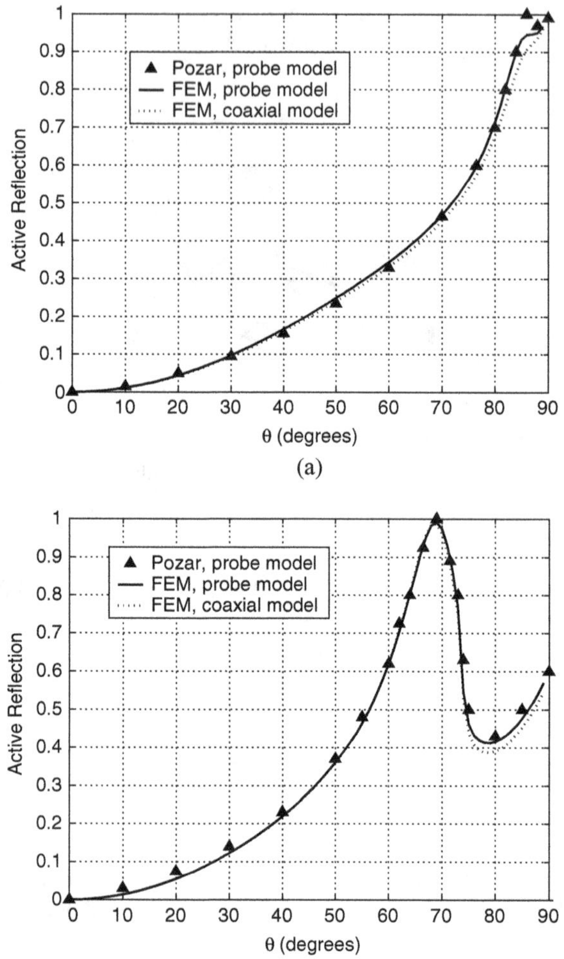

Figure 9.6 Active reflection coefficients for a microstrip patch array consisting of $0.3\lambda_0 \times 0.3\lambda_0$ rectangular patches ($T_x = T_y = 0.5\lambda_0$, $\varepsilon_r = 2.55$). (a) $h = 0.02\lambda_0$ and $s = 0.075\lambda_0$. (b) $h = 0.06\lambda_0$ and $s = 0.14\lambda_0$. (After Lou and Jin [7], Copyright © Wiley 2004.)

are compared with the moment-method solutions based on the same feed model [15]. Again, the two solutions agree very well. However, if we compare the calculations of the active input impedance versus frequency, we find that the simplified probe and precise coaxial feed models give significantly different results [7], especially at higher frequencies, and this difference is much more pronounced in the case of the thick substrate ($h = 0.1\lambda_0$) than in the case of the thin substrate ($h = 0.05\lambda_0$).

Next, we consider a circular patch array fed with coaxial lines. The periodic element spacings are given by $T_x = 34$ mm and $T_y = 36.1$ mm, and the patch radius is $R_o = 14.29$ mm. The dielectric substrate has a thickness of 0.79 mm and a relative

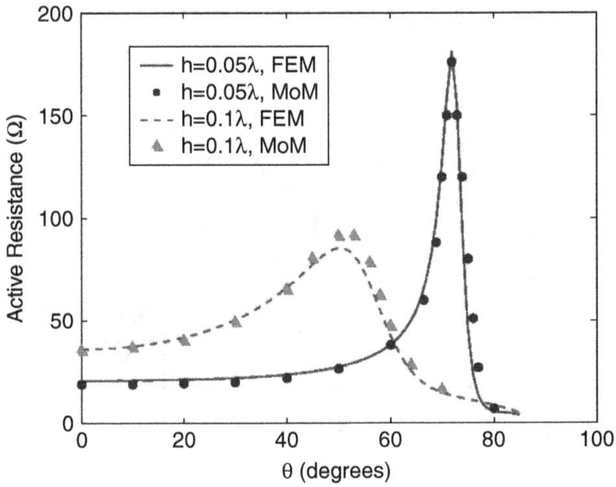

Figure 9.7 Active resistance versus scan angle for a microstrip patch array consisting of $0.25\lambda_0 \times 0.25\lambda_0$ rectangular patches ($T_x = T_y = 0.5\lambda_0$, $\varepsilon_r = 2.5$).

permittivity of 2.33. The outer and inner radii of the coaxial line used in the simulation are $r_o = 1.492$ mm and $r_i = 0.456$ mm, respectively, and it is filled with a dielectric having $\varepsilon_r = 2.024$. The S_{11} parameter measured at the coaxial port is shown in Figure 9.8. Both the magnitude and phase of the S_{11} parameter calculated agree with the results obtained by the generalized scattering matrix (GSM) combined with the finite element method [8] since in both simulations the coaxial feed model is used. These results also agree very well with those of experiments [8].

9.2 TIME-DOMAIN MODELING

As illustrated above, when analyzing an infinitely periodic antenna array using a unit cell, it is necessary to impose periodic boundary conditions which reflect a uniform phase shift between the antenna elements. This is straightforward in the frequency domain. It is, however, more challenging in the time domain when the antenna elements are excited by a pulse in time with an appropriate time shift to steer the main beam into a particular direction. The difficulty of enforcing the periodic boundary conditions for this case is due to the fact that the simple phase shift in the frequency domain translates into a time shift in the time domain. This time shift of the field at two parallel periodic surfaces requires future knowledge of the field at one of the two surfaces in order to enforce the periodic boundary conditions. Clearly, this is a major obstacle for a time-domain numerical method to overcome. Various methods have been suggested in the literature to tackle the problem of periodic boundary conditions in the time domain for the finite-difference time-domain (FDTD) analysis [16,17] and for time-domain integral-equation analysis [18]. One of the techniques originally

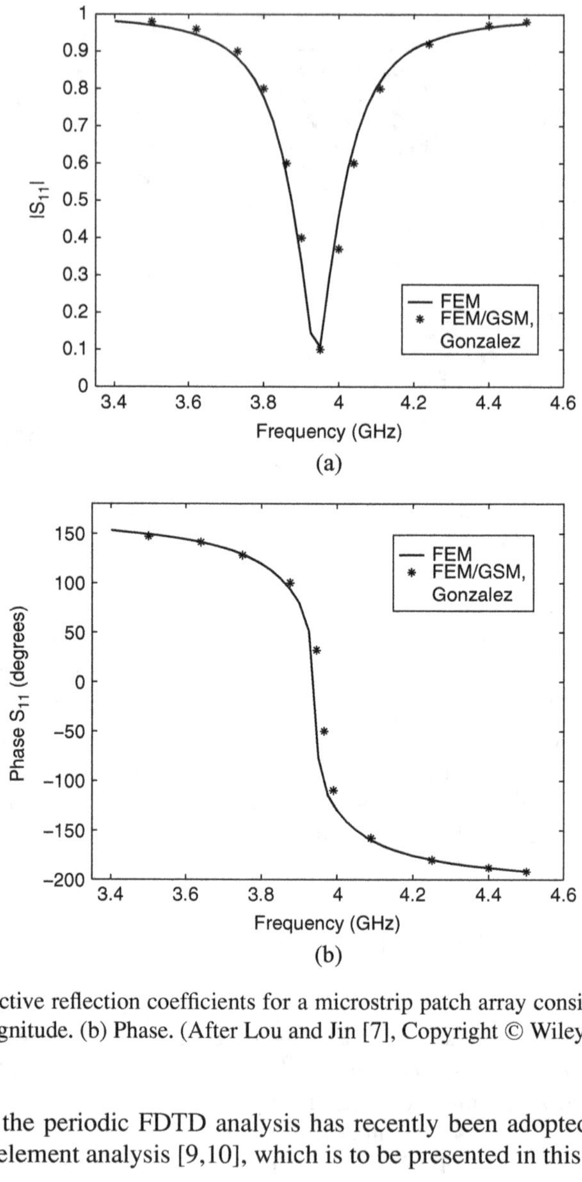

Figure 9.8 Active reflection coefficients for a microstrip patch array consisting of circular patches. (a) Magnitude. (b) Phase. (After Lou and Jin [7], Copyright © Wiley 2004.)

suggested for the periodic FDTD analysis has recently been adopted for the time-domain finite element analysis [9,10], which is to be presented in this section.

9.2.1 Transformed Field Variable

The method developed in Refs. 9 and 10 is based on transforming Maxwell's equations into a new set of equations involving transformed field variables instead of the regular field variables. By doing this, a transformed field variable at one periodic surface can be made equal to the transformed field variable at the parallel periodic surface for all time so that the periodic boundary conditions can easily be implemented.

To be more specific, we introduce the transformed field variable $\mathbf{P}_e$, which is related to the electric field vector by

$$\mathbf{E}(x, y, z; \omega) = \mathbf{P}_e(x, y, z; \omega) e^{-j(k_x^s x + k_y^s y)}. \tag{9.77}$$

By substituting (9.77) into (9.1), we obtain the periodic boundary conditions for $\mathbf{P}_e$ as

$$\mathbf{P}_e(x, y, z; \omega) = \begin{cases} \mathbf{P}_e(x + T_x, y, z; \omega) \\ \mathbf{P}_e(x, y + T_y, z; \omega). \end{cases} \tag{9.78}$$

Note that (9.78) remains the same in the time domain, which makes the periodic boundary conditions straightforward to enforce because there is no time delay between the fields at two parallel surfaces of the unit cell. By substituting the transformed field variable defined in (9.77) into the frequency-domain vector wave equation for the electric field,

$$\nabla \times \left(\overleftrightarrow{\mu}_r^{-1} \cdot \nabla \times \mathbf{E}\right) - k_0^2 \overleftrightarrow{\varepsilon}_r \cdot \mathbf{E} + j\omega\mu_0 \overleftrightarrow{\sigma} \cdot \mathbf{E} = -j\omega\mu_0 \mathbf{J}_{\text{imp}} - \nabla \times \left(\overleftrightarrow{\mu}_r^{-1} \cdot \mathbf{M}_{\text{imp}}\right) \tag{9.79}$$

and then transforming the resulting equation into the time domain, we obtain the following modified wave equation:

$$\begin{aligned} &\nabla \times \overleftrightarrow{\mu}_r^{-1} \cdot (\nabla \times \mathbf{P}_e) + \frac{1}{c^2} \overleftrightarrow{\varepsilon}_r \cdot \frac{\partial^2 \mathbf{P}_e}{\partial t^2} + \frac{1}{c^2} \hat{\mathbf{k}}_t^s \times \overleftrightarrow{\mu}_r^{-1} \cdot \left(\hat{\mathbf{k}}_t^s \times \frac{\partial^2 \mathbf{P}_e}{\partial t^2}\right) + \mu_0 \overleftrightarrow{\sigma} \cdot \frac{\partial \mathbf{P}_e}{\partial t} \\ &- \frac{1}{c} \nabla \times \overleftrightarrow{\mu}_r^{-1} \cdot \left(\hat{\mathbf{k}}_t^s \times \frac{\partial \mathbf{P}_e}{\partial t}\right) - \frac{1}{c} \hat{\mathbf{k}}_t^s \times \overleftrightarrow{\mu}_r^{-1} \cdot \left(\nabla \times \frac{\partial \mathbf{P}_e}{\partial t}\right) = \mathbf{G}(\mathbf{J}_{\text{imp}}, \mathbf{M}_{\text{imp}}) \end{aligned} \tag{9.80}$$

where $\mathbf{P}_e = \mathbf{P}_e(x, y, z; t)$, $\hat{\mathbf{k}}_t^s = \sin\theta_s \cos\phi_s \hat{x} + \sin\theta_s \sin\phi_s \hat{y}$, and $\mathbf{G}(\mathbf{J}_{\text{imp}}, \mathbf{M}_{\text{imp}})$ is the sum of the source terms:

$$\begin{aligned} \mathbf{G}(\mathbf{J}_{\text{imp}}, \mathbf{M}_{\text{imp}}) = &-\mu_0 \frac{\partial}{\partial t} \mathbf{J}_{\text{imp}}\left(t + \frac{\hat{\mathbf{k}}_t^s \cdot \mathbf{r}}{c}\right) - \nabla \times \left[\overleftrightarrow{\mu}_r^{-1} \cdot \mathbf{M}_{\text{imp}}\left(t + \frac{\hat{\mathbf{k}}_t^s \cdot \mathbf{r}}{c}\right)\right] \\ &+ \frac{\hat{\mathbf{k}}_t^s}{c} \times \left[\overleftrightarrow{\mu}_r^{-1} \cdot \frac{\partial}{\partial t} \mathbf{M}_{\text{imp}}\left(t + \frac{\hat{\mathbf{k}}_t^s \cdot \mathbf{r}}{c}\right)\right]. \end{aligned} \tag{9.81}$$

Equation (9.80) is to be solved over a unit cell with the associated truncation boundaries shown in Figure 9.2. With the coordinate system shown for the unit cell in Figure 9.2, we have the following periodic boundary conditions:

$$\mathbf{P}_e(x = 0, y, z; t) = \mathbf{P}_e(x = T_x, y, z; t) \tag{9.82}$$

$$\mathbf{P}_e(x, y = 0, z; t) = \mathbf{P}_e(x, y = T_y, z; t). \tag{9.83}$$

The weak-form representation for (9.80) can be formulated by taking the dot product with a testing function and integrating over the unit cell, yielding

$$\iiint_{V_{\mathrm{uc}}} \left\{ (\nabla \times \mathbf{T}) \cdot \overleftrightarrow{\mu}_r^{-1} \cdot (\nabla \times \mathbf{P}_e) + \frac{1}{c^2} \mathbf{T} \cdot \overleftrightarrow{\varepsilon}_r \cdot \frac{\partial^2 \mathbf{P}_e}{\partial t^2} - \frac{1}{c^2} (\hat{\mathbf{k}}_t^s \times \mathbf{T}) \cdot \overleftrightarrow{\mu}_r^{-1} \cdot \left(\hat{\mathbf{k}}_t^s \times \frac{\partial^2 \mathbf{P}_e}{\partial t^2} \right) \right.$$
$$\left. + \mu_0 \mathbf{T} \cdot \overleftrightarrow{\sigma} \cdot \frac{\partial \mathbf{P}_e}{\partial t} + \frac{1}{c} \left[\left(\hat{\mathbf{k}}_t^s \times \mathbf{T} \right) \cdot \overleftrightarrow{\mu}_r^{-1} \cdot \left(\nabla \times \frac{\partial \mathbf{P}_e}{\partial t} \right) - (\nabla \times \mathbf{T}) \cdot \overleftrightarrow{\mu}_r^{-1} \cdot \left(\hat{\mathbf{k}}_t^s \times \frac{\partial \mathbf{P}_e}{\partial t} \right) \right] \right\} dV$$
$$+ \iint_{S_{\mathrm{uc}} \cup S_T} \mathbf{T} \cdot \left[\hat{n} \times \overleftrightarrow{\mu}_r^{-1} \cdot \left(\nabla \times \mathbf{P}_e - \frac{1}{c} \hat{\mathbf{k}}_t^s \times \frac{\partial \mathbf{P}_e}{\partial t} \right) \right] dS = \iiint_{V_{\mathrm{uc}}} \mathbf{T} \cdot \mathbf{G}(\mathbf{J}_{\mathrm{imp}}, \mathbf{M}_{\mathrm{imp}})\, dV \tag{9.84}$$

where we have applied vector identities and Gauss's divergence theorem. An alternative and much easier approach to deriving (9.84) is to start with the weak-form representation of (9.79) using $\mathbf{T}e^{j(k_x^s x + k_y^s y)}$ as the testing function, substitute in (9.77), and then transform the resulting equation into the time domain.

The surface integral in (9.84) over the side surfaces S_{uc} can be used to implement the periodic boundary conditions, as we discussed in Section 9.1.1. The surface integral over the truncation surface S_T can be used to implement an absorbing boundary condition, as described in the next section.

9.2.2 Mesh Truncation Techniques

To complete the finite element formulation, we have to choose and implement a mesh truncation technique through the surface integral over S_T in (9.84). Although the use of the transformed field variable gives rise to two terms in this surface integral—one has the familiar form of $\hat{n} \times (\overleftrightarrow{\mu}_r^{-1} \cdot \nabla \times \mathbf{P}_e)$ and the other contains $\hat{n} \times \overleftrightarrow{\mu}_r^{-1} \cdot (\hat{\mathbf{k}}_t^s \times \partial \mathbf{P}_e / \partial t)$—both terms originate from $\hat{n} \times (\overleftrightarrow{\mu}_r^{-1} \cdot \nabla \times \mathbf{E})$. Therefore, the absorbing boundary conditions derived in Section 9.1.2 can be employed to derive the absorbing boundary conditions for the transformed field variable.

For simplicity, we first consider the use of the first-order absorbing boundary condition given in (9.63). By substituting (9.77) into (9.63), we obtain

$$\hat{z} \times (\nabla \times \mathbf{P}_e - jk_0 \hat{\mathbf{k}}_t^s \times \mathbf{P}_e) = \frac{jk_0}{\cos\theta_s} \hat{\mathbf{k}}_t^s (\hat{\mathbf{k}}_t^s \cdot \mathbf{P}_e) - jk_0 \cos\theta_s \hat{z} \times (\hat{z} \times \mathbf{P}_e) \qquad \forall z = z_c \tag{9.85}$$

which can be transformed into the time domain as

$$\hat{z} \times \left(\nabla \times \mathbf{P}_e - \frac{1}{c} \hat{\mathbf{k}}_t^s \times \frac{\partial \mathbf{P}_e}{\partial t} \right) = \frac{1}{c \cos\theta_s} \hat{\mathbf{k}}_t^s \left(\hat{\mathbf{k}}_t^s \cdot \frac{\partial \mathbf{P}_e}{\partial t} \right) - \frac{1}{c} \cos\theta_s \hat{z} \times \left(\hat{z} \times \frac{\partial \mathbf{P}_e}{\partial t} \right) \qquad \forall z = z_c. \tag{9.86}$$

Substituting (9.86) into the surface integral over S_T in (9.84) results in a successful implementation of the first-order absorbing boundary condition, which absorbs perfectly the fundamental Floquet mode in any direction. The resulting finite element matrix equation is given by

$$[T]\frac{d^2\{P_e\}}{dt^2} + [R + R^{\text{ABC}}]\frac{d\{P_e\}}{dt} + [S]\{P_e\} = \{f\} \tag{9.87}$$

where

$$T_{ij} = \frac{1}{c^2}\iiint_{V_{\text{uc}}} \left[\mathbf{N}_i \cdot \overleftrightarrow{\varepsilon}_r \cdot \mathbf{N}_j - \left(\hat{\mathbf{k}}_t^s \times \mathbf{N}_i\right) \cdot \overleftrightarrow{\mu}_r^{-1} \cdot \left(\hat{\mathbf{k}}_t^s \times \mathbf{N}_j\right)\right] dV \tag{9.88}$$

$$R_{ij} = \iiint_{V_{\text{uc}}} \Big\{\mu_0 \mathbf{N}_i \cdot \overleftrightarrow{\sigma} \cdot \mathbf{N}_j + \frac{1}{c}\big[\left(\hat{\mathbf{k}}_t^s \times \mathbf{N}_i\right) \cdot \overleftrightarrow{\mu}_r^{-1} \cdot (\nabla \times \mathbf{N}_j)$$

$$- (\nabla \times \mathbf{N}_i) \cdot \overleftrightarrow{\mu}_r^{-1} \cdot \left(\hat{\mathbf{k}}_t^s \times \mathbf{N}_j\right)\big]\Big\}\, dV \tag{9.89}$$

$$R_{ij}^{\text{ABC}} = \iint_{S_T} \left[\frac{\cos\theta_s}{c}(\hat{z} \times \mathbf{N}_i) \cdot (\hat{z} \times \mathbf{N}_j) + \frac{1}{c\cos\theta_s}\left(\hat{\mathbf{k}}_t^s \cdot \mathbf{N}_i\right)\left(\hat{\mathbf{k}}_t^s \cdot \mathbf{N}_j\right)\right] dS \tag{9.90}$$

$$S_{ij} = \iiint_{V_{\text{uc}}} (\nabla \times \mathbf{N}_i) \cdot \overleftrightarrow{\mu}_r^{-1} \cdot (\nabla \times \mathbf{N}_j)\, dV \tag{9.91}$$

$$f_i = \iiint_{V_{\text{uc}}} \mathbf{N}_i \cdot \mathbf{G}(\mathbf{J}_{\text{imp}}, \mathbf{M}_{\text{imp}})\, dV. \tag{9.92}$$

Equation (9.87) can be solved by time marching using the Newmark-beta integration method, as discussed in Chapter 2.

The first-order absorbing boundary condition given in (9.86) can absorb the fundamental Floquet mode propagating in any direction. For most practical applications where only the fundamental Floquet mode is excited, this mesh truncation is sufficiently accurate. An example is a well-designed phased array having no grating lobes. However, occasionally in certain periodic phased arrays and in many scattering applications, propagating higher-order modes may be important. More sophisticated boundary conditions for absorbing higher-order Floquet modes can be derived by taking these higher-order modes into account [10]. The most accurate one is the periodic boundary integral equation in (9.55), which in terms of the transformed field variable ($\mathbf{P}_e$) is given by

$$\hat{z} \times \left(\nabla \times \mathbf{P}_e - jk_0\hat{\mathbf{k}}_t^s \times \mathbf{P}_e\right) = \sum_{p=-\infty}^{\infty}\sum_{q=-\infty}^{\infty}\left[\frac{j\mathbf{k}_{tpq}}{k_{zpq}}(\mathbf{k}_{tpq} \cdot \tilde{\mathbf{P}}_{pq}) - jk_{zpq}\hat{z} \times (\hat{z} \times \tilde{\mathbf{P}}_{pq})\right]$$

$$\times\, e^{j(\kappa_{xp}x + \kappa_{yq}y)} \qquad \forall z \geq z_c \tag{9.93}$$

where $\kappa_{xp} = 2\pi p/T_x$, $\kappa_{yq} = 2\pi q/T_y$, and $\tilde{\mathbf{P}}_{pq}$ is given by (9.47). To transform this into the time domain, we first rewrite $\mathbf{k}_{tpq}(\mathbf{k}_{tpq} \cdot \tilde{\mathbf{P}}_{pq})$ as

$$\mathbf{k}_{tpq}(\mathbf{k}_{tpq} \cdot \tilde{\mathbf{P}}_{pq}) = \left(k_0^2 \hat{\mathbf{k}}_t^s \hat{\mathbf{k}}_t^s - k_0 \hat{\mathbf{k}}_t^s \boldsymbol{\kappa}_{pq} - k_0 \boldsymbol{\kappa}_{pq} \hat{\mathbf{k}}_t^s + \boldsymbol{\kappa}_{pq} \boldsymbol{\kappa}_{pq}\right) \cdot \tilde{\mathbf{P}}_{pq} \tag{9.94}$$

where $\boldsymbol{\kappa}_{pq} = \kappa_{xp}\hat{x} + \kappa_{yq}\hat{y}$. Hence, (9.93) can be written as

$$\hat{z} \times \left(\nabla \times \mathbf{P}_e - jk_0 \hat{\mathbf{k}}_t^s \times \mathbf{P}_e\right) = \sum_{p=-\infty}^{\infty} \sum_{q=-\infty}^{\infty} \Big\{\hat{\mathbf{k}}_t^s \hat{\mathbf{k}}_t^s \cdot W(\tilde{\mathbf{P}}_{pq}) + \left(\hat{\mathbf{k}}_t^s \boldsymbol{\kappa}_{pq} + \boldsymbol{\kappa}_{pq} \hat{\mathbf{k}}_t^s\right) \cdot X(\tilde{\mathbf{P}}_{pq})$$
$$+ \boldsymbol{\kappa}_{pq} \boldsymbol{\kappa}_{pq} \cdot Y(\tilde{\mathbf{P}}_{pq}) - \hat{z} \times [\hat{z} \times Z(\tilde{\mathbf{P}}_{pq})]\Big\} e^{j(\kappa_{xp}x + \kappa_{yq}y)} \tag{9.95}$$

where

$$W(\tilde{\mathbf{P}}_{pq}) = \frac{jk_0^2}{k_{zpq}} \tilde{\mathbf{P}}_{pq} \tag{9.96}$$

$$X(\tilde{\mathbf{P}}_{pq}) = -\frac{jk_0}{k_{zpq}} \tilde{\mathbf{P}}_{pq} \tag{9.97}$$

$$Y(\tilde{\mathbf{P}}_{pq}) = \frac{j}{k_{zpq}} \tilde{\mathbf{P}}_{pq} \tag{9.98}$$

$$Z(\tilde{\mathbf{P}}_{pq}) = jk_{zpq} \tilde{\mathbf{P}}_{pq}. \tag{9.99}$$

These four functions can be transformed into the time domain if we rewrite jk_{zpq} as

$$jk_{zpq} = \frac{\cos\theta_s}{c} \sqrt{a_{pq}^2 + (s + jb_{pq})^2} \tag{9.100}$$

where $s = j\omega$, $a_{pq} = \sqrt{c^2 \kappa_{pq}^2 / \cos^2\theta_s + b_{pq}^2}$, and $b_{pq} = c\hat{\mathbf{k}}_t^s \cdot \boldsymbol{\kappa}_{pq} / \cos^2\theta_s$, and use the Laplace transform pairs

$$\frac{1}{\sqrt{a^2 + (s+b)^2}} \leftrightarrow J_0(at) e^{-bt} u(t) \tag{9.101}$$

$$\sqrt{a^2 + (s+b)^2} \leftrightarrow \frac{\partial}{\partial t} + b\delta(t) + \frac{a}{t} J_1(at) e^{-bt} u(t) \tag{9.102}$$

and the operation rules

$$sF(s) \leftrightarrow \frac{\partial}{\partial t} f(t) \tag{9.103}$$

$$s^2 F(s) \leftrightarrow \frac{\partial^2}{\partial t^2} f(t) \tag{9.104}$$

along with the fact that multiplication of two functions in one domain results in a convolution in the other domain. In the equations above, J_0 is the zeroth-order Bessel function of the first kind, J_1 is the first-order Bessel function of the first kind, δ denotes the Dirac delta function, and u denotes the unit step function. The results are

$$W(\tilde{\mathbf{P}}_{pq}) \leftrightarrow \frac{1}{c\cos\theta_s}\left[\frac{\partial}{\partial t}\tilde{\mathbf{P}}_{pq}(t) + w_{pq}(t) * \tilde{\mathbf{P}}_{pq}(t)\right] \tag{9.105}$$

$$X(\tilde{\mathbf{P}}_{pq}) \leftrightarrow \frac{j}{\cos\theta_s} x_{pq}(t) * \tilde{\mathbf{P}}_{pq}(t) \tag{9.106}$$

$$Y(\tilde{\mathbf{P}}_{pq}) \leftrightarrow \frac{c}{\cos\theta_s} y_{pq}(t) * \tilde{\mathbf{P}}_{pq}(t) \tag{9.107}$$

$$Z(\tilde{\mathbf{P}}_{pq}) \leftrightarrow \frac{\cos\theta_s}{c}\left[\frac{\partial}{\partial t}\tilde{\mathbf{P}}_{pq}(t) + z_{pq}(t) * \tilde{\mathbf{P}}_{pq}(t)\right] \tag{9.108}$$

where

$$\begin{aligned} w_{pq}(t) = {} & \left[\left(2a_{pq}b_{pq} + \frac{a_{pq}}{t}\right) J_1(a_{pq}t) - (a_{pq}^2 + b_{pq}^2) J_0(a_{pq}t)\right] \\ & \times e^{-jb_{pq}t} u(t) - jb_{pq}\delta(t) \end{aligned} \tag{9.109}$$

$$x_{pq}(t) = \left[a_{pq}J_1(a_{pq}t) + jb_{pq}J_0(a_{pq}t)\right] e^{-jb_{pq}t} u(t) - \delta(t) \tag{9.110}$$

$$y_{pq}(t) = -J_0(a_{pq}t) e^{-jb_{pq}t} u(t) \tag{9.111}$$

$$z_{pq}(t) = \frac{a_{pq}}{t} J_1(a_{pq}t) e^{-jb_{pq}t} u(t) + jb_{pq}\delta(t). \tag{9.112}$$

By substituting the time-domain counterpart of (9.95) into the surface integral over S_T in (9.84), we obtain the finite element matrix equation

$$[T]\frac{d^2\{P_e\}}{dt^2} + [R + R^{\text{PRC}}]\frac{d\{P_e\}}{dt} + [S + S^{\text{PRC}} *]\{P_e\} = \{f\} \tag{9.113}$$

where the elements of $[T]$, $[R]$, $[S]$, and $\{f\}$ are given by (9.88), (9.89), (9.91), and (9.92), respectively, and the elements of $[R^{\text{PRC}}]$ and $[S^{\text{PRC}}]$ are given by

$$\begin{aligned} R_{ij}^{\text{PRC}} = {} & \frac{1}{T_x T_y} \sum_{p=-\infty}^{\infty} \sum_{q=-\infty}^{\infty} \left[\frac{1}{c\cos\theta_s} (\hat{\mathbf{k}}_t^s \cdot \tilde{\mathbf{N}}_{i,pq}^*)(\hat{\mathbf{k}}_t^s \cdot \tilde{\mathbf{N}}_{j,pq}) \right. \\ & \left. + \frac{\cos\theta_s}{c} (\hat{z} \times \tilde{\mathbf{N}}_{i,pq}^*) \cdot (\hat{z} \times \tilde{\mathbf{N}}_{j,pq}) \right] \end{aligned} \tag{9.114}$$

$$S_{ij}^{\text{PRC}} = \frac{1}{T_x T_y} \sum_{p=-\infty}^{\infty} \sum_{q=-\infty}^{\infty} \left\{ \frac{1}{\cos\theta_s} \left[\frac{1}{c} \left(\hat{\mathbf{k}}_t^s \cdot \tilde{\mathbf{N}}_{i,pq}^*\right)\left(\hat{\mathbf{k}}_t^s \cdot \tilde{\mathbf{N}}_{j,pq}\right) w_{pq}(t) \right.\right.$$
$$+ j\left[\left(\hat{\mathbf{k}}_t^s \cdot \tilde{\mathbf{N}}_{i,pq}^*\right)\left(\kappa_{pq} \cdot \tilde{\mathbf{N}}_{j,pq}\right) + \left(\kappa_{pq} \cdot \tilde{\mathbf{N}}_{i,pq}^*\right)\left(\hat{\mathbf{k}}_t^s \cdot \tilde{\mathbf{N}}_{j,pq}\right)\right] x_{pq}(t)$$
$$\left.\left. + c\left(\kappa_{pq} \cdot \tilde{\mathbf{N}}_{i,pq}^*\right)\left(\kappa_{pq} \cdot \tilde{\mathbf{N}}_{j,pq}\right) y_{pq}(t) \right] + \frac{\cos\theta_s}{c}\left(\hat{z} \times \tilde{\mathbf{N}}_{i,pq}^*\right)\left(\hat{z} \times \tilde{\mathbf{N}}_{j,pq}\right) z_{pq}(t) \right\} \tag{9.115}$$

in which

$$\tilde{\mathbf{N}}_{i,pq} = \frac{1}{T_x T_y} \iint_{S_T} \mathbf{N}_i \, e^{-j(\kappa_{xp} x + \kappa_{yq} y)} \, dS. \tag{9.116}$$

Note the difference between (9.116) and (9.60), although the same symbol is used [κ_{xp} and κ_{yq} are used in (9.116), whereas k_{xp} and k_{yq} were used in (9.60), in addition to the sign difference in the exponential function]. The detailed implementation of the formulation above is described in Ref. 10, which is quite complicated. However, the resulting method is highly effective, and numerical experiments show that highly accurate solutions are always obtained with a few higher-order Floquet modes included in the double summations in (9.114) and (9.115).

9.2.3 General Material Modeling

Whereas it is straightforward to model general dispersive, lossy materials in the frequency-domain simulation, the modeling in the time domain is much more challenging, as we have demonstrated in Chapter 2. The unit cell analysis of general materials in the time domain that is based on the transformed field variable is the topic of this section.

The finite element formulation for anisotropic, lossy, and dispersive electric and magnetic materials was described in Section 2.3.4. For convenience, the fundamental equations are restated here:

$$\nabla \times \mathbf{E}(t) = -\frac{\partial \mathbf{B}(t)}{\partial t} - \overleftrightarrow{\sigma}_m \cdot \mathbf{H}(t) \tag{9.117}$$

$$\nabla \times \mathbf{H}(t) = \frac{\partial \mathbf{D}(t)}{\partial t} + \overleftrightarrow{\sigma}_e \cdot \mathbf{E}(t) + \mathbf{J}_{\text{imp}}(t) \tag{9.118}$$

where $\mathbf{D}(t)$ is related to $\mathbf{E}(t)$ by the constitutive relation

$$\mathbf{D}(t) = \varepsilon_0 \overleftrightarrow{\varepsilon}_\infty \cdot \mathbf{E}(t) + \varepsilon_0 \overleftrightarrow{\chi}_e(t) * \mathbf{E}(t) \tag{9.119}$$

and $\mathbf{B}(t)$ is related to $\mathbf{H}(t)$ by the constitutive relation

$$\mathbf{B}(t) = \mu_0 \overleftrightarrow{\mu}_\infty \cdot \mathbf{H}(t) + \mu_0 \overleftrightarrow{\chi}_m(t) * \mathbf{H}(t). \tag{9.120}$$

Note that for nondispersive materials, $\overleftrightarrow{\varepsilon}_\infty \to \overleftrightarrow{\varepsilon}_r$, $\overleftrightarrow{\mu}_\infty \to \overleftrightarrow{\mu}_r$, and the susceptibility tensors are zero. The magnetic current term is omitted in (9.117) for simplicity.

The transformed field variable technique is now applied to (9.117)–(9.120) and the approach will be based on the **E–H** concept of Section 2.3.4. The electric field $\mathbf{E}(\omega)$ is replaced by $\mathbf{P}_e$, as defined in (9.77), and a similar transformation is introduced for the magnetic field $\mathbf{H}(\omega)$:

$$\mathbf{H}(x, y, z; \omega) = \mathbf{P}_h(x, y, z; \omega) e^{-j(k_x^s x + k_y^s y)}. \tag{9.121}$$

The time-domain equations (9.117) and (9.118) then become

$$\mu_0 \overleftrightarrow{\mu}_\infty \cdot \frac{\partial \mathbf{P}_h(t)}{\partial t} = -\nabla \times \mathbf{P}_e(t) + \frac{1}{c}\hat{\mathbf{k}}_t^s \times \frac{\partial \mathbf{P}_e}{\partial t} - \overleftrightarrow{\sigma}_m \cdot \mathbf{P}_h(t) - \mu_0 \dot{\overleftrightarrow{\chi}}_m(t) * \mathbf{P}_h(t) \tag{9.122}$$

$$\varepsilon_0 \overleftrightarrow{\varepsilon}_\infty \cdot \frac{\partial \mathbf{P}_e(t)}{\partial t} = \nabla \times \mathbf{P}_h(t) - \frac{1}{c}\hat{\mathbf{k}}_t^s \times \frac{\partial \mathbf{P}_h}{\partial t} - \overleftrightarrow{\sigma}_e \cdot \mathbf{P}_e(t) - \varepsilon_0 \dot{\overleftrightarrow{\chi}}_e(t) * \mathbf{P}_e(t) - \mathbf{J}_{\text{imp}}\left(t + \frac{\hat{\mathbf{k}}_t^s \cdot \mathbf{r}}{c}\right) \tag{9.123}$$

where the constitutive relations have been used. Differentiating (9.123) with respect to time and integrating its weighted expression over the volume of the unit cell leads to the weak-form equation

$$\iiint_{V_{\text{uc}}} \left\{ -(\nabla \times \mathbf{T}) \cdot \frac{\partial \mathbf{P}_h(t)}{\partial t} + \mathbf{T} \cdot \left[\frac{1}{c}\hat{\mathbf{k}}_t^s \times \frac{\partial^2 \mathbf{P}_h(t)}{\partial t^2}\right] + \varepsilon_0 \mathbf{T} \cdot \overleftrightarrow{\varepsilon}_\infty \cdot \frac{\partial^2 \mathbf{P}_e(t)}{\partial t^2} + \mathbf{T} \cdot \overleftrightarrow{\sigma}_e \cdot \frac{\partial \mathbf{P}_e(t)}{\partial t} + \varepsilon_0 \mathbf{T} \cdot \ddot{\overleftrightarrow{\chi}}_e(t) * \mathbf{P}_e(t) \right\} dV - \iint_{S_{\text{uc}} \cup S_T} \mathbf{T} \cdot \left[\hat{n} \times \frac{\partial \mathbf{P}_h(t)}{\partial t}\right] dS = -\iiint_{V_{\text{uc}}} \mathbf{T} \cdot \frac{\partial}{\partial t}\mathbf{J}_{\text{imp}}\left(t + \frac{\hat{\mathbf{k}}_t^s \cdot \mathbf{r}}{c}\right) dV. \tag{9.124}$$

Note that by differentiating (9.122) with respect to time, the second term in (9.124) can be written as

$$\mu_0 \frac{1}{c}\hat{\mathbf{k}}_t^s \times \frac{\partial^2 \mathbf{P}_h}{\partial t^2} = \frac{\hat{\mathbf{k}}_t^s}{c} \times \left\{ \overleftrightarrow{\mu}_\infty^{-1} \cdot \left[-\nabla \times \frac{\partial \mathbf{P}_e(t)}{\partial t} + \frac{1}{c}\hat{\mathbf{k}}_t^s \times \frac{\partial^2 \mathbf{P}_e(t)}{\partial t^2} - \overleftrightarrow{\sigma}_m \cdot \frac{\partial}{\partial t}\mathbf{P}_h(t) - \mu_0 \ddot{\overleftrightarrow{\chi}}_m * \frac{\partial^2}{\partial t^2}\mathbf{P}_h(t) \right] \right\} \tag{9.125}$$

where the time derivatives in the convolution have been transferred equivalently onto the fields.

To ultimately obtain an unconditionally stable numerical scheme that solves self-consistently for both $\mathbf{P}_e$ and $\mathbf{P}_h$, we add and subtract $\iiint_{V_{uc}} (\nabla \times \mathbf{T}) \cdot \overleftrightarrow{\mu}_{\infty}^{-1} \cdot (\nabla \times \mathbf{P}_e)\, dV$ to (9.124) and then substitute $\nabla \times \mathbf{P}_e$ in the subtracted term with the expression derived from (9.122). By doing so, the weak-form equation can then be written as

$$
\begin{aligned}
&\iiint_{V_{uc}} \Bigg\{ (\nabla \times \mathbf{T}) \cdot \overleftrightarrow{\mu}_{\infty}^{-1} \cdot (\nabla \times \mathbf{P}_e) + \mu_0\, \mathbf{T} \cdot \left[\frac{1}{c} \hat{\mathbf{k}}_t^s \times \frac{\partial^2 \mathbf{P}_h}{\partial t^2} \right] \\
&\quad - (\nabla \times \mathbf{T}) \cdot \left[\frac{\hat{\mathbf{k}}_t^s}{c} \times \frac{\partial \mathbf{P}_e}{\partial t} - \overleftrightarrow{\sigma}_m \cdot \mathbf{P}_h - \mu_0 \overleftrightarrow{\chi}_m * \frac{\partial \mathbf{P}_h}{\partial t} \right] \\
&\quad + \frac{1}{c^2} \mathbf{T} \cdot \overleftrightarrow{\varepsilon}_{\infty} \cdot \frac{\partial^2 \mathbf{P}_e}{\partial t^2} + \mu_0 \mathbf{T} \cdot \overleftrightarrow{\sigma}_e \cdot \frac{\partial \mathbf{P}_e}{\partial t} + \frac{1}{c^2} \mathbf{T} \cdot \overleftrightarrow{\chi}_e * \frac{\partial^2 \mathbf{P}_e}{\partial t^2} \Bigg\}\, dV \\
&\quad - \mu_0 \iint_{S_{uc} \cup S_T} \mathbf{T} \cdot \left[\hat{n} \times \frac{\partial \mathbf{P}_h}{\partial t} \right] dS = -\mu_0 \iiint_{V_{uc}} \mathbf{T} \cdot \frac{\partial}{\partial t} \mathbf{J}_{\text{imp}} \left(t + \frac{\hat{\mathbf{k}}_t^s \cdot \mathbf{r}}{c} \right) dV.
\end{aligned}
\tag{9.126}
$$

We now consider the finite element solution of (9.126) and restrict development to the recursive convolution formulation for the susceptibility tensors, as discussed in Chapter 2. Although (9.126) involves many terms, the solution is relatively straightforward. The basic strategy will be to solve a finite element equation for $\mathbf{P}_e$ that advances $\mathbf{P}_e$ to time $t = (n+1)\Delta t$, and solve a difference equation for $\mathbf{P}_h$ that advances $\mathbf{P}_h$ to time $t = (n+1/2)\Delta t$. However, for oblique scan angles, we will find that there will be time dependencies between $\mathbf{P}_e^{n+1}$ and $\mathbf{P}_h^{n+1/2}$ that will require special attention.

We begin by noting that for both the electric and magnetic transformed fields, the convolutions can be written as

$$
\overleftrightarrow{\chi}_m(t) * \left. \frac{\partial \mathbf{P}_h(t)}{\partial t} \right|_{t=n\Delta t} \cong \overleftrightarrow{\chi}_m^0 \cdot \left. \frac{\partial \mathbf{P}_h(t)}{\partial t} \right|_{t=n\Delta t} + \mathbf{\Phi}^n \tag{9.127}
$$

$$
\overleftrightarrow{\chi}_e(t) * \left. \frac{\partial^2 \mathbf{P}_e(t)}{\partial t^2} \right|_{t=n\Delta t} \cong \overleftrightarrow{\chi}_e^0 \cdot \left. \frac{\partial^2 \mathbf{P}_e(t)}{\partial t^2} \right|_{t=n\Delta t} + \mathbf{\Psi}^n \tag{9.128}
$$

where

$$
\mathbf{\Phi}^n = \sum_{k=1}^{n} \int_{(k-1/2)\Delta t}^{(k+1/2)\Delta t} \overleftrightarrow{\chi}_m(\tau) \cdot \dot{\mathbf{P}}_h^{n-k}\, d\tau = \frac{1}{\Delta t} \sum_{k=1}^{n} \overleftrightarrow{\chi}_m^k \cdot \left(\mathbf{P}_h^{n-k+1/2} - \mathbf{P}_h^{n-k-1/2} \right) \tag{9.129}
$$

$$
\mathbf{\Psi}^n = \sum_{k=1}^{n} \int_{(k-1/2)\Delta t}^{(k+1/2)\Delta t} \overleftrightarrow{\chi}_e(\tau) \cdot \ddot{\mathbf{P}}_e^{n-k}\, d\tau = \frac{1}{\Delta t^2} \sum_{k=1}^{n} \overleftrightarrow{\chi}_e^k \cdot \left(\mathbf{P}_e^{n-k+1} - 2\mathbf{P}_e^{n-k} + \mathbf{P}_e^{n-k-1} \right) \tag{9.130}
$$

in which

$$\overset{\leftrightarrow}{\chi}{}^{k}_{e,m} = \int_{(k-1/2)\Delta t}^{(k+1/2)\Delta t} \overset{\leftrightarrow}{\chi}_{e,m}(\tau)\,d\tau. \tag{9.131}$$

The fields are assumed to be identically zero at zero time, and causality of the susceptibility tensors is used in the evaluation of (9.131) for the case $k = 0$ required in (9.127) and (9.128). If the electric and magnetic susceptibilities can be represented by a pole expansion such that

$$\overset{\leftrightarrow}{\chi}_{e,m}(t) = \mathrm{Re}\{\overset{\leftrightarrow}{a}_{e,m}\, e^{-b_{e,m} t}\}\, u(t) \tag{9.132}$$

both (9.129) and (9.130) can be evaluated efficiently using the recursive relations

$$\hat{\mathbf{\Phi}}^{n} = \frac{1}{\Delta t} \overset{\leftrightarrow}{\chi}{}^{1}_{m} \cdot \left(\mathbf{P}_{h}^{n-1/2} - \mathbf{P}_{h}^{n-3/2}\right) + e^{-b_m \Delta t}\, \hat{\mathbf{\Phi}}^{n-1} \tag{9.133}$$

$$\hat{\mathbf{\Psi}}^{n} = \frac{1}{\Delta t^2} \overset{\leftrightarrow}{\chi}{}^{1}_{e} \cdot \left(\mathbf{P}_{e}^{n} - 2\mathbf{P}_{e}^{n-1} + \mathbf{P}_{e}^{n-2}\right) + e^{-b_e \Delta t}\, \hat{\mathbf{\Psi}}^{n-1} \tag{9.134}$$

and $\mathbf{\Phi}^n = \mathrm{Re}\{\hat{\mathbf{\Phi}}^n\}$, $\mathbf{\Psi}^n = \mathrm{Re}\{\hat{\mathbf{\Psi}}^n\}$. Such a recursive evaluation of the convolutions has been discussed in Chapter 2.

To formulate a time-marching equation for $\mathbf{P}_e$ from (9.126), we need to convert the term involving the second time derivative on $\mathbf{P}_h$ into an equivalent expression without such a second derivative. This can be accomplished by differentiating (9.127) with respect to time and then substituting into (9.125) to obtain the semidiscrete form

$$\begin{aligned} \mu_0 \frac{1}{c} \hat{\mathbf{k}}_t^s \times \left.\frac{\partial^2 \mathbf{P}_h}{\partial t^2}\right|_{t=n\Delta t} &= \frac{\hat{\mathbf{k}}_t^s}{c} \times \overset{\leftrightarrow}{\xi}{}^{-1} \cdot \left[-\nabla \times \left.\frac{\partial \mathbf{P}_e}{\partial t}\right|_{t=n\Delta t} + \frac{1}{c}\hat{\mathbf{k}}_t^s \times \left.\frac{\partial^2 \mathbf{P}_e}{\partial t^2}\right|_{t=n\Delta t} \right. \\ &\quad \left. - \overset{\leftrightarrow}{\sigma}_m \cdot \left.\frac{\partial \mathbf{P}_h}{\partial t}\right|_{t=n\Delta t} - \mu_0 \left.\frac{\partial \mathbf{\Phi}^n}{\partial t}\right|_{t=n\Delta t} \right] \end{aligned} \tag{9.135}$$

where $\overset{\leftrightarrow}{\xi} = \mathrm{Re}\{\overset{\leftrightarrow}{\chi}{}^{0}_{m}\} + \overset{\leftrightarrow}{\mu}_{\infty}$. Using (9.127) and (9.135), we can now write (9.126) as

$$\begin{aligned} \iiint_{V_{\mathrm{uc}}} \Bigg\{ & (\nabla \times \mathbf{T}) \cdot \overset{\leftrightarrow}{\mu}{}^{-1}_{\infty} \cdot (\nabla \times \mathbf{P}_e|_{t=n\Delta t}) \\ & + \mathbf{T} \cdot \frac{\hat{\mathbf{k}}_t^s}{c} \times \left[\overset{\leftrightarrow}{\xi}{}^{-1} \cdot \left(-\nabla \times \frac{\partial \mathbf{P}_e}{\partial t} + \frac{1}{c}\hat{\mathbf{k}}_t^s \times \frac{\partial^2 \mathbf{P}_e}{\partial t^2} - \overset{\leftrightarrow}{\sigma}_m \cdot \frac{\partial \mathbf{P}_h}{\partial t} - \mu_0 \frac{\partial \mathbf{\Phi}^n}{\partial t} \right)\Bigg|_{t=n\Delta t} \right] \\ & - (\nabla \times \mathbf{T}) \cdot \left[\left(\frac{\hat{\mathbf{k}}_t^s}{c} \times \frac{\partial \mathbf{P}_e}{\partial t} - \overset{\leftrightarrow}{\sigma}_m \cdot \mathbf{P}_h - \mu_0\, \mathrm{Re}\{\overset{\leftrightarrow}{\chi}{}^{0}_{m}\} \cdot \frac{\partial \mathbf{P}_h}{\partial t} \right)\Bigg|_{t=n\Delta t} - \mu_0 \mathbf{\Phi}^n \right] \end{aligned}$$

$$+\frac{1}{c^2}\mathbf{T}\cdot(\overleftrightarrow{\varepsilon}_\infty+\mathrm{Re}\{\overleftrightarrow{\chi}_e^0\})\cdot\left.\frac{\partial^2\mathbf{P}_e}{\partial t^2}\right|_{t=n\Delta t}+\mu_0\mathbf{T}\cdot\overleftrightarrow{\sigma}_e\cdot\left.\frac{\partial\mathbf{P}_e}{\partial t}\right|_{t=n\Delta t}+\frac{1}{c^2}\mathbf{T}\cdot\mathbf{\Psi}^n\Bigg\}\,dV$$

$$-\mu_0\iint_{S_{\mathrm{uc}}\cup S_T}\mathbf{T}\cdot\left[\hat{n}\times\left.\frac{\partial\mathbf{P}_h}{\partial t}\right|_{t=n\Delta t}\right]dS$$

$$=-\mu_0\iiint_{V_{\mathrm{uc}}}\mathbf{T}\cdot\frac{\partial}{\partial t}\mathbf{J}_{\mathrm{imp}}\left.\left(t+\frac{\hat{\mathbf{k}}_t^s\cdot\mathbf{r}}{c}\right)\right|_{t=n\Delta t}dV. \tag{9.136}$$

In this equation, $\mathbf{\Psi}^n$ is defined by (9.130) from which a recurrence relationship similar to (9.134) will be obtained to accommodate the case of electric dispersion. The situation of both magnetic dispersion and oblique scan/incident angles introduces a further complication because $\mathbf{P}_h$ and its time derivative at time step n will be dependent on $\mathbf{P}_e$ at time step $n+1$. A technique to resolve this is the following. By using the second-order central-difference approximation for the time derivatives, a semidiscrete equation for (9.122) is given by

$$\mathbf{P}_h^{n+1/2}=\overleftrightarrow{\mathbf{\Lambda}}_+^{-1}\cdot(\overleftrightarrow{\mathbf{\Lambda}}_-\cdot\mathbf{P}_h^{n-1/2})-\Delta t\,\overleftrightarrow{\mathbf{\Lambda}}_+^{-1}\cdot(\nabla\times\mathbf{P}_e|_{t=n\Delta t})-\mu_0\,\Delta t\,\overleftrightarrow{\mathbf{\Lambda}}_+^{-1}\cdot\mathbf{\Phi}^n$$
$$+\Delta t\,\overleftrightarrow{\mathbf{\Lambda}}_+^{-1}\cdot\left(\frac{1}{c}\hat{\mathbf{k}}_t^s\times\left.\frac{\partial\mathbf{P}_e}{\partial t}\right|_{t=n\Delta t}\right) \tag{9.137}$$

where

$$\overleftrightarrow{\mathbf{\Lambda}}_\pm=\mu_0\overleftrightarrow{\xi}\pm\frac{\overleftrightarrow{\sigma}_m\,\Delta t}{2}. \tag{9.138}$$

Note that for the case of oblique scan/incident angles, (9.137) is dependent on $\mathbf{P}_e$ at time $t=(n+1)\Delta t$ when a centered difference is used for $\partial\mathbf{P}_e/\partial t$. To resolve this challenge, we can write $\mathbf{P}_h^{n+1/2}$ as

$$\mathbf{P}_h^{n+1/2}=\tilde{\mathbf{P}}_h^{n+1/2}+\Delta t\,\overleftrightarrow{\mathbf{\Lambda}}_+^{-1}\cdot\left(\frac{1}{c}\hat{\mathbf{k}}_t^s\times\left.\frac{\partial\mathbf{P}_e}{\partial t}\right|_{t=n\Delta t}\right) \tag{9.139}$$

where

$$\tilde{\mathbf{P}}_h^{n+1/2}=\overleftrightarrow{\mathbf{\Lambda}}_+^{-1}\cdot(\overleftrightarrow{\mathbf{\Lambda}}_-\cdot\mathbf{P}_h^{n-1/2})-\Delta t\,\overleftrightarrow{\mathbf{\Lambda}}_+^{-1}\cdot(\nabla\times\mathbf{P}_e|_{t=n\Delta t})-\mu_0\,\Delta t\,\overleftrightarrow{\mathbf{\Lambda}}_+^{-1}\cdot\mathbf{\Phi}^n. \tag{9.140}$$

This decomposition permits the term involving $\partial\mathbf{P}_e/\partial t$ to be extracted from (9.139) and included into the finite element formulation for $\mathbf{P}_e$ based on (9.136), whereas the partial field $\tilde{\mathbf{P}}_h^{n+1/2}$ contributes as an additional forcing function to the right-hand side of (9.136). Note that after $\mathbf{P}_e^{n+1}$ is obtained, it must be included in (9.139) to finalize the advancement of $\mathbf{P}_h^{n+1/2}$.

Similarly, the time derivative $\partial \boldsymbol{\Phi}^n/\partial t$ in (9.136) can be written as

$$\left.\frac{\partial \boldsymbol{\Phi}^n}{\partial t}\right|_{t=n\Delta t} = \frac{1}{2\,\Delta t}(\boldsymbol{\Phi}^{n+1} - \boldsymbol{\Phi}^{n-1}). \tag{9.141}$$

From the recursive relationship (9.133), along with (9.139), we obtain

$$\begin{aligned}
\hat{\boldsymbol{\Phi}}^{n+1} &= \frac{1}{\Delta t}\overleftrightarrow{\chi}^{1}_{m}\cdot\left(\mathbf{P}_h^{n+1/2} - \mathbf{P}_h^{n-1/2}\right) + e^{-b_m\Delta t}\,\hat{\boldsymbol{\Phi}}^n \\
&= \frac{1}{\Delta t}\overleftrightarrow{\chi}^{1}_{m}\cdot\left(\tilde{\mathbf{P}}_h^{n+1/2} - \mathbf{P}_h^{n-1/2}\right) + e^{-b_m\Delta t}\,\hat{\boldsymbol{\Phi}}^n + \frac{\Delta t}{\Delta t}\,\overleftrightarrow{\chi}^{1}_{m}\cdot\overleftrightarrow{\boldsymbol{\Lambda}}_{+}^{-1}\cdot\left(\frac{1}{c}\hat{\mathbf{k}}_t^s \times \left.\frac{\partial \mathbf{P}_e}{\partial t}\right|_{t=n\Delta t}\right) \\
&= \tilde{\boldsymbol{\Phi}}^{n+1} + \overleftrightarrow{\chi}^{1}_{m}\cdot\overleftrightarrow{\boldsymbol{\Lambda}}_{+}^{-1}\cdot\left(\frac{1}{c}\hat{\mathbf{k}}_t^s \times \left.\frac{\partial \mathbf{P}_e}{\partial t}\right|_{t=n\Delta t}\right)
\end{aligned} \tag{9.142}$$

where

$$\tilde{\boldsymbol{\Phi}}^{n+1} = \frac{1}{\Delta t}\overleftrightarrow{\chi}^{1}_{m}\cdot\left(\tilde{\mathbf{P}}_h^{n+1/2} - \mathbf{P}_h^{n-1/2}\right) + e^{-b_m\Delta t}\,\hat{\boldsymbol{\Phi}}^n. \tag{9.143}$$

The time derivative $\partial \boldsymbol{\Phi}^n/\partial t$ can now be written in semidiscrete form as

$$\left.\frac{\partial \boldsymbol{\Phi}^n}{\partial t}\right|_{t=n\Delta t} = \frac{1}{2\,\Delta t}\mathrm{Re}\{\tilde{\boldsymbol{\Phi}}^{n+1} - \hat{\boldsymbol{\Phi}}^{n-1}\} + \frac{1}{2\Delta t}\,\mathrm{Re}\{\overleftrightarrow{\chi}^{1}_{m}\}\cdot\overleftrightarrow{\boldsymbol{\Lambda}}_{+}^{-1}\cdot\left(\frac{1}{c}\hat{\mathbf{k}}_t^s \times \left.\frac{\partial \mathbf{P}_e}{\partial t}\right|_{t=n\Delta t}\right) \tag{9.144}$$

which allows the extraction of the $\partial \mathbf{P}_e/\partial t$ term and then including it into the traditional finite element formulation for $\mathbf{P}_e$ based on (9.136).

Decompositions of the types described by (9.139) and (9.144) permit a standard Newmark-beta discretization of (9.136) for $\mathbf{P}_e$. The solution procedure for (9.136) for oblique scan/incident angles is initially to advance the partial terms $\tilde{\mathbf{P}}_h^{n+1/2}$ and $\tilde{\boldsymbol{\Phi}}^{n+1}$ from (9.140) and (9.143), respectively, and then use these results as additional source terms in the right-hand side of (9.136). The $\mathbf{P}_e^{n+1}$ calculated is then used in (9.139) and (9.142) to build the full solutions for $\mathbf{P}_h^{n+1/2}$ and $\boldsymbol{\Phi}^{n+1}$, respectively. This iterative procedure continues for the duration of the simulation. Extensive numerical experiments have indicated that this algorithm is unconditionally stable with regard to the time step, scan/incident angle, and many forms of material dispersion, although a theoretical proof is not currently available.

To verify the validity of the formulation described above and test the accuracy and stability of its numerical implementation, we consider the problem of plane-wave incidence on a uniform magnetically dispersive slab, as illustrated in Figure 9.9(a). The slab has a thickness of 10 cm, and its complex relative permeability is plotted in Figure 9.9(b). This slab can be modeled as a periodic structure with a uniform unit cell; hence, the problem can be solved numerically using the time-domain finite element formulation described in this section. The computed reflection and transmission

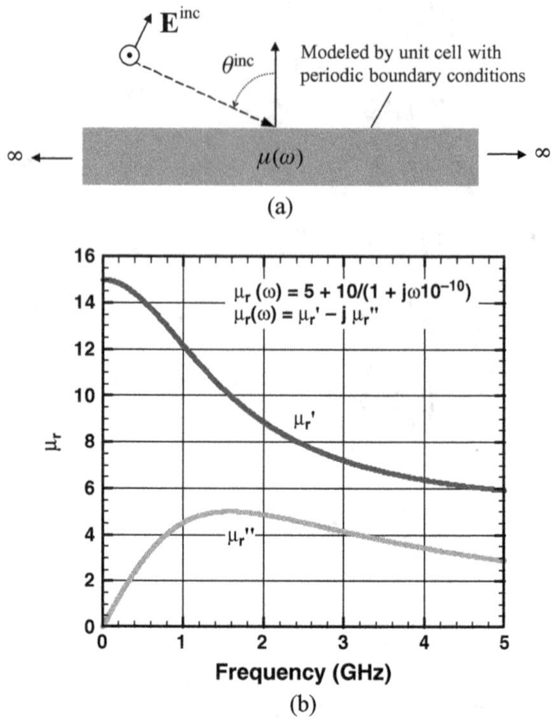

Figure 9.9 Magnetically dispersive slab with a thickness of 10 cm with an obliquely incident plane wave. (a) Geometry. (b) Dispersive permeability frequency profile of the slab.

coefficients are shown in Figures 9.10 and 9.11 for two angles of incidence and two polarizations. The results are compared with the exact solution, and it can be seen that the agreement is excellent for all the cases. Numerical examples for electrically dispersive and doubly dispersive dielectric slabs can be found in Ref. 19 along with an example that demonstrates the very high accuracy achieved with this formulation.

9.2.4 Application Examples

The numerical scheme described above can be implemented to analyze any infinitely periodic phased-array antenna. One issue that we need to be careful about is the handling of the excitation. When the antennas are excited by coaxial lines and the waveguide port boundary condition is employed to truncate the mesh at the coaxial port, the right-hand-side excitation term becomes $\mathbf{U}^{\text{inc}}(t + \hat{\mathbf{k}}_t^s \cdot \mathbf{r}/c)$ when the port boundary condition is converted for the transformed field variable and transformed into the time domain. A similar treatment is applied to the incident field for the plane-wave scattering analysis. Furthermore, once the transformed field variable is solved for, we have to convert it back to the electric field first and then calculate the radiation or scattering parameters. This conversion can be done in the frequency

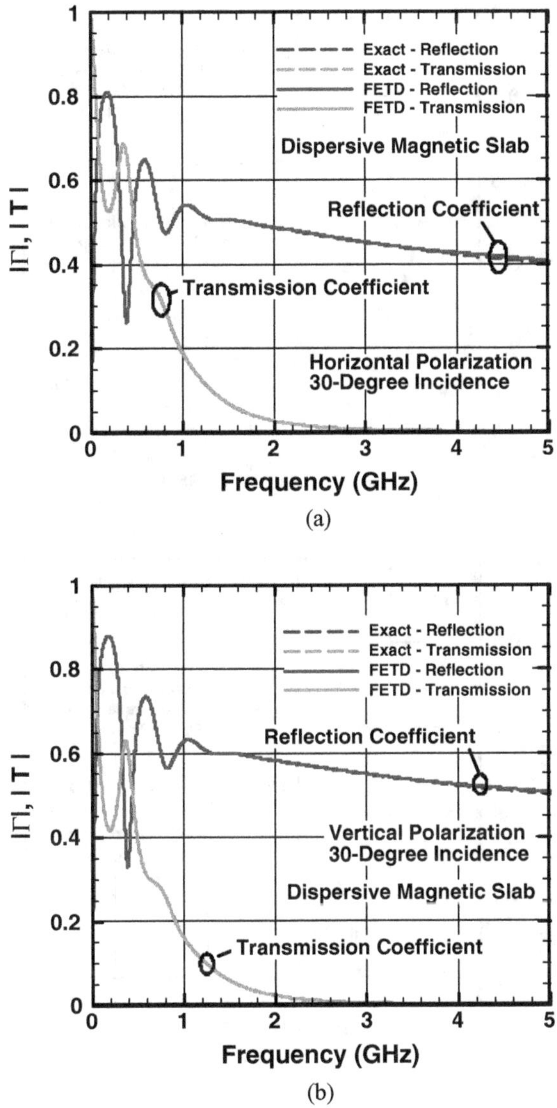

Figure 9.10 Reflection and transmission coefficients of a magnetically dispersive slab for incident angle $\theta^{\text{inc}} = 30°$. (a) Horizontal polarization. (b) Vertical polarization.

domain according to (9.77) or in the time domain according to the Fourier transform of (9.77), which yields

$$\mathbf{E}(x, y, z; t) = \mathbf{P}\left(x, y, z; t - \frac{\hat{\mathbf{k}}_t^s \cdot \mathbf{r}}{c}\right). \tag{9.145}$$

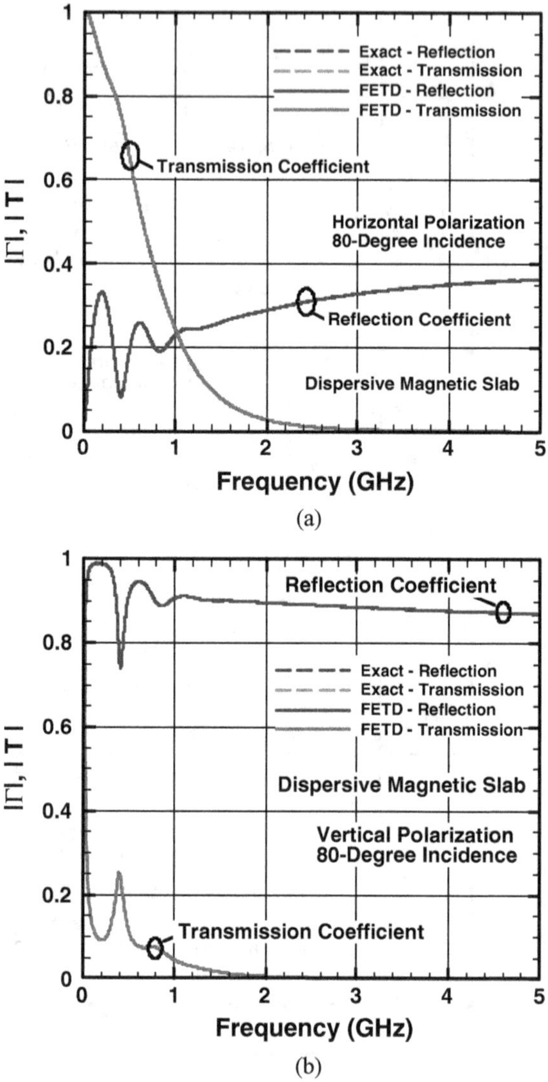

Figure 9.11 Reflection and transmission coefficients of a magnetically dispersive slab for incident angle $\theta^{\text{inc}} = 80°$. (a) Horizontal polarization. (b) Vertical polarization.

In this section we present four numerical examples to demonstrate the time-domain finite element analysis of infinite phased-array antennas. As the first example, consider a phased-array antenna based on a Vivaldi element shown in Figure 9.12. A dielectric substrate ($\varepsilon_r = 6$) with an etched conducting surface on one side is positioned normal to the ground plane such that it partially covers the aperture of a coaxial waveguide. The side of the substrate containing the etched conducting surface bisects the aperture of the coaxial waveguide, and the conducting surface is etched

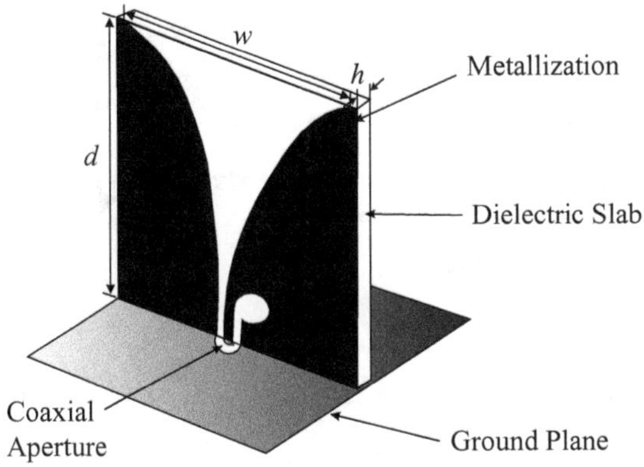

Figure 9.12 Unit cell of an infinite phased array of Vivaldi antennas.

such that it forms a short coplanar waveguide starting at the ground plane with the dimension of the two slots being equal to the width of the aperture of the coaxial waveguide. One of the slots of the coplanar waveguide transitions into a flared slotline, and the other slot is terminated by an open load formed by an etched circle of radius 2.5 mm. The dimensions of the unit cell are $T_x = 36$ mm and $T_y = 34$ mm, and the height of the substrate is 33.3 mm with a thickness of 1.27 mm. The coaxial waveguide is empty with the radii of the inner and outer conductors equal to 0.375 and 0.875 mm, respectively. The half-width of the flared slotline is given by $0.25e^{0.123z}$ mm, which gives a half width of 15 mm at the open end. The etched circle forming the open load of the other slot intersects the slot at a distance 3.5 mm from the ground plane. Note that the xz-plane that coincides with the center of the flared slotline divides the unit cell into two equal-sized parts. Figure 9.13 shows the simulated results for the magnitude of S_{11} as a function of frequency for broadside scanning $(\theta_s, \phi_s) = (0^\circ, 0^\circ)$ and oblique scanning $(\theta_s, \phi_s) = (45^\circ, 135^\circ)$. The results are computed using the time-domain finite element method [10] and are compared with another solution obtained using the commercial software suite HFSS, which is based on the frequency-domain finite element method. A prominent feature of the reflection coefficient is the sharp peak occurring at 4.6 GHz when the array is configured for broadside radiation. Such a phenomenon, referred to as the *impedance anomaly*, has been observed previously [20]. The anomaly can be modeled as the excitation of a certain resonant mode in the unit cell cavities, which are formed by the ground plane, the electric conducting surfaces on the sidewalls, and a magnetic conducting surface on the open aperture.

Simulation of scanning of periodic phased arrays to a very wide scan angle is often problematic for time-domain methods such as the FDTD because a significant reduction in the time step is required [21]. The periodic time-domain finite element method imposes no such restriction on the time step, due to its unconditional stability. This is demonstrated in the second example with the coaxially fed, doubly periodic monopole array shown in Figure 9.14. Scan performance over $0^\circ \leq \theta_s < 90^\circ$ in both the principal and diagonal planes is shown in Figure 9.15. Comparison data are based

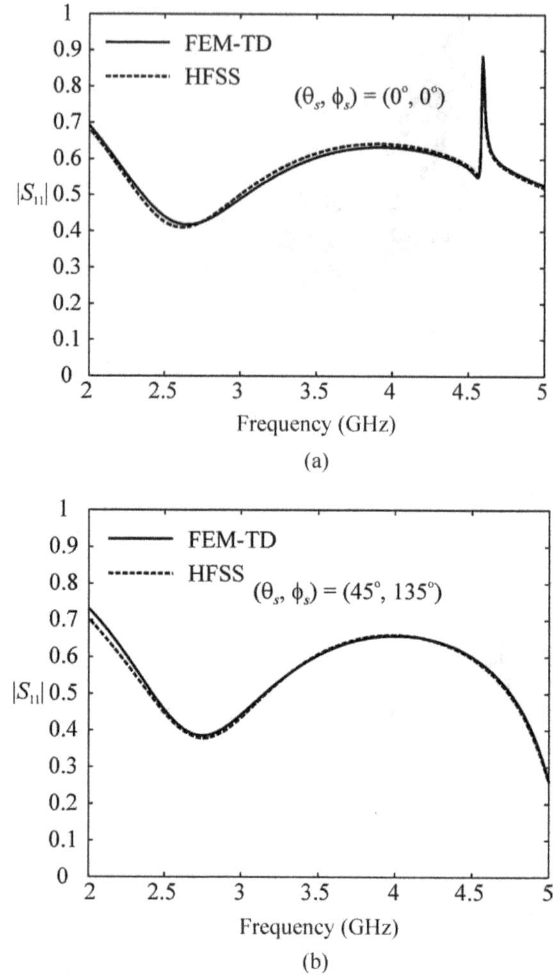

Figure 9.13 Reflection coefficient of a coaxial waveguide connected to an element in an infinite phased array of Vivaldi antennas. (After Petersson and Jin [10], Copyright © IEEE 2006.)

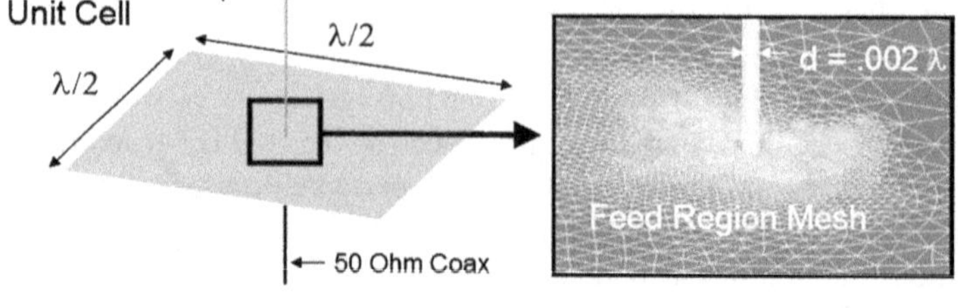

Figure 9.14 Unit cell geometry of a quarter-wavelength monopole array and the finite element surface mesh local to the feed region.

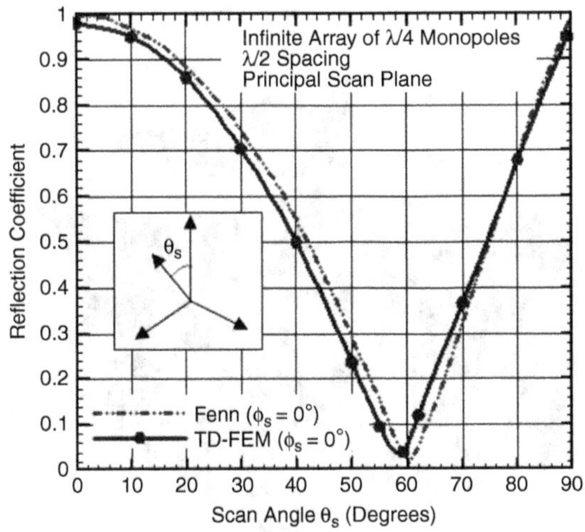

(a)

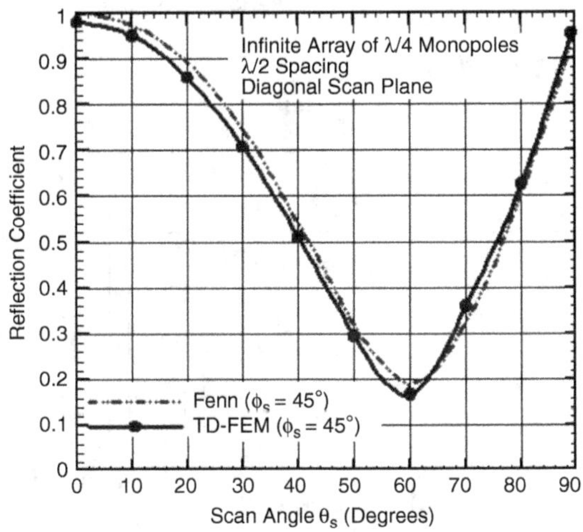

(b)

Figure 9.15 Scan performance of a quarter-wavelength monopole array. (a) In the principal ($\phi_s = 0°$) plane. (b) In the diagonal ($\phi_s = 45°$) plane.

on an approximate analytical solution that assumed a sinusoidal current distribution on the monopoles [22].

The third example concerns an infinite periodic array whose unit cell design is shown in Figure 9.16(a). The surface of the right-front corner is made semitransparent to show the internal structure of the unit cell. This is a novel, ultrawide-bandwidth array based on a dispersive magnetic substrate and a bowtie-shaped radiator. The two coaxial feeds are modeled using the waveguide port boundary condition. The insertion loss and VSWR for the broadside scan are shown in Figure 9.16(b),

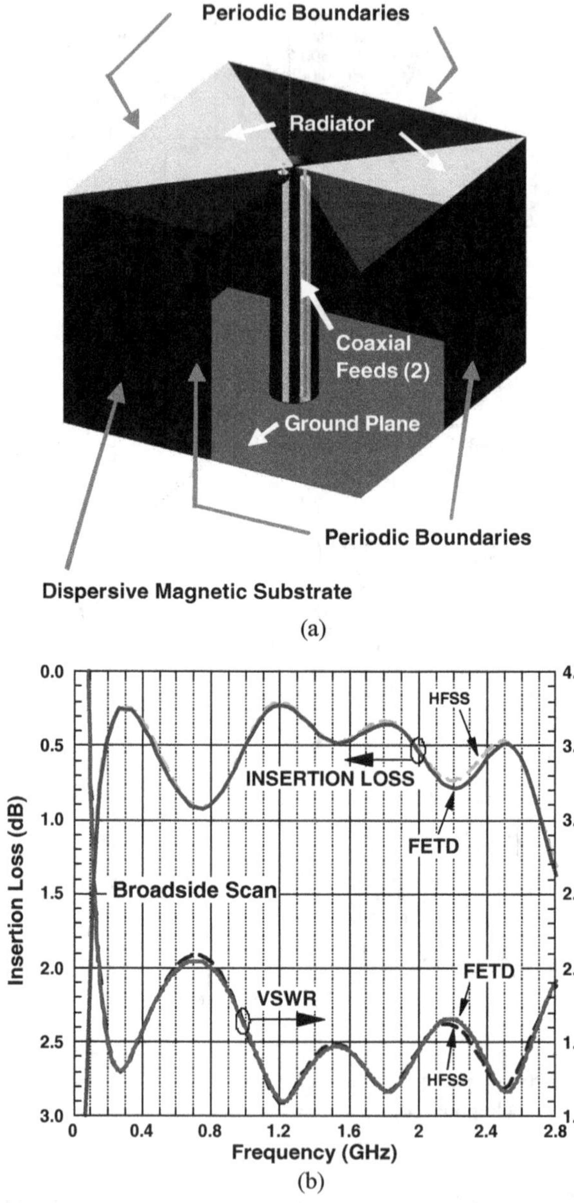

Figure 9.16 Ultrawideband phased array. (a) Unit cell geometry of the radiator and substrate. (b) Insertion loss and VSWR for the broadside scan of the ultrawideband phased array based on both time- and frequency-domain finite element solutions. [*See insert for color representation of* (a).]

where the periodic time-domain finite element solution is compared against the more traditional frequency-domain finite element solution (HFSS). The *insertion loss* is defined as the ratio of the power radiated to the power impressed into the two coaxial ports. Insertion loss contains both the absorptive and mismatch losses in the antenna. Note that this particular array design provides a 1-dB insertion loss bandwidth of approximately 20 : 1 based on this criterion. The results for the 30° H-plane scan are given in Figure 9.17. In addition to the insertion loss and VSWR, the return

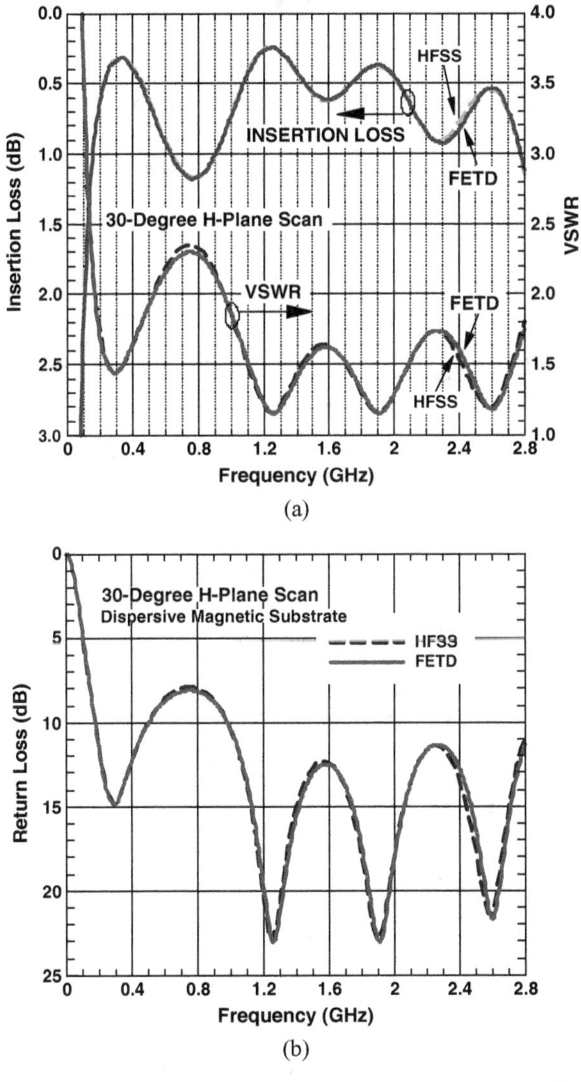

Figure 9.17 Performance of an ultrawideband phased array for the 30° H-plane scan based on both time- and frequency-domain finite element solutions. (a) Insertion loss and VSWR. (b) Return loss.

loss is also given, which shows the reflection observed at the feed port due to the impedance mismatch. The results for the E-plane scan can be found in Ref. 19. In all the cases, the periodic time-domain finite element solution agrees very well with the frequency-domain finite element calculation.

The last example demonstrates application of the periodic time-domain finite element method to the analysis of frequency-selective surfaces, which are often used in the design of antenna and array systems. The three-dimensional frequency-selective surface considered here consists of perfectly conducting rectangular patches periodically embedded in a dielectric slab with a relative permittivity of 2. The top

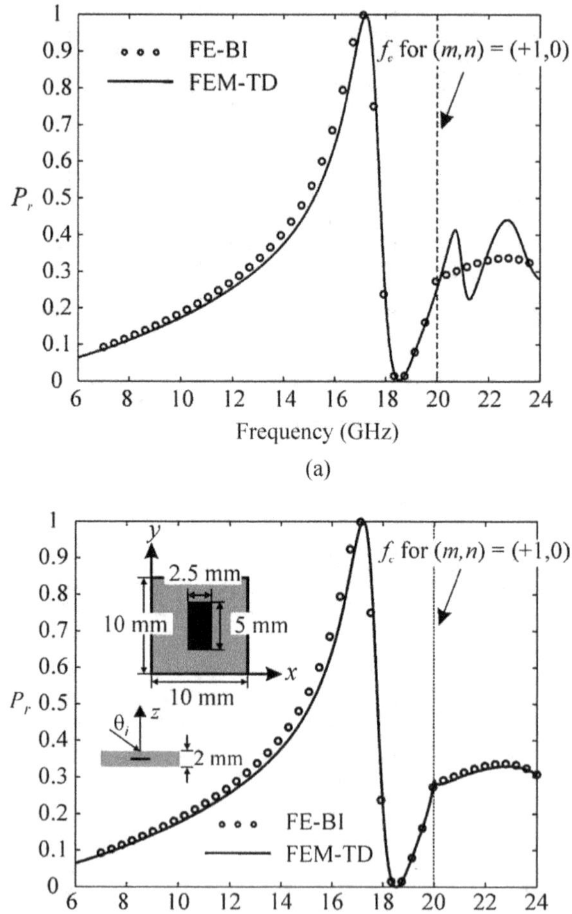

Figure 9.18 Power reflection coefficient due to a plane wave obliquely incident ($\theta^{\text{inc}}, \phi^{\text{inc}} = 30°, 0°$) on a frequency-selective surface. The electric field is parallel to the longest side of the PEC patches. (a) Results obtained using the first-order absorbing boundary condition. (b) Results obtained using the accurate periodic radiation condition. (After Petersson and Jin [9,10], Copyright © IEEE 2006.)

and side views of the unit cell are shown in Figure 9.18. The plane wave is obliquely incident with an angle of incidence $(\theta^{\text{inc}}, \phi^{\text{inc}}) = (30°, 0°)$, and the electric field is parallel to the longest side of the conducting patches. The problem is simulated by the time-domain finite element method using the first-order Floquet absorbing boundary condition (9.86) and then using the accurate periodic radiation condition given in (9.93). The results are shown as the solid curves in Figure 9.18(a) and (b), respectively, and are compared with the solutions obtained using the frequency-domain finite element method using the periodic boundary integral equation (9.56). It is evident from Figure 9.18 that the first-order absorbing boundary condition (9.86) is accurate below the cutoff frequency of the first higher-order Floquet mode, and the periodic radiation condition (9.93) remains accurate even above the cutoff frequency of the first higher-order Floquet mode.

9.3 APPROXIMATION TO FINITE ARRAYS

The analysis of the unit cell of an infinitely periodic phased array can be used to evaluate approximately the radiation patterns or the scattering patterns of a corresponding finite array. Once the field over a unit cell is obtained, we can find the field over every array element in the finite array by adding an appropriate phase shift and then computing the far-field patterns. This is equivalent to a simpler alternative in which one calculates the array factor based on the configuration of the finite array and then simply multiplies it by the far-field pattern of the unit cell. For a rectangular array having N_x elements in the x-direction and N_y elements in the y-direction, the array factor is given by [23]

$$AF(\theta, \phi) = \left| \frac{\sin[(N_x T_x/2)\Phi_x]}{\sin[(T_x/2)\Phi_x]} \right| \cdot \left| \frac{\sin[(N_y T_y/2)\Phi_y]}{\sin[(T_y/2)\Phi_y]} \right| \tag{9.146}$$

where $\Phi_x = k_0 \sin\theta \cos\phi - k_x^s$ and $\Phi_y = k_0 \sin\theta \sin\phi - k_y^s$. For a nonrectangular array, the array factor can be evaluated numerically using the expression

$$AF(\theta, \phi) = \sum_{i=1}^{N_P} e^{j(\Phi_x x_i + \Phi_y y_i)} \tag{9.147}$$

where N_P denotes the total number of the array elements and (x_i, y_i) denotes the center of the ith element.

We present three examples to demonstrate the performance of this simple approximation for finite arrays. The first example concerns a finite periodic 10×10 microstrip phased array in a ground plane with periodicity: $T_x = 5$ cm and $T_y = 5$ cm (Figure 9.19). Each 3-cm $\times$ 3-cm microstrip patch antenna resides in a square cavity with dimensions of 4.5 cm $\times$ 4.5 cm $\times$ 0.2 cm, and the substrate has a relative permittivity of 2.8. The scanning angle is $\phi_s = 0°$ and $\theta_s = 50°$. The radiation pattern in the xz-plane at 3 GHz is calculated using a rigorous analysis of the finite array and an approximate analysis based on the truncation of a corresponding infinite array. The results, given in Figure 9.20, show that both solutions agree very well for the

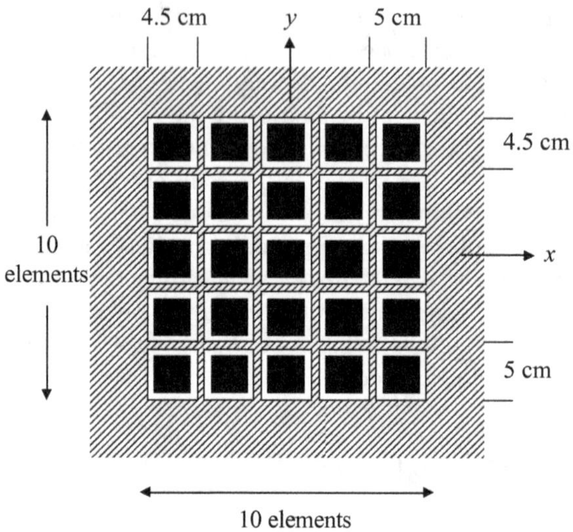

Figure 9.19 Finite periodic 10 × 10 microstrip phased array in a ground plane with periodicity: $T_x = 5$ cm and $T_y = 5$ cm. Each microstrip patch antenna of size 3 cm × 3 cm resides in a square cavity with the dimensions 4.5 cm × 4.5 cm × 0.2 cm, and the substrate has a relative permittivity of 2.8.

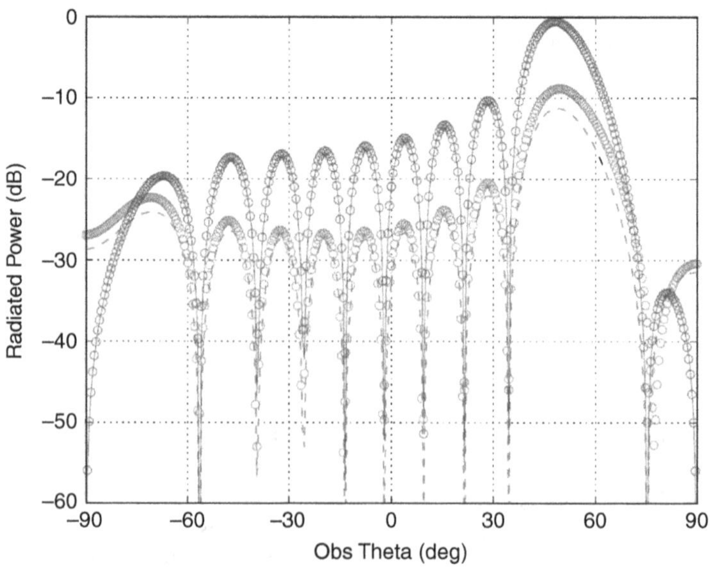

Figure 9.20 Radiation patterns of a 10 × 10 microstrip patch array in the xz-plane at 3 GHz. The scanning angle is set at $\phi_s = 0°$ and $\theta_s = 50°$. Solid line: co-polarization. Dashed line: cross-polarization. All lines: approximate solution based on the truncation of an infinite array. Circles: accurate solution for the finite array.

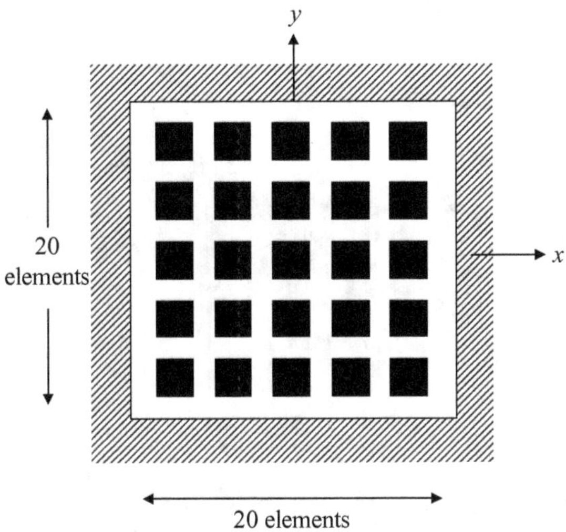

Figure 9.21 Finite periodic 20 × 20 microstrip phased array in a ground plane with periodicity: $T_x = 5$ cm and $T_y = 5$ cm. The microstrip patch antennas, of size 3 cm × 3 cm, are printed on one substrate of thickness 0.2 cm and relative permittivity 2.8. The entire array is housed in a square cavity with dimensions 100 cm × 100 cm × 0.2 cm.

co-polarization and differ by 3 dB for the cross-polarization in the main lobe. Similar agreement is obtained for the radiation pattern in the yz-plane.

The second example is a finite periodic 20 × 20 microstrip phased array in a ground plane with periodicity $T_x = 5$ cm and $T_y = 5$ cm (Figure 9.21). The microstrip patch antennas having a size of 3 cm × 3 cm are printed on one substrate having a thickness of 0.2 cm and a relative permittivity of 2.8. The entire array is housed in a square cavity with dimensions of 100 cm × 100 cm × 0.2 cm. The scanning angle is again set at $\phi_s = 0°$ and $\theta_s = 50°$. The radiation pattern in the xz-plane at 3 GHz is calculated using a rigorous analysis of the finite array, given in Figure 9.22(a), and an approximate analysis based on the truncation of a corresponding infinite array, displayed in Figure 9.22(b). Again both solutions agree very well for the co-polarization.

The third example considers the plane-wave scattering by a finite periodic 21 × 21 array of cavities in a ground plane with periodicity $T_x = 1$ cm and $T_y = 0.5$ cm (Figure 9.23). Each cavity has a dimension of 0.9 cm × 0.4 cm × 0.1 cm and is filled with air. A uniform plane wave is incident at $\phi^{\text{inc}} = 0°$ and $\theta^{\text{inc}} = 45°$. The VV-polarized bistatic radar cross section (RCS) in the xz-plane at 30 GHz is calculated using a rigorous analysis of the finite array and an approximate analysis based on the truncation of a corresponding infinite array. The results are given in Figure 9.24, which shows that the two solutions agree within 4 dB.

In all three examples, the difference between the accurate and approximate solutions is caused mainly by the edge effects because the fields on the elements close to the edges and corners of the array can be significantly different from the field on an element at the center of the array. Such an edge/corner effect can only be

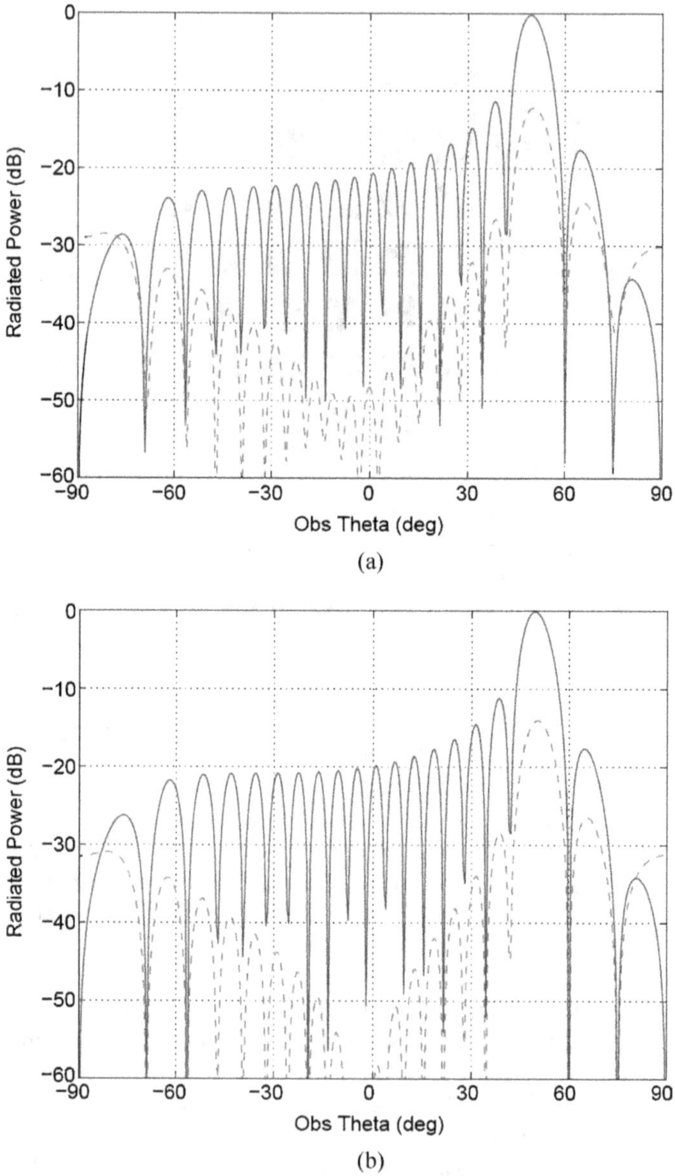

Figure 9.22 Radiation patterns of a 20 × 20 microstrip patch array in the xz-plane at 3 GHz. The scanning angle is set at $\phi_s = 0^\circ$ and $\theta_s = 50^\circ$. Solid line: co-polarization. Dashed line: cross-polarization. (a) Accurate solution for the finite array. (b) Approximate solution based on the truncation of an infinite array.

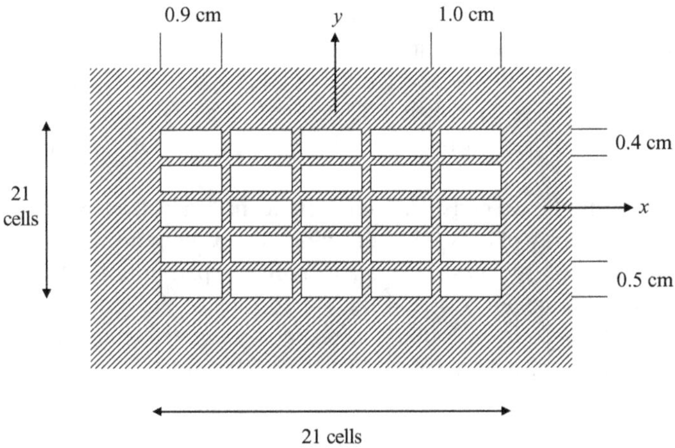

Figure 9.23 Finite periodic 21 × 21 cavity array in a ground plane with periodicity: $T_x = 1$ cm and $T_y = 0.5$ cm. Each cavity has dimensions 0.9 cm × 0.4 cm × 0.1 cm and is filled with air.

captured accurately by analyzing the finite array directly. Nevertheless, the simple, approximate solution described in this section is highly useful in practical array applications because of its fast solution time and reasonable accuracy. An intermediate step to evaluate the edge effects is to analyze an array that is finite in one direction and infinite in the other direction [24,25]. In this case, the computational domain is

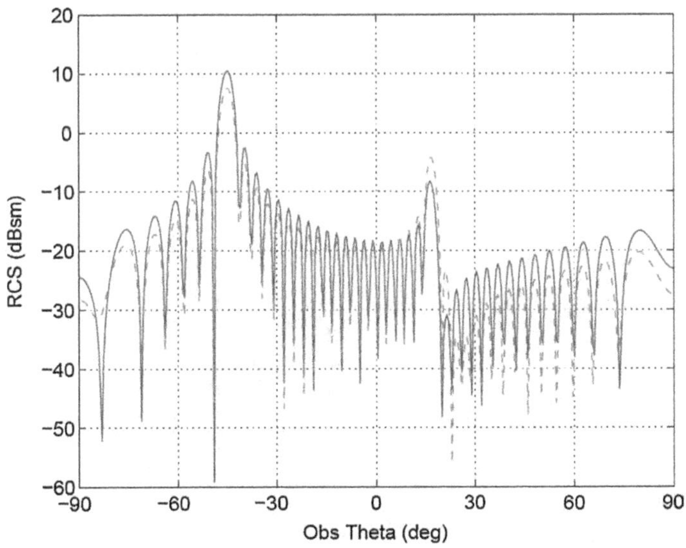

Figure 9.24 VV-polarized bistatic RCS in the xz-plane at 30 GHz for a uniform plane wave incident at $\phi^{\text{inc}} = 0°$ and $\theta^{\text{inc}} = 45°$. Solid line: approximate solution based on the truncation of an infinite array. Dashed line: accurate solution for the finite array.

reduced to one row of elements in the finite direction, and the periodic boundary and radiation conditions can be formulated to take into account the periodicity in the infinite direction. However, with the development of highly efficient numerical solutions for finite arrays, which are described in Chapter 10, the analysis of finite-by-infinite arrays becomes less important because it is still incapable of capturing the corner effect.

As the final example, we consider a finite frequency-selective surface (FSS) consisting of a 50 × 50 array of cross-shaped conducting patches printed on an infinitely large dielectric sheet of thickness 0.1 cm and a relative permittivity of 4.0. The finite array is shown in Figure 9.25(a), and the dimensions of the unit cell are given in Figure 9.25(b). The bistatic RCS in the xz-plane is computed using the approximate approach for the finite frequency-selective surface at 9 and 18 GHz with a plane-wave

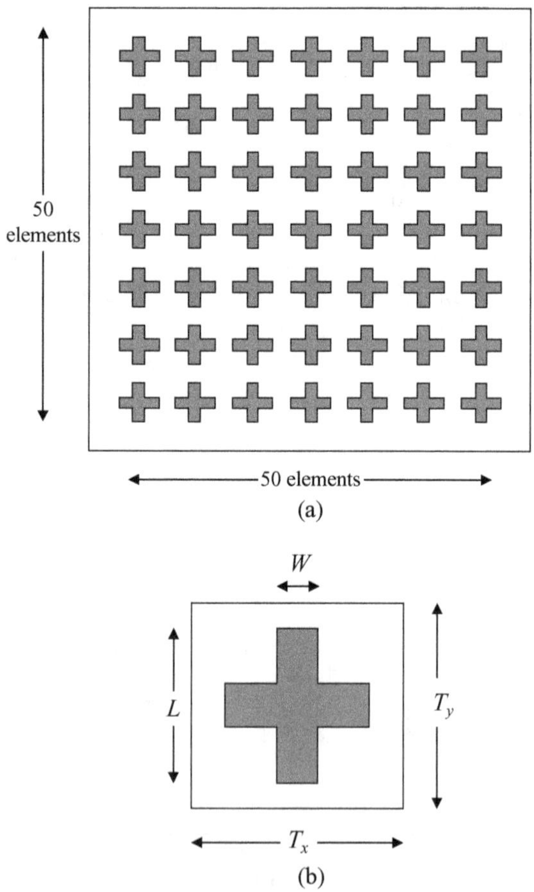

Figure 9.25 Finite frequency-selective surface consisting of a 50 × 50 array of cross-shaped conducting patches printed on an infinitely large dielectric sheet of thickness 0.1 cm and relative permittivity 4.0. (a) Finite array. (b) The unit cell ($T_x = T_y = 2$ cm, $L = 1.5$ cm, $W = 0.5$ cm).

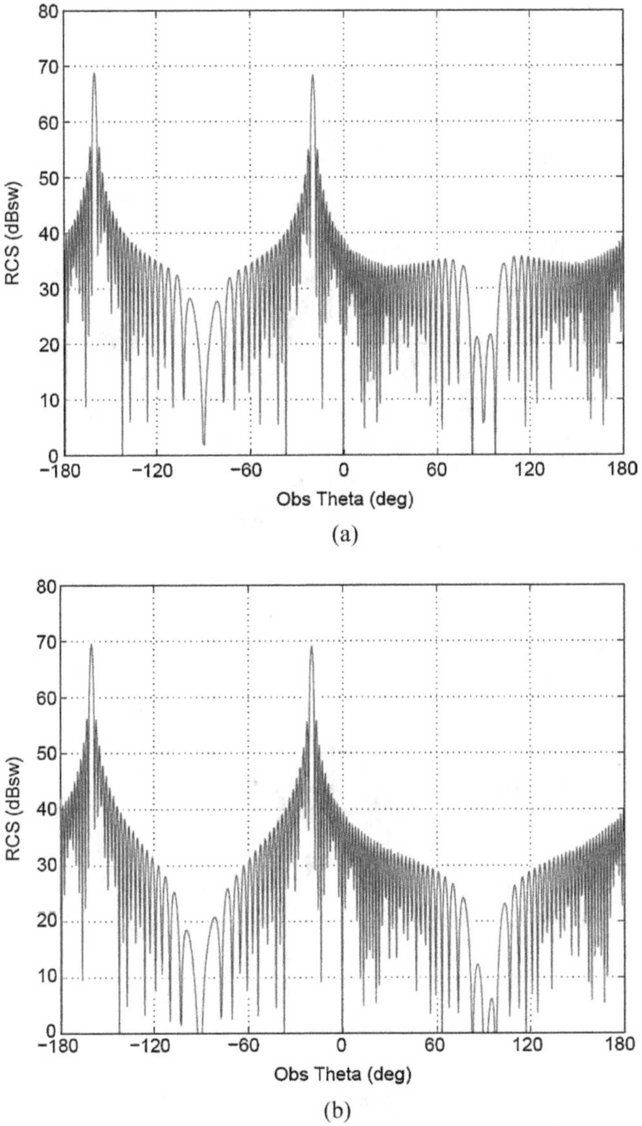

Figure 9.26 Bistatic RCS in the xz-plane for a finite FSS on an infinitely large dielectric slab at 9 GHz with a plane-wave incidence at $\phi^{\text{inc}} = 0^\circ$ and $\theta^{\text{inc}} = 20^\circ$. (a) HH polarization. (b) VV polarization.

incidence at $\phi^{\text{inc}} = 0^\circ$ and $\theta^{\text{inc}} = 20^\circ$. The results are displayed in Figures 9.26 and 9.27, respectively. It can be seen that at 9 GHz, there are two main peaks in the bistatic RCS, corresponding to the specular reflection and forward transmission. However, when the frequency is increased to 18 GHz, there are two additional peaks, which correspond to the first higher-order Floquet mode.

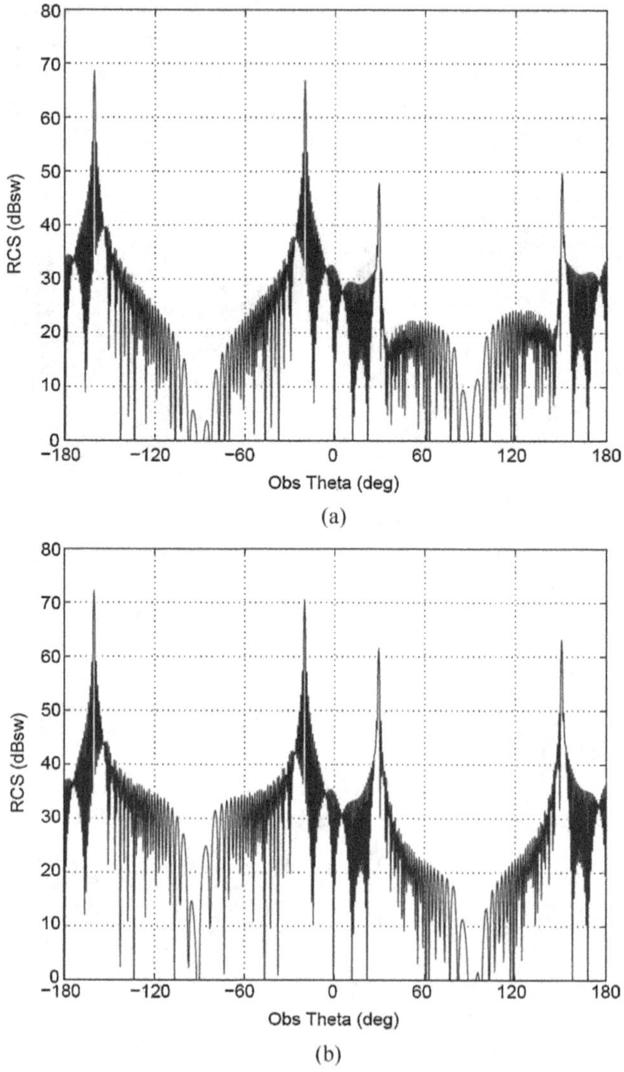

Figure 9.27 Bistatic RCS in the xz-plane for a finite FSS on an infinitely large dielectric slab at 18 GHz with a plane-wave incidence at $\phi^{\text{inc}} = 0^\circ$ and $\theta^{\text{inc}} = 20^\circ$. (a) HH polarization. (b) VV polarization.

9.4 SUMMARY

In this chapter we discussed the finite element analysis of three-dimensional doubly periodic infinite phased arrays. The periodicity of the problem reduces the computational domain to that of a unit cell. For the analysis in both the frequency and time domains, the main issues are to impose periodic boundary conditions on the side surfaces of the unit cell and to truncate the computational domain in the nonperiodic direction. For the latter, the simplest and most efficient approach is to employ the first-order Floquet absorbing boundary condition that can perfectly absorb the fundamental Floquet mode for any propagation direction, and the most accurate approach is to employ the exact periodic radiation condition that includes all Floquet modes. Although perfectly matched layers can also be used for mesh truncation, the performance is not as robust as the other two approaches since perfectly matched layers cannot be designed to absorb the fundamental Floquet mode in an arbitrary direction. Consequently, increasing the separation distance to a traditional perfectly matched layer may not add further improvement. To obtain possible improved performance, one has to employ more complicated perfectly matched layers, such as the complex frequency shifted and the second-order models, to absorb evanescent Floquet modes which are present when the frequency is close to or surpasses the cutoff frequency of the first Floquet mode.

Whereas the implementation of the finite element method in the frequency domain is rather straightforward once the periodic boundary conditions and mesh truncation are formulated, the implementation in the time domain is much more challenging. Although enforcement of the periodic boundary conditions can be simplified with introduction of the transformed field variable, implementation of the periodic radiation condition that includes higher-order Floquet modes for mesh truncation requires the time-consuming calculation of convolutions for each of the higher-order Floquet modes. Fortunately, numerical experiments showed that only the propagating Floquet modes and a few evanescent Floquet modes are needed to yield very accurate solutions. Moreover, for many practical applications, the first-order Floquet absorbing boundary condition that perfectly absorbs the fundamental Floquet mode in any propagation direction is sufficient to yield engineering accuracy without requiring the calculation of convolutions.

We have given a few examples to demonstrate use of the finite element method for the analysis of a variety of phased-array antennas. It is worthwhile to point out that phased-array antennas constitute only a small class of periodic structures that can be analyzed using the finite element method described here. Periodic structures have widespread applications in electrical and optical engineering. Other applications include frequency-selective surfaces, metamaterials (left-handed or double-negative materials), electromagnetic and photonic bandgap materials, photonic crystals, and propagation through walls containing metal reinforcing structures as well as periodic air holes in bricks. All these structures can be analyzed effectively with the method presented here. Although the method is formulated to deal with doubly periodic structures, it can be extended easily to handle either singly or triply periodic structures.

REFERENCES

1. R. J. Mailloux, *Phased Array Antenna Handbook*, 2nd ed. Norwood, MA: Artech House, 2005.
2. J. M. Jin and J. L. Volakis, "Scattering and radiation analysis of three-dimensional cavity arrays via a hybrid finite element method," *IEEE Trans. Antennas Propagat.*, vol. 41, pp. 1580–1586, Nov. 1993.

3. D. T. McGrath and V. P. Pyati, "Phased array antenna analysis with the hybrid finite element method," *IEEE Trans. Antennas Propagat.*, vol. 42, pp. 1625–1630, Dec. 1994.
4. E. W. Lucas and T. P. Fontana, "A 3-D hybrid finite element/boundary element method for the unified radiation and scattering analysis of general infinite periodic arrays," *IEEE Trans. Antennas Propagat.*, vol. 43, no. 2, pp. 145–153, Feb. 1995.
5. J. D'Angelo and I. Mayergoyz, "Phased array antenna analysis," in *Finite Element Software for Microwave Engineering*, G. Pelosi, P. Silvester, and T. Itoh, Eds. New York: Wiley, 1996.
6. T. F. Eibert, J. L. Volakis, D. R. Wilton, and D. R. Jackson, "Hybrid FE/BI modeling of 3-D doubly periodic structures utilizing triangular prismatic elements and MPIE formulation accelerated by the Ewald transformation," *IEEE Trans. Antennas Propagat.*, vol. 47, no. 5, pp. 843–851, May 1999.
7. Z. Lou and J. M. Jin, "Finite element analysis of phased array antennas," *Microwave Opt. Tech. Lett.*, vol. 40, no. 6, pp. 490–496, Mar. 2004.
8. M. A. Gonzalez de Aza, J. A. Encinar, J. Zapata, and M. Lambea, "Full-wave analysis of cavity-backed and probe-fed microstrip patch arrays by a hybrid mode-matching generalized scattering matrix and finite-element method," *IEEE Trans. Antennas Propagat.*, vol. 46, no. 2, pp. 234–242, Feb. 1998.
9. L. E. R. Petersson and J. M. Jin, "A three-dimensional time-domain finite element formulation for periodic structures," *IEEE Trans. Antennas Propagat.*, vol. 54, no. 1, pp. 12–19, Jan. 2006.
10. L. E. R. Petersson and J. M. Jin, "Analysis of periodic structures via a time-domain finite element formulation with a Floquet ABC," *IEEE Trans. Antennas Propagat.*, vol. 54, no. 3, pp. 933–944, Mar. 2006.
11. N. Amitay, V. Galindo, and C. Wu, *Theory and Analysis of Phased Array Antennas*. New York: Wiley, 1972.
12. D. Correia and J. M. Jin, "A simple and efficient implementation of CFS-PML in the FDTD analysis of periodic structures," *IEEE Microwave Wireless Components Lett.*, vol. 15, no. 7, pp. 487–489, July 2005.
13. D. Correia and J. M. Jin, "On the development of a higher-order PML," *IEEE Trans. Antennas Propagat.*, vol. 53, no. 12, pp. 4157–4163, Dec. 2005.
14. D. M. Pozar and D. H. Schaubert, "Analysis of an infinite array of rectangular microstrip patches with idealized probe feeds," *IEEE Trans. Antennas Propagat.*, vol. 32, pp. 1101–1107, Oct. 1984.
15. C. C. Liu, A. Hessel, and J. Shmoys, "Performance of probe-fed microstrip-patch element phased arrays," *IEEE Trans. Antennas Propagat.*, vol. 36, pp. 1501–1509, Nov. 1988.
16. M. E. Veysoglu, R. T. Shin, and J. A. Kong, "A finite-difference time-domain analysis of wave scattering from periodic structures: oblique incidence case," *J. Electromagn. Waves Appl.*, vol. 7, pp. 1595–1607, Dec. 1993.
17. J. Ren, O. P. Gandhi, L. R. Walker, J. Fraschilla, and C. R. Boerman, "Floquet-based FDTD analysis of two-dimensional phased array antennas," *IEEE Microwave Guided Wave Lett.*, vol. 4, pp. 109–111, Apr. 1994.
18. N.-W. Chen, B. Shanker, and E. Michielssen, "Integral-equation-based analysis of transient scattering from periodic perfectly conducting structures," *IEE Proc. Microwave Antennas Propagat.*, vol. 150, pp. 120–124, Apr. 2003.

19. D. Riley and J. M. Jin, "Finite-element time-domain analysis of electrically and magnetically dispersive periodic structures," *IEEE Trans. Antennas Propagat.*, vol. 56, no. 11, Nov. 2008.

20. H. Holter, T.-H. Chio, and D. H. Schaubert, "Elimination of impedance anomalies in single- and dual-polarized endfire tapered slot phased arrays," *IEEE Trans. Antennas Propagat.*, vol. 48, pp. 122–124, Jan. 2000.

21. J. Maloney, G. Smith, E. Thiele, O. Gandhi, N. Chavannes, and S. Hagness, "Modeling of antennas," in *Computational Electrodynamics: The Finite-Difference Time-Domain Method*, 3rd ed., A. Taflove and S. C. Hagness, Eds. Norwood, MA: Artech House, pp. 607–676, 2005.

22. A. J. Fenn, "Theoretical and experimental study of monopole phased array antennas," *IEEE Trans. Antennas Propagat.*, vol. 33, no. 10, pp. 1118–1126, Oct. 1985.

23. C. A. Balanis, *Antenna Theory: Analysis and Design*, 3rd ed. Hoboken, NJ: Wiley, 2005.

24. Z. Lou and J. M. Jin, "Higher-order finite element analysis of finite-by-infinite arrays," *Electromagnetics*, vol. 24, no. 7, pp. 497–514, Oct. 2004.

25. B. Munk, *Finite Antenna Arrays and FSS*. Hoboken, NJ: Wiley, 2003.

10 Finite Phased-Array Modeling

The modeling and analysis of a large finite phased-array antenna consisting of hundreds or even thousands of array elements is technically very challenging because of its large electrical size and often fine geometrical feed details. Traditionally, the design of a finite phased-array aperture is carried out by first approximating the finite array with a corresponding infinite array, where the analysis can be performed on a unit cell [1–3], and then approximately modeling the truncation effect on the radiated fields using the array theory, as demonstrated in Chapter 9. However, this design approach neglects the complex edge truncation effects seen at the element level of the array. Even though the radiation patterns for a finite array are often adequately predicted by the simple array theory based on a unit cell analysis, the impact of the edge truncation effects is particularly important with regard to understanding the variation in the input impedance seen by the outer elements of the array as a function of their distance from the edge of the array. Early in the phased-array design process, the beamformer and electronics required to perform full finite array measurements are rarely available. In addition, the active impedance of an element in an array is extremely difficult to measure and therefore such measurement is rarely performed. Consequently, accurate predictive methods not only for the radiation patterns for finite phased arrays but also for the active element impedances are of great practical importance in the design of modern phased-array antennas.

For arrays with one dimension much larger than the other, the truncation effect can be partially modeled using a finite-by-infinite array model [4]. For general arrays, an accurate numerical analysis of a finite array can be obtained by applying a conventional numerical method, such as the finite element method, directly to model the entire array. However, this application often results in a linear system with many millions or even billions of unknowns, whose solution requires a vast amount of memory and many floating-point operations. To fully model large finite arrays more efficiently, various fast algorithms have been developed, including the array decomposition method [5], which uses the fast Fourier transform (FFT) to evaluate the product of a block-Toeplitz matrix with a vector in an iterative solution for a finite periodic array; the characteristic basis function method [6], which forms larger basis functions using a collection of traditional basis functions to reduce the size of the moment-method system matrix; and the domain-decomposition methods [7–10], which divide the computational domain into subdomains and then recombine the

Finite Element Analysis of Antennas and Arrays, By Jian-Ming Jin and Douglas J. Riley

reduced systems for the subdomains to obtain a global solution. In this chapter we focus on the domain-decomposition methods for the finite element analysis and modeling of finite arrays because of their high efficiency and generality. Both the frequency- and time-domain formulations are presented along with many numerical examples.

10.1 FREQUENCY-DOMAIN MODELING

Among the various numerical schemes proposed for large-scale finite element simulations, the domain-decomposition method is one of the most efficient and versatile. Here we describe three such domain-decomposition algorithms which are especially well suited for simulating large finite-array antennas. The first method is the dual–primal finite element tearing and interconnecting method designed for electromagnetic analysis (referred to as FETI–DPEM1) [8]. This method is an extension of the dual–primal finite element tearing and interconnecting method for solving the Helmholtz equation in computational acoustics [11–14]. The second method is an improved formulation over the first method and is referred to as FETI–DPEM2. Unlike the first method, this improved formulation is numerically scalable with respect to the frequency of electromagnetic fields, which is critical for high-frequency applications. The third method is the nonconforming domain-decomposition method, which relaxes the requirement of periodic meshes and thereby simplifies mesh generation.

10.1.1 FETI–DPEM1 Formulation

The general principle of the FETI–DPEM1 method [8] is first to divide the entire computational domain into nonoverlapping subdomains, where an incomplete solution of the field is first evaluated using a direct solver for the finite element equations (in contrast to an *iterative solver*, a *direct solver* provides a complete factorization of the governing finite element system matrix in the subdomains). Next, the tangential continuity of electric and magnetic fields is enforced at the subdomain interfaces by using the Lagrange multipliers or dual variables. This yields a reduced-order interface problem, which can be solved using an iterative algorithm. The solution to the interface problem can then serve as the boundary condition for individual subdomains to evaluate the field inside the subdomains.

For an array type of geometry, the natural domain decomposition is to divide or *tear* the entire domain such that each array element becomes a subdomain. For an array that consists of $M \times N$ array elements, such a decomposition yields $M \times N$ subdomains, as sketched in Figure 10.1. Note that the method to be described is also applicable to skew arrays and hexagonal arrays. The rectangular array is used here simply for the purpose of illustration. To formulate the field in each subdomain, we need to prescribe an appropriate boundary condition at the surface of the subdomain. In addition to the standard Dirichlet, Neumann, and mixed boundary conditions for electrically conducting, magnetically conducting, and impedance surfaces, we

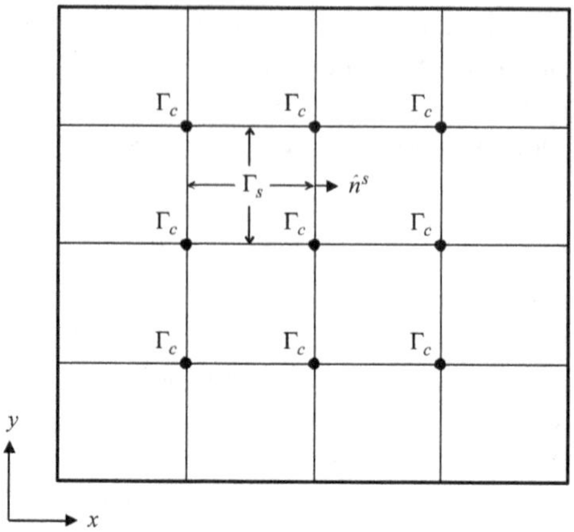

Figure 10.1 Decomposition of a computational domain into subdomains, with Γ_s denoting the surface of subdomain s that interfaces with neighboring subdomains and Γ_c denoting the corner edges shared by more than two subdomains.

assume that the field at the surfaces coincident with the exterior surface of the entire computational domain satisfies the first-order absorbing boundary condition

$$\hat{n} \times \left[\overleftrightarrow{\mu}_r^{-1} \cdot (\nabla \times \mathbf{E})\right] + jk_0\hat{n} \times [\overleftrightarrow{\varsigma} \cdot (\hat{n} \times \mathbf{E})] \approx 0 \qquad \mathbf{r} \in S_o \tag{10.1}$$

where $\overleftrightarrow{\varsigma}$ denotes the normalized admittance tensor chosen to minimize the reflection of the truncation boundary (for an isotropic medium, $\overleftrightarrow{\varsigma}$ becomes a scalar with a value of $\sqrt{\varepsilon_r/\mu_r}$). The first-order absorbing boundary condition is used here simply for the sake of simplicity since other more accurate truncation techniques can be employed as well. At the surfaces that border the adjacent subdomains, we assume that the field satisfies the Neumann boundary condition

$$\hat{n} \times \left(\overleftrightarrow{\mu}_r^{-1} \cdot \nabla \times \mathbf{E}\right) = -jk_0 Z_0 \hat{n} \times \mathbf{H} \tag{10.2}$$

where the tangential magnetic field is initially unknown. By following the finite element formulation described in Section 2.1, we obtain the finite element matrix equation for the subdomain as

$$[K^s]\{E^s\} = \{b^s\} - \{\lambda^s\} \tag{10.3}$$

where the superscript s denotes the subdomain number and the matrix and vectors are given by

$$[K^s] = \iiint_{V_s} \left[\nabla \times \{\mathbf{N}^s\} \cdot \overleftrightarrow{\mu}_r^{-1} \cdot \nabla \times \{\mathbf{N}^s\}^{\mathrm{T}} - k_0^2\{\mathbf{N}^s\} \cdot \overleftrightarrow{\varepsilon}_r \cdot \{\mathbf{N}^s\}^{\mathrm{T}}\right] dV$$
$$+ jk_0 \iint_{S_s \cap S_o} \hat{n}^s \times \{\mathbf{N}^s\} \cdot \overleftrightarrow{\varsigma} \cdot \hat{n}^s \times \{\mathbf{N}^s\}^{\mathrm{T}}\, dS \tag{10.4}$$

$$\{b^s\} = -\iiint_{V_s} \{\mathbf{N}^s\} \cdot \left[jk_0 Z_0 \mathbf{J}_{\text{imp}} + \nabla \times (\overleftrightarrow{\mu}_r^{-1} \cdot \mathbf{M}_{\text{imp}}) \right] dV \tag{10.5}$$

$$\{\lambda^s\} = -jk_0 Z_0 \iint_{S_s \cap \Gamma_s} \{\mathbf{N}^s\} \cdot (\hat{n}^s \times \mathbf{H})\, dS. \tag{10.6}$$

In the equations above, V_s denotes the volume of the subdomain, $S_s \cap S_o$ denotes the portion of the surface of the subdomain coincident with the exterior surface, $S_s \cap \Gamma_s$ denotes the remaining portion that interfaces with adjacent subdomains, and $\{\mathbf{N}^s\}$ is a column vector that stores all the vector basis functions associated with the subdomain.

Next, we split the unknowns in each subdomain into two parts. The first part consists of all the unknowns associated with the edges shared by more than two subdomains, which are called *corner edges*, and the vector storing these unknowns is denoted as $\{E_c^s\}$. The second part consists of all other unknowns in the subdomain, which include those inside the subdomain and on the surface of the subdomain, and the vector storing these unknowns is denoted as $\{E_r^s\}$. With this partition, (10.3) can be written as

$$\begin{bmatrix} K_{rr}^s & K_{rc}^s \\ K_{cr}^s & K_{cc}^s \end{bmatrix} \begin{Bmatrix} E_r^s \\ E_c^s \end{Bmatrix} = \begin{Bmatrix} b_r^s \\ b_c^s \end{Bmatrix} - \begin{Bmatrix} \lambda_r^s \\ \lambda_c^s \end{Bmatrix} \tag{10.7}$$

which can be split further into two equations as

$$\{E_r^s\} = [K_{rr}^s]^{-1}(\{b_r^s\} - \{\lambda_r^s\} - [K_{rc}^s]\{E_c^s\}) \tag{10.8}$$

$$([K_{cc}^s] - [K_{cr}^s][K_{rr}^s]^{-1}[K_{rc}^s])\{E_c^s\} = \{b_c^s\} - \{\lambda_c^s\} - [K_{cr}^s][K_{rr}^s]^{-1}(\{b_r^s\} - \{\lambda_r^s\}). \tag{10.9}$$

This formulation is carried out for each and every subdomain. It can be seen that the subdomain finite element systems are fully decoupled and the interaction between the subdomains is accounted for through $\{\lambda_r^s\}$ and $\{\lambda_c^s\}$.

The next step is to couple or *interconnect* the fields for all the subdomains. For this, we introduce two Boolean matrices to facilitate the mathematical formulation. The first one is a signed Boolean matrix, denoted as $[B_r^s]$, which extracts the unknowns on the interface from $\{E_r^s\}$ and places them in the global interface vector. Therefore, $[B_r^s]\{E_r^s\}$ represents the signed global vector that stores all the primal unknowns on the interface of subdomain s. Conversely, if we define the global interface vector $\{\lambda\}$ as

$$\{\lambda\} = -jk_0 Z_0 \iint_{\Gamma_b} \{\mathbf{N}^b\} \cdot (\hat{n} \times \mathbf{H})\, dS \tag{10.10}$$

where Γ_b denotes all the subdomain interfaces in the computational domain and $\{\mathbf{N}^b\}$ denotes all the vector basis functions associated with the unknowns on the interfaces but excluding those on the corner edges, $\{\lambda_r^s\}$ can be extracted as $\{\lambda_r^s\} = [B_r^s]^{\mathrm{T}}\{\lambda\}$. The second Boolean matrix, denoted as $[B_c^s]$, extracts $\{E_c^s\}$ from the global vector $\{E_c\}$ that stores all the unknowns on the corner edges; that is, $\{E_c^s\} = [B_c^s]\{E_c\}$. With the introduction of these two Boolean matrices, (10.8) and (10.9) can now be written as

$$[B_r^s]\{E_r^s\} = [B_r^s][K_{rr}^s]^{-1}(\{b_r^s\} - [B_r^s]^{\mathrm{T}}\{\lambda\} - [K_{rc}^s][B_c^s]\{E_c\}) \quad (10.11)$$

$$[B_c^s]^{\mathrm{T}}([K_{cc}^s] - [K_{cr}^s][K_{rr}^s]^{-1}[K_{rc}^s])[B_c^s]\{E_c\}$$
$$= [B_c^s]^{\mathrm{T}}(\{b_c^s\} - \{\lambda_c^s\} - [K_{cr}^s][K_{rr}^s]^{-1}(\{b_r^s\} - [B_r^s]^{\mathrm{T}}\{\lambda\})). \quad (10.12)$$

Note that the left-hand side of (10.11) consists of the unknown tangential electric fields on the subdomain interfaces.

To couple the fields over all the subdomains, we note that at the interface between two subdomains, say subdomains s and q, the electric field satisfies the tangential continuity condition

$$\hat{n}^s \times \mathbf{E}^s = -\hat{n}^q \times \mathbf{E}^q \qquad \text{or} \qquad \hat{n}^s \times \mathbf{E}^s + \hat{n}^q \times \mathbf{E}^q = 0. \quad (10.13)$$

This condition can be enforced by summing (10.11) over all the subdomains and setting it to zero, which yields

$$\sum_{s=1}^{N_s}[B_r^s]\{E_r^s\} = \sum_{s=1}^{N_s}[B_r^s][K_{rr}^s]^{-1}(\{b_r^s\} - [B_r^s]^{\mathrm{T}}\{\lambda\} - [K_{rc}^s][B_c^s]\{E_c\}) = 0 \quad (10.14)$$

where N_s denotes the total number of subdomains. This enforcement can only be done under the assumption that the meshes of two adjacent subdomains have to match at their interface. Equation (10.14) provides an incomplete system of equations for the unknown $\{\lambda\}$ and $\{E_c\}$. The remaining equations can be obtained by summing (10.12) over all the subdomains, resulting in

$$\sum_{s=1}^{N_s}[B_c^s]^{\mathrm{T}}([K_{cc}^s] - [K_{cr}^s][K_{rr}^s]^{-1}[K_{rc}^s])[B_c^s]\{E_c\}$$
$$= \sum_{s=1}^{N_s}[B_c^s]^{\mathrm{T}}(\{b_c^s\} - \{\lambda_c^s\} - [K_{cr}^s][K_{rr}^s]^{-1}(\{b_r^s\} - [B_r^s]^{\mathrm{T}}\{\lambda\})). \quad (10.15)$$

Because the tangential component of the magnetic field is continuous across the interface between the subdomains (assuming that there is no surface electric currents on the interface), based on the definition of $\{\lambda_c^s\}$ in (10.6), we have

$$\sum_{s=1}^{N_s} [B_c^s]^{\mathrm{T}} \{\lambda_c^s\} = 0. \tag{10.16}$$

Consequently, (10.15) can be written as

$$\sum_{s=1}^{N_s} [B_c^s]^{\mathrm{T}} ([K_{cc}^s] - [K_{cr}^s][K_{rr}^s]^{-1}[K_{rc}^s])[B_c^s]\{E_c\}$$
$$= \sum_{s=1}^{N_s} [B_c^s]^{\mathrm{T}} (\{b_c^s\} - [K_{cr}^s][K_{rr}^s]^{-1}(\{b_r^s\} - [B_r^s]^{\mathrm{T}}\{\lambda\})) \tag{10.17}$$

which can be coupled with (10.14) to form a complete system of equations for the solution of $\{\lambda\}$ and $\{E_c\}$. In (10.14) and (10.17), $\{\lambda\}$ is called the *dual variable* and $\{E_c\}$ is called the *primal variable*; hence, the formulation above is named as the *dual–primal method*.

Although (10.14) and (10.17) can be solved simultaneously for $\{\lambda\}$ and $\{E_c\}$, a more efficient approach is to eliminate $\{E_c\}$ in these two equations to find a system of equations for only $\{\lambda\}$. To show this clearly, we first rewrite (10.14) and (10.17), respectively, in compact form as

$$[F_{rr}]\{\lambda\} = \{d_r\} - [F_{rc}]\{E_c\} \tag{10.18}$$
$$[\tilde{K}_{cc}]\{E_c\} = \{\tilde{b}_c\} + [F_{rc}]^{\mathrm{T}}\{\lambda\} \tag{10.19}$$

where the matrices and vectors are defined as

$$[\tilde{K}_{cc}] = \sum_{s=1}^{N_s} [\tilde{K}_{cc}^s] = \sum_{s=1}^{N_s} ([B_c^s]^{\mathrm{T}}[K_{cc}^s][B_c^s] - ([K_{rc}^s][B_c^s])^{\mathrm{T}}[K_{rr}^s]^{-1}([K_{rc}^s][B_c^s])) \tag{10.20}$$

$$[F_{rr}] = \sum_{s=1}^{N_s} [F_{rr}^s] = \sum_{s=1}^{N_s} [B_r^s][K_{rr}^s]^{-1}[B_r^s]^{\mathrm{T}} \tag{10.21}$$

$$[F_{rc}] = \sum_{s=1}^{N_s} [F_{rc}^s] = \sum_{s=1}^{N_s} [B_r^s][K_{rr}^s]^{-1}[K_{rc}^s][B_c^s] \tag{10.22}$$

$$\{d_r\} = \sum_{s=1}^{N_s} \{d_r^s\} = \sum_{s=1}^{N_s} [B_r^s][K_{rr}^s]^{-1}\{b_r^s\} \tag{10.23}$$

$$\{\tilde{b}_c\} = \sum_{s=1}^{N_s} \{\tilde{b}_c^s\} = \sum_{s=1}^{N_s} [B_c^s]^{\mathrm{T}}(\{b_c^s\} - [K_{rc}^s]^{\mathrm{T}}[K_{rr}^s]^{-1}\{b_r^s\}). \tag{10.24}$$

In the above we used the symmetric property of the subdomain finite element matrix given by $[K_{cr}^s] = [K_{rc}^s]^{\mathrm{T}}$. Here, we assume that $\overleftrightarrow{\varepsilon}_r$, $\overleftrightarrow{\mu}_r$, and $\overleftrightarrow{\varsigma}$ are all symmetric; however, the formulation can be modified easily to deal with nonsymmetric cases. Although the expressions above seem complicated, every matrix and vector quantity is well defined and the summations can be carried out efficiently provided that the inverse of $[K_{rr}^s]$ is available. In particular, $[\tilde{K}_{cc}]$ is a highly sparse and symmetric matrix that relates all the unknowns on the corner edges in the entire computational domain; hence, (10.19) is referred to as the *coarse* or *corner-related problem*. Furthermore, $[F_{rr}]$ is also a symmetric matrix. Now, we can eliminate $\{E_c\}$ in (10.18) and (10.19) to find

$$([F_{rr}] + [F_{rc}][\tilde{K}_{cc}]^{-1}[F_{rc}]^{\mathrm{T}})\{\lambda\} = \{d_r\} - [F_{rc}][\tilde{K}_{cc}]^{-1}\{\tilde{b}_c\} \tag{10.25}$$

which can be solved for $\{\lambda\}$.

Equation (10.25), called the *global interface system of equations*, is much smaller than the size of the original finite element system since $\{\lambda\}$ is defined only on the subdomain interfaces. This interface equation can be solved iteratively using a Krylov subspace method such as the generalized minimum residual (GMRES) and the stabilized biconjugate gradient method (BiCGSTAB). In an iterative solution process, the system matrix is required only in matrix–vector multiplication. To carry out this multiplication efficiently, we can prefactorize all $[K_{rr}^s]$ $(s = 1,2,\ldots, N_s)$ and $[\tilde{K}_{cc}]$ so that the matrix–vector products involving $[K_{rr}^s]^{-1}$ and $[\tilde{K}_{cc}]^{-1}$ can be evaluated efficiently by a forward and a backward substitution. To accelerate the convergence of the iterative solution, we can construct a preconditioner, which is a good approximation of the inverse of $[F_{rr}]$. A simple approximation of the inverse of $[F_{rr}]$ is given by

$$\left[F_{rr}^{L}\right]^{-1} = \sum_{s=1}^{N_s} [B_r^s][K_{rr}^s][B_r^s]^{\mathrm{T}} \tag{10.26}$$

which is called the *lumped preconditioner* [15]. To simplify (10.26), we can first split $\{E_r^s\}$ into two parts, one for all the interior unknowns, $\{E_i^s\}$, and the other for the unknowns on the interface, $\{E_b^s\}$. The corresponding $[K_{rr}^s]$ can then be partitioned as

$$\left[K_{rr}^{s}\right] = \begin{bmatrix} K_{ii}^s & K_{ib}^s \\ K_{bi}^s & K_{bb}^s \end{bmatrix}. \tag{10.27}$$

Since $[B_r^s]$ extracts only those on the interface, (10.26) can be written as

$$\left[F_{rr}^{L}\right]^{-1} = \sum_{s=1}^{N_s} [B_r^s] \begin{bmatrix} 0 & 0 \\ 0 & K_{bb}^s \end{bmatrix} [B_r^s]^{\mathrm{T}} \tag{10.28}$$

which shows clearly that this lumped preconditioner neglects all the effects of the subdomain interior matrices. Although this preconditioner is computationally inexpensive to construct and works well for definite problems such as those arising from the analysis of static fields, it is not very effective and optimal for indefinite problems resulting from the analysis of electrodynamic problems. A much more effective preconditioner can be obtained by replacing $[K_{bb}^s]$ in (10.28) by the Schur complement $[S_{bb}^s] = [K_{bb}^s] - [K_{bi}^s][K_{ii}^s]^{-1}[K_{ib}^s]$, so that it becomes

$$[F_{rr}^D]^{-1} = \sum_{s=1}^{N_s} [B_r^s] \begin{bmatrix} 0 & 0 \\ 0 & S_{bb}^s \end{bmatrix} [B_r^s]^{\mathrm{T}}. \tag{10.29}$$

This preconditioner is called the *Dirichlet preconditioner* [15], and the matrix $[K_{ii}^s]$ is the same as the subdomain finite element matrix when the Dirichlet boundary condition is applied to the subdomain interface. Although this preconditioner is more expensive to construct, it is much more effective and mathematically optimal, especially for the high-frequency applications we are dealing with here.

Once the dual unknown $\{\lambda\}$ is obtained by solving (10.25), the primal unknown $\{E_c\}$ can then be computed by solving (10.19), which yields

$$\{E_c\} = [\tilde{K}_{cc}]^{-1}\left(\{\tilde{b}_c\} + [F_{rc}]^{\mathrm{T}}\{\lambda\}\right). \tag{10.30}$$

With the computed values of $\{\lambda\}$ and $\{E_c\}$, the electric field in each subdomain can be calculated using (10.8) or (10.11), which can be written as

$$\{E_r^s\} = [K_{rr}^s]^{-1}\left(\{b_r^s\} - [B_r^s]^{\mathrm{T}}\{\lambda\} - [K_{rc}^s][B_c^s]\{E_c\}\right). \tag{10.31}$$

Once the field is computed, all other engineering parameters can be extracted or evaluated based on their definitions.

From the formulation described above, it is obvious that the FETI–DPEM1 method is applicable to general electromagnetic problems with an arbitrary computational domain. However, the method is particularly suitable for the analysis of finite arrays because in a typical finite array, many array elements are identical. As a result, we only have to calculate and factorize a few different subdomain finite element matrices $[K_{rr}^s]$. For example, if all the array elements in a rectangular array are identical, there are nine different subdomain finite element matrices in the entire computational domain, which are illustrated in Figure 10.2. (Due to the application of exterior boundary conditions such as absorbing boundary conditions and perfectly matched layers, the finite element matrices for the subdomains at the edges and corners of the array are different from those in the interior.) All other subdomain finite element matrices can then be identified with one of these nine basic subdomains. Therefore, once these nine subdomain finite element matrices $[K_{rr}^s]$ are calculated and factorized, they can be used for all other subdomains in the iterative solution of the global interface equation and in the final calculation of the fields in each subdomain. This full exploitation of the geometrical repetitions in a finite array can

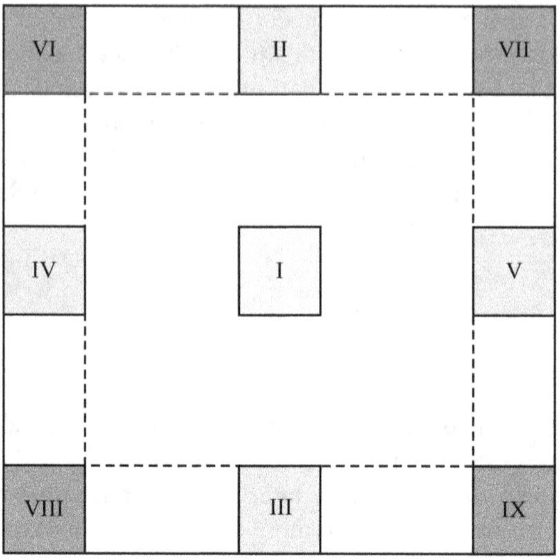

Figure 10.2 Nine subdomains having different subdomain finite element matrices because of the exterior boundary condition. The subdomain finite element matrix for any of other subdomains can be identified with one of the nine labeled here.

result in tremendous savings in computation time and memory. Note that if we extend the dual variable $\{\lambda\}$ over the exterior surface of the computational domain, all the subdomains would have the same finite element matrix for an array consisting of identical array elements. Consequently, only one $[K_{rr}^{s}]$, instead of nine, would have to be calculated and factorized. However, this would approximately double the size of the global interface system of equations, which would actually increase the total computation time. Also note that for an array that has identical interior array elements but different edge and corner elements, there are still only nine different subdomain finite element matrices to be calculated and factorized. The ability to treat different edge and corner elements is important for numerical simulations since it allows finite elements to be used to extend the location of the mesh truncation surface, thereby reducing the grid truncation error, and it is also important for investigating different practical designs, which may contain various edge and corner treatments that control undesirable edge effects.

When designing a numerical method for large-scale applications such as the simulation of large finite arrays considered here, it is very important to design the method such that it is numerically scalable. A numerically scalable algorithm can solve an n-times-larger problem in a constant time by using n times the number of parallel processors, or n times the computation time by using a fixed number of processors. From its formulation, it can be observed that the FETI–DPEM1 would be numerically scalable if the number of iterations for solving the global interface problem is independent of problem size. Numerical tests have shown that the number of iterations for solving (10.25) is indeed independent of the number of subdomains, the spatial size of the finite elements, and the wavenumber when the size of the subdomains is smaller than one-half wavelength. Such excellent scalability is due mainly to the construction of the corner-related coarse problem given by (10.19). In fact, it is this construction that distinguishes the dual–primal method from all other domain-decomposition methods. The reader can verify easily that one can still formulate a relatively simple domain-decomposition algorithm without separating the corner-related unknowns by following the basic procedure outlined in this section. The resulting global interface system propagates the residual error in its iterative solution from a subdomain to its neighboring subdomains. However, with the construction of the corner-related coarse problem, all the subdomains are coupled in this small system, and the residual error is propagated globally at each iteration. As a result, the low-frequency component of the residual error is effectively damped and the convergence of the iterative solution is improved significantly.

However, when the size of the subdomain exceeds one-half wavelength, the number of iterations for solving the global interface equation increases noticeably and the FETI–DPEM1 algorithm becomes numerically unscalable. This problem is caused by use of the Neumann boundary condition (10.2) at the subdomain interface. With this boundary condition, the subdomain finite element matrix in (10.4) is the same as that for a cavity formed by covering the interface with a perfect magnetic conductor, which can support resonant modes when the size exceeds one-half wavelength. Even though for phased-array antenna problems, where we have employed an absorbing boundary condition at the top surface (and perhaps also at the bottom surface if the bottom is open), there remain transverse resonant modes which are not affected by the absorbing boundary condition. For most practical array applications, this is not a serious problem because the array element size is usually designed to be smaller than one-half wavelength at the highest frequency of operation to prevent higher-order Floquet modes (grating lobes) from propagating. Even in applications when the array element size exceeds one-half wavelength, a possible solution is to subdivide an array element into smaller subdomains to control the number of iterations at the cost of increasing the size of the interface system and the time required to construct the finite element meshes. Another possible solution is simply to let the computation continue until it reaches convergence at the cost of a significantly longer computation time.

10.1.2 FETI–DPEM2 Formulation

The FETI–DPEM2 method [10] is designed to make the algorithm fully numerically scalable regardless of the electrical size of the subdomains or the frequency of electromagnetic fields, and this is accomplished with the introduction of two Lagrange multipliers [16]. As discussed in the preceding section, the FETI–DPEM1 becomes numerically unscalable because of the use of the Neumann boundary condition (10.2) at the subdomain interface. Therefore, instead of assuming (10.2) at the subdomain interface, we assume that the field satisfies a mixed boundary condition, or an impedance boundary condition, given by

$$\hat{n}^s \times \left[\overleftrightarrow{\mu}_r^{-1} \cdot (\nabla \times \mathbf{E}^s)\right] + jk_0 \hat{n}^s \times \left[\overleftrightarrow{\varsigma}^s \cdot (\hat{n}^s \times \mathbf{E}^s)\right] = \mathbf{\Lambda}^s \tag{10.32}$$

where the superscript s denotes the subdomain number, $\overleftrightarrow{\varsigma}^s$ denotes the normalized admittance tensor at the interface, and $\mathbf{\Lambda}^s$ denotes the dual unknown vector to be solved for. Equation (10.32) is also called the *Robin transmission condition* in mathematics. With this, the finite element matrix equation for the subdomain becomes

$$\begin{bmatrix} K_{rr}^s & K_{rc}^s \\ K_{cr}^s & K_{cc}^s \end{bmatrix} \begin{Bmatrix} E_r^s \\ E_c^s \end{Bmatrix} = \begin{Bmatrix} b_r^s \\ b_c^s \end{Bmatrix} - \begin{Bmatrix} \lambda_r^s \\ \lambda_c^s \end{Bmatrix} \tag{10.33}$$

where

$$\begin{aligned}[K_{rr}^s] = &\iiint_{V_s} \left[\nabla \times \{\mathbf{N}_r^s\} \cdot \overleftrightarrow{\mu}_r^{-1} \cdot \nabla \times \{\mathbf{N}_r^s\}^{\mathrm{T}} - k_0^2 \{\mathbf{N}_r^s\} \cdot \overleftrightarrow{\varepsilon}_r \cdot \{\mathbf{N}_r^s\}^{\mathrm{T}}\right] dV \\ &+ jk_0 \iint_{S_s} \hat{n}^s \times \{\mathbf{N}_r^s\} \cdot \overleftrightarrow{\varsigma}^s \cdot \hat{n}^s \times \{\mathbf{N}_r^s\}^{\mathrm{T}}\, dS\end{aligned} \tag{10.34}$$

$$\begin{aligned}[K_{rc}^s] = &\iiint_{V_s} \left[\nabla \times \{\mathbf{N}_r^s\} \cdot \overleftrightarrow{\mu}_r^{-1} \cdot \nabla \times \{\mathbf{N}_c^s\}^{\mathrm{T}} - k_0^2 \{\mathbf{N}_r^s\} \cdot \overleftrightarrow{\varepsilon}_r \cdot \{\mathbf{N}_c^s\}^{\mathrm{T}}\right] dV \\ &+ jk_0 \iint_{S_s \cap S_o} \hat{n}^s \times \{\mathbf{N}_r^s\} \cdot \overleftrightarrow{\varsigma}^s \cdot \hat{n}^s \times \{\mathbf{N}_c^s\}^{\mathrm{T}}\, dS\end{aligned} \tag{10.35}$$

$$\begin{aligned}[K_{cr}^s] = &\iiint_{V_s} \left[\nabla \times \{\mathbf{N}_c^s\} \cdot \overleftrightarrow{\mu}_r^{-1} \cdot \nabla \times \{\mathbf{N}_r^s\}^{\mathrm{T}} - k_0^2 \{\mathbf{N}_c^s\} \cdot \overleftrightarrow{\varepsilon}_r \cdot \{\mathbf{N}_r^s\}^{\mathrm{T}}\right] dV \\ &+ jk_0 \iint_{S_s \cap S_o} \hat{n}^s \times \{\mathbf{N}_c^s\} \cdot \overleftrightarrow{\varsigma}^s \cdot \hat{n}^s \times \{\mathbf{N}_r^s\}^{\mathrm{T}}\, dS\end{aligned} \tag{10.36}$$

$$\begin{aligned}[K_{cc}^s] = &\iiint_{V_s} \left[\nabla \times \{\mathbf{N}_c^s\} \cdot \overleftrightarrow{\mu}_r^{-1} \cdot \nabla \times \{\mathbf{N}_c^s\}^{\mathrm{T}} - k_0^2 \{\mathbf{N}_c^s\} \cdot \overleftrightarrow{\varepsilon}_r \cdot \{\mathbf{N}_c^s\}^{\mathrm{T}}\right] dV \\ &+ jk_0 \iint_{S_s \cap S_o} \hat{n}^s \times \{\mathbf{N}_c^s\} \cdot \overleftrightarrow{\varsigma}^s \cdot \hat{n}^s \times \{\mathbf{N}_c^s\}^{\mathrm{T}}\, dS\end{aligned} \tag{10.37}$$

$$\{b_r^s\} = -\iiint_{V_s} \{\mathbf{N}_r^s\} \cdot \left[jk_0 Z_0 \mathbf{J}_{\mathrm{imp}} + \nabla \times (\overleftrightarrow{\mu}_r^{-1} \cdot \mathbf{M}_{\mathrm{imp}})\right] dV \tag{10.38}$$

$$\{b_c^s\} = -\iiint_{V_s} \{\mathbf{N}_c^s\} \cdot \left[jk_0 Z_0 \mathbf{J}_{\mathrm{imp}} + \nabla \times (\overleftrightarrow{\mu}_r^{-1} \cdot \mathbf{M}_{\mathrm{imp}})\right] dV \tag{10.39}$$

$$\{\lambda_r^s\} = \iint_{S_s \cap \Gamma_s} \{\mathbf{N}_r^s\} \cdot \mathbf{\Lambda}^s\, dS + jk_0 \iint_{S_s \cap \Gamma_s} \hat{n}^s \times \{\mathbf{N}_r^s\} \cdot \overleftrightarrow{\varsigma}^s \cdot \hat{n}^s \times \{\mathbf{N}_c^s\}^{\mathrm{T}}\, dS \{E_c^s\} \tag{10.40}$$

$$\{\lambda_c^s\} = -jk_0 Z_0 \iint_{S_s \cap \Gamma_s} \{\mathbf{N}_c^s\} \cdot (\hat{n}^s \times \mathbf{H})\, dS. \tag{10.41}$$

Note that $[K_{rr}^s]$ is different from that in the preceding subsection since its surface integral is now over the entire surface of the subdomain because of the use of (10.32). It is important to note that although some notations used in this subsection are the same as those in the preceding subsection, the definitions may differ. Also note that the second terms in (10.35)–(10.37) do not exist for the current domain decomposition arrangement depicted in Figure 10.1 since $\hat{n}^s \times \{\mathbf{N}_c^s\}$ vanishes on the exterior surface S_o. They are included here to make the formulation applicable to more general cases. From (10.33) we obtain the electric fields on the subdomain interfaces as

$$[B_r^s]\{E_r^s\} = [B_r^s][K_{rr}^s]^{-1}(\{b_r^s\} - [B_r^s]^{\mathrm{T}}\{\lambda^s\} - [K_{rc}^s][B_c^s]\{E_c\}) \quad (10.42)$$

$$\begin{aligned}[B_c^s]^{\mathrm{T}}([K_{cc}^s] - [K_{cr}^s][K_{rr}^s]^{-1}[K_{rc}^s])[B_c^s]\{E_c\} \\ = [B_c^s]^{\mathrm{T}}(\{b_c^s\} - \{\lambda_c^s\} - [K_{cr}^s][K_{rr}^s]^{-1}(\{b_r^s\} - [B_r^s]^{\mathrm{T}}\{\lambda^s\})).\end{aligned} \quad (10.43)$$

These two equations are nearly the same as (10.11) and (10.12) except that $\{\lambda\}$ is now replaced by $\{\lambda^s\}$ because the dual variable $\mathbf{\Lambda}^s$ is related to each specific subdomain. Note that $[B_r^s]$ in (10.42) and (10.43) is defined differently from that in Section 10.1.1. Here, it denotes a Boolean matrix that simply extracts the unknowns on the interface from $\{E_r^s\}$ without sign assignment.

To couple the fields over all the subdomains, we enforce the continuity condition on the tangential components of the electric and magnetic fields across the subdomain interfaces. However, since the postulated boundary condition has the form of (10.32), the continuity condition has to be enforced in a weak sense. To illustrate this process clearly, consider two adjacent subdomains s and q; the continuity condition across their interface denoted by Γ_s^q is given by

$$\hat{n}^s \times \mathbf{E}^s = -\hat{n}^q \times \mathbf{E}^q \quad \text{and} \quad \hat{n}^s \times \mathbf{H}^s = -\hat{n}^q \times \mathbf{H}^q. \quad (10.44)$$

In formulating the field in subdomain s, we assumed the boundary condition given by (10.32), which can be written for the interface between subdomains s and q as

$$\hat{n}^s \times [\overleftrightarrow{\mu}_r^{-1} \cdot (\nabla \times \mathbf{E}_q^s)] + jk_0\hat{n}^s \times [\overleftrightarrow{\varsigma}^s \cdot (\hat{n}^s \times \mathbf{E}_q^s)] = \mathbf{\Lambda}_q^s \quad (10.45)$$

where $\mathbf{E}_q^s$ and $\mathbf{\Lambda}_q^s$ denote the associated quantities $\mathbf{E}^s$ and $\mathbf{\Lambda}^s$ on Γ_s^q. The corresponding boundary condition for formulating the field in subdomain q is given by

$$\hat{n}^q \times [\overleftrightarrow{\mu}_r^{-1} \cdot (\nabla \times \mathbf{E}_s^q)] + jk_0\hat{n}^q \times [\overleftrightarrow{\varsigma}^q \cdot (\hat{n}^q \times \mathbf{E}_s^q)] = \mathbf{\Lambda}_s^q. \quad (10.46)$$

By adding (10.45) and (10.46) and applying (10.44), we obtain

$$\mathbf{\Lambda}_q^s + \mathbf{\Lambda}_s^q = jk_0\hat{n}^s \times [(\overleftrightarrow{\varsigma}^s + \overleftrightarrow{\varsigma}^q) \cdot (\hat{n}^s \times \mathbf{E}_q^s)] \quad (10.47)$$

$$\mathbf{\Lambda}_q^s + \mathbf{\Lambda}_s^q = jk_0\hat{n}^q \times [(\overleftrightarrow{\varsigma}^s + \overleftrightarrow{\varsigma}^q) \cdot (\hat{n}^q \times \mathbf{E}_s^q)] \quad (10.48)$$

which weakly enforces the continuity conditions in (10.44). By testing it with $\{\mathbf{N}_r^s\}$ and integrating over Γ_s^q, (10.48) can be converted into a matrix equation as

$$\{\lambda_q^s\} + \{\lambda_s^q\} = [M_q^s]\{E_s^q\} \qquad s = 1, 2, \ldots, N_s \text{ and } q \in \text{neighbor}(s) \quad (10.49)$$

where

$$\{\lambda_q^s\} = \iint_{\Gamma_s^q} \{\mathbf{N}_r^s\} \cdot \mathbf{\Lambda}_q^s \, dS + jk_0 \iint_{\Gamma_s^q} \hat{n}^s \times \{\mathbf{N}_r^s\} \cdot \overleftrightarrow{\varsigma}^s \cdot \hat{n}^s \times \{\mathbf{N}_c^s\}^{\mathrm{T}} \, dS \{E_c^s\} \quad (10.50)$$

$$\{\lambda_s^q\} = \iint_{\Gamma_s^q} \{\mathbf{N}_r^s\} \cdot \mathbf{\Lambda}_s^q \, dS + jk_0 \iint_{\Gamma_s^q} \hat{n}^q \times \{\mathbf{N}_r^s\} \cdot \overleftrightarrow{\varsigma}^q \cdot \hat{n}^q \times \{\mathbf{N}_c^q\}^{\mathrm{T}} \, dS \{E_c^q\}$$
$$(10.51)$$

$$[M_q^s] = jk_0 \iint_{\Gamma_s^q} \hat{n}^s \times \{\mathbf{N}_r^s\} \cdot (\overleftrightarrow{\varsigma}^s + \overleftrightarrow{\varsigma}^q) \cdot \hat{n}^q \times \{\mathbf{N}_r^q\}^{\mathrm{T}} \, dS. \quad (10.52)$$

Although $\{\lambda_q^s\}$ and $\{\lambda_s^q\}$ have long and complicated expressions, we do not have to be concerned with these specific forms, because we will find that they will be solved directly and then employed to calculate the fields in the subdomains using the first equation in (10.33).

Equation (10.49) can be employed to connect the discrete dual variables and the fields in (10.42) and (10.43) across all the interfaces. To assemble the subdomains, we introduce several projection matrices. Assume that $\{\lambda\} = [\{\lambda_r^1\}, \{\lambda_r^2\}, \ldots, \{\lambda_r^{N_s}\}]^{\mathrm{T}}$. A projection Boolean matrix $[Q^s]$ is introduced to extract $\{\lambda_r^s\}$ from $\{\lambda\}$; that is,

$$\{\lambda_r^s\} = [Q^s]\{\lambda\}. \quad (10.53)$$

Another projection Boolean matrix $[T_q^s]$ is introduced to extract $\{E_q^s\}$ from $\{E_b^s\}$, which stores the discrete fields on the interface of subdomain s. Hence,

$$\{E_q^s\} = [T_q^s]\{E_b^s\}, \qquad \{\lambda_q^s\} = [T_q^s]\{\lambda_r^s\}. \quad (10.54)$$

Consequently, we have

$$\{\lambda\} = \sum_{s=1}^{N_s} [Q^s]^{\mathrm{T}} \{\lambda_r^s\}, \qquad \{\lambda_r^s\} = \sum_{q \in \text{neighbor}(s)} [T_q^s]^{\mathrm{T}} \{\lambda_q^s\}. \quad (10.55)$$

With the aid of (10.53)–(10.55), (10.49) can be written as

$$\{\lambda_q^s\} + [T_s^q]\{\lambda_r^q\} = [M_q^s][T_s^q][B_r^q]\{E_r^q\}. \quad (10.56)$$

By eliminating $\{E_r^q\}$ using (10.42), this can be written further as

$$\{\lambda_q^s\} + ([T_s^q] + [P_s^q][F_{rr}^q])\{\lambda_r^q\} + [P_s^q][F_{rc}^q]\{E_c\} = [P_s^q]\{d_r^q\} \quad (10.57)$$

where $[P_s^q] = [M_q^s][T_s^q]$, and $[F_{rr}^q]$, $[F_{rc}^q]$, and $\{d_r^q\}$ are defined similarly to (10.21)–(10.23).

Now we are ready to assemble the global system of equations. We first premultiply (10.57) by $[T_q^s]^{\mathrm{T}}$, then sum over all the neighboring subdomains for q, and finally, sum over all the subdomains for s. Doing so, we find the global interface-related system to be

$$[\tilde{F}_{rr}]\{\lambda\} + [\tilde{F}_{rc}]\{E_c\} = \{\tilde{d}_r\} \tag{10.58}$$

where

$$[\tilde{F}_{rr}] = [I] + \sum_{s=1}^{N_s} [Q^s]^{\mathrm{T}} \sum_{q \in \text{neighbor}(s)} [T_q^s]^{\mathrm{T}} ([T_s^q] + [P_s^q][F_{rr}^q])[Q^q] \tag{10.59}$$

$$[\tilde{F}_{rc}] = \sum_{s=1}^{N_s} [Q^s]^{\mathrm{T}} \sum_{q \in \text{neighbor}(s)} [T_q^s]^{\mathrm{T}} [P_s^q][F_{rc}^q] \tag{10.60}$$

$$\{\tilde{d}_r\} = \sum_{s=1}^{N_s} [Q^s]^{\mathrm{T}} \sum_{q \in \text{neighbor}(s)} [T_q^s]^{\mathrm{T}} [P_s^q]\{d_r^q\}. \tag{10.61}$$

In (10.59), $[I]$ denotes the identity matrix. The global corner-related system can be obtained by assembling (10.43) over all the subdomains as

$$[\tilde{K}_{cc}]\{E_c\} = \{\tilde{b}_c\} + [\tilde{F}_{cr}]\{\lambda\} \tag{10.62}$$

where the expressions of $[\tilde{K}_{cc}]$ and $\{\tilde{b}_c\}$ are given by (10.20) and (10.24), respectively, and $[\tilde{F}_{cr}]$ is given by

$$[\tilde{F}_{cr}] = \sum_{s=1}^{N_s} [F_{rc}^s]^{\mathrm{T}} [Q^s] \tag{10.63}$$

with $[F_{rc}^s]$ given by (10.22). Equations (10.58) and (10.62) form a complete system of equations for the solution of the dual unknown $\{\lambda\}$ and the primal unknown $\{E_c\}$. We can also eliminate $\{E_c\}$ to find a reduced system

$$([\tilde{F}_{rr}] + [\tilde{F}_{rc}][\tilde{K}_{cc}]^{-1}[\tilde{F}_{cr}])\{\lambda\} = \{\tilde{d}_r\} - [\tilde{F}_{rc}][\tilde{K}_{cc}]^{-1}\{\tilde{b}_c\} \tag{10.64}$$

which can be solved for $\{\lambda\}$ by using a Krylov subspace method such as GMRES or BiCGSTAB. In general, BiCGSTAB is more efficient for this solution [10]. Once $\{\lambda\}$ is available, $\{E_c\}$ can be computed by solving (10.62). When both $\{\lambda\}$ and $\{E_c\}$ are computed, the field in each subdomain can be calculated using the first equation of

(10.33), which can be written as

$$\{E_r^s\} = [K_{rr}^s]^{-1}(\{b_r^s\} - [B_r^s]^{\mathrm{T}}[Q^s]\{\lambda\} - [K_{rc}^s][B_c^s]\{E_c\}). \tag{10.65}$$

Compared to the FETI–DPEM1 formulation, the global interface system (10.64) in the FETI–DPEM2 formulation is twice as large since two sets of dual variables are defined on the subdomain interfaces. However, the amount of computation and computer memory needed for an iterative solution is about the same because of the identity matrix in (10.59), which requires no computation and storage. More interestingly, because of the identity matrix in (10.59) and the global corner-related system (10.62), the final global interface system (10.64) is very well conditioned and can be solved efficiently without using a preconditioner. Numerical tests have shown [10] that FETI–DPEM2 maintains the numerical scalability of FETI–DPEM1 with regard to the number of iterations required for solving (10.64) being independent of the number of subdomains and the spatial size of the finite elements. More important, it is also numerically scalable with respect to the frequency of the electromagnetic field, which was a critical property lacking in the FETI–DPEM1 formulation. This is because $[K_{rr}^s]$ in (10.34) corresponds to the finite element matrix for subdomain s with all six of its surfaces made resistive through the application of the mixed boundary condition (10.45) and the radiation condition (10.1). As a result, $[K_{rr}^s]$ cannot support any resonant modes with real-valued frequencies. This is the major difference between the two FETI–DPEM formulations. The introduction of two sets of dual variables in the FETI–DPEM2 is simply the cost to achieve this important distinction.

For the numerical implementation of the FETI–DPEM2 algorithm, we have to choose the values for $\overleftrightarrow{\varsigma}^s$ and $\overleftrightarrow{\varsigma}^q$ to be used in (10.52). In theory, the best choice is such that they minimize the reflection by the interface associated with the transmission conditions (10.45) and (10.46). If the material at the interface is isotropic, $\overleftrightarrow{\varsigma}^s$ and $\overleftrightarrow{\varsigma}^q$ become scalars and a good choice for their values is given by $\sqrt{\varepsilon_r/\mu_r}$. If the material property has a jump at the interface, $\overleftrightarrow{\varsigma}^s$ will use the value on the side of subdomain s and $\overleftrightarrow{\varsigma}^q$ will use the value on the side of subdomain q. However, since (10.45) and (10.46) are not true physical boundary conditions and are introduced simply to weakly enforce the field continuity conditions, the choice of the values for $\overleftrightarrow{\varsigma}^s$ and $\overleftrightarrow{\varsigma}^q$ is not critical for the numerical solution. We can simply set them to unity for all practical applications.

10.1.3 Nonconforming Domain Decomposition

In the two FETI–DPEM formulations described above, the finite element surface meshes on the two sides of an interface are required to be the same or to conform to each other. This requirement is caused by the use of a single set of the primal variable $\{E_c\}$ on the corner edges and by the elimination of the discrete electric fields on the interfaces. In practical applications, generating conforming meshes at an interface may not be an easy task for some mesh generators. In that case it is desirable to have a domain-decomposition algorithm that can handle different surface meshes, or nonconforming surface meshes, at the subdomain interfaces. Such an algorithm is often called a *nonconforming domain-decomposition algorithm.*

In fact, if we are willing to sacrifice the corner-related coarse system of equations and introduce two sets of discrete electric fields on the subdomain interfaces, we can extend the FETI–DPEM formulations to deal with nonconforming interface meshes. Here, we describe a well-tested algorithm [7,9] which combines the alternating Schwarz nonoverlapping domain-decomposition method [17,18] with the concept of "cement" elements [19]. Like all other domain-decomposition methods, this algorithm first decomposes a computational domain into smaller subdomains and then calculates the fields in all the subdomains iteratively with some form of enforcement of the field continuity through the subdomain interfaces. While direct enforcement of the field continuity has been found to lead to convergence problems when solving the wave equations (similar to FETI–DPEM1), the weak enforcement through Robin-type transmission conditions effectively ensures the convergence of the iterative process (similar to FETI–DPEM2). At the interface between subdomains s and q, the Robin-type transmission conditions are given by

$$\begin{aligned}&\hat{n}^s \times \left[\overleftrightarrow{\mu}_r^{-1} \cdot (\nabla \times \mathbf{E}^s)\right] + jk_0\hat{n}^s \times (\hat{n}^s \times \mathbf{E}^s)\\&\quad = -\hat{n}^q \times \left[\overleftrightarrow{\mu}_r^{-1} \cdot (\nabla \times \mathbf{E}^q)\right] + jk_0\hat{n}^q \times (\hat{n}^q \times \mathbf{E}^q)\end{aligned} \tag{10.66}$$

$$\begin{aligned}&\hat{n}^q \times \left[\overleftrightarrow{\mu}_r^{-1} \cdot (\nabla \times \mathbf{E}^q)\right] + jk_0\hat{n}^q \times (\hat{n}^q \times \mathbf{E}^q)\\&\quad = -\hat{n}^s \times \left[\overleftrightarrow{\mu}_r^{-1} \cdot (\nabla \times \mathbf{E}^s)\right] + jk_0\hat{n}^s \times (\hat{n}^s \times \mathbf{E}^s).\end{aligned} \tag{10.67}$$

It can be seen easily that the sum of the two equations yields the tangential electric field continuity condition

$$\hat{n}^s \times (\hat{n}^s \times \mathbf{E}^s) = \hat{n}^q \times (\hat{n}^q \times \mathbf{E}^q) \tag{10.68}$$

and subtraction of the two equations yields the tangential magnetic field continuity condition

$$\hat{n}^s \times \left[\overleftrightarrow{\mu}_r^{-1} \cdot (\nabla \times \mathbf{E}^s)\right] = -\hat{n}^q \times \left[\overleftrightarrow{\mu}_r^{-1} \cdot (\nabla \times \mathbf{E}^q)\right]. \tag{10.69}$$

Hence, (10.66) and (10.67) can be considered to be equivalent to (10.68) and (10.69).

With the formulation of the transmission conditions at the subdomain interfaces, the iterative process for the alternating Schwarz nonoverlapping domain-decomposition method is given by

$$\nabla \times \left(\overleftrightarrow{\mu}_r^{-1} \cdot \nabla \times \mathbf{E}_{(i)}^s\right) - k_0^2\overleftrightarrow{\varepsilon}_r \cdot \mathbf{E}_{(i)}^s = -jk_0Z_0\mathbf{J}_{\text{imp}} - \nabla \times \left(\overleftrightarrow{\mu}_r^{-1} \cdot \mathbf{M}_{\text{imp}}\right) \qquad \mathbf{r} \in V_s \tag{10.70}$$

$$\hat{n} \times \left[\overleftrightarrow{\mu}_r^{-1} \cdot \left(\nabla \times \mathbf{E}_{(i)}^s\right)\right] + jk_0\hat{n} \times \left[\overleftrightarrow{\varsigma} \cdot \left(\hat{n} \times \mathbf{E}_{(i)}^s\right)\right] \approx 0 \qquad \mathbf{r} \in S_s \cap S_o \tag{10.71}$$

$$\hat{n}^s \times \left[\overleftrightarrow{\mu}_r^{-1} \cdot \left(\nabla \times \mathbf{E}^s_{(i)}\right)\right] + jk_0 \hat{n}^s \times \left(\hat{n}^s \times \mathbf{E}^s_{(i)}\right)$$
$$= -\hat{n}^q \times \left[\overleftrightarrow{\mu}_r^{-1} \cdot \left(\nabla \times \mathbf{E}^q_{(i-1)}\right)\right] + jk_0 \hat{n}^q \times \left(\hat{n}^q \times \mathbf{E}^q_{(i-1)}\right) \qquad \mathbf{r} \in S_s \cap \Gamma^q_s \tag{10.72}$$

in conjunction with the appropriate boundary conditions inside the subdomain, and here the subscript (i) denotes the iteration number. In other words, the iteration process starts with an initial guess for the field and then solves the boundary-value problem defined in (10.70)–(10.72) for each subdomain ($s = 1, 2, \ldots, N_s$). The tangential fields computed at the interfaces are then passed to the neighboring subdomains to construct the right-hand side of the transmission condition, which is used for the calculation in the next iteration. This iterative process continues until the solution converges to certain accuracy.

Although the implementation of (10.72), which is nothing more than a mixed boundary condition, is well established in the finite element method, construction of its right-hand side is less straightforward. To facilitate the finite element implementation and, more important, to handle nonconforming surface meshes at the interfaces illustrated in Figure 10.3, a cement element method has been proposed [7,9] that introduces an additional quantity, called the *cement variable*, over the subdomain interfaces. This quantity is defined as

$$\mathbf{j}^s = \hat{n}^s \times \left[\overleftrightarrow{\mu}_r^{-1} \cdot (\nabla \times \mathbf{E}^s)\right] \tag{10.73}$$

which is related to the equivalent surface electric current in electromagnetics. With this cement variable, the boundary-value problem for subdomain s at the ith iteration becomes

$$\nabla \times \left(\overleftrightarrow{\mu}_r^{-1} \cdot \nabla \times \mathbf{E}^s_{(i)}\right) - k_0^2 \overleftrightarrow{\varepsilon}_r \cdot \mathbf{E}^s_{(i)} = -jk_0 Z_0 \mathbf{J}_{\text{imp}} - \nabla \times \left(\overleftrightarrow{\mu}_r^{-1} \cdot \mathbf{M}_{\text{imp}}\right) \qquad \mathbf{r} \in V_s \tag{10.74}$$

$$\hat{n} \times \left[\overleftrightarrow{\mu}_r^{-1} \cdot \left(\nabla \times \mathbf{E}^s_{(i)}\right)\right] + jk_0 \hat{n} \times \left[\overleftrightarrow{\varsigma} \cdot \left(\hat{n} \times \mathbf{E}^s_{(i)}\right)\right] \approx 0 \qquad \mathbf{r} \in S_s \cap S_o \tag{10.75}$$

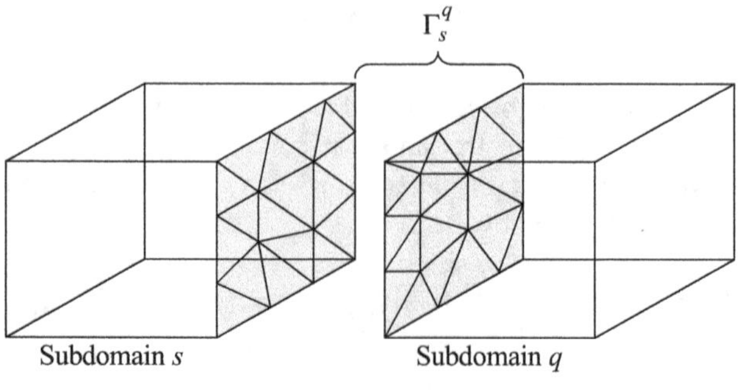

Figure 10.3 Subdomains with nonconforming meshes at the interface.

$$\hat{n}^s \times \left[\overleftrightarrow{\mu}_r^{-1} \cdot \left(\nabla \times \mathbf{E}_{(i)}^s\right)\right] = \mathbf{j}_{(i)}^s \qquad \mathbf{r} \in S_s \cap \Gamma_s \tag{10.76}$$

$$\mathbf{j}_{(i)}^s + jk_0\hat{n}^s \times \left(\hat{n}^s \times \mathbf{E}_{(i)}^s\right) = -\mathbf{j}_{(i-1)}^q + jk_0\hat{n}^q \times \left(\hat{n}^q \times \mathbf{E}_{(i-1)}^q\right) \qquad \mathbf{r} \in S_s \cap \Gamma_s^q. \tag{10.77}$$

The finite element discretization of this boundary-value problem defined by (10.74)–(10.76) yields

$$\begin{bmatrix} K_{ii}^s & K_{ib}^s & 0 \\ K_{bi}^s & K_{bb}^s & B_{bb}^s \end{bmatrix} \begin{Bmatrix} E_i^s \\ E_b^s \\ j^s \end{Bmatrix}_{(i)} = \begin{Bmatrix} b_i^s \\ b_b^s \end{Bmatrix} \tag{10.78}$$

where the combined $[K^s]$ matrix is given by (10.4), the combined $\{b^s\}$ is given by (10.5), and

$$\left[B_{bb}^s\right] = \iint_{S_s \cap \Gamma_s} \left\{\mathbf{N}_b^s\right\} \cdot \hat{n}^s \times \left\{\mathbf{N}_b^s\right\}^{\mathrm{T}} dS. \tag{10.79}$$

In (10.78), the subscript i denotes the quantities inside the subdomain and the subscript b denotes the quantities on the surface (including the corner edges) of the subdomain. By testing it with $\hat{n}^s \times \{\mathbf{N}_b^s\}$, integrating over Γ_s^q, and summing over all the interfaces between subdomain s and its neighboring subdomains, the transmission condition (10.77) can be converted into a matrix equation as

$$\begin{bmatrix} B_{bb}^{s\ \mathrm{T}} & C_{bb}^s \end{bmatrix} \begin{Bmatrix} E_b^s \\ j^s \end{Bmatrix}_{(i)} = \sum_{q \in \mathrm{neighbor}(s)} \begin{bmatrix} U_s^q & V_s^q \end{bmatrix} \begin{Bmatrix} E_b^q \\ j^q \end{Bmatrix}_{(i-1)} \tag{10.80}$$

where

$$\left[C_{bb}^s\right] = \frac{j}{k_0} \iint_{S_s \cap \Gamma_s} \hat{n}^s \times \left\{\mathbf{N}_b^s\right\} \cdot \hat{n}^s \times \left\{\mathbf{N}_b^s\right\}^{\mathrm{T}} dS \tag{10.81}$$

$$\left[U_s^q\right] = \iint_{S_s \cap \Gamma_s^q} \hat{n}^s \times \left\{\mathbf{N}_b^s\right\} \cdot \left\{\mathbf{N}_b^q\right\}^{\mathrm{T}} dS \tag{10.82}$$

$$\left[V_s^q\right] = \frac{1}{jk_0} \iint_{S_s \cap \Gamma_s^q} \hat{n}^s \times \left\{\mathbf{N}_b^s\right\} \cdot \hat{n}^q \times \left\{\mathbf{N}_b^q\right\}^{\mathrm{T}} dS. \tag{10.83}$$

Equations (10.78) and (10.80) can now be combined to form a complete system for subdomain s as

$$\begin{bmatrix} K_{ii}^s & K_{ib}^s & 0 \\ K_{bi}^s & K_{bb}^s & B_{bb}^s \\ 0 & B_{bb}^{s\ \mathrm{T}} & C_{bb}^s \end{bmatrix} \begin{Bmatrix} E_i^s \\ E_b^s \\ j^s \end{Bmatrix}_{(i)} = \begin{Bmatrix} b_i^s \\ b_b^s \\ g^s \end{Bmatrix}_{(i-1)} \tag{10.84}$$

where

$$\{g^s\} = \sum_{q \in \text{neighbor}(s)} \begin{bmatrix} U_s^q & V_s^q \end{bmatrix} \begin{Bmatrix} E_b^q \\ j^q \end{Bmatrix}. \tag{10.85}$$

Equation (10.84) can be solved using either the Jacobi or the Gauss–Seidel method [7,9].

For very large arrays modeled with millions, or perhaps billions, of unknowns, calculating all the fields in all the subdomains is not only time consuming but also memory intensive. In this case we can first reduce (10.84) into a smaller system that involves only the fields and the cement variables on the subdomain surfaces. This can be done by extracting $[E_b^s, j^s]^{\mathrm{T}}$ from (10.84) as

$$\begin{Bmatrix} E_b^s \\ j^s \end{Bmatrix}_{(i)} = \begin{bmatrix} 0 & I_{bb}^s & 0 \\ 0 & 0 & I_{bb}^s \end{bmatrix} \begin{Bmatrix} E_i^s \\ E_b^s \\ j^s \end{Bmatrix}_{(i)}$$

$$= \begin{bmatrix} 0 & I_{bb}^s & 0 \\ 0 & 0 & I_{bb}^s \end{bmatrix} \begin{bmatrix} K_{ii}^s & K_{ib}^s & 0 \\ K_{bi}^s & K_{bb}^s & B_{bb}^s \\ 0 & B_{bb}^{s\,\mathrm{T}} & C_{bb}^s \end{bmatrix}^{-1} \left(\begin{Bmatrix} b_i^s \\ b_b^s \\ 0 \end{Bmatrix} + \begin{Bmatrix} 0 \\ 0 \\ g^s \end{Bmatrix}_{(i-1)} \right) \tag{10.86}$$

where $[I_{bb}^s]$ denotes the identity matrix. This equation can be placed into a compact form as

$$\{v^s\}_{(i)} = \left[R_b^s\right]\{u^s\}_{(i)} = \left[R_b^s\right][A^s]^{-1}(\{\tilde{b}^s\} + \left[R_j^s\right]^{\mathrm{T}}\{g^s\}_{(i-1)}) \tag{10.87}$$

where

$$\{v^s\} = \begin{Bmatrix} E_b^s \\ j^s \end{Bmatrix}, \qquad \{u^s\} = \begin{Bmatrix} E_i^s \\ E_b^s \\ j^s \end{Bmatrix}, \qquad \{\tilde{b}^s\} = \begin{Bmatrix} b_i^s \\ b_b^s \\ 0 \end{Bmatrix} \tag{10.88}$$

$$\left[R_b^s\right] = \begin{bmatrix} 0 & I_{bb}^s & 0 \\ 0 & 0 & I_{bb}^s \end{bmatrix}, \qquad \left[R_j^s\right] = [0 \quad 0 \quad I_{bb}^s] \tag{10.89}$$

$$[A^s] = \begin{bmatrix} K_{ii}^s & K_{ib}^s & 0 \\ K_{bi}^s & K_{bb}^s & B_{bb}^s \\ 0 & B_{bb}^{s\,\mathrm{T}} & C_{bb}^s \end{bmatrix}. \tag{10.90}$$

It is now clear that we can precompute $[R_b^s][A^s]^{-1}\{\tilde{b}^s\}$ and $[R_b^s][A^s]^{-1}[R_j^s]^{\mathrm{T}}$ so that in subsequent iterations the calculation of (10.87) is reduced to a simple matrix–vector multiplication. Furthermore, for identical subdomains, the matrix $[R_b^s][A^s]^{-1}[R_j^s]^{\mathrm{T}}$ is identical in each subdomain, which makes the method highly attractive for array-type problems, similar to the FETI–DPEM methods.

Although the Gauss–Seidel method is simple to implement, its convergence is rather slow and in some applications it may not converge [20]. This problem can be alleviated by replacing the stationary iterative method with a Krylov subspace method. For this, we have to convert (10.87) into a system with all the unknown variables moved to the left-hand side. Let us first denote the global unknown vector as $\{v\} = [\{v^1\}, \{v^2\}, \ldots, \{v^{N_s}\}]^{\mathrm{T}}$ and introduce a Boolean matrix $[Q^s]$ to extract $\{v^s\}$ from $\{v\}$. Hence,

$$\{v^s\} = [Q^s]\{v\}, \qquad \{v\} = \sum_{s=1}^{N_s} [Q^s]^{\mathrm{T}}\{v^s\}. \tag{10.91}$$

With this, we can premultiply (10.87) by $[Q^s]^{\mathrm{T}}$ and sum the result over all the subdomains to find the final system of equations as

$$([I] - [S])\{v\} = \{f\} \tag{10.92}$$

where

$$[S] = \sum_{s=1}^{N_s} [Q^s]^{\mathrm{T}}\big[R_b^s\big][A^s]^{-1}\big[R_j^s\big]^{\mathrm{T}} \sum_{q\in\text{neighbor}(s)} \big[D_s^q\big][Q^q] \tag{10.93}$$

$$\{f\} = \sum_{s=1}^{N_s} [Q^s]^{\mathrm{T}}\big[R_b^s\big][A^s]^{-1}[\tilde{b}^s] \tag{10.94}$$

in which $[D_s^q] = [\,U_s^q \quad V_s^q\,]$. Equation (10.92) can now be solved using a Krylov subspace method such as GMRES or BiCGSTAB. The convergence of the iterative solution can be accelerated by using the symmetric Gauss–Seidel (SGS) preconditioner [20] or the block symmetric successive over relaxation (SSOR) preconditioner [21], which reduces to the SGS preconditioner when the relaxation factor is set to 1. Once $\{v\}$ is obtained, the field inside each subdomain can be calculated by solving either the first set of equations in (10.78), the entire system of equations in (10.78), or the system of equations in (10.84), where the invertibility is guaranteed.

10.1.4 Application Examples

In this section we present three numerical examples to demonstrate the finite element simulation of large finite antenna arrays using the domain-decomposition methods. The first example considered is a broadband single-polarized Vivaldi antenna array having 31 × 31 array elements, which are identical to the one shown in Figure 9.12. Figure 10.4(a) shows the magnitude of the active reflection coefficient (defined by using the characteristic impedance of the feeding coaxial line as the reference

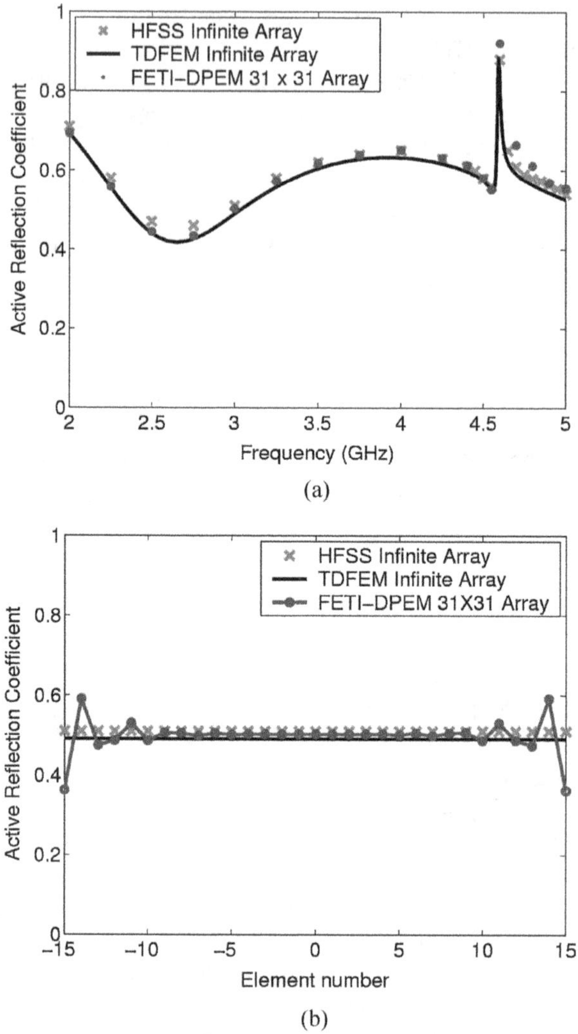

Figure 10.4 Active reflection coefficient for a 31 × 31 Vivaldi antenna array. (a) For the center element as a function of frequency. (b) For the midrow elements at 3 GHz. (After Li and Jin [8], Copyright © IEEE 2006.)

impedance) as a function of frequency for the center element. The result, obtained using the FETI–DPEM1, is compared with those for the corresponding infinite array from the time-domain finite element method and HFSS [8]. Good agreement can be observed between the three different methods over a wide frequency band.

To show the edge effects of the finite array as compared with the corresponding infinite array, the active reflection coefficient for the elements from the central row of the array is computed at 3 GHz and the result is shown in Figure 10.4(b) with respect to the element position. As can be expected, the infinite-array approximation

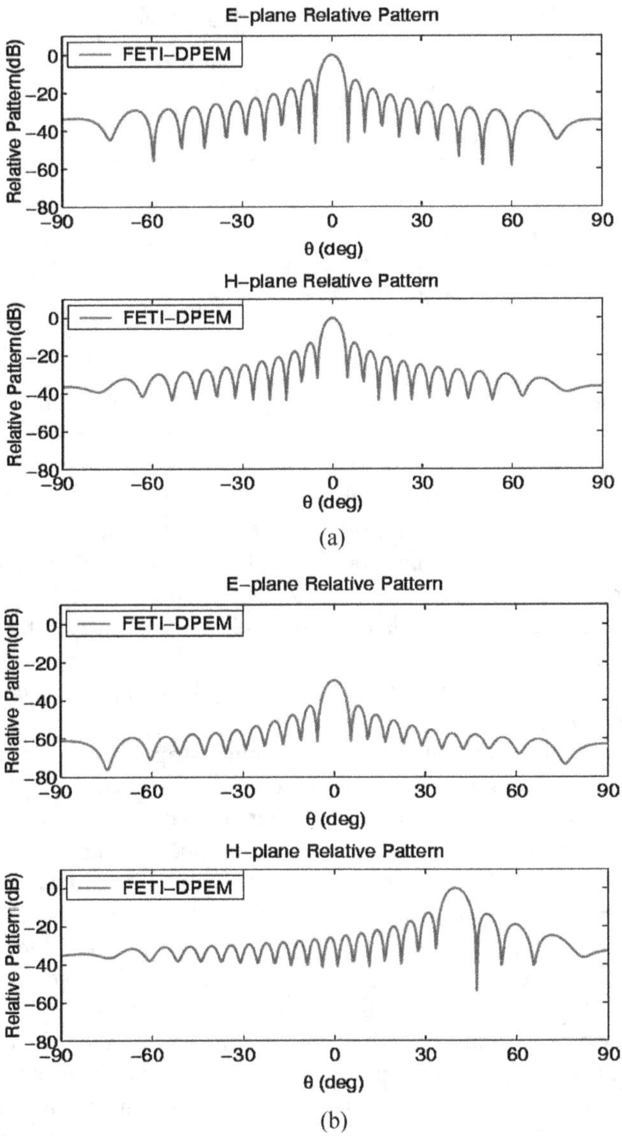

Figure 10.5 Radiation patterns of a 31 × 31 Vivaldi antenna array at 3 GHz. (a) Broadside scan. (b) $\theta_s = 40^\circ$ and $\phi_s = 0^\circ$. (After Li and Jin [8], Copyright © IEEE 2006.)

is accurate only for the interior elements, and the edge effects are prominent for the outer elements. Finally, the radiation patterns of the arrays are shown in Figure 10.5 for two excitations. For this simulation, nearly 40 million unknowns were used and the computation required 541 MB of memory and 12 minutes of computation time on a 1.5-GHz Itanium II processor using the FETI–DPEM1 method.

TABLE 10.1 Data for Various Vivaldi Antenna Array Simulations

Size of Array	Total Number of Unknowns	Number of Dual Unknowns	Memory Used (MB)	Interface Solution Time (h:min:s)	Number of Iterations	Total Solution Time (h:min:s)
10 × 10	5.7M	0.1M	500	00:00:10	19	00:06:36
50 × 50	100.9M	1.9M	685	00:03:02	18	00:20:23
100 × 100	396.2M	7.8M	2,100	00:10:54	17	01:12:30
200 × 200	1,572.2M	31.5M	6,700	01:09:36	20	04:16:35
300 × 300	3,528.4M	70.8M	15,000	05:27:15	20	12:27:14

To further demonstrate the capability of the FETI–DPEM1 method, a series of Vivaldi antenna arrays have been simulated on a 1.5-GHz Itanium II processor. The array size ranges from 10 × 10 to 300 × 300. Table 10.1 gives the total number of unknowns, the number of dual unknowns, the number of iterations for solving the global interface equation using GMRES with a restart number of 5, the memory requirements, the computation time for solving the interface equation, and the total simulation time. Note that for the 300 × 300 array, the total number of unknowns exceeds 3.5 billion, and the number of dual unknowns is about 70.8 million. A computation time of 5.5 hours was required to solve the global interface equation and 12.5 hours was required to simulate the entire problem, including computing the fields in all the subdomains.

The second example deals with a 51 × 51 microstrip patch antenna array on a cylindrical surface having a radius of 454.5 cm, as illustrated in Figure 10.6. The patch antenna array consists of 2601 square metallic patches, each having a size of 3.0 cm × 2.0 cm printed periodically on a dielectric substrate of thickness 0.08 cm, a relative permittivity of 2.17, and a relative permeability of 1.0. The center-to-center distance between adjacent patches is 4.5 cm in the ϕ-direction and 3.5 cm in the z-direction. The coaxial feeds have a filling material of $\varepsilon_r = 2.0$, an inner radius $r_i = 0.2$ mm, and an outer radius $r_o = 0.6$ mm, and are offset by $s = 0.38$ cm in the ϕ-direction. The entire array is flush-mounted on the metallic cylindrical surface, which is assumed to be infinitely long along the z-direction. In the simulation by the FETI–DPEM2 [22], the patch array is excited uniformly at the frequency of 3.3 GHz, which corresponds to the resonant frequency of a single patch antenna. The radiation pattern in the $\theta = 90°$ plane is plotted in Figure 10.7 and is compared with that on a planar platform. The radiation patterns are normalized with respect to the results from the planar array calculation. The active reflection coefficients for the central line of elements in both the E- and H-planes are shown in Figure 10.8. It is shown that the shape of the host cylinder has little effect on the reflection coefficient and thus on the input impedance of the patch antenna array, in contrast to the significant effect on the radiation patterns. For this simulation, the total number of the primal unknowns is 120.3 million and the dimension of the interface equation is 4.76 million. The BiCGSTAB was employed to solve the global interface equation, which converged with the residual error below 10^{-3} within 70 iterations. The entire simulation required

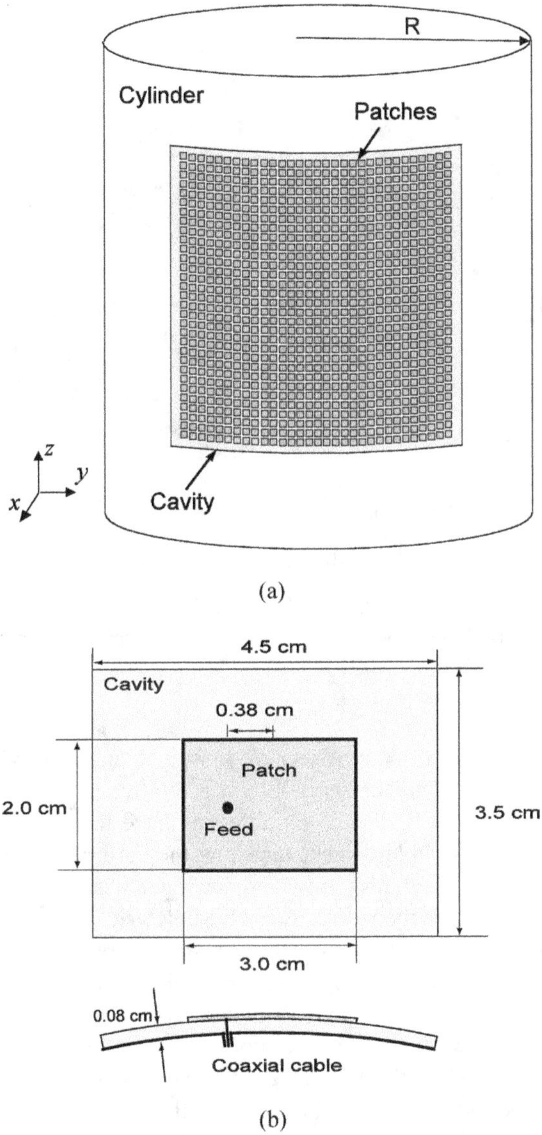

Figure 10.6 Geometry of a 51 × 51 patch antenna array flush-mounted on an infinitely long metallic cylinder. (a) The patch antenna array on a platform. (b) The array element configuration. (After Jin et al [22], Copyright © IEEE 2008.)

1.6 GB of memory and 66 minutes of computation time on a 1.5-GHz Itanium II processor.

The third example is concerned with the mutual coupling between two identical microstrip patch arrays placed in the near-field region of each other. This example, where the configuration is shown in Figure 10.9, is modified from a similar example

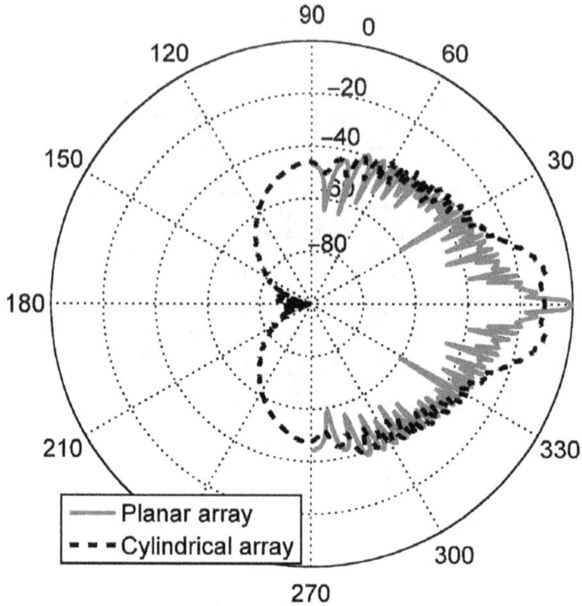

Figure 10.7 Normalized radiation patterns for the 51 × 51 patch antenna array at 3.3 GHz. (After Jin et al. [22], Copyright © IEEE 2008.)

presented in Ref. 23. Each array consists of 9 × 9 patch antennas, which are printed on a dielectric substrate of thickness 0.08 cm, a relative permittivity of 2.17, and a relative permeability of 1.0. The patch antennas have the same configuration as those in the preceding example, except that now they are placed on a flat surface. The center-to-center distance between two adjacent patches is $T_x = 4.5$ cm in the x-direction and $T_y = 3.5$ cm in the y-direction. The dielectric substrate has a dimension of 42 cm and 33 cm in the x- and y-directions, respectively. Each array is housed in a 42-cm × 33-cm × 0.08-cm cavity recessed on an infinitely large ground plane and the two arrays are separated by 34.5 cm. To characterize the mutual coupling between the two arrays, we first calculate the total normalized voltage induced at the feeds of the 81 patch antennas of the receiving array shown on the right-hand side of Figure 10.9 as a function of the scanning angle of the transmitting array on the left-hand side. Direct calculation is very time consuming because of the large number of scanning angles to be considered. A more efficient approach [23] is to use the following expression for the normalized voltage received at the element on the mth column and nth row of the receiving array:

$$V_{mn} = \frac{1}{N_x N_y} \sum_{p=1}^{N_x} \sum_{q=1}^{N_y} S_{mn,pq} e^{-jk_0 p T_x U_t} e^{-jk_0 q T_y V_t} \tag{10.95}$$

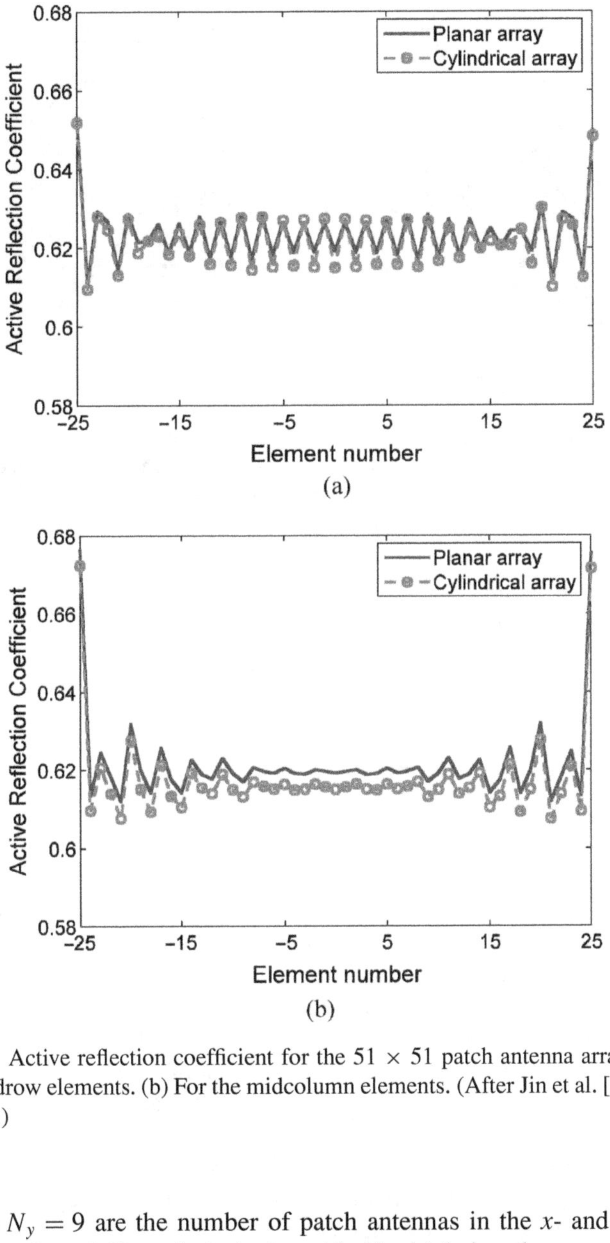

Figure 10.8 Active reflection coefficient for the 51 × 51 patch antenna array at 3.3 GHz. (a) For the midrow elements. (b) For the midcolumn elements. (After Jin et al. [22], Copyright © IEEE 2008.)

where $N_x = N_y = 9$ are the number of patch antennas in the x- and y-directions, $U_t = \sin\theta_t \cos\phi_t$ and $V_t = \sin\theta_t \sin\phi_t$, with (θ_t, ϕ_t) being the scanning angle of the transmitting array, and $S_{mn,pq}$ are the scattering parameters, which represent the transmission between two elements in the transmitting and receiving arrays. The scattering parameters $S_{mn,pq}$ can be calculated by exciting only the element on the pth column and qth row of the transmitting array. Because of the symmetry in the

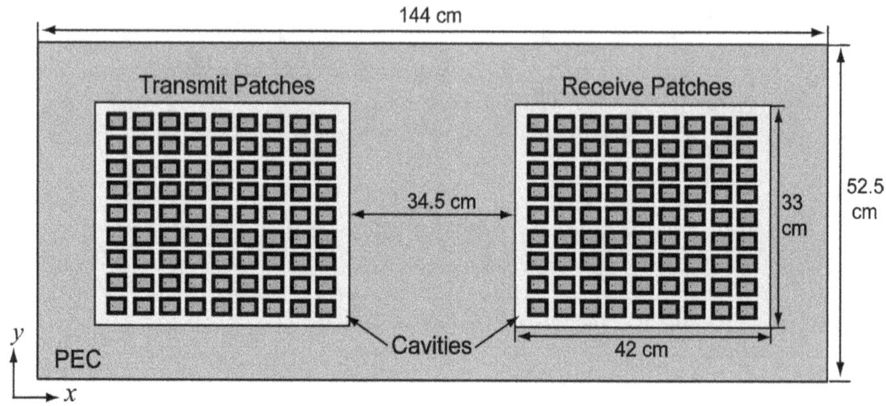

Figure 10.9 Geometry of two 9 × 9 microstrip patch antenna arrays placed on a ground plane. Each array is housed in a shallow cavity recessed in the ground plane so that it is flush-mounted. (*See insert for color representation of figure.*)

y-direction, this calculation can be accomplished by simulating the problem with 45 different excitations. Once the normalized received voltages are computed, the total normalized power coupled from the transmitting array to the receiving array can be calculated as

$$P_r = \left| \sum_{m=1}^{N_x} \sum_{n=1}^{N_y} V_{mn} e^{-jk_0 m T_x U_r} e^{-jk_0 n T_y V_r} \right|^2 \tag{10.96}$$

where $U_r = \sin\theta_r \cos\phi_r$ and $V_r = \sin\theta_r \sin\phi_r$, with (θ_r, ϕ_r) being the scanning angle of the receiving array. Figure 10.10 plots the normalized power as a function of U_r and V_r for the broadside scan of the transmitting array at 3.3 GHz [24]. By the reciprocity theorem, this also represents the normalized power as a function of U_t and V_t when the receiving array operates at the broadside scan. It can be seen that the maximum mutual coupling occurs at the endfire scan, where $U_r = \pm 1$ and $V_r = 0$, and the maximum normalized coupled power is about -40 dB. The simulation was carried out using the FETI–DPEM2 with approximately 24 million primal unknowns and 645,000 dual unknowns. The BiCGSTAB was employed to solve the global interface equation and the convergence was declared when the residual was reduced below 10^{-3}. The average number of iterations for the 45 excitations was 57, and the entire simulation used 1.2 GB of memory and 5 hours of computation time.

For the nonconforming domain-decomposition algorithm described in Section 10.1.3, many examples have been presented in Refs. 7 and 9 using the Gauss–Seidel method and in Refs. 20 and 21 using the Krylov subspace method for solving the global interface system of equations.

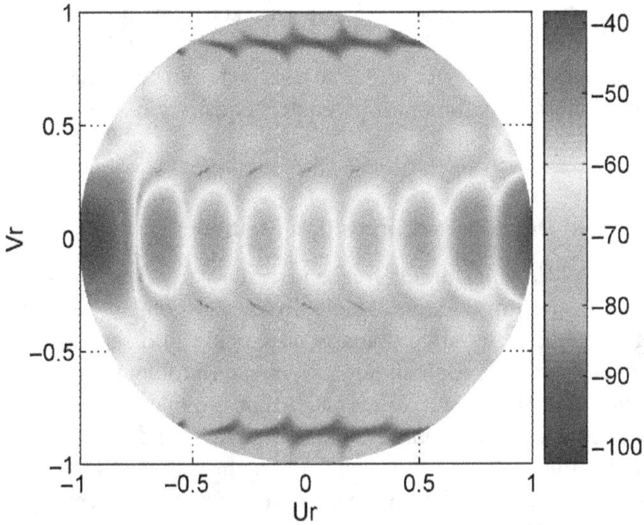

Figure 10.10 Normalized power (in decibels) coupled from the transmitting array with broadside scan to the receiving array as a function of the scan angle of the receiving array. (*See insert for color representation of figure.*)

10.2 TIME-DOMAIN MODELING

Time-domain analysis can be particularly beneficial when modeling broadband phased arrays; however, developing an efficient time-domain formulation for a large finite array is not as straightforward as the analogous frequency-domain formulation described above. Whereas the system matrix is solved only once in a frequency-domain simulation, it is solved at each time step in a time-domain simulation. Hence, the matrix solving is more critical in a time-domain simulation, and a tremendous amount of research effort has been directed into this aspect. To minimize the total computational time, solving the linear system using a direct solver is preferable in the time domain since the factorization can be reused at each time step. For large finite arrays, however, the computer time and memory storage required by a direct solver become impractically large for the factorization of a sparse matrix resulting from a global finite element discretization, although the computer time for each time step becomes more manageable once the matrix is factorized. As a result, the time-domain finite element method often resorts to an iterative solver for large problems to avoid excessive memory use and computer time for the matrix factorization. However, the convergence of an iterative solver heavily depends on the properties of the system matrix and the preconditioner it employs.

In this section we discuss two domain-decomposition schemes for simulating large finite arrays using the time-domain finite element method. The first, the dual-field domain-decomposition method, employs a direct solver to deal with the finite element matrix for each subdomain and computes the electric and magnetic fields based on a leapfrogging time-marching scheme [25,26]. The second method maintains the basic formulation of the time-domain finite element method, as described in Chapter 2, but employs the domain-decomposition idea to distribute the computation over parallel processors for the iterative solution of the matrix equation in each time step.

10.2.1 Dual-Field Domain-Decomposition Method

Similar to frequency-domain simulations, one strategy to solve large-scale problems more efficiently in the time domain is to employ the domain-decomposition idea by dividing the original global computational domain into several smaller subdomains. With a reduced size, each smaller subdomain problem can then be factored and solved using a sparse direct solver such that the overall computational complexity can be reduced as compared to the original single-domain (global) problem. To further reduce the computational time, the subdomain problems can be distributed onto a parallel computing system and solved concurrently. To this end we can use the FETI–DPEM or any other domain-decomposition method described earlier. A drawback of this method is that the global interface problem, which is usually solved by an iterative solver, needs to be solved at each time step. A more efficient time-domain domain-decomposition scheme that does not require solving a global problem has recently been proposed [25,26]. This method, referred to as the *dual-field domain-decomposition* (DFDD) *time-domain finite element method*, solves the dual-field second-order vector wave equations in each subdomain and couples the adjacent subdomains explicitly using the equivalent surface currents on the subdomain interfaces. Since adjacent subdomains are coupled explicitly at each time step, a global interface problem does not need to be formulated and solved. At every time step, each subdomain problem is solved efficiently using its local prefactorized matrix.

To describe the dual-field domain-decomposition method for the analysis of a finite array, we consider the domain decomposition shown in Figure 10.1. In each subdomain, the electric and magnetic fields satisfy the second-order vector wave equations

$$\nabla \times \left[\overleftrightarrow{\mu}^{-1} \cdot \nabla \times \mathbf{E}^s(t)\right] + \overleftrightarrow{\varepsilon} \cdot \frac{\partial^2 \mathbf{E}^s(t)}{\partial t^2} = -\frac{\partial \mathbf{J}^s_{\text{imp}}}{\partial t} - \nabla \times \left(\overleftrightarrow{\mu}^{-1} \cdot \mathbf{M}^s_{\text{imp}}\right) \qquad (10.97)$$

$$\nabla \times \left[\overleftrightarrow{\varepsilon}^{-1} \cdot \nabla \times \mathbf{H}^s(t)\right] + \overleftrightarrow{\mu} \cdot \frac{\partial^2 \mathbf{H}^s(t)}{\partial t^2} = -\frac{\partial \mathbf{M}^s_{\text{imp}}}{\partial t} + \nabla \times \left(\overleftrightarrow{\varepsilon}^{-1} \cdot \mathbf{J}^s_{\text{imp}}\right) \qquad (10.98)$$

where the superscript s denotes the subdomain number. For simplicity, we assume that the medium is lossless and nondispersive; however, the formulation can be extended to deal with lossy and dispersive media by following the approaches described in Chapter 2. Also for simplicity, we assume the following first-order absorbing boundary conditions for the outer surface of the entire computational domain:

$$\hat{n} \times \left[\overleftrightarrow{\mu}^{-1} \cdot \nabla \times \mathbf{E}(t) \right] + Y_0 \hat{n} \times \left[\overleftrightarrow{\varsigma} \cdot \hat{n} \times \frac{\partial \mathbf{E}(t)}{\partial t} \right] \approx 0 \qquad \mathbf{r} \in S_o \tag{10.99}$$

$$\hat{n} \times \left[\overleftrightarrow{\varepsilon}^{-1} \cdot \nabla \times \mathbf{H}(t) \right] + Z_0 \hat{n} \times \left[\overleftrightarrow{\zeta} \cdot \hat{n} \times \frac{\partial \mathbf{H}(t)}{\partial t} \right] \approx 0 \qquad \mathbf{r} \in S_o \tag{10.100}$$

where $\overleftrightarrow{\varsigma}$ and $\overleftrightarrow{\zeta}$ denote the normalized admittance and impedance tensors chosen to minimize the reflection of the truncation boundary. (For an isotropic medium, $\overleftrightarrow{\varsigma}$ and $\overleftrightarrow{\zeta}$ become scalars with values of $\sqrt{\varepsilon_r/\mu_r}$ and $\sqrt{\mu_r/\varepsilon_r}$, respectively.)

By taking the dot product with the vector basis function $\{\mathbf{N}^s\}$, integrating over the volume of the subdomain, and then invoking the divergence theorem, the weak-form representations of (10.97) and (10.98) can be written as

$$\begin{aligned}
&\iiint_{V_s} \left[(\nabla \times \{\mathbf{N}^s\}) \cdot \overleftrightarrow{\mu}^{-1} \cdot (\nabla \times \mathbf{E}^s) + \{\mathbf{N}^s\} \cdot \overleftrightarrow{\varepsilon} \cdot \frac{\partial^2 \mathbf{E}^s}{\partial t^2} \right] dV \\
&\quad + Y_0 \iint_{S_s \cap S_o} (\hat{n} \times \{\mathbf{N}^s\}) \cdot \overleftrightarrow{\varsigma} \cdot \left(\hat{n} \times \frac{\partial \mathbf{E}^s}{\partial t} \right) dS \\
&= \iint_{S_s \cap \Gamma_s} (\hat{n}^s \times \{\mathbf{N}^s\}) \cdot \overleftrightarrow{\mu}^{-1} \cdot (\nabla \times \mathbf{E}^s)\, dS \\
&\quad - \iiint_{V_s} \{\mathbf{N}^s\} \cdot \left[\frac{\partial \mathbf{J}^s_{\text{imp}}}{\partial t} + \nabla \times \left(\overleftrightarrow{\mu}^{-1} \cdot \mathbf{M}^s_{\text{imp}} \right) \right] dV
\end{aligned} \tag{10.101}$$

$$\begin{aligned}
&\iiint_{V_s} \left[(\nabla \times \{\mathbf{N}^s\}) \cdot \overleftrightarrow{\varepsilon}^{-1} \cdot (\nabla \times \mathbf{H}^s) + \{\mathbf{N}^s\} \cdot \overleftrightarrow{\mu} \cdot \frac{\partial^2 \mathbf{H}^s}{\partial t^2} \right] dV \\
&\quad + Z_0 \iint_{S_s \cap S_o} (\hat{n} \times \{\mathbf{N}^s\}) \cdot \overleftrightarrow{\zeta} \cdot \left(\hat{n} \times \frac{\partial \mathbf{H}^s}{\partial t} \right) dS \\
&= \iint_{S_s \cap \Gamma_s} (\hat{n}^s \times \{\mathbf{N}^s\}) \cdot \overleftrightarrow{\varepsilon}^{-1} \cdot (\nabla \times \mathbf{H}^s)\, dS \\
&\quad - \iiint_{V_s} \{\mathbf{N}^s\} \cdot \left[\frac{\partial \mathbf{M}^s_{\text{imp}}}{\partial t} - \nabla \times \left(\overleftrightarrow{\varepsilon}^{-1} \cdot \mathbf{J}^s_{\text{imp}} \right) \right] dV
\end{aligned} \tag{10.102}$$

where V_s denotes the volume of subdomain s and Γ_s denotes its interface with adjacent subdomains. At the subdomain interface Γ_s, an application of Maxwell's equations yields

$$\hat{n}^s \times \left[\overleftrightarrow{\mu}^{-1} \cdot \nabla \times \mathbf{E}^s(t) \right] = -\hat{n}^s \times \frac{\partial \mathbf{H}^s(t)}{\partial t} \qquad \mathbf{r} \in \Gamma_s \tag{10.103}$$

$$\hat{n}^s \times \left[\overleftrightarrow{\varepsilon}^{-1} \cdot \nabla \times \mathbf{H}^s(t) \right] = \hat{n}^s \times \frac{\partial \mathbf{E}^s(t)}{\partial t} \qquad \mathbf{r} \in \Gamma_s \tag{10.104}$$

under the assumption that there is no source on Γ_s (a condition that can be removed easily). Hence, (10.101) and (10.102) become

$$\iiint_{V_s} \left[(\nabla \times \{\mathbf{N}^s\}) \cdot \overleftrightarrow{\mu}^{-1} \cdot (\nabla \times \mathbf{E}^s) + \{\mathbf{N}^s\} \cdot \overleftrightarrow{\varepsilon} \cdot \frac{\partial^2 \mathbf{E}^s}{\partial t^2} \right] dV$$
$$+Y_0 \iint_{S_s \cap S_0} (\hat{n} \times \{\mathbf{N}^s\}) \cdot \overleftrightarrow{\varsigma} \cdot \left(\hat{n} \times \frac{\partial \mathbf{E}^s}{\partial t} \right) dS$$
$$= \iint_{S_s \cap \Gamma_s} (\hat{n}^s \times \{\mathbf{N}^s\}) \cdot \left(\hat{n}^s \times \frac{\partial \mathbf{J}^s}{\partial t} \right) dS$$
$$- \iiint_{V_s} \{\mathbf{N}^s\} \cdot \left[\frac{\partial \mathbf{J}^s_{\text{imp}}}{\partial t} + \nabla \times \left(\overleftrightarrow{\mu}^{-1} \cdot \mathbf{M}^s_{\text{imp}} \right) \right] dV \tag{10.105}$$

$$\iiint_{V_s} \left[(\nabla \times \{\mathbf{N}^s\}) \cdot \overleftrightarrow{\varepsilon}^{-1} \cdot (\nabla \times \mathbf{H}^s) + \{\mathbf{N}^s\} \cdot \overleftrightarrow{\mu} \cdot \frac{\partial^2 \mathbf{H}^s}{\partial t^2} \right] dV$$
$$+ Z_0 \iint_{S_s \cap S_0} (\hat{n} \times \{\mathbf{N}^s\}) \cdot \overleftrightarrow{\zeta} \cdot \left(\hat{n} \times \frac{\partial \mathbf{H}^s}{\partial t} \right) dS$$
$$= \iint_{S_s \cap \Gamma_s} (\hat{n}^s \times \{\mathbf{N}^s\}) \cdot \left(\hat{n} \times \frac{\partial \mathbf{M}^s}{\partial t} \right) dS$$
$$- \iiint_{V_s} \{\mathbf{N}^s\} \cdot \left[\frac{\partial \mathbf{M}^s_{\text{imp}}}{\partial t} - \nabla \times \left(\overleftrightarrow{\varepsilon}^{-1} \cdot \mathbf{J}^s_{\text{imp}} \right) \right] dV \tag{10.106}$$

where $\mathbf{J}^s$ and $\mathbf{M}^s$ denote the equivalent surface currents, which are defined by

$$\mathbf{J}^s = \hat{n}^s \times \mathbf{H}^s, \qquad \mathbf{M}^s = \mathbf{E}^s \times \hat{n}^s \qquad \mathbf{r} \in \Gamma_s. \tag{10.107}$$

From (10.105) and (10.106), it can be observed that the electric field in subdomain s is produced by the impressed source $(\mathbf{J}^s_{\text{imp}}, \mathbf{M}^s_{\text{imp}})$ and the equivalent surface electric current $\mathbf{J}^s$ on Γ_s, whereas the magnetic field in subdomain s is produced by the impressed source $(\mathbf{J}^s_{\text{imp}}, \mathbf{M}^s_{\text{imp}})$ and the equivalent surface magnetic current $\mathbf{M}^s$ on Γ_s. Therefore, if $\mathbf{J}^s$ is obtained, the electric field can be evaluated, and similarly, if $\mathbf{M}^s$ is available, the magnetic field can be calculated.

The finite element discretization of (10.105) and (10.106) yields the matrix equations

$$[T^s_e] \frac{d^2\{E^s\}}{dt^2} + [R^s_e] \frac{d\{E^s\}}{dt} + [S^s_e]\{E^s\} = \{f^s_e\} + \frac{d\{j^s\}}{dt} \tag{10.108}$$

$$[T^s_h] \frac{d^2\{H^s\}}{dt^2} + [R^s_h] \frac{d\{H^s\}}{dt} + [S^s_h]\{H^s\} = \{f^s_h\} + \frac{d\{m^s\}}{dt} \tag{10.109}$$

where

$$\left[T_e^s\right] = \iiint_{V_s} \{\mathbf{N}^s\} \cdot \overleftrightarrow{\varepsilon} \cdot \{\mathbf{N}^s\}^{\mathrm{T}} \, dV \tag{10.110}$$

$$\left[R_e^s\right] = Y_0 \iint_{S_s \cap S_0} \hat{n} \times \{\mathbf{N}^s\} \cdot \overleftrightarrow{\varsigma} \cdot \hat{n} \times \{\mathbf{N}^s\}^{\mathrm{T}} \, dS \tag{10.111}$$

$$\left[S_e^s\right] = \iiint_{V_s} \nabla \times \{\mathbf{N}^s\} \cdot \overleftrightarrow{\mu}^{-1} \cdot \nabla \times \{\mathbf{N}^s\}^{\mathrm{T}} \, dV \tag{10.112}$$

$$\left\{f_e^s\right\} = -\iiint_{V_s} \{\mathbf{N}^s\} \cdot \left[\frac{\partial \mathbf{J}_{\mathrm{imp}}^s}{\partial t} + \nabla \times \left(\overleftrightarrow{\mu}^{-1} \cdot \mathbf{M}_{\mathrm{imp}}^s\right)\right] dV \tag{10.113}$$

$$\{j^s\} = \iint_{S_s \cap \Gamma_s} \hat{n}^s \times \{\mathbf{N}^s\} \cdot (\hat{n}^s \times \mathbf{J}^s) \, dS \tag{10.114}$$

and

$$\left[T_h^s\right] = \iiint_{V_s} \{\mathbf{N}^s\} \cdot \overleftrightarrow{\mu} \cdot \{\mathbf{N}^s\}^{\mathrm{T}} \, dV \tag{10.115}$$

$$\left[R_h^s\right] = Z_0 \iint_{S_s \cap S_0} \hat{n} \times \{\mathbf{N}^s\} \cdot \overleftrightarrow{\varsigma} \cdot \hat{n} \times \{\mathbf{N}^s\}^{\mathrm{T}} \, dS \tag{10.116}$$

$$\left[S_h^s\right] = \iiint_{V_s} \nabla \times \{\mathbf{N}^s\} \cdot \overleftrightarrow{\varepsilon}^{-1} \cdot \nabla \times \{\mathbf{N}^s\}^{\mathrm{T}} \, dV \tag{10.117}$$

$$\left\{f_h^s\right\} = -\iiint_{V_s} \{\mathbf{N}^s\} \cdot \left[\frac{\partial \mathbf{M}_{\mathrm{imp}}^s}{\partial t} - \nabla \times \left(\overleftrightarrow{\varepsilon}^{-1} \cdot \mathbf{J}_{\mathrm{imp}}^s\right)\right] dV \tag{10.118}$$

$$\{m^s\} = \iint_{S_s \cap \Gamma_s} \hat{n}^s \times \{\mathbf{N}^s\} \cdot (\hat{n}^s \times \mathbf{M}^s) \, dS. \tag{10.119}$$

It remains to discretize (10.108) and (10.109) in the time domain so that the fields can be updated step by step in time. Since the updating of the electric field depends on $\{j^s\}$, which in turn depends on the magnetic field on the interface, and the updating of the magnetic field depends on $\{m^s\}$, which in turn depends on the electric field on the interface, the electric and magnetic fields should be discretized so that they are staggered one-half of a time step with respect to each other. In other words, we can discretize the electric field on integer time indices ($n = 0, 1, 2, \ldots, N$) and the magnetic field on half-integer indices ($n = \frac{1}{2}, \frac{3}{2}, \frac{5}{2}, \ldots, N + \frac{1}{2}$). Once the fields are discretized, we apply the Newmark-beta method to integrate (10.108) and (10.109), which yields

$$\{E^s\}^{n+1} = \left[A_e^s\right]^{-1}\Bigg(\left[B_e^s\right]\{E^s\}^n - \left[C_e^s\right]\{E^s\}^{n-1} + \left\{f_e^s\right\}^n + \frac{\{j^s\}^{n+1/2} - \{j^s\}^{n-1/2}}{\Delta t}\Bigg) \tag{10.120}$$

$$\{H^s\}^{n+3/2} = \left[A_h^s\right]^{-1}\Bigg(\left[B_h^s\right]\{H^s\}^{n+1/2} - \left[C_h^s\right]\{H^s\}^{n-1/2} + \left\{f_h^s\right\}^{n+1/2} + \frac{\{m^s\}^{n+1} - \{m^s\}^n}{\Delta t}\Bigg) \tag{10.121}$$

where

$$\left[A_{e,h}^s\right] = \frac{1}{(\Delta t)^2}\left[T_{e,h}^s\right] + \frac{1}{2\Delta t}\left[R_{e,h}^s\right] + \frac{1}{4}\left[S_{e,h}^s\right] \tag{10.122}$$

$$\left[B_{e,h}^s\right] = \frac{2}{(\Delta t)^2}\left[T_{e,h}^s\right] - \frac{1}{2}\left[S_{e,h}^s\right] \tag{10.123}$$

$$\left[C_{e,h}^s\right] = \frac{1}{(\Delta t)^2}\left[T_{e,h}^s\right] - \frac{1}{2\Delta t}\left[R_{e,h}^s\right] + \frac{1}{4}\left[S_{e,h}^s\right]. \tag{10.124}$$

In the equations above, we assumed that $\beta = 1/4$ in the application of the Newmark-beta method. Equations (10.120) and (10.121) can be used to compute the electric and magnetic fields in a leapfrog fashion.

To compute the electric field in subdomain s using (10.120), we require $\{j^s\}^{n-1/2}$ and $\{j^s\}^{n+1/2}$ in addition to the electric fields at the previous two steps and the impressed source. Similarly, to compute the magnetic field in subdomain s using (10.121), we require $\{m^s\}^n$ and $\{m^s\}^{n+1}$ in addition to the magnetic fields at the previous two steps and the impressed source. From (10.107), (10.114), and (10.119), it can be seen that $\{j^s\}$ is related to the tangential magnetic field at the subdomain interface and $\{m^s\}$ is related to the tangential electric field at the subdomain interface. It is these two quantities that couple the fields across the subdomains. At the interface between subdomains s and q, the field continuity conditions are given by (10.44), which can also be written as

$$\mathbf{J}^s(\mathbf{r}) = \hat{n}^s \times \mathbf{H}^q(\mathbf{r}) \qquad \text{and} \qquad \mathbf{M}^s(\mathbf{r}) = \mathbf{E}^q(\mathbf{r}) \times \hat{n}^s \qquad \mathbf{r} \in \Gamma_s^q. \tag{10.125}$$

Therefore, when $\{H^q\}^{n+1/2}$ is computed, we can calculate $\{j^s\}^{n+1/2}$ as

$$\{j^s\}^{n+1/2} = -\sum_{q \in \text{neighbor}(s)} \iint_{\Gamma_s^q} \hat{n}^s \times \{\mathbf{N}^s\} \cdot \mathbf{H}_s^{q(n+1/2)}\, dS \tag{10.126}$$

and when $\{E^q\}^{n+1}$ is computed, we can calculate $\{m^s\}^{n+1}$ as

$$\{m^s\}^{n+1} = \sum_{q \in \text{neighbor}(s)} \iint_{\Gamma_s^q} \hat{n}^s \times \{\mathbf{N}^s\} \cdot \mathbf{E}_s^{q(n+1)} \, dS \tag{10.127}$$

where $\mathbf{E}_s^q$ and $\mathbf{H}_s^q$ denote the electric and magnetic fields on the subdomain q side of Γ_s^q. With the calculation of $\{j^s\}^{n+1/2}$ and $\{m^s\}^{n+1}$, the time-marching process can now be carried out to compute the electric and magnetic fields step by step. Given the initial field values and the impressed source, we first calculate $\{E^s\}^{(1)}$ for each subdomain ($s = 1, 2, \ldots, N_s$) using (10.120). Once this calculation is completed, we calculate $\{m^s\}^{(1)}$ using (10.127), which is then used to compute $\{H^s\}^{(3/2)}$ for each subdomain using (10.121). Once $\{H^s\}^{(3/2)}$ is computed, we calculate $\{j^s\}^{(3/2)}$, which is then employed to compute $\{E^s\}^{(2)}$ for each subdomain. This process continues until either the fields vanish or they reach the steady state, depending on the type of impressed source used for the simulation.

If we examine (10.126) and (10.127) carefully, we find that the interfacing between the subdomains in the dual-field domain-decomposition method is very simple. Once we compute the electric field in each subdomain, all we do is to pass its tangential component on the interface to the neighboring subdomains and receive the corresponding tangential component computed in the neighboring subdomain for the calculation of the magnetic field in the next half-step. Similarly, once we compute the magnetic field in each subdomain, we pass its tangential component on the interface to the neighboring subdomains and receive the corresponding tangential component computed in the neighboring subdomain for the calculation of the electric field in the next half-step. Surprisingly, such a simple field-exchanging scheme works very effectively and yields a highly stable time-marching process. Although the field continuity conditions are not enforced explicitly at the subdomain interfaces and there are two values for the tangential field at any instant (associated with the two neighboring subdomains), these two values converge to one when the finite element discretization and the time-step size are sufficiently fine [25]. From the physical point of view, the field-exchanging scheme models the phenomenon described by the well-known Huygens' principle. The field generated in a subdomain becomes the part of the source for the field in the neighboring subdomains and hence should be passed to the neighboring subdomains in the form of a surface equivalent current. Numerical analysis and tests show that this interfacing scheme maintains the overall accuracy of the final numerical solution [25,26].

The simple yet robust subdomain interfacing scheme via an explicit field exchanging described above does not come without cost. Although we have employed the Newmark-beta method to discretize the time domain in each subdomain, and consequently, the time marching in each subdomain using (10.120) and (10.121) is unconditionally stable, the global time marching for the entire domain becomes only conditionally stable because the subdomain fields are exchanged at the interfaces explicitly.

A stability analysis [25,26] shows that the global time marching is stable when the time-step size satisfies the stability condition given by

$$\Delta t < \frac{2}{c_0} \frac{1}{\rho([L])} \tag{10.128}$$

where $c_0 = 1/\sqrt{\mu_0 \varepsilon_0}$ and $\rho([L])$ denotes the spectral radius of $[L]$, which is defined as

$$[L] = \begin{bmatrix} M_e & 0 \\ 0 & M_h \end{bmatrix}^{-1} \begin{bmatrix} 0 & P \\ P & 0 \end{bmatrix} \tag{10.129}$$

in which

$$[M_e] = \iiint_V \{\mathbf{N}\} \cdot \overleftrightarrow{\varepsilon}_r \cdot \{\mathbf{N}\}^{\mathrm{T}} \, dV \tag{10.130}$$

$$[M_h] = \iiint_V \{\mathbf{N}\} \cdot \overleftrightarrow{\mu}_r \cdot \{\mathbf{N}\}^{\mathrm{T}} \, dV \tag{10.131}$$

$$[P] = \iint_\Gamma \{\mathbf{N}\} \cdot \hat{n} \times \{\mathbf{N}\}^{\mathrm{T}} \, dS. \tag{10.132}$$

In the expressions above, $\{\mathbf{N}\}$ denotes the column vector consisting of all the basis functions in the computational domain and Γ denotes all the subdomain interfaces. It is evident that $[P]$ has nonzero entries only for basis functions residing on the subdomain interfaces. Therefore, $[L]$ depends only on the spatial discretization directly related to the interfaces. If the computational domain is meshed into tetrahedral elements, the stability condition depends only on the tetrahedral elements directly connected to the interfaces. It does not depend on the size of tetrahedral elements inside each subdomain. Therefore, we can use very small elements inside each subdomain to resolve small geometrical details without any adverse impact on the stability condition. This stability condition has been fully verified in numerical tests [25,26]. For equilateral tetrahedral elements with a side length of h that are connected to the interface, the estimated stability criterion is $\Delta t < 0.3h/c$ when the first-order basis functions are employed, where c denotes the speed of light in the medium. Therefore, it is rather straightforward to determine the maximum time-step size based on the size of elements attached to the interfaces. When higher-order basis functions are used, the estimate above can be scaled by the order of the basis functions.

From its formulation it is clear that the dual-field domain-decomposition method can be applied not only to array-type problems, but also to arbitrary geometries. For arbitrary geometries, the efficiency of the method is due mainly to the decomposition of an extremely large finite element system for the entire computational domain into many smaller systems for the subdomains. Because of the exponential increases in the computer time and memory requirements for the factorization of a matrix with an increasing dimension (even if the matrix is sparse), decomposing a large matrix into many smaller ones greatly reduces the computer resources required for the matrix factorization. The highest efficiency can be achieved when the domain decomposition is applied to the extreme case where every finite element is treated as a subdomain, resulting in a completely explicit algorithm [27]. The drawback for this limiting case is that the maximum time-step size is now dictated by the smallest finite element in the entire computational domain since every element is now connected directly to the subdomain interfaces. This is usually undesirable for practical applications. A more balanced approach is to include small finite elements inside subdomains and treat each large finite element as a subdomain so that a high efficiency can be obtained without significantly reducing the maximum time-step size. This approach can be considered as a hybrid explicit–implicit numerical algorithm [27].

Based on the discussions above, it is clear that when the dual-field domain-decomposition method is implemented on a serial computer, the matrix factorization time and memory requirements can be reduced significantly when the computational domain is decomposed into subdomains. The computational complexity for the factorization and time marching approaches $O(N)$ as the number of subdomains increases. When the method is implemented on a parallel computer, we can achieve a superlinear speedup for the matrix factorization, a superlinear reduction in the memory requirements, and a nearly linear speedup (corresponding to a parallel efficiency of nearly 100%) for time marching [25].

When the dual-field domain-decomposition method is applied to large finite array problems, additional savings can be obtained by exploiting the geometrical repetitions of an array structure. As discussed earlier, when the computational domain is divided such that each array element becomes a subdomain, there are only nine subdomains that have different finite element matrices (Figure 10.2). For each of the nine basic subdomains, we can assemble the finite element matrix and then factorize it and store the result in memory at the beginning of the simulation. During the time marching, each of the array elements is updated using the appropriate factorized matrix according to its position in the array. Thus, the dual-field domain decomposition requires the storage of only nine factorized matrices regardless of the actual size of the array. Considering that matrix factorization typically dominates the memory consumption, a significant reduction in the memory requirements can be achieved by using this method for large arrays. The same argument can be made for the factorization time. However, the total computation time required by the time-marching process remains unaffected. The reason for this is that even though the system matrices are the same for the subdomains with identical geometry, the electric fields throughout these subdomains must be updated individually at every time step.

10.2.2 Domain Decomposition for Iterative Solutions

As discussed in the preceding subsection, the dual-field domain-decomposition method achieves its high performance by decomposing the finite element system of equations into many smaller subsystems and exploiting the repetitions of the subsystems for array-type applications. The only drawback is the compromise of the otherwise unconditional stability of the time-domain finite element solution because of the explicit, and therefore conditionally stable, field-exchanging scheme at the subdomain interfaces. For problems having small geometries next to or across the subdomain interfaces that have to be modeled with small finite elements, the time-step size must be reduced accordingly to maintain the stability of the time-marching process. For such problems it may be more efficient to apply directly the traditional time-domain finite element method described in Chapter 2, or its hybridization with the finite-difference time-domain (FDTD) algorithm discussed in Chapter 4, and then employ the domain-decomposition idea for an iterative solution of the global finite element matrix equation in each time step on a parallel computer. A domain-decomposition strategy based on this approach is described in this subsection.

An appealing aspect of the traditional finite element time-domain (FETD) method is its global unconditional stability when the Newmark-beta time-integration method is applied to the wave equation. Because unconditionally stable time-marching algorithms must necessarily be implicit, a system of equations needs to be solved at each time step, and this can be accomplished by either a direct or an iterative solution procedure. As mentioned previously, a direct solver is appealing for time-domain applications because the matrix equation typically only needs to be factored once and then applied at each time step. However, because the factorization of a sparse matrix generally leads to a relatively dense matrix, this solution approach is applicable only to small discretization domains, such as those used with the dual-field domain-decomposition method described in the preceding subsection. Iterative solution techniques can be applied to much larger discretization domains, although practical limits still remain with regard to the number of the degrees of freedom that can be solved efficiently. Indeed, this aspect was one of the primary reasons for adopting the hybrid FETD–FDTD technique described in Chapter 4. In addition, generating a single three-dimensional unstructured finite element grid that may contain tens, or perhaps hundreds, of millions of cells can be challenging for mesh generation software. Consequently, there is a need for a domain-decomposition method that globally preserves the properties of the original FETD method, and is also applicable to extremely large unstructured finite element grids. The iterative domain-decomposition (IDD) approach can be used to address both of these issues effectively.

Similar to the dual-field domain-decomposition method, we consider a single global domain with volume V that is subdivided into N_s subdomains such that V_s denotes the volume of subdomain s and Γ_s denotes its interface with adjacent subdomains. In the IDD method, the basic strategy will be to construct finite element matrices for each subdomain that correspond identically to the equivalent edges associated with the global matrices that characterize the original single domain discretization. Because each subdomain can be gridded in isolation with the IDD approach, the assembled finite element matrix entries for the interface edges will be incomplete, and hence will require a computer message-passing procedure to extract the necessary matrix information from the neighboring subdomains. The IDD method fundamentally assumes that the cell edges on Γ_s are aligned between the neighboring subdomains and hence is considered to be a conformal domain-decomposition technique. In addition, because the matrix entries for the interface edges are completed by using the local matrix information from neighboring subdomains, it is also an

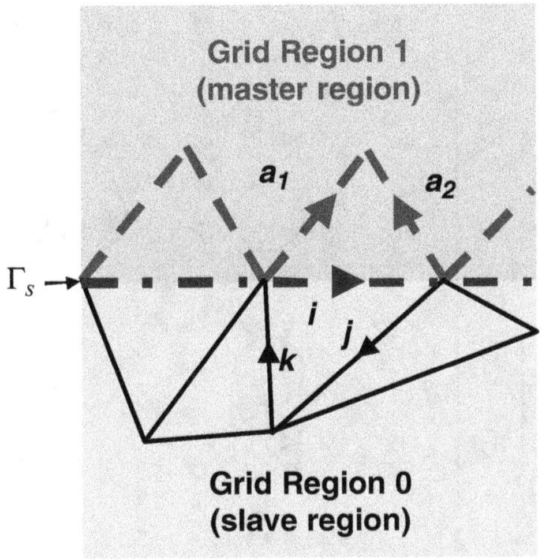

Figure 10.11 Edges associated with the finite element cells in the interface region between two subdomains for application of the iterative domain-decomposition method. Finite element cells within the master region are used to augment the finite element matrices associated with the slave region.

overlapping domain-decomposition technique. There are two benefits of the IDD approach for time-domain applications: (1) Because the assembled matrices for each subdomain are identical to their respective portions of an equivalent discretization of the original single domain application, the unconditional stability properties for the traditional FETD method apply to the decomposed volume, and (2) limitations associated with traditional mesh generation software can be avoided because the subdomains can be gridded independently.

To implement the IDD method, we consider the interface between the two subdomains illustrated in Figure 10.11. The implementation strategy is based on defining a master–slave relationship for the two subdomains relative to the interface Γ_s. Because each subdomain is assumed to be gridded independently, additional grid information will be required by either the master or slave regions to formulate the global solution. More specifically, the proper assembly of the finite element matrices based on vector edge basis functions will require the local cell topology for all the cells that share edges and faces along the subdomain interface. Because the interface edges associated with both the master and slave regions are missing many of these required cells, the additional matrix information associated with these cells is acquired by a message-passing procedure.

To examine more carefully how the matrix entries are augmented, we consider a portion of an assembled finite element matrix for the slave subdomain as illustrated in Figure 10.12, where the diagonal entry for edge i that resides on the interface has been highlighted. It is noted that this submatrix would correspond to a portion of the $[T]$, $[R]$, or $[S]$ matrices that are associated with a Newmark-beta time-marching

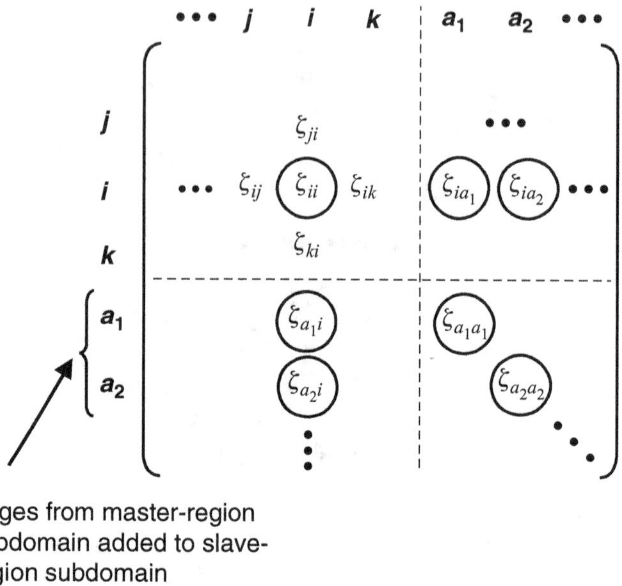

Figure 10.12 Portion of an assembled finite element matrix for the slave-region subdomain showing augmented matrix entries due to contributions from cells residing in the master-region subdomain. The regional locations of the edges $i,\ j,\ k,\ a_1$, and a_2 are defined in Figure 10.11. The modified entries are denoted by circles.

scheme for the wave equation, such as the one described by (2.36). The highlighted *ii* matrix entry has contributions from both the slave and master regions, where the additive contribution due to the master region is acquired by message passing.

As described above, a message-passing procedure can be used to complete the matrix entries for the interface edges associated with the slave region. However, additional edge-based information will be required to fully time advance the fields associated with these interface edges, which can be seen by examining the time-marching scheme described by (2.36). More specifically, field data at the previous time step for the master region edges that are adjacent to but not connected directly to the interface will be needed by the interface edges associated with the slave region. Thus, the slave region finite element matrices require further augmentation by these master region offset edges, and the corresponding matrix additions for these edges are also illustrated in Figure 10.12, where the labels a_1 and a_2 have been used. A message-passing procedure is again used to acquire this matrix data from the master region. With the matrix entries for the interface edges corrected, along with the augmentation due to the offset edges, the slave region matrices are now suitable for time marching. No additional modifications are required for the master region matrices. The final solution procedure is to pass the electric field coefficients associated with the interface edges in the slave region to the master region, and pass the offset interface edge coefficients from the master region to the slave region.

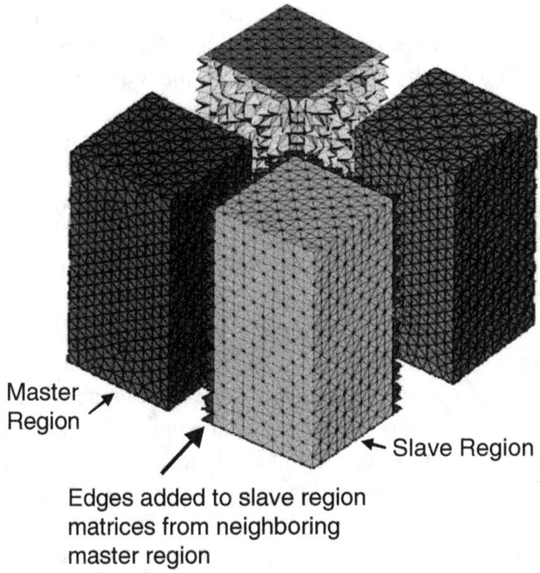

Figure 10.13 Two-dimensional decomposition of a three-dimensional tetrahedral mesh for application of the IDD technique. The master region edges that are used to augment the slave region are highlighted. (*See insert for color representation of figure.*)

The approach described above corresponds directly to a one-dimensional domain decomposition. However, the strategy is also used for two- and three-dimensional decompositions by additionally considering the edges and corners between the subdomains. A simple two-dimensional decomposition of a three-dimensional tetrahedral grid that highlights the overlapping edges between the master and slave regions is shown in Figure 10.13.

A variety of software packages currently exist to partition a global domain into an arbitrary number of subdomains [28,29], and the procedure above to construct the finite element matrices for these subdomains is directly applicable. However, because traditional grid partitioning algorithms operate on the global mesh, practical limitations may exist with regard to the total number of cells that can be simultaneously partitioned. An alternative approach is to independently grid the various subdomains, subject to conformal surface meshes on the interfaces, and this is why the IDD technique was formulated in a general manner. Assuming that a sufficient number of distributed computer processors are available, this later approach removes any restriction on the overall number of finite elements that can be constructed and solved, although the load balancing efficiency of the subdomains may not be optimal for certain applications.

As mentioned previously, iterative solution techniques can benefit from well-designed preconditioners. For the FETD method, the incomplete *LU* (ILU) factorization technique leads to an efficient preconditioner for the global system matrix. By judiciously increasing the matrix fill-in for the ILU, the number of iterations required for a defined solution residual can be reduced, albeit at the expense of increased computer memory storage requirements as well as a larger number of floating-point operations per iteration. As the variation between the maximum and minmum finite element edge lengths increases, which typically occurs when resolving the fine details of antenna feeds, a proportional increase in the ILU fill-in can effectively keep the required number of iterations relatively low, with a typical iteration count being less than 10, even with edge-length ratios exceeding 100 : 1. As the fill-in is further increased, the ILU technique will eventually reduce to a full *LU* factorization and the iterative solution method will converge in a single iteration, thereby rendering a functional equivalence between iterative and direct solution techniques.

The ILU preconditioner can also be applied to the individual subdomains associated with a domain decomposition. However, when applied at the subdomain level the ILU performance can become degraded relative to its application to the global matrix. This is because the dot products associated with an iterative solution method must be accumulated globally, even though only those edges within a particular subdomain are included in the matrix–vector multiplications. Consequently, application of a local ILU preconditioner at the subdomain level may not provide a sufficiently robust preconditioner relative to the global level. The result may be an increase in the number of iterations required to reach a defined solution residual for the partitioned domain relative to a single global domain. The development of robust preconditioners for the massively parallel application of iterative solution methods remains an important research topic in computational sciences. A promising class of preconditioners for this purpose is based on multigrid concepts [30].

The iterative domain-decomposition technique described in this subsection is applicable to arbitrary three-dimensional geometry, although the method is particularly beneficial to applications with repetitive subdomains, such as phased arrays and frequency-selective surfaces. For example, by applying a one-dimensional domain decomposition of a finite phased array, it is only necessary to grid three rows of the array: specifically, the front and back rows, as well as a single row from the center region. Each row is then mapped to a computer processor, with the center row simply being translated the appropriate number of times to construct the desired physical depth of the array. With a two-dimensional decomposition strategy it is only necessary to build nine zones of the finite array, where these zones also correspond to those shown in Figure 10.2. For a massively parallel computer implementation, each of these unit cells, with the appropriate spatial translations, could be placed on a separate processor and solved concurrently, with the result being numerically equivalent to treating the entire array as a single domain. An example of a 25 × 25 ultrawideband phased array using the one-dimensional IDD approach is discussed in the following section.

10.2.3 Application Examples

In this section we present three examples simulated using the time-domain finite element method: The first two employ the dual-field domain-decomposition method

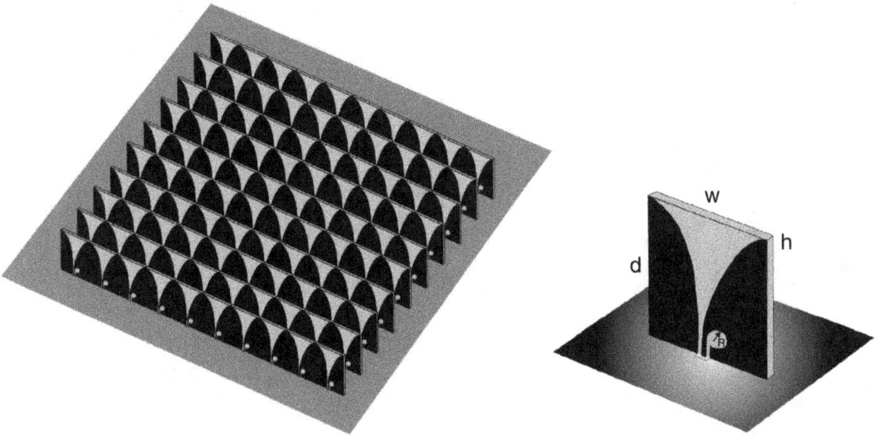

(a) (b)

Figure 10.14 (a) Geometry of a 10 × 10 Vivaldi array. (b) The unit cell ($w = 40$ mm, $d = 55$ mm; substrate thickness $h = 1.5$ mm, relative permittivity $\varepsilon_{r,\text{sub}} = 3.0$; stripline width is 2 mm, relative permittivity $\varepsilon_{r,\text{strip}} = 1.0$). (After Lou and Jin [26], Copyright © Elsevier 2007.) [*See insert for color representation of* (a).]

and the third uses the iterative domain-decomposition method for the solution of the matrix equation in each time step.

The first example deals with the characterization of a broadband phased-array antenna, which consists of 10 × 10 Vivaldi elements illustrated in Figure 10.14(a). The unit cell configuration is shown in Figure 10.14(b). This configuration is similar to the one shown in Figure 9.12, but now having different dimensions and the feed is a stripline from below the ground plane. The spacing between the array elements is 40 mm in both the x- and y-directions. The separation corresponds to one-half wavelength at 3.75 GHz. An infinite ground plane is assumed and the stripline feed is modeled as a TEM port. The simulation employs 144 subdomains, and the total number of unknowns is approximately 3.3 million. The voltage standing-wave ratio (VSWR) parameter calculated at the feed port of the center element is shown in Figure 10.15 for two scan angles. A prominent feature of the VSWR curve is the two sharp peaks occurring at 3.92 and 5.32 GHz when the array is configured for broadside radiation. As discussed in Chapter 9, such phenomena, referred to as the *impedance anomalies*, have been observed in the frequency-domain analysis of the infinite array of Vivaldi antennas [31,32]. The anomalies can be modeled as the excitation of certain resonant modes in the unit cell cavities, which are formed by the ground plane, the electric conducting surfaces on the sidewalls, and a magnetic conducting surface on the open aperture. The resonant frequencies predicted by the cavity model are $f_{10} = 3.99$ GHz and $f_{11} = 5.55$ GHz, which are similar to the resonant frequencies obtained by the numerical simulation. Note that the cavity model does not take into account the thin dielectric substrate. If the effect of the dielectrics is considered, the resonant frequencies predicted would become slightly lower. The radiation patterns

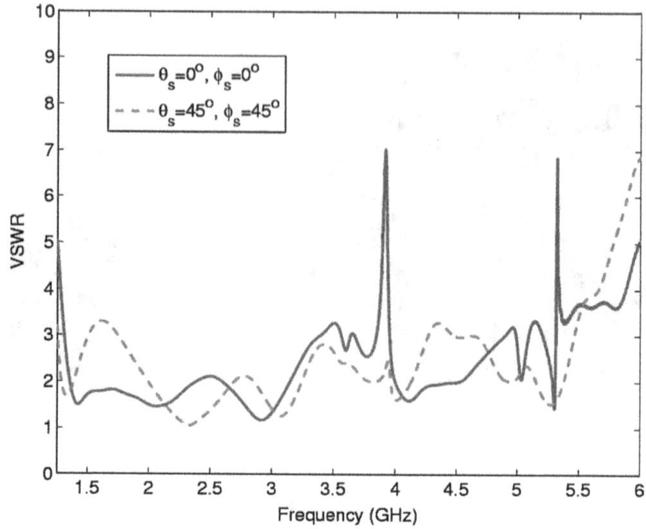

Figure 10.15 VSWR calculated at the feed port of the center element for a 10 × 10 Vivaldi array. (After Lou and Jin [26], Copyright © Elsevier 2007.)

for a few scan angles are displayed in Figures 10.16 and 10.17 for both E- and H-planes.

The second example was not designed for any practical applications; rather, it was designed as one of the test cases to verify the dual-field domain-decomposition method and demonstrate the unique characteristics of electromagnetic bandgap structures. The example consists of a monopole placed in the center of a 9 × 9 square array of dielectric cylinders on an infinitely large ground plane, as illustrated in Figure 10.18. The periodic length of the array is denoted as a in both directions, which is often referred to as the *lattice constant*. The dielectric cylinders have a height of $h = 1.5a$, a radius of $r = 0.18a$, and a relative permittivity of 11.56. The monopole has a height of $h' = 0.75a$ and a radius of $r' = 0.1a$ and is fed by a coaxial waveguide below the ground plane with an inner radius of $r_i = 0.1a$ and an outer radius of $r_o = 0.23a$. The monopole replaces the dielectric cylinder at the center and is excited with a modulated Gaussian pulse with a central frequency of $f_0 = 0.4c_0/a$. This problem was simulated using the dual-field domain-decomposition method in the time domain [26] with approximately 1.4 million unknowns, and the reflection coefficient calculated at the coaxial port in the time domain is plotted in Figure 10.19(a). The reflection coefficient as a function of frequency obtained using the Fourier transform is given in Figure 10.19(b) and is compared with the prediction by the FETI–DPEM1 in the frequency domain [33]. We chose to compare the predictions of the reflection coefficient because this parameter is related directly to the near field and is more sensitive to errors in the numerical computation. Radiation patterns, in contrast, are rather insensitive to the errors in the near-field computation and are not ideal to verify a solution. Figure 10.19(b) shows excellent agreement between two

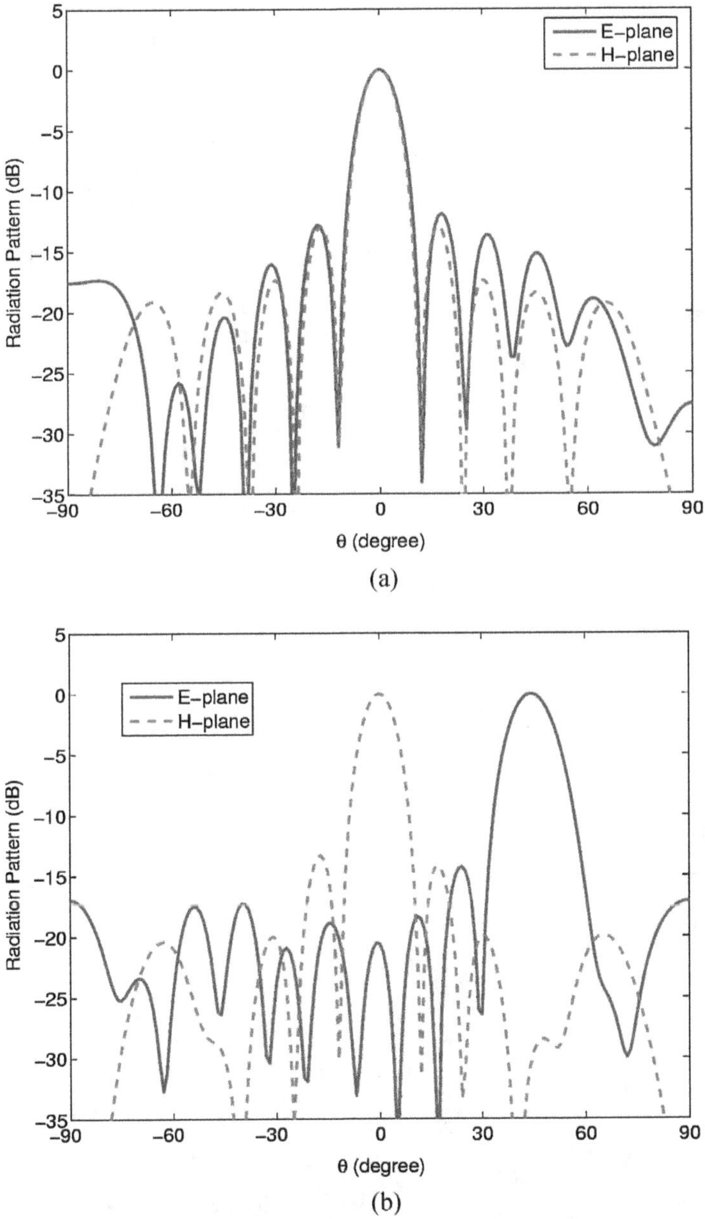

Figure 10.16 E- and H-plane radiation patterns for a 10 × 10 Vivaldi array. (a) Broadside scan. (b) $\theta_s = 45^\circ$ and $\phi_s = 90^\circ$. (After Lou and Jin [26], Copyright © Elsevier 2007.)

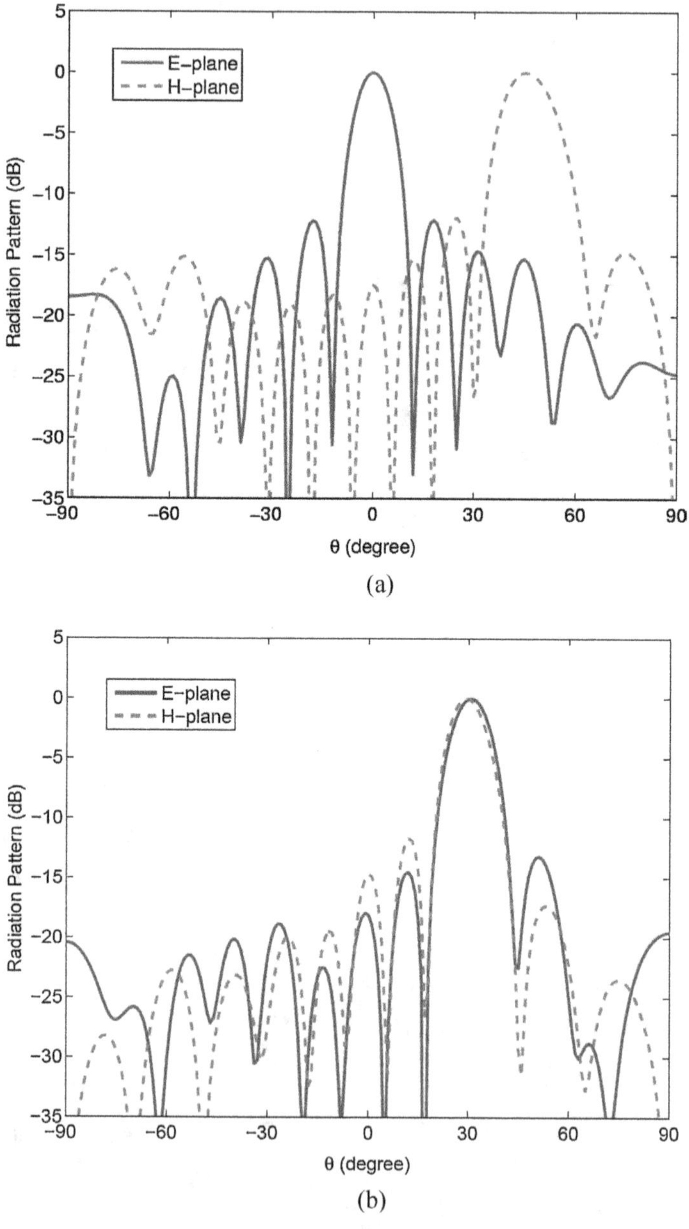

(a)

(b)

Figure 10.17 *E*- and *H*-plane radiation patterns for a 10 × 10 Vivaldi array. (a) $\theta_s = 45^\circ$ and $\phi_s = 0^\circ$. (b) $\theta_s = 45^\circ$ and $\phi_s = 45^\circ$. (After Lou and Jin [26], Copyright © Elsevier 2007.)

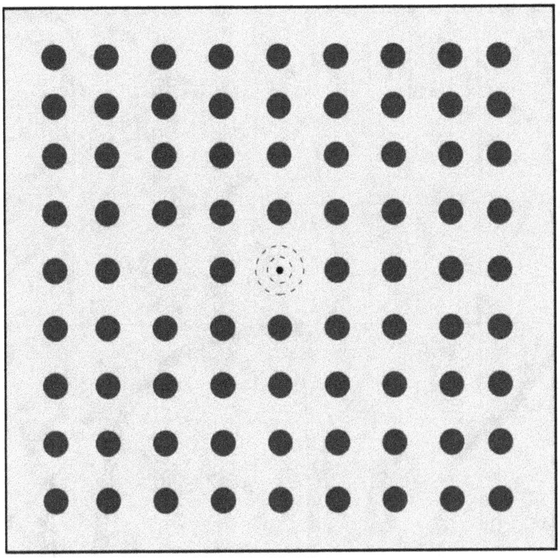

Figure 10.18 Monopole placed at the center of a 9 × 9 array of dielectric cylinders on an infinitely large ground plane.

predictions even though the two simulations are based on different formulations and simulated in different domains. The slight frequency shift is caused by the insufficient mesh density in the dual-field domain-decomposition calculation. A bandgap is observed clearly around $f = 0.35c_0/a$, where the electromagnetic field is unable to propagate through the lattice, and the total power is reflected back into the feed line.

The third example is a 25 × 25 ultrabroadband finite array as illustrated in Figure 10.20. The design of this array is based on the unit cell shown in Figure 9.16, where the radiating elements are connected bowties. The substrate of the array is based on a single-pole Debye dispersive magnetic material. The hybrid FETD–FDTD method described in Chapter 4 was used for this application, where the FETD region contained the array. The finite array was spatially partitioned by using a one-dimensional domain decomposition. Because of the repetitive nature of the array elements in the center region, it was only necessary to create a 25 × 3 finite element grid that would be used to characterize the entire 25 × 25 array. Specifically, the center row of the 25 × 3 grid was translated to fill the complete center section of the array structure, as shown in Figure 10.20. The 25 rows of the array were each mapped to a separate distributed computer processor and solved concurrently by using the iterative domain-decomposition technique described in Section 10.2.2. The global FETD domain consisted of approximately 50 million unknowns, and its solution required approximately 12 seconds per time step using 2.6-GHz Opteron processors, with a steady state being achieved in 3600 time steps. The broadside realized gain, which includes the mismatch and absorptive losses for each element in the array, is shown in

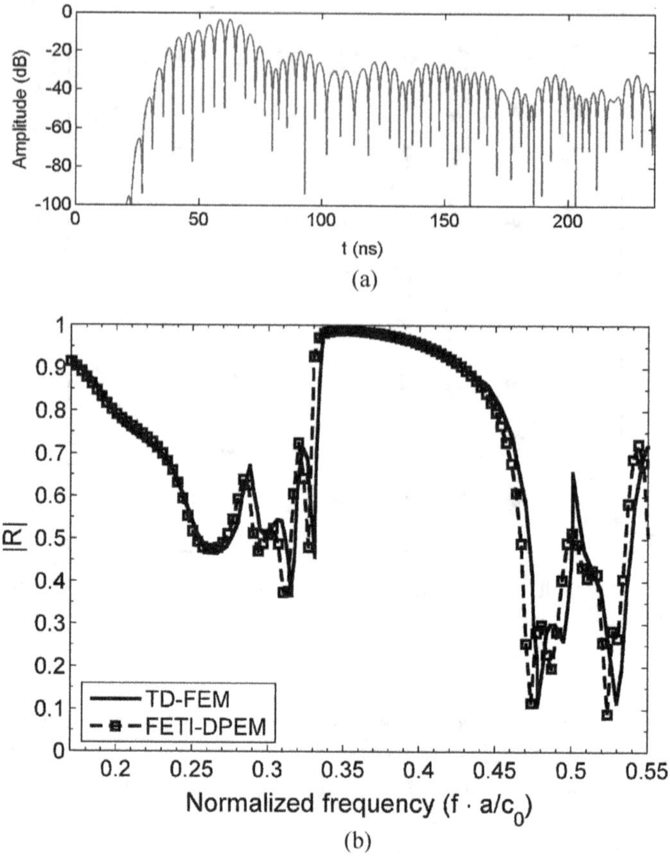

Figure 10.19 Reflection coefficient observed at the coaxial port. (a) The amplitude in time computed by the dual-field domain-decomposition method. (b) The magnitude as a function of the normalized frequency computed by the dual-field domain-decomposition method and the FETI–DPEM1 method. (After Li and Jin [33], Copyright © Optical Society of America 2007.)

Figure 10.21. The performance of this array is compared to that of the ideal aperture (i.e., $4\pi A/\lambda^2$, where $A = 0.25$ m^2 denotes the aperture area).

10.3 SUMMARY

In this chapter we discussed the finite element modeling and analysis of large finite antenna arrays, which is often considered to be one of grand challenges in computational electromagnetics because of the size of the resulting numerical system and the importance of arrays to both commercial and military applications. The complete and rigorous analysis of a finite array has many practical benefits. First, it provides

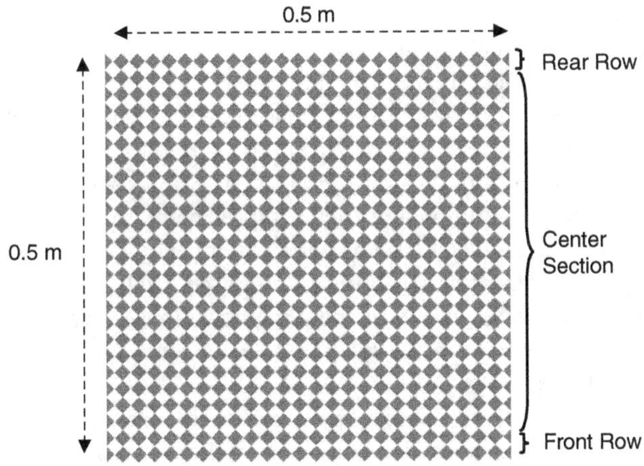

Figure 10.20 Ultrabroadband, 25 × 25 bowtie phased array etched on a dispersive magnetic substrate with a ground plane. The unit cell is based on Figure 9.16.

a reference solution to assess the accuracy of more efficient approximate solutions, such as those based on the unit cell analysis, array theory, and asymptotic approaches. Second, it provides critical information about the edge truncation effect, including edge diffraction and the variation of the active input impedances for the elements near the edges of the array, which is difficult to measure. Third, the ability to apply arbitrary excitations to the radiating elements permits the study of many problems

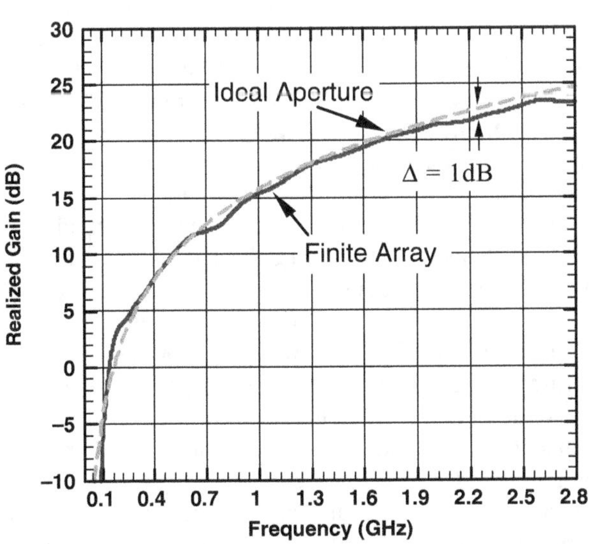

Figure 10.21 Realized gain for a 25 × 25 connected bowtie array compared to that of an ideal aperture.

that cannot be handled using an infinite array model that requires uniform excitation. Some of these problems include the characterization of mutual coupling between individual array elements and mutual coupling between subarrays.

Similar to the analysis of individual antennas, the finite element analysis of large finite antenna arrays can be accomplished in either the frequency or the time domain, both of which were covered in this chapter. The basic approach was to employ the idea of domain decomposition, where each array element, or a collection of array elements, is considered as a subdomain. For the frequency-domain analysis, many domain-decomposition algorithms have been proposed, such as those based on the Schwarz methods and the Schur complement methods. In this chapter we described three methods that have been fully developed, tested extensively, and are well suited for the analysis of finite arrays.

The first frequency-domain method was based on the dual–primal finite element tearing and interconnecting method adapted for solving electromagnetic field problems. This method distinguished itself from other domain-decomposition methods in the construction of a global corner-related coarse problem, which was coupled with the global interface system of equations. The global coarse problem related the subdomain fields over the entire computational domain and propagated the residual error globally in each iteration for solving the reduced interface equation. As a result, the method exhibited an excellent numerical scalability with respect to the number of subdomains, the finite element mesh density, and the wavenumber when the subdomain size is smaller than one-half wavelength. The second method was an improved version of the first method, where the fields across the subdomain interfaces were coupled through weak enforcement of the Robin-type transmission conditions. In this method, two sets of Lagrange multipliers or dual variables were introduced over each subdomain interface. Again, a global corner-related coarse problem was constructed together with the global reduced interface system of equations. Because of the use of the Robin-type transmission conditions, the second method became numerically scalable regardless of the frequency of the electromagnetic fields, which is an important characteristic that is particularly attractive for high-frequency applications. The third method was designed to be applicable when the subdomains do not have matching meshes at their interface, which could alleviate the requirement of periodic meshes and thus simplify mesh generation. In this method, two sets of primal variables (related to the tangential electric fields) and two sets of Lagrange multipliers (related to the tangential magnetic fields) were introduced over the subdomain interfaces. The subdomain interior unknowns were eliminated first to formulate a system of equations for the interface unknowns, which were then solved using either the Gauss–Seidel iteration method or a Krylov subspace method with a symmetric Gauss–Seidel preconditioner. All three methods can effectively exploit the geometrical repetitions in a finite array and are capable of simulating large finite arrays even on a desktop computer.

Compared with the frequency-domain analysis, the time-domain analysis of large finite antenna arrays is more challenging because the fields over the entire computational domain have to be updated in each time step. In this chapter we described two strategies to perform the time-domain finite element modeling of a large finite array using the idea of domain decomposition. The first strategy was based on a recently developed novel technique, the dual-field domain-decomposition method. In this method the vector wave equations for both the electric and magnetic fields were discretized using the finite element method in each subdomain. A leapfrog algorithm was then designed that allowed the electric and magnetic fields to be updated in an alternating fashion. The fields across the subdomains were coupled explicitly by exchanging the surface fields at the interface between adjacent subdomains. With this domain decomposition, the finite element matrix for the entire computational domain was decomposed into many smaller matrices, whose factorization requires significantly less computer time and memory, thus permitting the analysis of large-scale electromagnetic problems. The geometrical repetitions in a finite array could also be utilized since most subdomains have the identical finite element matrices. However, since the fields at the subdomain interfaces were exchanged explicitly, the dual-field domain-decomposition algorithm was only conditionally stable, with the maximum time-step size dictated by the finite element discretization of the subdomain interfaces. The second strategy was to apply the traditional time-domain finite element method and employ the domain-decomposition idea for the iterative solution of the large finite element matrix equation in each time step. This approach preserved the unconditional stability of the time-domain finite element solution. Because of the need to update the fields over the entire computational domain in each time step, the efficiency of time-domain finite element domain-decomposition techniques may be reduced relative to their frequency-domain counterparts, particularly if only narrowband information is required. However, the ability of the finite element time-domain method to perform broadband characterizations and to model nonlinear devices, components, and media (a feature that has not yet been explored extensively) can be highly attractive for the future design of advanced actively driven broadband phased arrays.

REFERENCES

1. D. T. McGrath and V. P. Pyati, "Phased array antenna analysis with the hybrid finite element method," *IEEE Trans. Antennas Propagat.*, vol. 42, pp. 1625–1630, Dec. 1994.
2. E. W. Lucas and T. P. Fontana, "A 3-D hybrid finite element/boundary element method for the unified radiation and scattering analysis of general infinite periodic arrays," *IEEE Trans. Antennas Propagat.*, vol. 43, no. 2, pp. 145–153, Feb. 1995.
3. Z. Lou and J. M. Jin, "Finite element analysis of phased array antennas," *Microwave Opt. Tech. Lett.*, vol. 40, no. 6, pp. 490–496, Mar. 2004.
4. Z. Lou and J. M. Jin, "Higher-order finite element analysis of finite-by-infinite arrays," *Electromagnetics*, vol. 24, no. 7, pp. 497–514, Oct. 2004.
5. R. Kindt, K. Sertel, E. Topsakal, and J. Volakis, "Array decomposition method for the accurate analysis of finite arrays," *IEEE Trans. Antennas Propagat.*, vol. 51, pp. 1364–1372, June 2003.
6. J. Yeo, V. Prakash, and R. Mittra, "Efficient analysis of a class of microstrip antennas using the characteristic basis function method (CBFM)," *Microwave Opt. Tech. Lett.*, vol. 39, pp. 456–464, Dec. 2003.

7. S.-C. Lee, M. N. Vouvakis, and J.-F. Lee, "A non-overlapping domain decomposition method with non-matching grids for modeling large finite antenna arrays," *J. Comput. Phys.*, vol. 203, pp. 1–21, Feb. 2005.

8. Y. J. Li and J. M. Jin, "A vector dual–primal finite element tearing and interconnecting method for solving 3-D large-scale electromagnetic problems," *IEEE Trans. Antennas Propagat.*, vol. 54, no. 10, pp. 3000–3009, Oct. 2006.

9. M. N. Vouvakis, Z. Cendes, and J.-F. Lee, "A FEM domain decomposition method for photonic and electromagnetic band gap structures," *IEEE Trans. Antennas Propagat.*, vol. 54, pp. 721–733, Feb. 2006.

10. Y. J. Li and J. M. Jin, "A new dual–primal domain decomposition approach for finite element simulation of 3D large-scale electromagnetic problems," *IEEE Trans. Antennas Propagat.*, vol. 55, no. 10, pp. 2803–2810, Oct. 2007.

11. C. Farhat and F.-X. Roux, "A method of finite element tearing and interconnecting and its parallel solution algorithm," *Int. J. Numer. Methods Eng.*, vol. 32, pp. 1205–1227, Oct. 1991.

12. C. Farhat, A. Macedo, and M. Lesoinne, "A two-level domain decomposition method for the iterative solution of high frequency exterior Helmholtz problems," *Numer. Math.*, vol. 85, pp. 283–308, 2000.

13. C. Farhat, M. Lesoinne, P. Le Tallec, K. Pierson, and D. Rixen, "FETI-DP: a dual–primal unified FETI method. I: A faster alternative to the two-level FETI method," *Int. J. Numer. Methods Eng.*, vol. 50, pp. 1523–1544, 2001.

14. C. Farhat, P. Avery, R. Tezaur, and J. Li, "FETI-DPH: a dual–primal domain decomposition method for acoustic scattering," *J. Comput. Acoustics*, vol. 13, no. 3, pp. 499–524, Sept. 2005.

15. D. Rixen and C. Farhat, "A simple and efficient extension of a class of substructure based preconditioners to heterogeneous structural mechanics problems," *Int. J. Numer. Methods Eng.*, vol. 44, pp. 489–516, Feb. 1999.

16. C. Farhat, A. Macedo, M. Lesoinne, F.-X. Roux, and F. Magoules, "Two-level domain decomposition methods with Lagrange multipliers for the fast iterative solution of acoustic scattering problems," *Comput. Methods Appl. Mech. Eng.*, vol. 184, pp. 213–239, Apr. 2000.

17. B. Després, P. Joly, and J. E. Roberts, "A domain decomposition method for the harmonic Maxwell equations," in *Iterative Methods in Linear Algebra*. Amsterdam: Elsevier, pp. 475–484, 1992.

18. B. Stupfel and M. Mognot, "A domain decomposition method for the vector wave equation," *IEEE Trans. Antennas Propagat.*, vol. 48, no. 5, pp. 653–660, May 2000.

19. Y. Achdou, C. Japhet, Y. Maday, and F. Nataf, "A new cement to glue nonconforming grids with Robin interface conditions: the finite volume case," *Numer. Math.*, vol. 92, pp. 593–620, 2002.

20. K. Zhao, V. Rawat, S.-C. Lee, and J.-F. Lee, "A domain decomposition method with non-conformal meshes for finite periodic and semi-periodic structures," *IEEE Trans. Antennas Propagat.*, vol. 55, no. 9, pp. 2559–2570, Sept. 2007.

21. Z.-Q. Lu, X. An, and W. Hong, "A fast domain decomposition method for solving three-dimensional large-scale electromagnetic problems," *IEEE Trans. Antennas Propagat.*, vol. 56, no. 8, pp. 2200–2210, Aug. 2008.

22. J. M. Jin, Z. Lou, Y. J. Li, N. Riley, and D. Riley, "Finite element analysis of complex antennas and arrays," *IEEE Trans. Antennas Propagat.,* vol. 56, no. 8, pp. 2222–2240, Aug. 2008.

23. C. Smith, M. Little, B. Porter, and M. N. Vouvakis, "Analysis of co-planar phased array coupling using finite element domain decomposition," *IEEE AP-S Int. Symp. Dig.*, pp. 3524–3527, Honolulu, HI, June 2007.

24. Y. Li and J. M. Jin, "Analysis of antenna array mutual coupling effects using the parallelized FETI–DPEM method," *USNC/URSI Natl. Radio Sci. Mtg. Dig.*, p. 331.3, San Diego, CA, July 2008.

25. Z. Lou and J. M. Jin, "A novel dual-field time-domain finite-element domain-decomposition method for computational electromagnetics," *IEEE Trans. Antennas Propagat.,* vol. 54, no. 6, pp. 1850–1862, June 2006.

26. Z. Lou and J. M. Jin, "A dual-field domain-decomposition method for the time-domain finite-element analysis of large finite arrays," *J. Comput. Phys.*, vol. 222, no. 1, pp. 408–427, Mar. 2007.

27. Z. Lou and J. M. Jin, "A new explicit time-domain finite-element method based on element-level decomposition," *IEEE Trans. Antennas Propagat.,* vol. 54, no. 10, pp. 2990–2999, Oct. 2006.

28. "ParMETIS: parallel graph partitioning and fill-reducing matrix ordering," University of Minnesota, Minneapolis, MN, http://glaros.dtc.umn.edu/gkhome/metis/parmetis/overview.

29. "Aztec: a parallel iterative package for linear systems," Sandia National Laboratories, Albuquerque, NM, http://www.cs.sandia.gov/CRF/aztec1.html.

30. "Trilinos," Sandia National Laboratories, Albuquerque, NM, http://trilinos.sandia.gov/index.html.

31. Z. Lou and J. M. Jin, "Finite element analysis of phased array antennas," *Microwave Opt. Tech. Lett.,* vol. 40, no. 6, pp. 490–496, Mar. 2004.

32. H. Holter, T.-H. Chio, and D. H. Schaubert, "Elimination of impedance anomalies in single- and dual-polarized endfire tapered slot phased arrays," *IEEE Trans. Antennas Propagat.*, vol. 48, pp. 122–124, Jan. 2000.

33. Y. J. Li and J. M. Jin, "Fast full-wave analysis of large-scale three-dimensional photonic crystal devices," *J. Opt. Soc. Am. B*, vol. 24, pp. 2406–2415, Sept. 2007.

11 Antenna–Platform Interaction Modeling

In practice, antennas are integrated into platforms, such as airframes, missiles, satellites, ships, or structural supports for commercial wireless applications. Mounting an antenna on a platform frequently introduces distortion in its expected radiation pattern and causes coupling between the antenna and the platform. The distortion in the radiation pattern may reduce the desired coverage for effective communications or alternatively, compromise the accuracy of postprocessing algorithms used for separating and locating signals. The term *installed performance* typically refers to the performance of the antenna in the presence of the platform. The existence of mutual coupling between multiple onboard antennas—caused by space waves, surface waves, and scattering by the platform—can further degrade the performance of antenna systems through co-site interference, where the term *co-site* is used to describe two or more antenna systems that are operating simultaneously and is related to the underlying electronics as well as the phenomenology of the electromagnetic coupling. In addition, because antenna systems utilizing amplitude and phase received at various antenna ports to locate signals may experience degraded accuracy, it is important to calibrate direction finding and angle-of-arrival systems by using analytical or numerical predictions of the anticipated amplitude and phase received at the antenna ports as a function of angle, frequency, and polarization. For these reasons, it is important to develop accurate numerical prediction tools to characterize the radiation patterns and mutual coupling of antennas mounted on a complex, and often electrically large, platform.

Predicting the installed performance of antenna systems has historically been based on the use of ray-optics techniques, such as physical optics (PO), geometrical optics (GO), and their extensions, which include the geometrical theory of diffraction (GTD) and the uniform theory of diffraction (UTD). In these approaches, predicted or measured antenna patterns based on the antenna in isolation can be used as a source for eminating rays which then interact with the platform and thereby lead to predictions for the effect of the platform on the radiated antenna patterns. Provided that the frequency is sufficiently high that optics-based assumptions are valid, these techniques have been proven to be powerful and useful. However, these methods can have challenges with the effects of complex materials, shadow boundaries, and near-field physics.

Finite Element Analysis of Antennas and Arrays, By Jian-Ming Jin and Douglas J. Riley

To provide a more rigorous analysis of installed antenna performance, especially in the lower-frequency bands, the use of fast multipole–accelerated boundary integral formulations has recently become popular. Because these acceleration techniques can reduce the traditional moment-method computational complexity of $O(N^3)$ down to $O(N \log N)$, the full-wave analysis of, for example, moderate-size airframes in the S-band (2–4 GHz) frequency region is tractable with small distributed parallel computer systems. In contrast to a purely ray optics–based analysis, the accelerated boundary integral approaches use detailed models of the antennas, and therefore account rigorously for their near-field physics, which can be important in a co-site analysis. In addition, the hybridization of boundary integrals and ray-optical methods is also widely used for the analysis of installed antenna performance. However, accommodating material regions that cannot be characterized by a surface impedance representation, or by a known Green's function, will reduce the performance and hence the application domain of the boundary integral methods. To address this issue, fast multipole–accelerated volumetric integral equations that can accommodate inhomogeneous regions of complex materials are currently an active research topic in computational electromagnetics.

A third approach to an installed antenna performance analysis is based on the direct use of differential equation methods, such as the finite element and finite-difference methods, as well as their hybridization with the other two approaches discussed above. By adopting advanced domain-decomposition techniques, and because of the excellent ability of finite element and finite-difference techniques to accommodate complex materials, this approach can provide increased generality with regard to the overall complexity of an antenna–platform interaction analysis. Therefore, in this chapter we discuss the finite element modeling of antennas mounted on a platform, using two approaches. The first is a rigorous coupled analysis in which the effect of the platform is included in both the computation of the radiation patterns and the calculation of the input impedance and mutual coupling in the multiantenna case. The second is an approximate and decoupled approach in which antennas are first analyzed in isolation from the platform to determine their near-field equivalent surface currents or their radiation patterns, and then the effect of the platform is considered in computation of the radiation patterns by using either a numerical or an asymptotic technique. The two approaches will leverage some of the best known domain-decomposition concepts, as well as accelerated boundary integral equation formulations and ray-optical techniques.

11.1 COUPLED ANALYSIS

The most accurate approach to analyzing antennas installed on a platform is to treat both the antennas and the platform as a single electromagnetic system and apply any first-principle method for its analysis. For example, we can enclose the entire system with an artificial surface and use the finite element method in conjunction with an absorbing boundary condition to compute the electromagnetic fields. Alternatively, we can combine the finite element method and the finite-difference time-domain

(FDTD) method so that the FDTD can be used to model any regular regions inside the artificial surface. If the platform is made of perfectly conducting surfaces, a more efficient approach is to tailor the finite element–boundary integral (FE-BI) method specifically to this application so that the induced surface currents on the platform can be formulated accurately with a boundary integral equation. In this section we discuss these three approaches briefly together with their technical challenges and solutions.

11.1.1 FETI–DPEM with Domain Decomposition

As mentioned earlier, one of the simplest approaches to analyzing antennas installed on a platform is to apply the finite element method directly to the entire problem. The efficiency of this approach depends highly on the size and shape of the platform. When the platform is electrically large and has an irregular shape, the resulting finite element system can be extremely large, involving a few hundred millions or perhaps billions of unknowns. Even though the finite element matrix is highly sparse, solving such a large linear system of equations is challenging for both direct and iterative solvers. A solution to this problem is to use a domain-decomposition technique that leverages the increasing power of parallel computing.

Among various domain-decomposition schemes, the electromagnetic dual–primal finite element tearing and interconnecting (FETI–DPEM) method described in Chapter 10 for the finite-array analysis is one of the most efficient. If we examine the formulation of the FETI–DPEM carefully, we find that this method is not restricted to only finite-array geometries and, in fact, can be applied to arbitrary geometries as well. There are two differences between its applications to an array and to an arbitrary geometry. The first is that the array geometry provides a natural decomposition of the entire domain into subdomains, whereas for an arbitrary domain, one has to design an algorithm to divide the entire domain automatically into a desired number of subdomains. Each subdomain should have an approximately equal number of unknowns, and the number of unknowns on the subdomain interfaces should be kept to a minimum to minimize the size of the final reduced-order interface system. Fortunately, partitioning algorithms have been researched actively over the past several years, and a variety of software tools are now available [1]. The second difference is that in an array problem many subdomains are identical and hence have the same finite element matrix. When this matrix is assembled and factorized for one subdomain, it can be used for the remaining identical subdomains. In contrast, for an arbitrary problem the subdomains are generally different and a finite element matrix has to be assembled and factorized for each subdomain, which could greatly increase the computational time of the FETI–DPEM. However, with the availability of massively parallel computing systems, the computation associated with the subdomains can be distributed effectively to different processors for parallel computation. No interprocessor communication is required with this step of the FETI–DPEM technique because the finite element matrix for each subdomain is completely independent of other subdomains, and coupling between the subdomains occurs later through solution of the global interface problem. Once the interface equation is solved, the calculation of the fields in the subdomains is again completely independent of each other and can be distributed to different processors. Consequently, the parallel FETI–DPEM is highly efficient.

The approach described above has recently been implemented for the analysis of arbitrary electromagnetic scattering and radiation problems [2]. For example, the problem discussed in Section 7.1.2, which was solved with the time-domain finite element method, has also been simulated using the FETI–DPEM. This problem consists of five monopole antennas mounted on a finite ground plane, and its geometrical details are given in Figure 7.2. The finite element discretization results in 1.4 million unknowns. When the computational domain is decomposed into 66 subdomains, the dimension of the interface system is 134,000. The computation is distributed onto 16 processors. Figures 11.1 and 11.2 show calculated input and

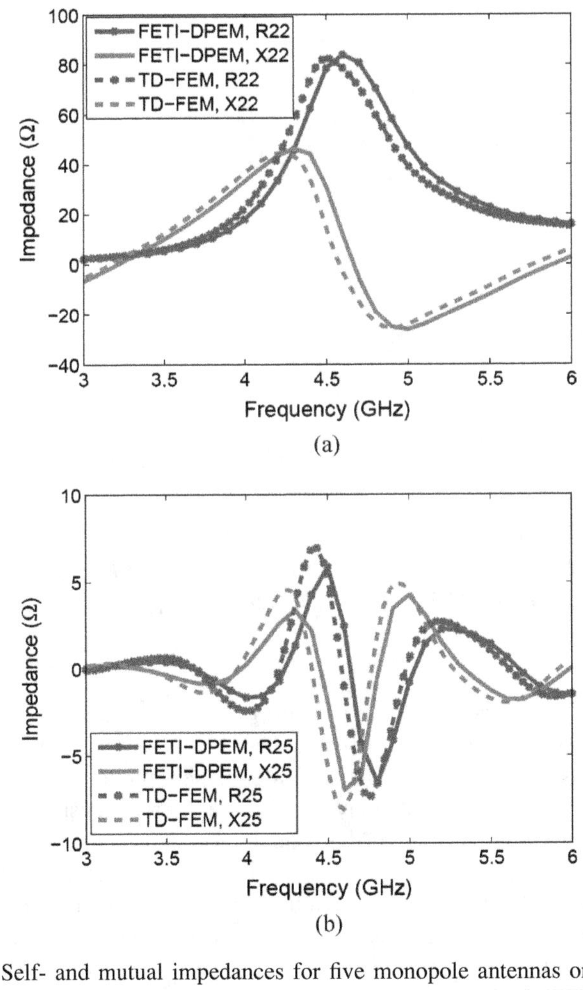

Figure 11.1 Self- and mutual impedances for five monopole antennas on the finite plate shown in Figure 7.2. (a) Z_{22}. (b) Z_{25}. (After Li and Jin [2], Copyright © IEEE 2008.)

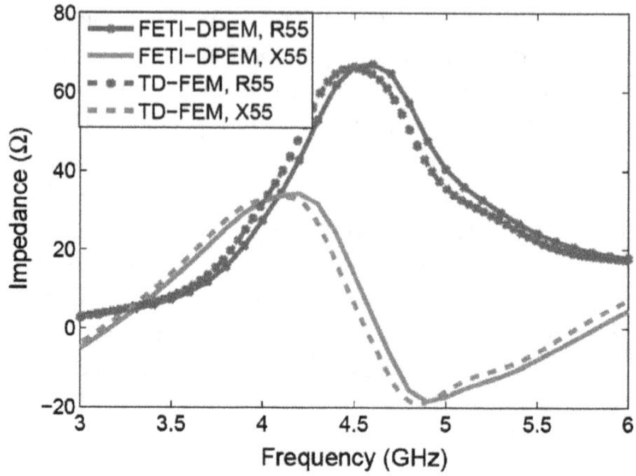

(a)

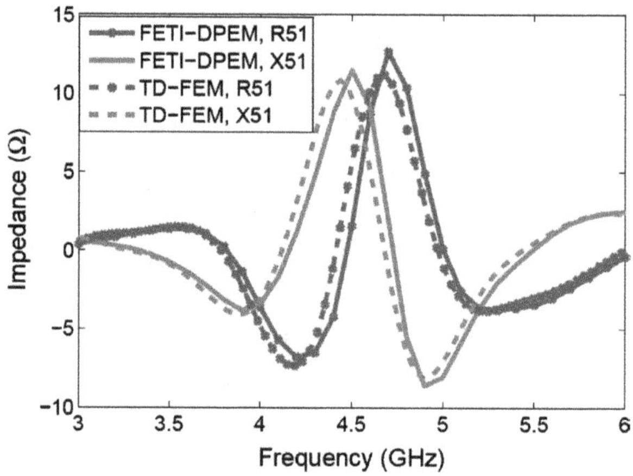

(b)

Figure 11.2 Self- and mutual impedances for five monopole antennas on the finite plate shown in Figure 7.2. (a) Z_{55}. (b) Z_{51}. (After Li and Jin [2], Copyright © IEEE 2008.)

mutual impedances observed at the input port of the SMA connector, which is 1.08 cm below the surface of the ground plane. The results are compared with the previous calculations using the time-domain finite element method, which are given in Figure 7.3. Note that the results displayed in Figure 7.3 are measured at the aperture of the coaxial feed coincident with the ground plane, and these results have been converted to those shown in Figures 11.1 and 11.2 by using transmission-line theory. Considering the fact that the calculation of the input and mutual impedances depends directly on the fields local to the antenna feeds, agreement between the two solutions is good. The calculated radiation patterns also agree well with those displayed in Figure 7.4.

11.1.2 Hybrid FETD–FDTD with Domain Decomposition

As noted in the preceding subsection, the major challenge in the finite element analysis of antennas on a large platform is to solve the extremely large number of unknowns associated with the volumetric finite element mesh that fills the large space between the platform and the truncation surface. This challenge can be partially resolved by using the hybrid FETD–FDTD technique described in Chapter 4. With this technique, the finite element method is used to model the antennas and the curved surfaces of the platform, while the remainder of the computational domain is discretized with the efficient FDTD method. This approach can significantly reduce the size of the finite element region and hence the finite element matrix. Consequently, both computer memory requirements and the computational time can be reduced significantly. In addition, further computational benefits may be obtained by applying parallel computing algorithms to both the finite element and FDTD regions of the computational volume. Because the finite element subdomains are contained within the FDTD background grid, a variety of parallelization strategies become possible, and a few of these are described in this subsection.

Distributed parallelization techniques for the FDTD method are mature and very efficient [3,4]. One-, two-, and three-dimensional decompositions of FDTD grids that contain billions of cells have been mapped successfully onto parallel computers with hundreds of processors [5]. One of the more common parallelization strategies for the FDTD method is similar to the iterative domain decomposition (IDD) that was described in Section 10.2.2 for the time-domain finite element technique. However, because the traditional FDTD method is an explicit time-stepping algorithm, there is no need to augment the matrix entries for those edges that lie on the interface between subdomains, and therefore implementation of parallel FDTD algorithms is considerably simplified.

Figure 11.3 illustrates a possible one-dimensional domain decomposition along the z-direction for the FDTD algorithm. The approach is based on a one-cell overlapping region in which a message-passing procedure is used to exchange the boundary

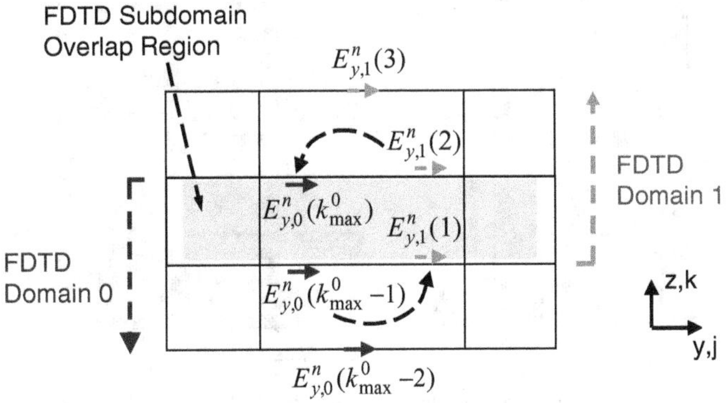

Figure 11.3 Interface region between two FDTD subdomains, showing an interchange of electric fields.

electric fields. In the FDTD domain 0, the electric field component E_y at local position $z_0 = k^0_{\max}\,\Delta z$ is denoted by $E_{y,0}(k^0_{\max})$, where $k^0_{\max}$ denotes the number of cells in the z-direction in domain 0. Similarly, the E_y component in domain 1 at local position $z_1 = 0\Delta z$ is denoted by $E_{y,1}(1)$. Because domain 0 terminates at the local index $k^0_{\max}$, $E^n_{y,0}(k^0_{\max})$ must be updated from domain 1 through the mapping $E^n_{y,1}(2) \rightarrow E^n_{y,0}(k^0_{\max})$. Similarly, domain 1 requires the update mapping $E^n_{y,0}(k^0_{\max} - 1) \rightarrow E^n_{y,1}(1)$. This procedure can easily be extended to two- and three-dimensional FDTD domain decompositions. Because advanced parallel computers often dedicate separate processors to the message-passing communication, it is possible to time-advance the field components while communicating the field data mappings. Consequently, parallel efficiencies exceeding 90% are often obtainable, even with a large number of subdomains and computer processors.

The hybrid FETD–FDTD technique can be applied with either a single FETD region or multiple FETD regions inset into the FDTD background grid. The inclusion of multiple FETD regions can be useful for predicting installed antenna performance because the interaction effects between these regions can easily be studied through simple spatial translations of their locations within the supporting FDTD mesh. A simple parallel decomposition would place each of the FETD regions on a separate processor and then further adopt an FDTD decomposition such as that described above. Alternatively, the iterative domain-decomposition technique described in Section 10.2.2 could be used for the distributed parallel processing of each of the FETD regions, where each region could have a different number of subdomains. For example, a two-dimensional representation of four FETD regions inset into an FDTD grid is illustrated in Figure 11.4. Each FETD region contains a cylindrical monopole antenna along with its coaxial feeding system. The monopoles reside on a metallic

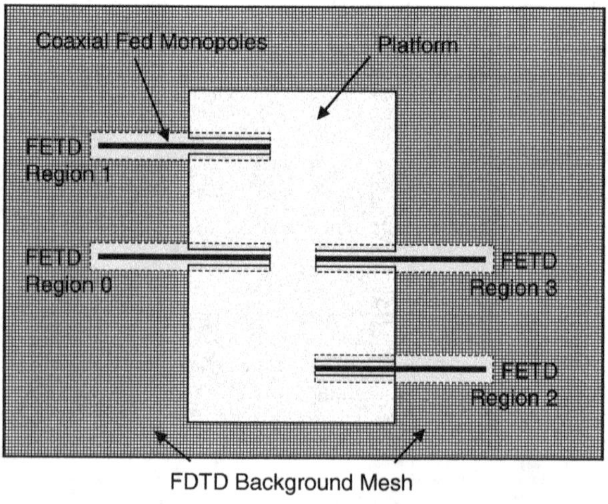

Figure 11.4 Four coaxial-fed monopoles on a simple platform. Each monopole is enclosed in an unconditionally stable FETD region that is inset into an FDTD uniform background grid. Independent domain decompositions could be applied to each region of the global domain.

platform that resides primarily within the FDTD grid, although the metallic surface of the platform extends into the FETD regions as shown in Figure 11.4. A practical parallel computer implementation could partition each FETD monopole region into several finite element subdomains, as well as partition the background FDTD region into several FDTD subdomains. Note that the total number of FETD subdomains does not have to equal the total number of FDTD subdomains, although this will be the case in certain decompositions. Because the IDD technique preserves the local unconditional stability of the FETD regions, the parallel FETD–FDTD method retains the original properties of the serial formulation.

To illustrate the application of the parallel FETD–FDTD method, we consider the problem illustrated in Figure 11.5, where two monopole antennas are mounted on

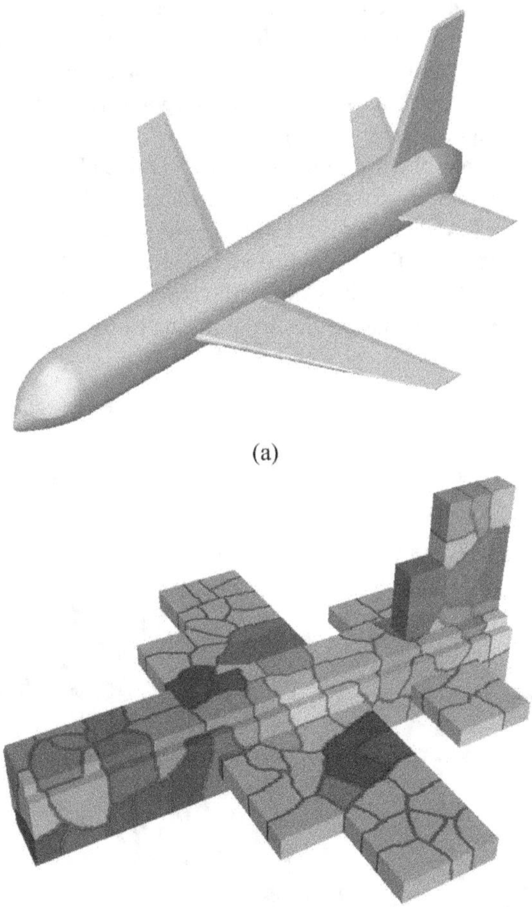

(a)

(b)

Figure 11.5 (a) Solid model of a 15-m-long airframe. (b) Enclosure of the finite element discretization region. (After Riley et al. [6], Copyright © IEEE 2002.) (*See insert for color representation of figure.*)

the front edges of the wings of an airframe [6]. The 15-m-long airframe is illustrated in Figure 11.5(a), and its surface is discretized by 95,000 triangular patches. The entire airframe, including the antennas, is enclosed by a rectangular container that defines the unstructured finite element region as shown in Figure 11.5(b). This single finite element region is then decomposed into 256 subdomains for 256 processors by using a technique based the Kernighan–Lin partitioning algorithm [7], and the total number of unknowns in this unstructured grid region is approximately 2.7 million. By using the techniques described in Chapter 4, the unstructured grid is interfaced with a rectangular structured FDTD grid with approximately 12 million FDTD cells, including those in the perfectly matched layers that provide the grid truncation. The FDTD grid is further partitioned and its calculation is distributed onto 256 processors by using a simple three-dimensional block decomposition process. The surface current on the airframe at a time instant of 45 ns (900 time steps) for the monopoles driven by a Gaussian-modulated 500-MHz sinusoid is shown in Figure 11.6. At this frequency, the airframe is approximately 25 wavelengths long. Surface current visualizations can provide a useful diagnostic tool to identify coupling paths for specific platform-installed antenna configurations.

An additional example based on the hybrid FETD–FDTD technique is shown in Figure 11.7(a). This geometry consists of two surface-mounted, dual-frequency and circularly polarized, stacked patch GPS antennas mounted on a missile platform that is approximately 40 wavelengths in length and 20 wavelengths in circumference. The two antennas are located on opposite sides of the platform. A domain-decomposition approach that isolates the patch antennas from the platform yet retains full coupling

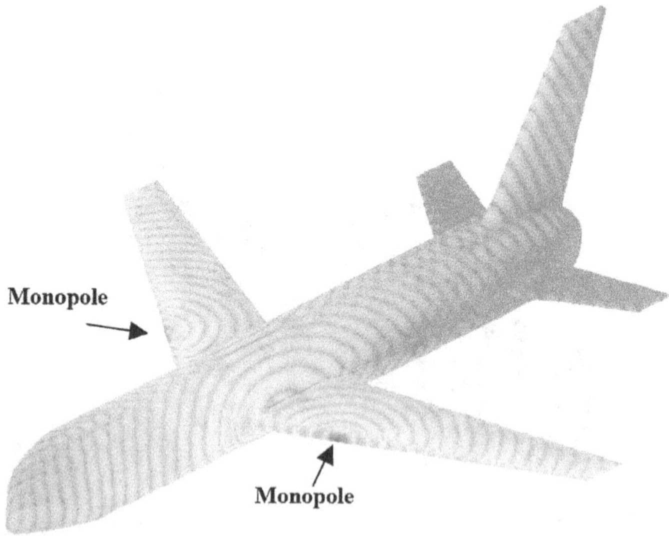

Figure 11.6 Surface electric current distribution due to radiation by two monopole antennas mounted on a simple airplane model. (After Riley et al. [6], Copyright © IEEE 2002.) (*See insert for color representation of figure.*)

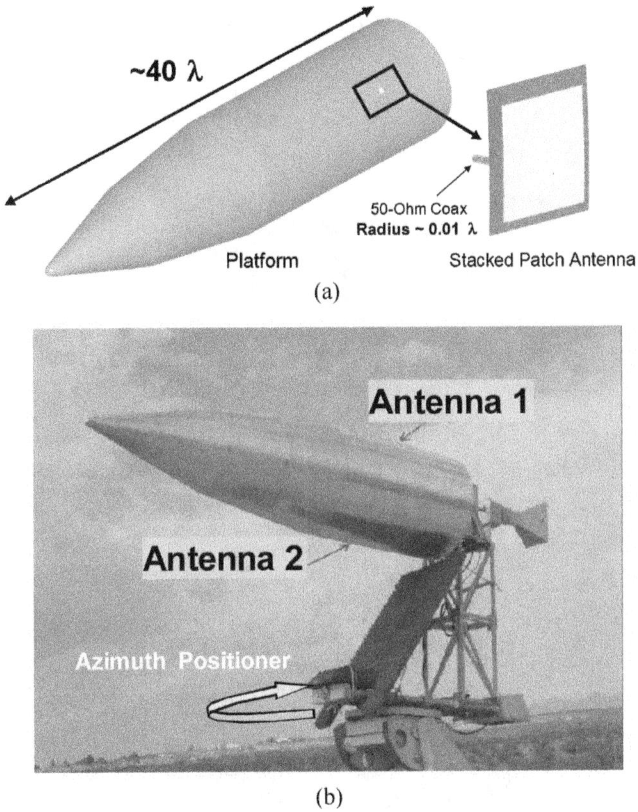

Figure 11.7 Two dual-frequency coaxially fed stacked patch antennas mounted on an electrically large platform. (a) Geometry. (b) Measurement setup.

is used, where the basic strategy is similar to that shown in Figure 11.4. The time-domain finite element method is used to model the antennas and a thin volumetric layer placed around the perimeter of the platform. The finite element region is transitioned to an FDTD grid that is terminated with perfectly matched layers using the technique described in Chapter 4. The radiation patterns of the antennas on the platform were measured using the setup shown in Figure 11.7(b). The measurements were performed by the Physical Science Laboratory, New Mexico State University. Predictions and measurements for the right-hand circularly polarized (RHCP) gain pattern at approximately 1.5 GHz are shown in Figure 11.8. The cut plane is through the antennas and along the axis of the platform. Good agreement is seen for all angles, with the exception of the tail section. The positioner shown in Figure 11.7(b) that was used to support and rotate the physically large platform adversely affected the measurements for observation angles in this region. It is noteworthy that RF predictions often overcome practical measurement challenges, such as blockage by positioners or ground reflections when antennas are mounted on the bottom of airplanes and measured on the runway.

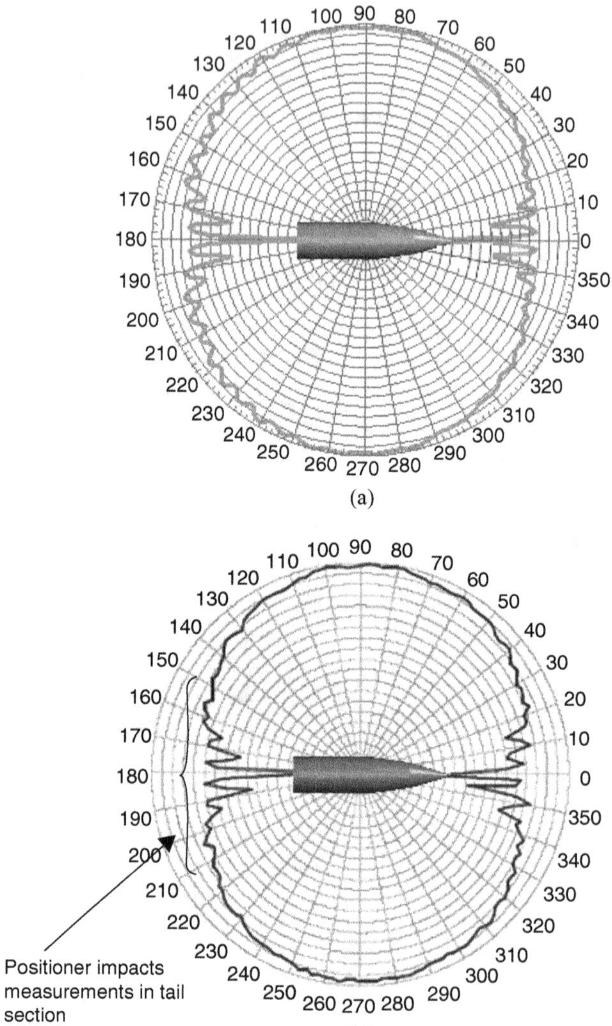

Figure 11.8 Predicted and measured RHCP gain for a cut plane taken axially through the antennas. (a) Prediction. (b) Measurement. Equivalent decibel scales with 2-dB increments. The platform positioner affected the data measured for observation angles in the tail section.

As discussed in Chapter 2, one of the major sources of error in the finite element analysis of electrically large wave propagation problems is the grid dispersion error. The same is true for the FDTD analysis. For both the FDTD technique and the finite element method based on first-order basis functions, the dispersion error is proportional to $(h/\lambda)^2$, where h denotes the elemental edge length and λ denotes the wavelength. Because the dispersion error is accumulative, the total dispersion error becomes larger over electrically large distances. Therefore, when the finite element method and the FDTD are applied to electrically large geometries, the grid density must be increased accordingly to control the dispersion error. For example, when the computational domain is increased by a factor of 10, the element size must be reduced by a factor of 3 to maintain a similar overall dispersion error. However, it must be pointed out that although both structured and unstructured grids have dispersion error proportional to $(h/\lambda)^2$, the magnitude of the error in an unstructured grid is typically much smaller than that in a structured grid because the inherent randomness of an unstructured grid tends to mitigate the accumulation of the dispersion errors [8].

11.1.3 Hybrid FE–BI Method with FMM Acceleration

One of the most accurate methods for the analysis of complex antennas that are mounted on a platform with electromagnetically impenetrable surfaces is the finite element method combined with boundary integral equations, which is commonly known as the FE-BI method and was described in Section 3.3. In the FE-BI method, the finite element technique is used to model the details of the antennas, while boundary integral equations are employed to model the effects of the platform. Since boundary integral equations model wave propagation by using a Green's function, they are free of the spatial dispersion errors that are associated with the volumetric finite element or finite-difference methods. This characteristic, combined with the facts that they require only surface discretization and satisfy the Sommerfeld radiation condition rigorously, makes the FE-BI method an attractive approach for modeling antennas on a platform.

Formulation of the FE-BI method for the analysis of antennas on a platform is similar to the approach described in Section 3.3. To illustrate it briefly, we consider the problem shown in Figure 11.9, where two antennas are mounted on a perfect electrically conducting (PEC) platform. To formulate an FE-BI solution, the first step is to separate the finite element region from the exterior region where the fields are formulated by a surface integral equation. For conformal antennas, the aperture

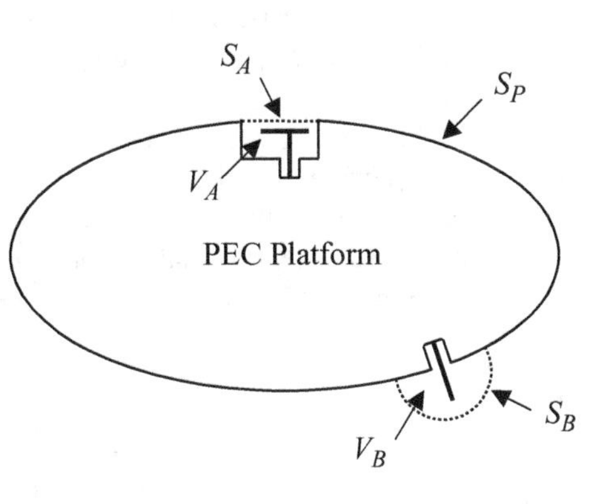

Figure 11.9 Two antennas mounted on an electrically conducting platform.

provides a natural separation surface, which is denoted as S_A. For nonconformal antennas, we can introduce an arbitrary surface, such as a hemispherical surface, to enclose the antenna. This surface is denoted as S_B in Figure 11.9. The remainder of the conducting surface of the platform is denoted by S_P. Both tangential electric and magnetic fields exist on S_A and S_B; however, only tangential magnetic fields exist on S_P, which are, of course, related to the induced surface electric currents. Therefore, for the problem illustrated in Figure 11.9, the coupled FE-BI system corresponding to (3.125) is given by

$$\begin{bmatrix} K_{V_A,V_A} & K_{V_A,S_A} & 0 & 0 & 0 & 0 & 0 \\ K_{S_A,V_A} & K_{S_A,S_A} & B_{S_A,S_A} & 0 & 0 & 0 & 0 \\ 0 & 0 & 0 & K_{V_B,V_B} & K_{V_B,S_B} & 0 & 0 \\ 0 & 0 & 0 & K_{S_B,V_B} & K_{S_B,S_B} & B_{S_B,S_B} & 0 \\ 0 & P_{S_A,S_A} & Q_{S_A,S_A} & 0 & P_{S_A,S_B} & Q_{S_A,S_B} & Q_{S_A,S_P} \\ 0 & P_{S_B,S_A} & Q_{S_B,S_A} & 0 & P_{S_B,S_B} & Q_{S_B,S_B} & Q_{S_B,S_P} \\ 0 & P_{S_P,S_A} & Q_{S_P,S_A} & 0 & P_{S_P,S_B} & Q_{S_P,S_B} & Q_{S_P,S_P} \end{bmatrix} \begin{Bmatrix} E_{V_A} \\ E_{S_A} \\ \overline{H}_{S_A} \\ E_{V_B} \\ E_{S_B} \\ \overline{H}_{S_B} \\ \overline{H}_{S_P} \end{Bmatrix} = \begin{Bmatrix} b_{V_A} \\ b_{S_A} \\ b_{V_B} \\ b_{S_B} \\ b_{S_A}^{\text{inc}} \\ b_{S_B}^{\text{inc}} \\ b_{S_P}^{\text{inc}} \end{Bmatrix}. \tag{11.1}$$

This system is obtained by combining the finite element matrix equations for the fields in the antenna regions V_A and V_B with the boundary matrix equation discretized from the combined-field integral equation, and the resulting matrices are similar to those defined in Section 3.3. To be more specific, the K matrices are defined by (3.92) with the exception that the volume integral is now defined over either V_A or V_B, and the B matrices are defined by (3.93) except that the surface integral is now defined over either S_A or S_B. The P and Q matrices are defined by (3.123) and (3.124), respectively, except that the surface integration in (3.123) and (3.124) is carried out over the surface defined by the first subscript and the surface integration in (3.109) and (3.110) is carried out over the surface defined by the second subscript. For the right-hand-side vector in (11.1), the top four vectors are due to internal sources and they are given by (2.19), with the exception that the volume integral is now defined over either V_A or V_B. In a scattering analysis, the bottom three vectors are due to incident field excitations on S_A, S_B, and S_P, respectively; however, in an antenna radiation analysis, these terms vanish.

The K and B matrices are highly sparse, whereas the P and Q matrices are fully populated. Therefore, the system matrix in (11.1) consists of both dense and sparse matrices. Depending on the size of the problem, (11.1) can be solved either by a direct solver when the matrix size is sufficiently small, or otherwise by iteration. If a direct solver is employed, it is preferable to reduce (11.1) into a system of a smaller size,

$$\begin{bmatrix} Q'_{S_A,S_A} & Q'_{S_A,S_B} & Q_{S_A,S_P} \\ Q'_{S_B,S_A} & Q'_{S_B,S_B} & Q_{S_B,S_P} \\ Q'_{S_P,S_A} & Q'_{S_P,S_B} & Q_{S_P,S_P} \end{bmatrix} \begin{Bmatrix} \overline{H}_{S_A} \\ \overline{H}_{S_B} \\ \overline{H}_{S_P} \end{Bmatrix} = \begin{Bmatrix} b'_{S_A} \\ b'_{S_B} \\ b'_{S_P} \end{Bmatrix} \tag{11.2}$$

where

$$\begin{bmatrix} Q'_{S_A,S_A} \\ Q'_{S_B,S_A} \\ Q'_{S_P,S_A} \end{bmatrix} = \begin{bmatrix} Q_{S_A,S_A} \\ Q_{S_B,S_A} \\ Q_{S_P,S_A} \end{bmatrix} - \begin{bmatrix} P_{S_A,S_A} \\ P_{S_B,S_A} \\ P_{S_P,S_A} \end{bmatrix} \begin{bmatrix} 0_{S_A,V_A} & I_{S_A,S_A} \end{bmatrix} \begin{bmatrix} K_{V_A,V_A} & K_{V_A,S_A} \\ K_{S_A,V_A} & K_{S_A,S_A} \end{bmatrix}^{-1} \begin{bmatrix} 0_{V_A,S_A} \\ B_{S_A,S_A} \end{bmatrix} \tag{11.3}$$

$$\begin{bmatrix} Q'_{S_A,S_B} \\ Q'_{S_B,S_B} \\ Q'_{S_P,S_B} \end{bmatrix} = \begin{bmatrix} Q_{S_A,S_B} \\ Q_{S_B,S_B} \\ Q_{S_P,S_B} \end{bmatrix} - \begin{bmatrix} P_{S_A,S_B} \\ P_{S_B,S_B} \\ P_{S_P,S_B} \end{bmatrix} \begin{bmatrix} 0_{S_B,V_B} & I_{S_B,S_B} \end{bmatrix} \begin{bmatrix} K_{V_B,V_B} & K_{V_B,S_B} \\ K_{S_B,V_B} & K_{S_B,S_B} \end{bmatrix}^{-1} \begin{bmatrix} 0_{V_B,S_B} \\ B_{S_B,S_B} \end{bmatrix} \tag{11.4}$$

$$\begin{aligned} \begin{Bmatrix} b'_{S_A} \\ b'_{S_B} \\ b'_{S_P} \end{Bmatrix} = \begin{Bmatrix} b^{\text{inc}}_{S_A} \\ b^{\text{inc}}_{S_B} \\ b^{\text{inc}}_{S_P} \end{Bmatrix} &- \begin{bmatrix} P_{S_A,S_A} \\ P_{S_B,S_A} \\ P_{S_P,S_A} \end{bmatrix} \begin{bmatrix} 0_{S_A,V_A} & I_{S_A,S_A} \end{bmatrix} \begin{bmatrix} K_{V_A,V_A} & K_{V_A,S_A} \\ K_{S_A,V_A} & K_{S_A,S_A} \end{bmatrix}^{-1} \begin{Bmatrix} b_{V_A} \\ b_{S_A} \end{Bmatrix} \\ &- \begin{bmatrix} P_{S_A,S_B} \\ P_{S_B,S_B} \\ P_{S_P,S_B} \end{bmatrix} \begin{bmatrix} 0_{S_B,V_B} & I_{S_B,S_B} \end{bmatrix} \begin{bmatrix} K_{V_B,V_B} & K_{V_B,S_B} \\ K_{S_B,V_B} & K_{S_B,S_B} \end{bmatrix}^{-1} \begin{Bmatrix} b_{V_B} \\ b_{S_B} \end{Bmatrix}. \end{aligned} \tag{11.5}$$

In these equations, subscripts have been added to both the null and identity matrices to indicate their dimensions. The calculation of (11.3)–(11.5) requires solving two purely sparse, symmetric finite element matrices, which can be accomplished efficiently by using sparse direct solvers. This direct solution procedure is especially attractive for the scattering analysis of antennas mounted on a platform.

Equation (11.1) can alternatively be solved using an iterative solver. The iterative solution approach is preferred if the number of surface unknowns on S_A, S_B, and S_P is sufficently large such that a direct solution is no longer feasible, or if we are only interested in just a few right-hand-side vectors. However, to obtain an efficient iterative solution, there are two challenges that we must overcome. The first is slow convergence because the matrix in (11.1) typically has a large condition number. This problem can be handled by using a preconditioned iterative solver. A simple preconditioner can be obtained from the inverse of the coefficient matrix simply by retaining the K matrices and discarding all other matrices. The use of this preconditioner is equivalent to solving (11.2) iteratively. A more robust preconditioner can be constructed by retaining both the K and B matrices and replacing the fully populated P and Q matrices with the corresponding sparse matrices defined by (3.128) and (3.129). As noted in Section 3.3, this highly effective preconditioner is actually the same as the inverse of the finite element system matrix obtained by replacing the combined-field integral equation with the first-order absorbing boundary condition [8]. Since this system matrix is purely sparse, it can be constructed and factorized efficiently by using a sparse direct solver. Alternatively, the associated preconditioning equation can be solved iteratively with an incomplete LU preconditioner. This results in a two-loop iteration algorithm, in which the outer loop is to solve (11.1) iteratively and the inner loop is to solve the preconditioning equation iteratively in each iteration of the outer loop.

The second challenge to obtaining an efficient iterative solution is the time-consuming evaluation of the matrix–vector multiplications involving the fully populated P and Q matrices. This challenge can be overcome by using a fast algorithm such as the adaptive integral method or the fast multipole method [9–12]. For a surface-based application, the fast multipole method is a better choice. To understand the basic idea of the fast multipole method, we first recognize that matrix–vector products involving the P and Q matrices represent interactions between the basis and testing functions. Therefore, P_{ij} and Q_{ij} as defined by (3.123) and (3.124) can be considered as the fields radiated by the basis function $\mathbf{g}_j$ and received by the testing function $\mathbf{t}_i(\mathbf{r})$. To calculate such interactions directly $O(N^2)$ operations are required, where N denotes the total number of basis or testing functions involved. The basic idea of the fast multipole method [9] for obtaining a fast evaluation of these matrix–vector products is first to divide the entire surface $S_A \cup S_B \cup S_P$ into groups. The addition theorem is then used to translate the field radiated by different basis functions within a group into a single center, and this process is called *aggregation*. By doing this, the number of radiation centers is reduced significantly from the number of basis functions to the number of groups. Similarly, for each group, the field radiated by all other group centers can be first received by its center and then redistributed to the testing functions belonging to the group. This process is called *disaggregation*. It has been shown that this one-level fast multipole method can reduce the memory requirement and computational complexity to $O(N^{1.5} \log N)$. To further reduce the memory requirement and computational complexity, we can extend the one-level algorithm to multiple levels. In this multilevel algorithm, which is known as the multilevel fast multipole algorithm [10], the entire object is first enclosed in a large box, which is divided into eight smaller boxes. Each subbox is then recursively subdivided into smaller boxes until the edge length of the smallest boxes is about one-half wavelength. For a testing and basis function spatially residing in the same or nearby smallest boxes, their interaction is calculated in a direct manner. However, when they reside in different boxes such that they are in the far field with respect to each other their interaction is calculated by the fast multipole method, as described above. The level of the boxes on which the fast multipole method is applied depends upon the spatial separation between the testing and basis functions. The efficiency of the algorithm is obviously enhanced by increasing the ratio of the far-field interactions to the near-field interactions, and this ratio can be affected by the manner in which a particular application is discretized. A detailed description of the multilevel fast multipole algorithm and its application to the FE-BI method may be found in Refs. 10–12.

The FE-BI formulation described here can be extended to deal with nonmetallic platforms such as those coated with radar-absorbing materials [13]. It can also be replicated into a time-domain setting [14]. The following example is taken from Ref. 13 to demonstrate the capability of the method. The example consists of a microstrip patch antenna housed in a cavity that resides on a platform consisting of a conducting circular cylinder and a conducting plate (wing). The microstrip

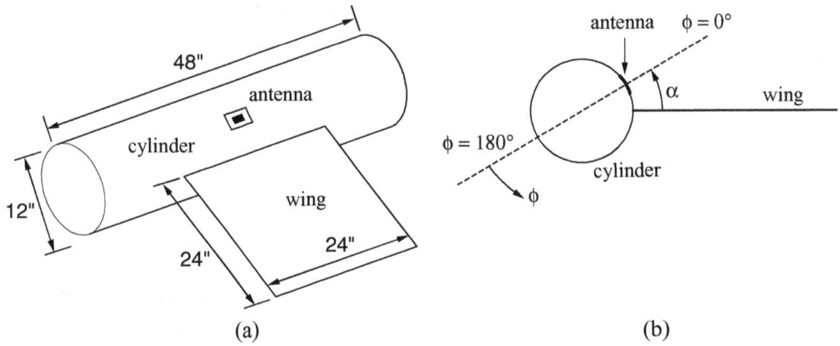

Figure 11.10 Microstrip patch antenna placed on a cylinder with a wing. (a) Entire structure. (b) Cross-sectional view. (After Liu and Jin [13], Copyright © Wiley 2003.)

patch antenna is designed to operate at 3.3 GHz, and its longer edge is aligned with the cylinder's axis. The entire structure is shown in Figure 11.10, and the detailed information about the patch antenna is given in Figure 11.11. The normalized radiation pattern in the H-plane is shown in Figure 11.12 for two cases: one has the patch antenna placed $\alpha = 28^\circ$ from the wing and the other is for $\alpha = 45^\circ$. It is

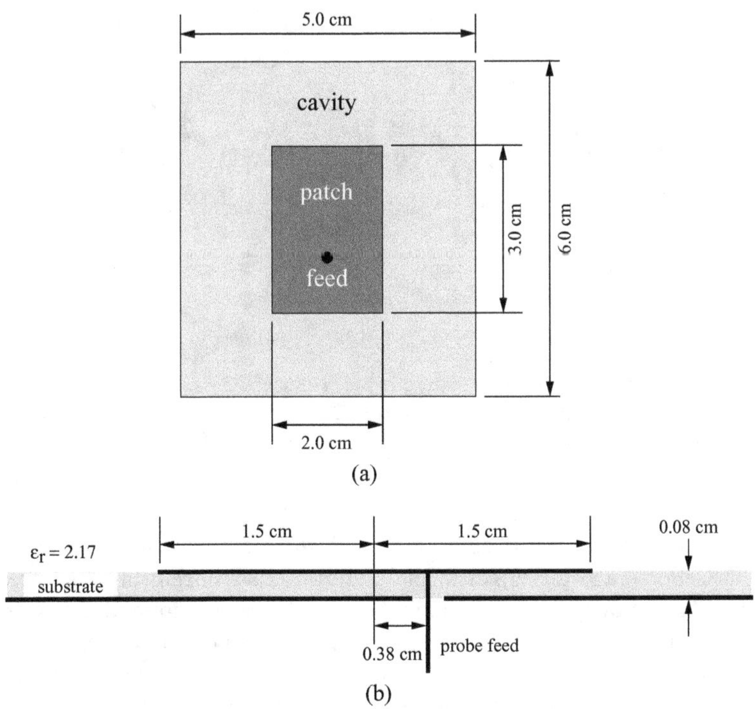

Figure 11.11 Configuration of a microstrip patch antenna. (a) Top view. (b) Side view. (After Liu and Jin [13], Copyright © Wiley 2003.)

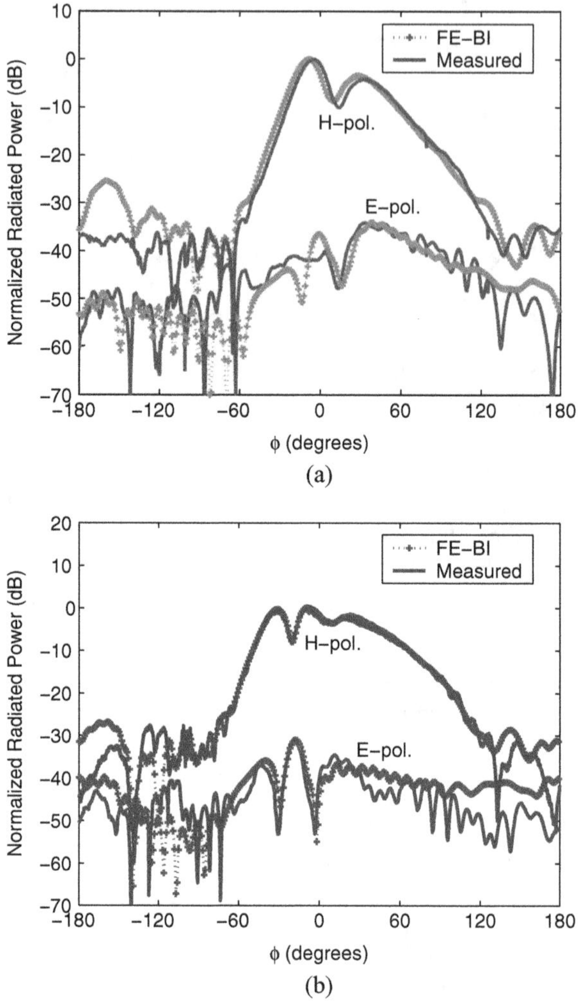

Figure 11.12 Normalized radiation pattern in the H-plane at 3.3 GHz for a microstrip patch antenna placed on a cylinder with a wing. (a) $\alpha = 28°$. (b) $\alpha = 45°$. (After Liu and Jin [13], Copyright © Wiley 2003.)

seen that the numerical results agree well with the measurement for both co- and cross-polarizations, and the effect of the platform is significant in both cases. For these numerical simulations, the total number of unknowns is approximately 60,000. Because of the use of a highly effective preconditioner, only 14 iterations are required to obtain a converged solution, and the memory used was on the order of 1 GB, due to the application of the multilevel fast multipole algorithm. This geometry has also been simulated using the time-domain FE-BI method accelerated with the time-domain adaptive integral algorithm, which yielded similar results [14].

The FE-BI formulation presented in this section requires that the finite element meshes for the antenna regions match exactly at interfaces S_A and S_B with the surface mesh used to discretize the boundary integral equation. This requirement may create challenges for some mesh generators, especially when the finite element regions and the exterior surfaces are meshed separately. The requirement can be relaxed by adopting a variety of domain-decomposition schemes. For example, when the finite element mesh does not match the exterior surface mesh at the interface, we can introduce two sets of equivalent surface currents (which are related to the tangential fields) over the interface [15]. One set is used in the finite element formulation of the interior field, and the other set is used in the boundary integral equation to formulate the exterior field. These two sets of equivalent surface currents can then be related by using the Robin-type transmission condition and field continuity conditions. They can also be expanded independently using different basis functions. In addition to the regular subdomain basis functions, we can use macro basis functions or entire-domain basis functions defined over the interface, and the field coupling at the interface can be accomplished using the generalized scattering matrix approach [16,17].

11.2 DECOUPLED ANALYSIS

In the preceding section we described techniques for a rigorous, coupled analysis of antenna–platform interactions. These techniques are very accurate, although they are computationally intensive and their implementation can be tedious. To simplify the analysis, various approximate numerical schemes have been developed. In a recent paper by Yilmaz et al. [14], one of the applications analyzed involves the radiation of a Vivaldi antenna installed on the bottom of an airplane (see Figure 5.21). The results show that although the platform has a significant effect on the radiation patterns, its effect on the input impedance is very small and can be neglected for practical purposes. This is to be expected because when the field radiated by the antenna is reflected and diffracted by the platform, the radiation patterns can be changed significantly. However, for the platform to affect the input impedance the platform has to reflect a significant amount of power back to the antenna, and the antenna has to couple the power reflected to the input port. This is unlikely to happen in many practical applications. Therefore, one of the primary goals for the analysis of antenna–platform interactions is to evaluate the effect of the platform on the radiation patterns instead of the effect of the platform on the input impedance. This evaluation can be carried out in an approximate sense by a decoupling process, which can simplify the analysis significantly. In this decoupled analysis, we first analyze the antenna by focusing on the antenna and its nearby structures to obtain the current distribution on the antenna or the radiated near field, and then compute the far-field radiation pattern by letting the current radiate in the presence of the platform or by letting the near field propagate in the presence of the platform. The computation of the radiation pattern can be carried out by either a numerical method or an asymptotic technique such as physical optics or the geometrical theory of diffraction. In this section we describe this decoupled analysis in more detail.

11.2.1 Near-Field Calculation

The first step in the decoupled analysis is either to calculate the current distribution on an isolated antenna or calculate its radiated near field. Since the finite element method works with the field directly, it is more convenient to calculate the near field. The calculation can be done using the finite element method with the computational domain truncated with an absorbing boundary condition, a perfectly matched layer, or a boundary integral equation, as discussed in Chapters 3 and 4. For a conformal antenna on a flat surface, such as antenna A in Figure 11.9, we can use the FE-BI method with the half-space Green's function discussed in Section 3.3.3. For other types of antennas, such as antenna B in Figure 11.9, we can use either an absorbing boundary condition or a perfectly matched layer to enclose the antenna locally. From the finite element solution, we can extract the tangential electric fields on S_A and S_B, which can then be converted into the equivalent surface magnetic currents, which are denoted here as $\mathbf{M}_{S_A}$ and $\mathbf{M}_{S_B}$, respectively.

11.2.2 Far-Field Evaluation by Numerical Methods

The second step is to calculate the far-field radiation based on the near fields or the equivalent surface magnetic currents obtained in the first step. For this, we let $S_o = S_A \cup S_B \cup S_P$ in (3.107) and (3.108) and note that $\hat{n}' \times \mathbf{E} = 0$ on S_P, to find

$$\mathbf{E}(\mathbf{r}) = \mathbf{E}^{\text{ex}}(\mathbf{r}) - jk_0 \oiint_{S_A \cup S_B \cup S_P} \left[\bar{\mathbf{J}}_s(\mathbf{r}')G_0(\mathbf{r}, \mathbf{r}') + \frac{1}{k_0^2} \nabla' \cdot \bar{\mathbf{J}}_s(\mathbf{r}')\, \nabla G_0(\mathbf{r}, \mathbf{r}') \right] dS' \qquad \mathbf{r} \in V_\infty \tag{11.6}$$

$$\bar{\mathbf{H}}(\mathbf{r}) = \bar{\mathbf{H}}^{\text{ex}}(\mathbf{r}) - \oiint_{S_A \cup S_B \cup S_P} \bar{\mathbf{J}}_s(\mathbf{r}') \times \nabla G_0(\mathbf{r}, \mathbf{r}')\, dS' \qquad \mathbf{r} \in V_\infty \tag{11.7}$$

where $\bar{\mathbf{H}} = Z_0 \mathbf{H}$, $\bar{\mathbf{J}}_s = Z_0 \mathbf{J}_s$, and

$$\mathbf{E}^{\text{ex}}(\mathbf{r}) = \iint_{S_A} \mathbf{M}_{S_A}(\mathbf{r}') \times \nabla G_0(\mathbf{r}, \mathbf{r}')\, dS' + \iint_{S_B} \mathbf{M}_{S_B}(\mathbf{r}') \times \nabla G_0(\mathbf{r}, \mathbf{r}')\, dS' \tag{11.8}$$

$$\begin{aligned} \bar{\mathbf{H}}^{\text{ex}}(\mathbf{r}) = &-jk_0 \iint_{S_A} \left[\mathbf{M}_{S_A}(\mathbf{r}')G_0(\mathbf{r}, \mathbf{r}') + \frac{1}{k_0^2} \nabla' \cdot \mathbf{M}_{S_A}(\mathbf{r}')\, \nabla G_0(\mathbf{r}, \mathbf{r}') \right] dS' \\ &- jk_0 \iint_{S_B} \left[\mathbf{M}_{S_B}(\mathbf{r}')G_0(\mathbf{r}, \mathbf{r}') + \frac{1}{k_0^2} \nabla' \cdot \mathbf{M}_{S_B}(\mathbf{r}')\, \nabla G_0(\mathbf{r}, \mathbf{r}') \right] dS'. \end{aligned} \tag{11.9}$$

Since $\mathbf{M}_{S_A}$ and $\mathbf{M}_{S_B}$ are known, $\mathbf{E}^{\text{ex}}$ and $\bar{\mathbf{H}}^{\text{ex}}$ can be calculated, which represent the fields radiated by $\mathbf{M}_{S_A}$ and $\mathbf{M}_{S_B}$ in free space. This field induces surface electric currents $\bar{\mathbf{J}}_s$ on S_A, S_B, and S_P, which modify the total radiated field due to the presence of the platform. Note that because S_A and S_B are not true conducting surfaces, the surface electric currents on those surfaces are actually equivalent currents, whereas the surface current on S_P is a truly induced current. Equations (11.6) and (11.7)

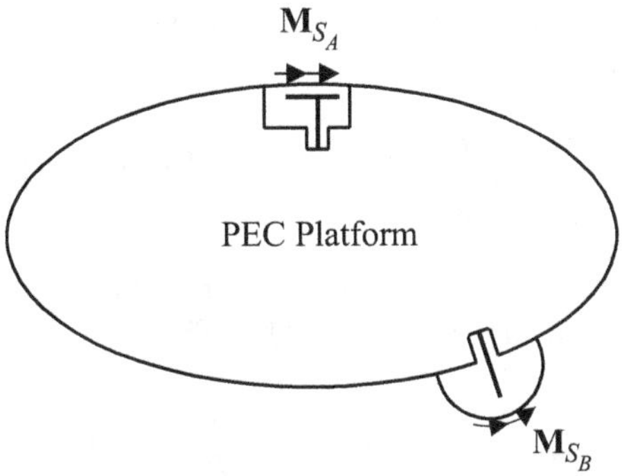

Figure 11.13 Equivalent problem for calculating the exterior fields.

are the mathematical expressions of the surface equivalence principle, which states that the exterior radiated field can be obtained by letting $\mathbf{M}_{S_A}$ and $\mathbf{M}_{S_B}$ radiate in the presence of the platform with both S_A and S_B replaced by perfect electrically conducting surfaces. The equivalent problem for Figure 11.9 is shown in Figure 11.13.

It remains to find $\overline{\mathbf{J}}_s$ on S_A, S_B, and S_P, and for this we apply (11.6) and (11.7) to S_A, S_B, and S_P to obtain

$$\hat{n} \times \mathcal{L}(\overline{\mathbf{J}}_s) = \hat{n} \times \mathbf{E}^{\text{ex}}(\mathbf{r}) + \begin{cases} \mathbf{M}_{S_A}(\mathbf{r}) & \mathbf{r} \in S_A \\ \mathbf{M}_{S_B}(\mathbf{r}) & \mathbf{r} \in S_B \\ 0 & \mathbf{r} \in S_P \end{cases} \tag{11.10}$$

$$\tfrac{1}{2}\overline{\mathbf{J}}_s(\mathbf{r}) + \hat{n} \times \tilde{\mathcal{K}}(\overline{\mathbf{J}}_s) = \hat{n} \times \overline{\mathbf{H}}^{\text{ex}}(\mathbf{r}) \qquad \mathbf{r} \in S_A, S_B, S_P. \tag{11.11}$$

These two integral equations can be discretized and solved using a procedure similar to the one described in Section 3.3, which also provides definitions for the operators $\mathcal{L}$ and $\tilde{\mathcal{K}}$. Once $\overline{\mathbf{J}}_s$ is obtained numerically, the exterior field can be calculated by using (11.6) and (11.7).

The forward calculation approach described above can be used to evaluate any exterior field, including the far field. It has the advantage of calculating the radiated field in all directions once $\overline{\mathbf{J}}_s$ is obtained by solving (11.10) and (11.11). However, the evaluation of $\mathbf{E}^{\text{ex}}$ and $\overline{\mathbf{H}}^{\text{ex}}$ in (11.10) and (11.11) must be done carefully because of the singularity that occurs when $\mathbf{r} = \mathbf{r}'$. If we are interested only in the far field, this can be obtained by solving a plane-wave scattering problem and utilizing the reciprocity theorem [18]. In this approach, an infinitesimally small current element is placed at the observation point. Because this current element is located far from the platform, the wave it produces can be regarded as a plane wave when incident on the platform. More specifically, let the field radiated by $\mathbf{M}_{S_A}$ and $\mathbf{M}_{S_B}$ be denoted as $\mathbf{E}^{\text{rad}}$, the field produced by the current element in the presence of the platform (with both S_A and S_B replaced by perfect electrically conducting surfaces) be denoted as

$\mathbf{H}_D$, and the current density of the current element be denoted as $\mathbf{J}_D$. According to the reciprocity theorem [19], we have

$$\iiint \mathbf{E}^{\mathrm{rad}} \cdot \mathbf{J}_D \, dV = -\iint_{S_A} \mathbf{H}_D \cdot \mathbf{M}_{S_A} \, dS - \iint_{S_B} \mathbf{H}_D \cdot \mathbf{M}_{S_B} \, dS. \tag{11.12}$$

Assuming that the current element has a dipole moment of Il and is located at (r, θ, ϕ) with an orientation in the $\hat{a}$-direction, then (11.12) becomes

$$\hat{a} \cdot \mathbf{E}^{\mathrm{rad}}(r, \theta, \phi) = -\frac{1}{Il} \iint_{S_A} \mathbf{H}_D \cdot \mathbf{M}_{S_A} \, dS - \frac{1}{Il} \iint_{S_B} \mathbf{H}_D \cdot \mathbf{M}_{S_B} \, dS. \tag{11.13}$$

To find the θ- and ϕ-components of the radiated field, we let $\hat{a} = \hat{\theta}$ and $\hat{\phi}$, respectively, and then use the expression of the field produced by a current element to find

$$E_\theta^{\mathrm{rad}}(r, \theta, \phi) = \frac{jk_0 Z_0 e^{-jk_0 r}}{4\pi r} \left[\iint_{S_A} \mathbf{H}^\theta \cdot \mathbf{M}_{S_A} \, dS + \iint_{S_B} \mathbf{H}^\theta \cdot \mathbf{M}_{S_B} \, dS \right] \tag{11.14}$$

$$E_\phi^{\mathrm{rad}}(r, \theta, \phi) = \frac{jk_0 Z_0 e^{-jk_0 r}}{4\pi r} \left[\iint_{S_A} \mathbf{H}^\phi \cdot \mathbf{M}_{S_A} \, dS + \iint_{S_B} \mathbf{H}^\phi \cdot \mathbf{M}_{S_B} \, dS \right] \tag{11.15}$$

where $\mathbf{H}^\theta$ and $\mathbf{H}^\phi$ denote the magnetic field induced on the surface of the platform by an incident plane wave with unit magnitude and with the electric field polarized in the $\hat{\theta}$- and $\hat{\phi}$-directions, respectively. Equations (11.14) and (11.15) can also be written in terms of the induced surface current as

$$E_\theta^{\mathrm{rad}}(r, \theta, \phi) = \frac{jk_0 e^{-jk_0 r}}{4\pi r} \left[\iint_{S_A} \overline{\mathbf{J}}_s^\theta \cdot \mathbf{E}_{S_A} \, dS + \iint_{S_B} \overline{\mathbf{J}}_s^\theta \cdot \mathbf{E}_{S_B} \, dS \right] \tag{11.16}$$

$$E_\phi^{\mathrm{rad}}(r, \theta, \phi) = \frac{jk_0 e^{-jk_0 r}}{4\pi r} \left[\iint_{S_A} \overline{\mathbf{J}}_s^\phi \cdot \mathbf{E}_{S_A} \, dS + \iint_{S_B} \overline{\mathbf{J}}_s^\phi \cdot \mathbf{E}_{S_B} \, dS \right] \tag{11.17}$$

where $\mathbf{E}_{S_A}$ and $\mathbf{E}_{S_B}$ are the electric fields calculated in the first step, as described in Section 11.2.1, and $\overline{\mathbf{J}}_s^{\theta,\phi} = Z_0 \hat{n} \times \mathbf{H}^{\theta,\phi}$.

The induced surface currents $\overline{\mathbf{J}}_s^\theta$ and $\overline{\mathbf{J}}_s^\phi$ can be obtained by solving the following integral equations using the moment-method procedure:

$$\hat{n} \times \mathcal{L}(\overline{\mathbf{J}}_s) = \hat{n} \times \mathbf{E}^{\mathrm{inc}}(\mathbf{r}) \qquad \mathbf{r} \in S_A, S_B, S_P \tag{11.18}$$

$$\tfrac{1}{2}\overline{\mathbf{J}}_s(\mathbf{r}) + \hat{n} \times \tilde{\mathcal{K}}(\overline{\mathbf{J}}_s) = \hat{n} \times \overline{\mathbf{H}}^{\mathrm{inc}}(\mathbf{r}) \qquad \mathbf{r} \in S_A, S_B, S_P. \tag{11.19}$$

Comparing these two equations with (11.10) and (11.11), we find that their left-hand sides are the same, and therefore the moment-method matrices will also be the same. The difference between (11.10), (11.11) and (11.18), (11.19) is due to the right-hand sides. In (11.18) and (11.19), the right-hand sides are simply the tangential components of the incident field. By using the $\hat{\theta}$- and $\hat{\phi}$-polarized incident fields, we can solve (11.18) and (11.19) for $\overline{\mathbf{J}}_s^{\theta}$ and $\overline{\mathbf{J}}_s^{\phi}$, respectively. Note that in this reverse calculation approach, we have to solve a scattering problem for each observation angle in order to calculate the radiation patterns, whereas in the previous approach using (11.10) and (11.11) the moment-method matrix only needs to be solved once. Therefore, this approach is efficient only when a direct solver is used to solve the moment-method matrix equation.

11.2.3 Far-Field Evaluation by Asymptotic Techniques

Because many platforms are electrically large, calculation of the far fields using a numerical method, as discussed in the preceding subsection, can be time consuming. A more efficient alternative is to use an asymptotic technique such as physical optics or the geometrical theory of diffraction [19]. In this subsection we describe the shooting- and bouncing-ray method, which is a combination of geometrical optics and physical optics.

Similar to the far-field evaluation by a numerical method, there are two approaches to using the shooting- and bouncing-ray method to compute the far-field radiation. One approach is first to compute the radiated field over a small hemispherical surface covering the antenna, as discussed in Section 11.2.1. This field is then converted into many outgoing rays which shoot along the radial directions [20]. Each ray is then traced as it bounces across the platform, and here the bounces are governed by geometrical optics. At each and every intersection point of the ray with the platform, a physical-optics type of integration is performed to determine the ray contribution to the radiated far field, which is given by

$$\mathbf{H}^{\mathrm{PO}}(\mathbf{r}) = 2\iint_{\sigma_{\mathrm{ray}}} \nabla G_0(\mathbf{r}, \mathbf{r}') \times [\hat{n}' \times \mathbf{H}^{\mathrm{ray}}(\mathbf{r}')]\, dS' \tag{11.20}$$

where σ_{ray} denotes the footprint of the ray and $\mathbf{H}^{\mathrm{ray}}$ is the magnetic field of the ray. The final result is the summation of the contributions from all rays. This process is illustrated in Figure 11.14, where only one ray is traced and calculated for clarity. As mentioned earlier, this forward calculation has the advantage of computing the radiated field in all directions simultaneously. However, to obtain accurate results, the field on the hemispherical surface surrounding the antenna must be divided into many rays, and in tracing each ray its divergence factor must be calculated and tracked, as described in Ref. 19. This approach has been implemented into the commercial AntFarm [21] antenna simulation toolkit.

The second approach is to use the reciprocity theorem to calculate the radiated field by initially carrying out the scattering analysis. The theoretical formulation is the same as that described in the preceding subsection. The equations used to calculate the radiated field are given by (11.16) and (11.17). The difference is in the calculation of $\overline{\mathbf{J}}_s^{\theta}$ and $\overline{\mathbf{J}}_s^{\phi}$, which are now calculated by using the shooting- and bouncing-ray method. For this approach, the incident wave is represented by a dense grid of parallel rays, typically about 10 to 20 rays per wavelength, which shoot

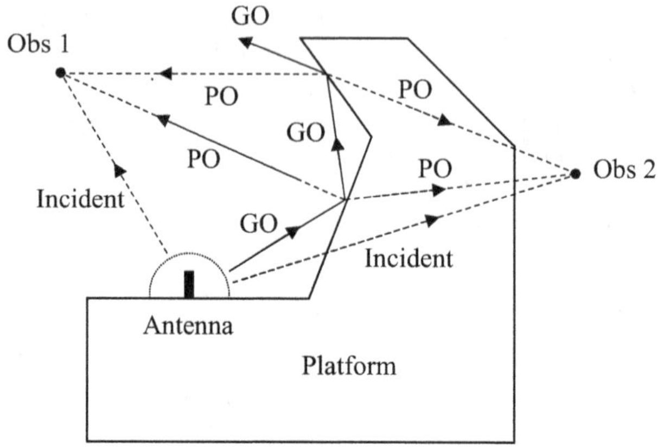

Figure 11.14 Shooting- and bouncing-ray method, which combines the geometrical optics (GO) and physical optics (PO) to calculate radiated fields at two observation points. The rays (solid lines) are traced according to GO, and a PO integration is performed at each and every hit point to evaluate the contribution to the fields at the observation points (dashed lines). For the sake of clarity, the edge diffraction is not included here.

from the observation direction and bounce across the platform (with both S_A and S_B replaced by perfect electrically conducting surfaces). As the rays hit S_A and S_B, their magnetic fields are recorded, from which the induced currents (either $\overline{\mathbf{J}}_s^{\theta}$ or $\overline{\mathbf{J}}_s^{\phi}$, depending on the polarization of the incident field) are computed, which are then used in (11.16) and (11.17) for calculation of the radiated far field.

Although the principle of the shooting- and bouncing-ray method, as illustrated in Figure 11.15, is simple and the method has been implemented in the XPATCH [22] scattering simulation code, its use for the calculation of $\overline{\mathbf{J}}_s^{\theta}$ and $\overline{\mathbf{J}}_s^{\phi}$ needs to be implemented carefully to obtain accurate results, because S_A and S_B are part of the platform. First, it is recognized that the major contribution to $\overline{\mathbf{J}}_s^{\theta}$ and $\overline{\mathbf{J}}_s^{\phi}$ is the direct illumination. This field can be calculated efficiently by launching a ray from the field point on S_A and S_B toward the observation direction. If the ray is not intercepted by the platform, the field point is illuminated directly and a direct incident field is then added to the total magnetic field at that point. This calculation is rather straightforward. It is the calculation of the field reflected by the platform to S_A or S_B that requires extra care. In practical applications, some platforms may have curved surfaces instead of planar surfaces. When a curved surface is represented by a faceted model, each facet actually represents a small portion of the curved surface, although each facet itself is either flat or planar. Thus, when a ray hits one of these curved facets, its reflected ray tube must either converge (when reflected from a concave surface), diverge (when reflected from a convex surface), or do a combination of the two (when reflected from a saddle point) in order to model the physics accurately. Neglecting the effect of the curved facets on ray tubes will lead to an erroneous calculation of the near field, although its effect on the far-field calculation is insignificant because the latter

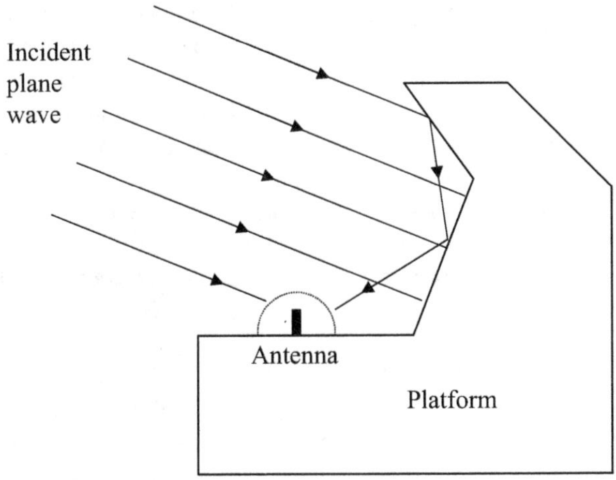

Figure 11.15 Shooting- and bouncing-ray method to calculate the field over a surface that encloses an antenna. A grid of parallel rays is shooting from the observation direction. As the rays are bounced around the platform, some hit the surface enclosing the antenna and contribute to the surface magnetic field. The edge diffraction is not included here for the sake of clarity.

is calculated by integrating the surface field as described above. For example, the reflected ray tubes may fail to cover the entire space and have large gaps between the adjacent ray tubes. A simple method to solve this problem is to multiply each ray tube by a diverging factor determined at the hit point of the center of the tube so that each ray will diverge as it leaves the curved surface. This method works satisfactorily for a surface having a constant curvature such as a cylindrical and spherical surface. However, most curved surfaces have different curvatures at different points. For such surfaces, the simple method described above leads to gaps or false overlapping between adjacent ray tubes, resulting in errors in the field calculation. To solve this problem, we trace the four corner rays of a ray tube separately instead of tracing the centerline of a ray tube and determining the reflecting direction and diverging factor based on the hit point of the center of the ray tube [23]. Each corner ray is reflected into the direction based on its hit point, and the reflected ray tube is then formed by connecting the four corners. The diverging factor used to calculate the field is determined from the ratio of the cross section of the ray tube at the field point to its original size.

The procedure described above works well in most cases; however, some challenges arise due to the finite round-off errors inherent in the numerical computations. Ideally, the grid of ray tubes launched into the problem domain is intended to cover the space uniformly, but in cases where a point lies either on or very near the border of two ray tubes, it can often be either double-counted (hit by two adjacent ray tubes) or missed completely. This can lead to inaccurate field calculations. The solution to this problem is to redefine the ray tube and introduce ray-tube basis functions [23]. Instead of having a grid of barely touching ray tubes with a constant magnitude within each ray tube, the new ray tubes are defined with twice the width and twice the height, so that the ray tubes are overlapping. To ensure a uniform incident field, a ray-tube basis function is then introduced such that at any point in space the contributions from all overlapping tubes add up to unity. This basis function can be written as $f(u, v) = (1 - |u|)(1 - |v|)$ for a ray tube that extends over $-1 \leq u \leq 1$ and $-1 \leq v \leq 1$ in normalized ray-tube coordinates. Clearly, the magnitude is 1 at the center of the ray tube and decreases to zero at the edges of the tube. With this modification, we can accurately calculate the field reflected to S_A and S_B by the platform.

When S_A and S_B reside in the shadow region of the incident and reflected fields, the calculation described above will yield $\overline{\mathbf{J}}_s^{\theta} = \overline{\mathbf{J}}_s^{\phi} = 0$. To obtain a more accurate value for $\overline{\mathbf{J}}_s^{\theta}$ and $\overline{\mathbf{J}}_s^{\phi}$ for this case, we can include the contribution from the fields diffracted by the edges on the platform. To show how the edge diffraction is calculated, we first break an arbitrary edge into short segments, then calculate the diffracted field from each segment, and finally, add the contributions from all the segments to find the diffracted field from the entire edge. The diffracted field from a short edge segment is computed using the incremental length diffraction coefficients, and the following description is adapted from Ref. 24. Consider an arbitrary wedge as illustrated in Figure 11.16, where a wave is incident upon the edge from the $\hat{s}'$-direction and we are interested in calculating the diffracted field at point P. According to the method

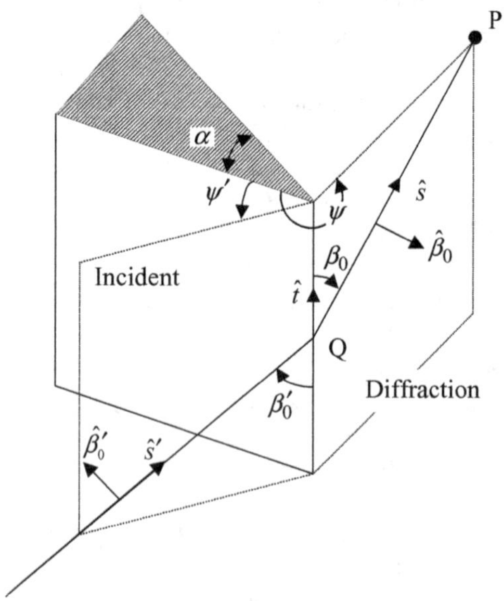

Figure 11.16 Geometry of edge diffraction. Note that ψ' and ψ are measured from the illuminated face of the edge so that $0 < \psi' < \pi$ and $0 < \psi < n\pi$, where $\alpha = (2 - n)\pi$.

of incremental length diffraction coefficients [25,26], the electric field diffracted by a short edge segment of length l located at Q is given by

$$\begin{Bmatrix} E_\theta^d(P) \\ E_\phi^d(P) \end{Bmatrix} = -\frac{1}{s}\begin{bmatrix} d_{\theta\theta} & d_{\theta\phi} \\ d_{\phi\theta} & d_{\phi\phi} \end{bmatrix}\begin{Bmatrix} E_\theta^{\text{inc}}(Q) \\ E_\phi^{\text{inc}}(Q) \end{Bmatrix} e^{-jk_0 s} \tag{11.21}$$

where s denotes the distance from the segment at Q to the field point at P, and $E_\theta^{\text{inc}}(Q)$ and $E_\phi^{\text{inc}}(Q)$ denote the components of the incident field at Q. Furthermore, d_{uv} $(u, v = \theta, \phi)$, the incremental length diffraction coefficients, are given by

$$d_{uv} = \frac{l}{4\pi} e^{jk_0 l(\hat{s}-\hat{s}')\cdot\hat{t}}[(\hat{u}\cdot\hat{t})(\hat{t}\cdot\hat{v})D_s + \hat{u}\cdot(\hat{t}\times\hat{s})(\hat{t}\times\hat{s}')\cdot\hat{v}D_h]\,\text{sinc}\frac{k_0 l(\hat{s}-\hat{s}')\cdot\hat{t}}{2} \tag{11.22}$$

where sinc $x = \sin x/x$ and $D_{s,h}$ are diffraction coefficients defined in terms of the Keller–Ufimtsev diffraction coefficients [26]

$$D_{s,h} = 2e^{j\pi/4}\sqrt{\frac{2\pi k_0}{\sin\beta_0'\sin\beta_0}}D_{s,h}^{\text{dif}}. \tag{11.23}$$

If the interior wedge angle is denoted as $\alpha = (2-n)\pi$, the $D_{s,h}^{\text{dif}}$ are then [19,27]

$$D_{s,h}^{\text{dif}} = -\frac{e^{-j\pi/4}}{2n\sqrt{2\pi k_0\sin\beta_0'\sin\beta_0}}(D_1 \mp D_2) \tag{11.24}$$

where

$$D_1 = \cot\frac{\pi+(\psi-\psi')}{2n}F[k_0 L g^+(\psi-\psi')] + \cot\frac{\pi-(\psi-\psi')}{2n}F[k_0 L g^-(\psi-\psi')] \tag{11.25}$$

$$D_2 = \cot\frac{\pi+(\psi+\psi')}{2n}F[k_0 L g^+(\psi+\psi')] + \cot\frac{\pi-(\psi+\psi')}{2n}F[k_0 L g^-(\psi+\psi')]. \tag{11.26}$$

In the equations above, the function $F[\bullet]$ involves a Fresnel integral, which is defined by

$$F[x] = 2j\sqrt{x}e^{jx}\int_{\sqrt{x}}^{\infty} e^{-jt^2}\,dt \tag{11.27}$$

and the other parameters are given by $L = s\sin\beta_0'\sin\beta_0$ and

$$g^{\pm}(X) = 1 + \cos(X - 2n\pi N^{\pm}) \tag{11.28}$$

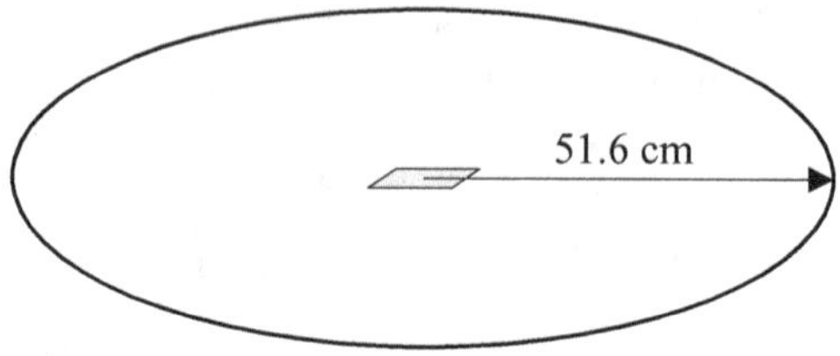

Figure 11.17 Microstrip patch antenna mounted on a finite circular ground plane.

where $N^{\pm}$ are the integers that most closely satisfy $2n\pi N^{\pm} = X \pm \pi$. Once the diffracted electric field is calculated from (11.21), the diffracted magnetic field can be found through the use of Maxwell's equations.

The edge-diffraction computation described above can be used in the forward calculation as well. For most practical applications, only the first-order edge diffraction, which is the diffraction from the edges directly illuminated by the incident wave, needs to be considered. The edge diffraction by multiply bounced and diffracted rays typically has a rather small contribution and can usually be neglected, which makes the numerical implementation more manageable and the computation more efficient.

To demonstrate the performance of the reverse calculation approach described above, we provide two examples [24]. The first example is for a cavity-backed microstrip patch antenna placed at the center of a finite, circular ground plane, as illustrated in Figure 11.17. The patch antenna has the configuration shown in Figure 11.11, which resonates at 3.3 GHz. The computed E- and H-plane radiation patterns are shown in Figure 11.18 and compared with the radiation patterns of the same antenna residing on an infinitely large ground plane. The difference between the two patterns is due primarily to edge diffraction, which is more pronounced in the E-plane pattern.

The second example is the waveguide-fed trihedral shown in Figure 11.19. The incident waveguide mode is TE_{10} and the frequency is 9.0 GHz. The computed E- and H-plane radiation patterns are compared with measured data in Figure 11.20, showing good agreement. For this geometry, the E-plane radiation pattern can be calculated by using a two-dimensional technique if the vertical side plate is removed, and the H-plane radiation pattern can be calculated by using a two-dimensional technique if the bottom plate is removed. The two-dimensional results are given in Ref. 18 and generally agree well with the three-dimensional results, with the exceptions that the variation is missing in the region from 40° to 90° in the H-plane pattern and from −90° to −40° in the E-plane pattern. These variations are caused primarily by the diffraction from the side edges, which are not modeled in the two-dimensional calculations. The fact that these variations are predicted correctly in the three-dimensional analysis is a clear indication of the validity of the edge diffraction calculation described in this subsection.

Although accurate radiation patterns can be computed efficiently by using the method discussed in this subsection, we emphasize that asymptotic (ray optics)

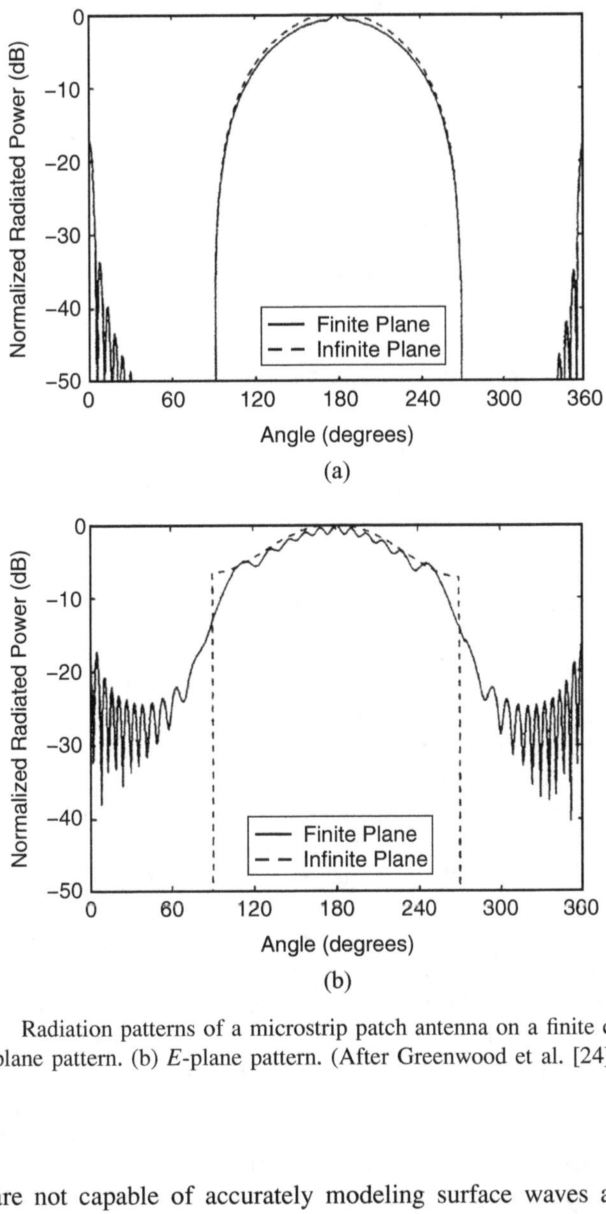

Figure 11.18 Radiation patterns of a microstrip patch antenna on a finite circular ground plane. (a) H-plane pattern. (b) E-plane pattern. (After Greenwood et al. [24], Copyright © Wiley 1996.)

techniques are not capable of accurately modeling surface waves and near-field scattering for the general case of complex, noncanonical geometry. Consequently, the approach described above cannot accurately simulate the mutual coupling of antennas on general platforms when surface wave and near-field physics are important, which is often the case when there is no line of sight between the antennas. In this situation, to carry out the analysis it is necessary to use a first-principle method such as the one described in the preceding subsection.

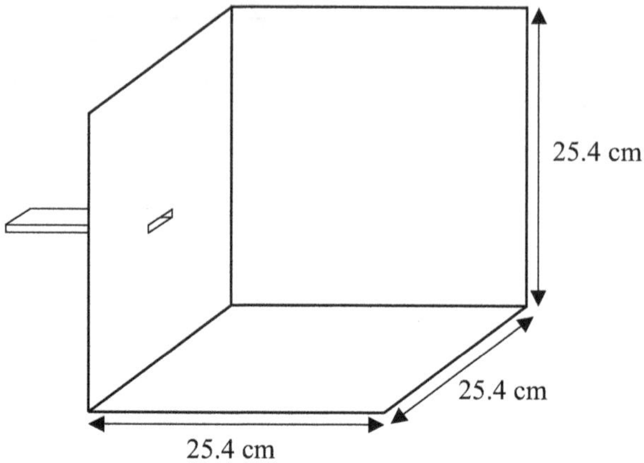

Figure 11.19 Waveguide-fed trihedral made of three 25.4-cm × 25.4-cm square conducting plates. The waveguide opening has the dimensions 2.286 cm × 1.016 cm.

11.2.4 Direct and Iterative Improvements

The decoupled analysis approach discussed in this chapter incorporates the effect of the platform on the radiation pattern but ignores the effect of the platform on the antenna itself. An improvement to this approach is to incorporate the first- or higher-order effect of the platform on the antenna by considering the reflected field by the platform as an incident field on the antenna. There are two ways to accomplish this. If the analysis employs the FE-BI method, this can be accomplished by modifying the Green's function in the boundary integral equation to include the presence of the platform [28,29], similar to the formulation of the hybrid moment and asymptotic methods for antenna analysis [30]. For example, in Ref. 28 the Green's function is modified by using the geometrical theory of diffraction [31] to account for edge diffraction, and in Ref. 29 the modification is made by using the more accurate uniform theory of diffraction [32]. Alternatively, we can first let the antenna radiate in free space and then compute the field reflected back to the antenna by the platform using an asymptotic technique. The analysis is then repeated by considering the reflected field as the excitation to the antenna. This iterative process can be repeated until a converged solution is obtained. In practice, a single iteration is often sufficient unless the antenna is placed in a highly resonant platform. This approach has been tested for scattering analyses by using the shooting- and bouncing-ray method as the asymptotic technique [23], and for antenna radiation analyses by using physical optics and the uniform theory of diffraction [33]. For example, in Ref. 33 the analysis of antennas is carried out iteratively by using the finite element method in conjunction with a Robin boundary condition calculated by boundary integral equations. Specifically, the field reflected and diffracted by electrically large conducting objects is calculated by using physical optics and the uniform theory of diffraction, and this contribution is then added to the field computed by the boundary integral equations.

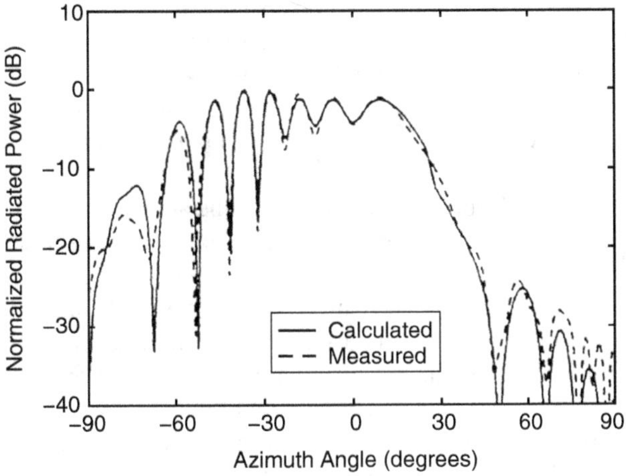

(a)

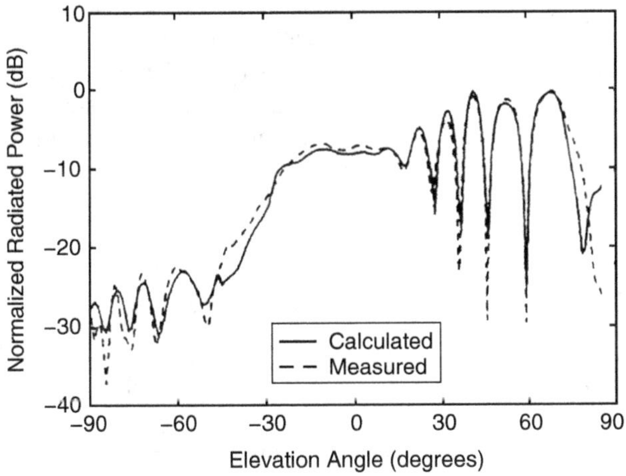

(b)

Figure 11.20 Radiation patterns of a waveguide-fed trihedral at 9.0 GHz. (a) H-plane pattern. (b) E-plane pattern. (After Greenwood et al. [24], Copyright © Wiley 1996.)

11.3 SUMMARY

In this chapter we discussed the numerical modeling of the interactions between antennas and the host platform. In particular, we described two approaches to analyzing the effect of the platform on the performance of the installed antennas. The first approach models the entire problem as a single object and applies a numerical method for a rigorous, coupled analysis. Three numerical methods were described, including the domain-decomposition implementation of the finite element method, the hybrid finite element and finite-difference time-domain technique, and the hybrid finite element and boundary integral method accelerated using the fast multipole algorithm. Because the fully coupled analysis is based on first-principle techniques, it can be used to accurately calculate both the antenna radiation patterns and the mutual coupling between the antennas installed on platforms having arbitrary geometry and materials. However, the primary drawback of this approach is that the computation is intensive.

The second approach decouples the analysis of the antennas from that of the platform by first simulating the isolated antennas to obtain their near fields, and then calculating the far-field radiation patterns in the presence of the platform by using these near fields as the excitation mechanism. The far-field calculation can be carried out by using either a numerical method such as the moment method or an asymptotic technique. In both cases, the far field can be calculated in either a forward or a reverse manner. The forward calculation is more efficient, although its implementation is more complicated. The reverse calculation based on the reciprocity theorem is simpler, although it can be more computationally time consuming. The application of asymptotic techniques was described in detail by using the shooting- and bouncing-ray method as an example. When an asymptotic technique is employed, the decoupled analysis becomes very efficient and can handle large platforms. The only drawback is that it cannot accurately characterize the mutual coupling between antennas, especially those which are not placed in the line of sight. Finally, we described briefly approaches to incorporate multiple interactions between the antennas and the platform that can improve the accuracy of the decoupled analysis.

REFERENCES

1. C. Farhat, S. Lanteri, and H. D. Simon, "TOP/DOMDEC: a software tool for mesh partitioning and parallel processing," *Comput. Syst. Eng.*, vol. 6, no. 1, pp. 13–26, Feb. 1995.
2. Y. J. Li and J. M. Jin, "Parallel implementation of the FETI–DPEM method for 3-D electromagnetic simulations," *IEEE AP-S Int. Symp. Dig.*, p. 306.7, San Diego, CA, July 2008.
3. S. K. Mazumdar, J. E. Lumpp, and S. D. Gedney, "Performance modeling of the finite-difference time-domain method on high performance parallel systems," *Appl. Comput. Electromagn. J.*, vol. 13, no. 2, pp. 147–159, 1998.
4. L. Taricone and A. Esposito, *Grid Computing for Electromagnetics*. Boston: Artech House, Chap. 4, 2004.
5. U. Andersson and G. Ledfelt, "Large scale FD-TD: a billion cells," *Proc. 15th Annu. Rev. Prog. Appl. Comput. Electromagn.* (*ACES*), Monterey, CA, pp. 572–577, Mar. 1999.
6. D. Riley, N. Riley, M. Pasik, J. Kotulski, and C. Turner, "Analysis of airframe-mounted antennas using parallel and hybridized finite-element time-domain methods," *IEEE AP-S Int. Symp. Dig.*, vol. 2, pp. 168–171, June 2002.
7. G. L. Hennigan and J. N. Shadid, "NEMESIS I: a set of functions for describing unstructured finite-element data on parallel computers," User's Manual, Sandia National Laboratories, Albuquerque, NM, Dec. 2, 1998.

8. J.-M. Jin, *The Finite Element Method in Electromagnetics*, 2nd ed. Hoboken, NJ: Wiley, 2002.
9. R. Coifman, V. Rokhlin, and S. Wandzura, "The fast multipole method for the wave equation: a pedestrian prescription," *IEEE Antennas Propagat. Mag.*, vol. 35, pp. 7–12, June 1993.
10. J. M. Song and W. C. Chew, "Multilevel fast-multipole algorithm for solving combined field integral equations of electromagnetic scattering," *Microwave Opt. Tech. Lett.*, vol. 10, no. 1, pp. 14–19, Sept. 1995.
11. X. Q. Sheng, J. M. Jin, J. M. Song, C. C. Lu, and W. C. Chew, "On the formulation of hybrid finite element and boundary integral methods for 3-D scattering," *IEEE Trans. Antennas Propagat.*, vol. 46, no. 3, pp. 303–311, Mar. 1998.
12. W. C. Chew, J. M. Jin, E. Michielssen, and J. M. Song, Eds., *Fast and Efficient Algorithms in Computational Electromagnetics*. Norwood, MA: Artech House, 2001.
13. J. Liu and J. M. Jin, "Analysis of conformal antennas on a complex platform," *Microwave Opt. Tech. Lett.*, vol. 36, no. 2, pp. 139–142, Jan. 2003.
14. A. E. Yilmaz, Z. Lou, E. Michielssen, and J. M. Jin, "A single-boundary, implicit, and FFT-accelerated time-domain finite element–boundary integral solver," *IEEE Trans. Antennas Propagat.*, vol. 55, no. 5, pp. 1382–1397, May 2007.
15. M. Vouvakis, K. Zhao, S. M. Seo, and J. F. Lee, "A domain decomposition approach of non-conformal couplings between finite and boundary elements for unbounded electromagnetic problems in R^3," *J. Comput. Phys.*, vol. 225, no. 1, pp. 975–994, July 2007.
16. J. Rubio, M. A. Gonzalez, and J. Zapata, "Generalized-scattering-matrix analysis of a class of finite arrays of coupled antennas by using 3-D FEM and spherical mode expansion," *IEEE Trans. Antennas Propagat.*, vol. 53, no. 3, pp. 1133–1144, Mar. 2005.
17. A. Barka and P. Caudrillier, "Domain decomposition method based on generalized scattering matrix for installed performance of antennas on aircraft," *IEEE Trans. Antennas Propagat.*, vol. 55, no. 6, pp. 1833–1842, June 2007.
18. J. M. Jin, J. A. Berrie, R. Kipp, and S. W. Lee, "Calculation of radiation patterns of microstrip antennas on cylindrical bodies of arbitrary cross section," *IEEE Trans. Antennas Propagat.*, vol. 45, no. 1, pp. 126–132, Jan. 1997.
19. C. A. Balanis, *Advanced Engineering Electromagnetics*. New York: Wiley, Chap. 13, 1989.
20. T. Özdemir, M. W. Nurnberger, J. L. Volakis, R. Kipp, and J. Berrie, "A hybridization of finite-element and high-frequency methods for pattern prediction for antennas on aircraft structures," *IEEE Antennas Propagat. Mag.*, vol. 38, pp. 28–37, June 1996.
21. "AntFarm^{TM} Antenna Simulation Toolkit," SAIC-DEMACO, Champaign, IL, http://www.saic.com/products/software/antfarm/.
22. D. Andresh, M. Hazlett, S. W. Lee, D. D. Reeves, D. P. Sullivan, and Y. Chu, "XPATCH: a high-frequency electromagnetic-scattering prediction code and environment for complex three-dimensional objects," *IEEE Antennas Propagat. Mag.*, vol. 36, pp. 65–69, Feb. 1994.
23. J. M. Jin, F. Ling, S. T. Carolan, J. M. Song, W. C. Gibson, W. C. Chew, C. C. Lu, and R. Kipp, "A hybrid SBR/MoM technique for analysis of scattering from small protrusions on a large conducting body," *IEEE Trans. Antennas Propagat.*, vol. 46, no. 9, pp. 1349–1356, Sept. 1998.

24. A. D. Greenwood, S. S. Ni, J. M. Jin, and S. W. Lee, "Hybrid FEM/SBR method to compute the radiation pattern from a microstrip patch antenna in a complex geometry," *Microwave Opt. Tech. Lett.*, vol. 13, no. 2, pp. 84–87, Oct. 1996.
25. K. M. Mitzner, "Incremental length diffraction coefficients," Tech. Rep. AFAL-TR-73-296, Aircraft Division, Northrop Corporation, Los Angeles, CA, Apr. 1974.
26. S. W. Lee and S. K. Jeng, "NcPTD-1.2: a high frequency near-field RCS computation code based on the physical theory of diffraction," Tech. Rep. DI-SC-90-012, DEMACO, Champaign, IL, June 1991.
27. S. K. Jeng and S. W. Lee, "Two new diffraction coefficients for physical theory of diffraction," Electromagn. Lab. Sci. Rep. 94-2, University of Illinois, Urbana-Champaign, IL, Jan. 1994.
28. C. J. Reddy, M. D. Deshpande, C. R. Cockrell, and F. B. Beck, "Radiation characteristics of cavity-backed aperture antennas in finite ground plane using the hybrid FEM/MoM technique and geometrical theory of diffraction," *IEEE Trans. Antennas Propagat.*, vol. 44, no. 10, pp. 1327–1333, Oct. 1996.
29. M. Alaydrus, V. Hansen, and T. F. Eibert, "Hybrid^2: combining the three-dimensional hybrid finite element–boundary integral technique for planar multilayered media with the uniform geometrical theory of diffraction," *IEEE Trans. Antennas Propagat.*, vol. 50, no. 1, pp. 67–74, Jan. 2002.
30. G. A. Thiele, "Overview of selected hybrid methods in radiating system analysis," *Proc. IEEE*, vol. 80, pp. 66–78, Jan. 1992.
31. J. B. Keller, "Geometrical theory of diffraction," *J. Opt. Soc. Am.*, vol. 52, no. 2, pp. 116–130, Feb. 1962.
32. R. G. Kouyoumjian and P. H. Pathak, "A uniform geometrical theory of diffraction for an edge in a perfectly conducting surface," *Proc. IEEE*, vol. 62, pp. 1448–1461, Nov. 1974.
33. R. Fernandez-Recio, L. E. Garcia-Castillo, I. Gomez-Revuelto, and M. Salazar-Palma, "Fully coupled multi-hybrid FEM-PO/PTD-UTD method for the analysis of radiation problems," *IEEE Trans. Magn.*, vol. 43, pp. 1341–1344, Apr. 2007.

12 Numerical and Practical Considerations

After reading Chapters 1 to 11, readers should have a solid understanding of the basic principles, technical issues, and solution technologies for the finite element analysis of complex antennas and arrays. In this chapter we discuss various numerical and practical considerations regarding the choice of simulation techniques, the applicability of frequency- or time-domain formulations, ensuring numerical convergence, the application of domain decomposition combined with parallel computing, and finally, some observations on the verification and validation of predictions.*

12.1 CHOICE OF SIMULATION TECHNOLOGIES

Over the past several decades, many electromagnetic simulation techniques have been developed for antenna analysis. A wide variety of radio-frequency (RF) prediction tools are available to antenna engineers, including integral and differential equation–based solvers implemented in both frequency and time domains, as well as high-frequency solvers that are useful when the physical size of the geometry is electrically very large. Every validated analysis approach has a place in the antenna engineer's suite of tools. The discussion in this section is intended to support antenna design by identifying specifically when frequency- and time-domain finite element formulations are a preferred choice of analytic tool. Due to the many competing considerations, it is difficult to provide hard and fast rules; therefore, only general guidance is provided.

The attraction of the finite element solvers is their generality. With a good mesh generator, the finite element method can model arbitrarily complex geometries that have very general material parameters. As an example, a wideband sinuous antenna, etched on a thick conical dielectric shell and fed by coaxial lines, would be a good candidate for the time-domain finite element method, or possibly the frequency-domain finite element method, while it may be challenging for many other solution methods. The reasons for this are the presence of the material regions along with a

*This chapter is expanded from Section 7 of the book chapter "Finite Element Analysis and Modeling of Antennas" in *Modern Antenna Handbook*, C. Balanis, Ed., Wiley, Hoboken, NJ, 2008.

very wide range of element sizes, as well as the absence of a known Green's function that describes the conical substrate. The price that one pays for the generality with finite element approaches is the increased time required to build volumetric meshes compared to the surface meshes generally used by integral equation–based moment methods.

When selecting an analysis approach, it is important to understand both the capabilities and limitations of the techniques and codes that are available. Capabilities and limitations may depend either on a code's specific implementation or on the fundamental underlying theory. For instance, although theory exists to allow thin wires, thin slots, and thin materials to be accurately included in the finite element analysis tools without directly meshing these geometrical details, a particular computer code may not have implemented that theory. Moreover, although it may, in principle, be possible to mesh a particular antenna geometry, the specific mesh generator integrated into a software package may be challenged by an especially complex geometry. In practice, limitations are often experienced as excessively long setup and/or run times.

Although any of a variety of analysis tools may be able to provide accurate solutions, it is important to choose the tool that provides the necessary accuracy within reasonable computer execution times. For instance, although predicting coupling between two monopoles on a large structure may be possible using either frequency- or time-domain finite element methods, run times for high-frequency methods may be orders of magnitude faster and should be used if acceptable accuracy can be demonstrated. However, it is important to keep in mind that high-frequency methods may not provide sufficient accuracy when one antenna is deep in the shadow region, and in such a case, the finite element method or other numerical methods might be required to predict coupling to antennas in regions of the structure where there is no line of sight.

It is also important to realize that the appropriate choice of tool for a specific problem changes slowly as new capabilities are formulated within the various classes of solvers. For example, simulation of wide-angle scanning of phased arrays to nearly 90° has recently become possible in the time-domain finite element method, making it an excellent choice for an analysis of wideband phased arrays. Another example is the acceleration of integral-equation solvers using the fast multipole method. This acceleration now makes integral-equation solvers a practical choice for applications that are moderately electrically large. In addition, research currently being performed to add general materials to the fast multipole method will probably further increase the class of problems to which integral-equation solvers will apply in the future. Finally, as discussed in Chapters 10 and 11, advancements in domain-decomposition techniques for the finite element method now permit the method to efficiently address electrically large applications such as large finite arrays in both the time and frequency domains.

12.2 FREQUENCY- VERSUS TIME-DOMAIN SIMULATION TOOLS

When beginning a simulation, it is useful to have a desired goal in mind to narrow the solution space and guide the simulation approach. The antenna engineer's expectation of analysis tools often differs from one application to another. For instance, when predicting impedance and radiation patterns of an existing antenna design, physical insight may not be needed, whereas if a novel class of antenna is being developed, physical insight is of great importance. Therefore, before choosing a simulation tool, some of the first questions to ask are: What information is needed from the predictions? Does one need simply to predict only impedance and radiation patterns of an existing antenna? Or is the goal of the analysis to explain why a particular antenna has specific measured characteristics, such as a high cross-polarization value?

When it is necessary to fully understand the underlying physics, both the frequency- and time-domain responses of the antenna provide useful and different insights. A frequency-domain simulation provides direct visualizations of how field lines and Poynting vectors in the near-field region vary spatially at a given frequency. For example, a visualization of strong field lines coupling the vertical polarization port of a printed antenna to the horizontal polarization port might guide design improvements in cross-polarization performance of that antenna. Alternatively, time-domain predictions provide additional insight into where physical phenomena occur within an antenna model, in the same way that time-domain reflectometry measurements provide knowledge of the locations from where signals are reflected by discontinuities along a transmission line. For example, edge reflections from finite arrays can be clearly observed as reflected waves move across an array and enter individual radiator feeds. Distances to reflections and other physical phenomena that cause the impedance of wideband antennas to vary with frequency are easily recognized. Knowledge of those distances guides the antenna designer by providing valuable insight into the locations of the reflections as well as the coupling to adjacent sources, where both affect wideband impedance.

A good example of the use of a frequency-domain finite element solution method is for the case of electrically small resonant antennas, because this method does not suffer numerical challenges due to the small electrical size until perhaps the zero-frequency limit is approached. On the other hand, time-domain approaches may not be well suited to electrically small antennas because these resonant structures often require exceedingly long run times to reach a steady solution state.

The analysis of wideband phased arrays, however, is an excellent application for the time-domain finite element method because of its efficiency for this type of application. Wideband predictions are obtained with a single simulation, often in less time than is required to sample the frequency band using a frequency-domain finite element method. In addition, due to its unconditional stability, the time-domain finite element method provides reasonable run times even at wide scan angles when performing a unit cell analysis. Because phased arrays are designed to operate without grating lobes, a well-designed infinite phased array radiates a single plane wave within its intended band of operation. The Floquet absorbing boundary conditions described in Chapter 9 for both time- and frequency-domain finite element methods, which are specialized to absorb energy nearly perfectly when a plane wave is incident from a known scan angle up to nearly 90°, can significantly improve the prediction of scan impedance, particularly over wide bandwidths. Furthermore, the fine frequency resolution available in a time-domain simulation through the FFT ensures that sharp resonant anomalies, sometimes found in flared-notch arrays and log-periodic antennas, are captured.

In both frequency- and time-domain formulations, various interpolation and extrapolation strategies can be useful. For example, frequency interpolation has been implemented in some frequency-domain finite element solvers to increase efficiency when many frequencies are of interest [1–3]. Similarly, a postprocessing extrapolation of the time-domain response into the late time may be possible [4]. This extrapolation improves efficiency of time-domain methods for highly resonant antennas, which would otherwise require long run times to obtain accurate antenna performance characteristics. While such methods extend the usefulness of a particular code, additional care is required to ensure that these techniques are applicable and properly used. In addition, visualizations of all field quantities may not be available for the interpolated frequencies or time steps. However, by adopting various interpolation and extrapolation strategies, it is often possible to extend the utility of a given method so that it can address a design problem that is slightly outside the primary strengths of the method.

12.3 FAST FREQUENCY SWEEP

Many antenna designs require computation of frequency responses over a broad band rather than at one or a few isolated frequencies. Such calculations can be very time consuming when a traditional frequency-domain numerical method is used because a matrix equation must be solved repeatedly at many frequencies. The number of calculations is proportional to the electrical size and the resonance of the problem and can be large for many applications. Therefore, there is a need to find approximate solution techniques that can simulate a frequency response efficiently over a broad band. This can be accomplished by the method of asymptotic waveform evaluation [1]. In this method, the unknown solution vector, the right-hand vector, and the system matrix are first expanded into a Taylor series at a chosen frequency. The expansion coefficients of the solution vector are then determined by moment matching. The Taylor series of the solution vector is then converted into a Padé rational function to broaden the radius of convergence. With this approach, we obtain a solution that is accurate at frequencies near the point of expansion. The accuracy of the solution decreases gradually when the frequency moves away from the point of expansion. In many practical applications, a solution is required over a specified frequency band, where one point of expansion may not be sufficient, and multiple points of expansion become necessary. These points can be selected automatically using a simple binary search algorithm [1]. In this algorithm, the asymptotic waveform evaluation method is first applied to the two ends of the frequency band, denoted by $f_{\min}$ and $f_{\max}$, respectively. The two corresponding solutions are then compared at the midfrequency point of the band, denoted as f_{mid}. If the difference exceeds a user-determined tolerance, the asymptotic waveform evaluation method is applied at the midpoint f_{mid}, and its solution is then compared with that of the start point, $f_{\min}$, at their midpoint, $(f_{\min} + f_{\text{mid}})/2$, and also compared with that of the end point,

f_{max}, at their midpoint, $(f_{mid} + f_{max})/2$, to check the convergence. This process is applied recursively until convergence is obtained over the entire band. This method can significantly reduce the number of frequency points where the matrix equation has to be solved and is capable of capturing sharp resonances. A few examples for antenna analysis can be found in Refs. 2 and 3.

12.4 NUMERICAL CONVERGENCE

A requirement for a numerical method is that it must be used within its range of applicability. When a numerical method is applied outside its intended domain, it is possible that poor results will be obtained, which may then lead the user to conclude that a particular numerical technique, or a simulation package, is of limited or no practical utility. In light of the enormous amount of physical insight that can be gained from modern numerical methods, and the range of the design space that can be explored only by simulation, this would be an unfortunate conclusion. Therefore, it is very important to use a numerical method within its range of applicability. One of the most probable reasons for obtaining poor results with numerical solution methods is the use of a mesh that is not sufficiently fine to resolve the field variation, which then yields unconverged results. Such unconverged results are particularly troubling because they often appear to be physically plausible but are, in fact, inaccurate. For problems where the field variation is due primarily to wave propagation, a general rule of thumb for finite element and finite-difference formulations is to use 20 points per wavelength to obtain a reasonably accurate solution. However, this rule has to be used with care because it does not guarantee any specific accuracy—different problems often have different accuracy considerations in their finite element solutions. For example, an accurate computation of the input impedance of an antenna may require a much finer mesh than that required for the antenna patterns. In addition, a finer mesh may be required to model wave propagation accurately over large electrical distances.

To understand this further, it is necessary to address briefly the issue related to the accuracy of a finite element solution. In the finite element method presented in this book, the field is spatially interpolated within each finite element using a set of basis functions. Since the basis functions have a finite interpolation order, such as the first-order basis in (2.15), this interpolation often cannot represent an arbitrary field exactly. As a result, there is an interpolation error in a finite element solution. For static problems, this is usually the major source of error. For wave-related problems, the imperfect interpolation causes waves to propagate on a finite element mesh at a slightly different speed than the true speed, which in the frequency domain leads to an error in the phase of the complex field. This error, often referred to as *grid dispersion error*, is particularly detrimental to the solution of wave-related problems. This is because phase errors can accumulate as waves propagate over a finite element or finite-difference mesh, and the larger the computational domain in terms of wavelength, the larger the final accumulated phase error, which can eventually render a solution inaccurate. As a result, even with a fixed mesh density, the accuracy of a finite element solution will be different for different problems, and the accuracy typically decreases as the size of the computational domain increases in terms of wavelength. Theoretical as well as numerical investigations [5] show that for the first-order basis functions defined in (2.15), the grid dispersion error is proportional to n^{-2}, where n denotes the number of points per wavelength. As a result, one can always reduce the error in a finite element solution by increasing the mesh density. For example, when the mesh density is doubled, say, from 10 to 20 points per wavelength, the error will be reduced by a factor of 4.

A much more effective approach to reducing the grid dispersion error is to employ higher-order basis functions in the finite element formulation, as discussed in Section 2.5. It has been shown [5] that the grid dispersion error is proportional to n^{-2p}, where p denotes the order of the basis functions employed. Therefore, for a mesh density of 10 points per wavelength, if the order of basis functions is increased from the first to second (which will yield a similar number of unknowns to a mesh with 20 points per wavelength), the dispersion error will be reduced by a factor of 100, which results in much faster convergence. Of course, this can only be done with codes in which higher-order basis functions have been implemented.

Time-domain formulations require slightly more attention to convergence criteria, because the solution must converge in both space and time. Both the traditional leapfrog time-integration technique used by the finite-difference time-domain (FDTD) method and the Newmark-beta formulation used with the finite element time-domain (FETD) method have a temporal truncation error proportional to Δt^2. With explicit, and therefore, conditionally stable time-stepping methods such as the FDTD technique, the Courant condition on a uniform mesh with grid size Δ constrains the time step according to $c\Delta t \leq \Delta/\sqrt{3}$ for numerical stability purposes. As noted above, if we assume a spatial sampling of 20 points per wavelength at the highest frequency of interest, then $\Delta t \leq 1/(20\sqrt{3}f_{\max})$ and we see that we have a relatively high temporal sampling even at the highest frequencies. Consequently, sufficient temporal resolution is normally not an issue with the traditional FDTD method. However, with unconditionally stable techniques, such as the FETD method based on the Newmark-beta formulation, additional care is required in the selection of the time step, because numerical stability no longer constrains how large the time step can be with these methods. Therefore, accuracy limits the largest practical time step and the user must choose it appropriately. Unconditionally stable formulations are particularly beneficial to practical applications having a mix of both coarse and fine geometrical details. This is because the fine geometrical details will not force a further reduction in the time step, which is a significant benefit that is not possible with a conditionally stable formulation. For example, for wideband antennas such as spirals, the complex and geometrically small details associated with the feed region will have no impact on the time step associated with an unconditionally stable formulation. As a simple guideline, defining the time step for the unconditionally stable FETD method according to $c\Delta t \leq h_{\max}$, where $h_{\max}$ denotes the maximum spatially resolved edge length on the grid, has been found to provide a good balance of accuracy and efficiency. It is also important with time-domain simulations to let the solution reach a steady state, either through direct time marching or possibly by using a time extrapolation method [4]. Premature truncation of the transient response will give rise to inaccurate artifacts in the transformed frequency spectrum.

Finally, we note that besides the field variation caused by wave propagation, geometry singularities such as edges, corners, and sharp tips and material discontinuities can also cause fields to vary rapidly. These issues should also be considered in the finite element discretization by using finer meshes around these singularities and discontinuities. Of course, all numerical solution methods are subject to these considerations and the discussions here are therefore not unique to finite element formulations.

In summary, it is recommended that numerical simulations be run with either multiple discretizations/mesh densities, or an increased order of the basis functions, to check for a change in physical quantities of interest so that a converged solution is ensured. Note that different physical quantities may require different mesh densities; for example, a low mesh density may be sufficient for calculating antenna radiation patterns, whereas a much higher mesh density, or a higher-order basis function, may be required for near-field quantities such as the input impedance.

12.5 DOMAIN DECOMPOSITION AND PARALLEL COMPUTING

As discussed in Chapters 10 and 11, it is often beneficial to partition an electrically large application regionally, a process generally known as *domain decomposition*. Within the regions, different solution algorithms and/or parallel processing techniques are applied. For example, it is often useful to decompose certain applications into regions where first-principle techniques, such as finite element methods and the method of moments, are applied, and regions where asymptotic techniques, such as physical or geometrical optics, are applied. A domain decomposition of this type can be solved effectively on a serial computer. Alternatively, the various regions can be further partitioned into subdomains that are distributed onto a parallel computing system and solved concurrently. Because domain-decomposition strategies can enable the solution of important and practical applications that could not otherwise be solved, the development and implementation of robust and efficient domain-decomposition methods are of great practical importance and therefore an active research area in computational science. A few commercial RF prediction codes have adopted certain levels of distributed parallel computing strategies, and these range from the simple concurrent processing of multiple frequencies associated with a single finite element mesh to the true decomposition of finite element or finite-difference meshes distributed across multiple computer processors.

For an efficient solution on a parallel computer that provides good scaling with an increasing number of processors, it is important to have equally distributed workloads, a process typically referred to as *load balancing*. As discussed previously, because of the repetitive nature of the large finite-array problem, this type of application can easily be decomposed such that the computational work is nearly ideally balanced across a very large number of processors working in parallel. For more general three-dimensional applications, there are at least two fundamental challenges associated with very large scale finite element simulations. First, the direct creation of unstructured grids that consist of tens or even hundreds of millions of elements can be challenging. Grids of large magnitude are often constructed in pieces which are then "stitched" together to form the global mesh. Because grids of realistic geometry will typically have elements with widely varying dimensions, this leads to the second challenge of logically decomposing this global mesh such that the computational work levels are uniformly balanced across the distributed computing platform. To simplify this decomposition process, a variety of techniques are currently available [6–10].

12.6 VERIFICATION AND VALIDATION OF PREDICTIONS

The term *verification* addresses the following question: Are the equations for the theoretical model being solved correctly? In comparison, the term *validation* addresses the question: Are all the required physics included in the theoretical model? These two questions have significant differences in the assessment of results obtained through predictive methods. Verification of a solution is established by comparing the result against similar results obtained from other codes or solution techniques that have been applied to an equivalent system of equations and boundary conditions, and this process is often well defined. However, to answer the question posed by validation requires careful measurements. Although many examples could be cited for electromagnetics, a practical example of a formulation that may perform extremely well from the verification perspective, yet possibly perform poorly from the validation perspective, is with regard to the surface impedance boundary condition (IBC). In this case, various numerical implementations of the IBC may be applied, for example, to a spherical geometry and then compared against the exact solution based on the Mie series subject to the similar IBC. Such a comparison was performed in Chapter 6, where good correlation from a verification perspective was shown when the numerical solution was based on the finite element time-domain technique. However, as pointed out in Chapter 6, the scalar IBC is an approximate technique that represents a three-dimensional volumetric material region by a two-dimensional surface along with an assumption of normally incident waves. Although this and similar approximations may be sufficient in many practical applications, comparison with measurements is valuable in determining the limitations of the approximate models, particularly when they are incorporated into otherwise rigorous solution methods.

Balancing the fidelity of the solution method with the domain knowledge, such as the geometrical details and material definitions, required by the application is also important, and an interesting example based on this concept is the following. The radar cross section (RCS) solution of, for example, the metallic double ogive discussed in Chapter 4 can be calculated precisely using either integral or differential equation formulations. In fact, the resulting RCS predictions could possibly be used in calibration of the measurement facility since they are not subject to certain measurement errors, such as those due to positioners. Because the geometry is defined precisely, excellent correlation would be expected from both verification and validation perspectives, and this was demonstrated in Chapter 4. However, if we desired a similar RCS analysis when the metallic double ogive was simply buried in a medium such as sand, the overall fidelity of the analysis may become significantly different, even if the same solution techniques are applied. The distinction, of course, lies in one's ability to characterize precisely the material properties of the sand, as well as its surface irregularities. In many practical applications, acquiring sufficient domain knowledge can be challenging, particularly if various system components come from different manufacturers and information is required over wide bandwidths for which measurements may not exist. Consequently, it is important to understand the overall errors that may be present in an RF assessment when selecting a particular solution strategy, which include both the uncertainties in the input parameters and the errors in the particular numerical formulation.

12.7 SUMMARY

Finite element methods in both the frequency and time domains have advanced significantly in recent years and now represent a powerful technology for the solution of complex antennas and phased arrays both in isolation and in the presence of platforms. In addition to the many theoretical and numerical formulations described in detail throughout the book, there are many practical considerations for an accurate numerical simulation. In this chapter we discussed some of these considerations, which include choosing the basic solution technology, such as finite element or integral equation methods, the applicability of frequency- and time-domain formulations, and guidelines for ensuring numerical convergence. Also discussed briefly was the application of parallel computing and domain-decomposition concepts, and finally some observations on the verification, validation, and overall error assessment associated with the application of numerical methods to predict physical electromagnetic phenomena.

REFERENCES

1. M. A. Kolbehdari, M. Srinivasan, M. S. Nakhla, Q. J. Zhang, and R. Achar, "Simultaneous time and frequency domain solutions of EM problems using finite element and CFH techniques," *IEEE Trans. Microwave Theory Tech.*, vol. 44, pp. 1526–1534, Sept. 1996.
2. D. Jiao and J. M. Jin, "Fast frequency-sweep analysis of cavity-backed microstrip patch antennas," *Microwave Opt. Tech. Lett.*, vol. 22, no. 6, pp. 389–393, Sept. 1999.
3. D. Jiao and J. M. Jin, "Fast frequency-sweep analysis of microstrip antennas on a dispersive substrate," *Electron. Lett.*, vol. 35, no. 14, pp. 1122–1123, July 1999.
4. W. L. Ko and R. Mittra, "A combination of FDTD and Prony's methods for analyzing microwave integrated circuits," *IEEE Trans. Microwave Theory Tech.*, vol. 39, pp. 2176–2181, Dec. 1991.
5. J.-M. Jin, *The Finite Element Method in Electromagnetics*, 2nd ed. Hoboken, NJ: Wiley, 2002.
6. W. Gropp, E. Lusk, and T. Sterling, *Beowulf Cluster Computing with Linux*, 2nd ed. Cambridge, MA: MIT Press, 2003.

7. D. Riley, N. Riley, M. Pasik, J. Kotulski, and C. Turner, "Analysis of airframe-mounted antennas using parallel and hybridized finite-element time-domain methods," *IEEE AP-S Int. Symp. Dig.*, vol. 3, pp. 168–171, June 2002.
8. "ParMETIS: parallel graph partitioning and fill-reducing matrix ordering," University of Minnesota, Minneapolis, MN, http://glaros.dtc.umn.edu/gkhome/metis/parmetis/overview.
9. "Aztec: a parallel iterative package for linear systems," Sandia National Laboratories, Albuquerque, NM, http://www.cs.sandia.gov/CRF/aztec1.html.
10. "Trilinos," Sandia National Laboratories, Albuquerque, NM, http://trilinos.sandia.gov/index.html.

INDEX

www.ingramcontent.com/pod-product-compliance
Lightning Source LLC
LaVergne TN
LVHW011556200926
848484LV00001B/322